Selected Papers from IIKII 2019 conferences in Symmetry

Selected Papers from IIKII 2019 conferences in Symmetry

Special Issue Editors

TeenHang Meen
Charles Tijus
Jih-Fu Tu

MDPI • Basel • Beijing • Wuhan • Barcelona • Belgrade • Manchester • Tokyo • Cluj • Tianjin

Special Issue Editors

TeenHang Meen
National Formosa University
Taiwan

Charles Tijus
Université Paris 8
France

Jih-Fu Tu
St. John's University
Taiwan

Editorial Office
MDPI
St. Alban-Anlage 66
4052 Basel, Switzerland

This is a reprint of articles from the Special Issue published online in the open access journal *Symmetry* (ISSN 2073-8994) (available at: https://www.mdpi.com/journal/symmetry/special_issues/IIKII_2019_conferences).

For citation purposes, cite each article independently as indicated on the article page online and as indicated below:

LastName, A.A.; LastName, B.B.; LastName, C.C. Article Title. *Journal Name* **Year**, *Article Number*, Page Range.

ISBN 978-3-03936-240-0 (Pbk)
ISBN 978-3-03936-241-7 (PDF)

Contents

About the Special Issue Editors

Teen-Hang Meen, Dr., was born in Tainan, Taiwan on August 1, 1967. He received his BS degree from Department of Electrical Engineering, National Cheng Kung University (NCKU), Tainan, Taiwan in 1989, his MS degree and PhD from Institute of Electrical Engineering, National Sun Yat-Sen University (NSYSU), Kaohsiung, Taiwan in 1991 and 1994, respectively. He was Chairman of the Department of Electronic Engineering from 2005 to 2011 at National Formosa University, Yunlin, Taiwan. He received prestigious research awards from National Formosa University in both 2008 and 2014. Currently, he is a distinguished professor at the Department of Electronic Engineering, National Formosa University, Yunlin, Taiwan. He is also the president of the International Institute of Knowledge Innovation and Invention (IIKII) and the Chair of the IEEE Tainan Section Sensors Council. He has published more than 100 SCI, SSCI and EI papers in recent years.

Charles Tijus, Dr., has worked on visual perception for two years, with Adam Reeves (Vision Lab, Northeastern University, Boston, USA) as a research assistant. Charles Tijus is the Director of the Cognitions Humaine et Artificielle Laboratory, founded with M. Bui and F. Jouen, the CHArt, a cognitive science laboratory (problem solving, understanding, robotics and cognitive ergonomics), and "Laboratoire des Usages des Techniques d'Information Numériques", with D. Boullier; a new cognitive ergonomics living LAB laboratory, (LUTIN), which is a "USERLAB" (something like the Audience Research Facility, Boston), located at the "Cité des Sciences et de l'Industrie", La Villette. LUTIN is a platform for usability observations and experiments. LUTIN owns most of the analytical equipment required for its work. It provides access to various shared equipment, such as eye-tracking systems, evoked potentials systems, physiological recording systems, video recording and analysi. The advantages of LUTIN are participants for observations, technologies for observation and experimentation, cognitive simulation, interface between disciplines, and links with industries. LUTIN has close relationships with hospitals, industries, and professional teams and users. It offers services and advice for the adequate conception and use of information technologies. Charles Tijus develops a contextual categorization-based approach in order to study the cognitive processes of understanding: thinking, reasoning, decision-making and learning in early child development and adults. How people develop abilities and competencies is the major concern when it comes to adaptive behavior. The methods comprise empirical research, eye-tracking, event-related potentials (ERPs): N400, and computer models for cognitive simulations. The current interdisciplinary and collaborative research (cognitive psychology, neuroscience and computer science) is on problem solving and operative language, as well as figurative language understanding, and on cognitive robotics and other smart cognitive technologies. Charles Tijus was one of the 2014 IBM Faculty Award recipients.

Jih-Fu Tu, Dr., received his PhD in Computer Science from Preston University, USA. He is also the professor of the Industrial Engineering and Management Department at St. John's University in Taiwan. He was a technology consultant for WanWell Cop. He is interested in the computer architecture of multiprocessor systems and multithreaded processors, discrete event systems (DES), VLSI design, the I/O devices design of a computer, and AIoT. He was also the peer reviewer of The Journal of Supercomputing, Computers and Electrical Engineering, Microsystem Technologies, and the Committee of International Conference, etc. He has published over 20 journal papers, 5 authored books, 7 edited books, and 30 papers in international conference proceedings.

Editorial

Selected Papers from IIKII 2019 Conferences in Symmetry

Teen-Hang Meen [1,*], Charles Tijus [2] and Jih-Fu Tu [3,*]

[1] Department of Electronic Engineering, National Formosa University, Yunlin 632, Taiwan
[2] Director of the Cognitions Director of the Cognitions Humaine et Artificielle Laboratory, University Paris 8, 93526 Paris, France; tijus@univ-paris8.fr
[3] Department of Industrial Engineering and Management, St. John's University, New Taipei City 25135, Taiwan
* Correspondence: thmeen@nfu.edu.tw (T.-H.M.); tu@mail.sju.edu.tw (J.-F.T.)

Received: 21 April 2020; Accepted: 21 April 2020; Published: 26 April 2020

Abstract: The International Institute of Knowledge Innovation and Invention (IIKII) is an institute that promotes the exchange of innovations and inventions, and establishes a communication platform for international innovations and researches. In 2019, IIKII cooperated with the Institute of Electrical and Electronics Engineers (IEEE) Tainan Section Sensors Council to hold IEEE conferences such as IEEE ICIASE 2019, IEEE ECBIOS 2019, IEEE ICKII 2019, ICUSA-GAME 2019, and IEEE ECICE 2019. This Special Issue entitled "Selected Papers from IIKII 2019 conferences" aims to select excellent papers from IIKII 2019 conferences, including symmetry in physics, chemistry, biology, mathematics, and computer science, etc. It selected 21 excellent papers from 750 papers presented in IIKII 2019 conferences on the topic of symmetry. The main goals of this Special Issue are to encourage scientists to publish their experimental and theoretical results in as much detail as possible, and to discover new scientific knowledge relevant to the topic of symmetry.

Keywords: physics symmetry; mathematics symmetry; computer Science

1. Introduction

Symmetry in language refers to a sense of harmonious and beautiful proportion and balance. In mathematics, "symmetry" has a more precise definition, where an object is invariant to any of the various transformations, including reflection, rotation, or scaling. Mathematical symmetry may be observed with respect to the passage of time; as a spatial relationship; through geometric transformations; through other kinds of functional transformations; and as an aspect of abstract objects, theoretic models, and even knowledge itself. Recently, the symmetry theorem and simulation have been widely applied in engineering to improve the developments of new technologies.

In addition, the International Institute of Knowledge Innovation and Invention (IIKII, http://www.iikii.org) is an institute that promotes the exchange of innovations and inventions, and establishes a communication platform for international innovations and researches. In 2019, IIKII cooperated with the Institute of Electrical and Electronics Engineers (IEEE) Tainan Section Sensors Council to hold IEEE conferences such as IEEE ICIASE 2019, IEEE ECBIOS 2019, IEEE ICKII 2019, ICUSA-GAME 2019, and IEEE ECICE 2019. This Special Issue entitled "Selected Papers from IIKII 2019 conferences" aims to select excellent papers from IIKII 2019 conferences, including symmetry in physics, chemistry, biology, mathematics, and computer science, etc. It selected 21 excellent papers from 750 papers presented in IIKII 2019 conferences on the topic of symmetry. The main goals of this Special Issue are to encourage scientists to publish their experimental and theoretical results in as much detail as possible, and to discover new scientific knowledge relevant to the topic of symmetry.

2. The Topic of Symmetry

This special issue selected 21 excellent papers from 750 papers presented in IIKII 2019 conferences on the topic of symmetry. The published papers are introduced as follows: Kubíček et al. reported "Proposal of Technological Geographic Information System (GIS) Support as Part of Resident Parking in Large Cities–Case Study, City of Brno" [1]. The aim of this study is to design and optimize the integrated collection of image data localized by satellite Global Satellite Navigation Systems (GNSS) technologies in the GIS environment to support the resident parking system, including an evaluation of its effectiveness. To achieve this goal, a residential parking monitoring system was designed and implemented, based on dynamic monitoring of the parking state using a vehicle equipped with a digital camera system and Global Satellite Navigation Systems (GNSS) technology for measuring the vehicle position, controlled by spatial and attribute data flow from static and dynamic spatial databases in the Geographic Information System (GIS), which integrate the whole monitoring system. The control algorithm of a vehicle passing through the street network works on the basis of graph theory with a defined recurrence interval for the same route, taking into account other parameters such as the throughput of the street network at a given time, its traffic signs, and the usual level of traffic density. Statistics after one year of operation show that the proposed system significantly increased the economic yield from parking areas from the original 30% to 90% and reduced the overall violation of parking rules to only 10%. It further increased turnover and, thus, the possibility of short-term parking for visitors, and it also ensured availability of parking for residents in the historical center of Brno and surrounding monitored areas.

Lee et al. reported the "Design and Implementation of Virtual Private Storage Framework Using Internet of Things Local Networks" [2]. This paper presents a virtual private storage framework (VPSF) using Internet of Things (IoT) local networks. The VPSF uses the extra storage space of sensor devices in an IoT local network to store users' private data, while guaranteeing expected network lifetime, by partitioning the storage space of a sensor device into data and system volumes and, if necessary, logically integrating the extra data volumes of the multiple sensor devices to virtually build a single storage space. When user data need to be stored, the VPSF gateway divides the original data into several blocks and selects the sensor devices in which the blocks will be stored based on their residual energy. The blocks are transmitted to the selected devices using the modified speedy block-wise transfer (BlockS) option of the constrained application protocol (CoAP), which reduces communication overhead by retransmitting lost blocks without a retransmission request message. To verify the feasibility of the VPSF, an experimental implementation was conducted using the open-source software libcoap. The results demonstrate that the VPSF is an energy-efficient solution for virtual private storage because it averages the residual energy amounts for sensor devices within an IoT local network and reduces their communication overhead.

Kwon et al. reported "Failure Prediction Model Using Iterative Feature Selection for Industrial Internet of Thing" [3]. This paper presents a failure prediction model using iterative feature selection, which aims to accurately predict the failure occurrences in industrial Internet of Things (IIoT) environments. In general, vast amounts of data are collected from various sensors in an IIoT environment, and they are analyzed to prevent failures by predicting their occurrence. However, the collected data may include data irrelevant to failures and thereby decrease the prediction accuracy. To address this problem, the authors propose a failure prediction model using iterative feature selection. To build the model, the relevancy between each feature (i.e., each sensor) and the failure was analyzed using the random forest algorithm, to obtain the importance of the features. Then, feature selection and model building were conducted iteratively. In each iteration, a new feature was selected considering the importance and added to the selected feature set. The failure prediction model was built for each iteration via the support vector machine (SVM). Finally, the failure prediction model having the highest prediction accuracy was selected. The experimental implementation was conducted using open-source R. The results showed that the proposed failure prediction model achieved high prediction accuracy.

Lan et al. reported "Symmetric Modeling of Communication Effectiveness and Satisfaction for Communication Software on Job Performance" [4]. Users in the Taiwanese community send messages or share information through communication software that leads to more dependence from business. Various business problems have been solved and job performance has increased through the diversified functions on communication software. Thus, this research supposed that staff are willing to continuously use communication software LINE (a new communication app that allows one to make FREE voice calls and send FREE messages), and they agree that the varied functions of the communication software would mean that information delivery more symmetrically affects their job performance. According to the research outcomes, communication effectiveness significantly influenced communication satisfaction and job performance, and communication satisfaction significantly influenced job performance. As organizational communication must be conducted through media that disseminate information, and different media have different communication effects, the relationship between communication effectiveness and job performance was completely mediated by communication satisfaction. The research suggested companies or organizations use LINE as a symmetric communication method to not only help employees improve their job performance, but also help enterprises achieve their goals or raise their profit, or even steady development for enterprises.

Li et al. reported "Homomorphic Encryption-Based Robust Reversible Watermarking for 3D Model" [5]. Robust reversible watermarking in an encrypted domain is a technique that preserves privacy and protects copyright for multimedia transmission in the cloud. In general, most models of buildings and medical organs are constructed by three-dimensional (3D) models. A 3D model shared through the internet can be easily modified by an unauthorized user, and in order to protect the security of 3D models, a robust reversible 3D models watermarking method based on homomorphic encryption is necessary. In this study, a 3D model is divided into non-overlapping patches, and the vertex in each patch is encrypted by using the Paillier cryptosystem. On the cloud side, in order to utilize the addition and multiplication homomorphism of the Paillier cryptosystem, three direction values of each patch are computed for constructing the corresponding histogram, which is shifted to embed the watermark. For obtaining watermarking robustness, the robust interval is designed in the process of histogram shifting. The watermark can be extracted from the symmetrical direction histogram, and the original encrypted model can be restored by histogram shifting. Moreover, the process of watermark embedding and extraction are symmetric. Experimental results show that compared to the existing watermarking methods in encrypted 3D models, the quality of the decrypted model is improved. Moreover, the proposed method is robust to common attacks, such as translation, scaling, and Gaussian noise.

Zhang et al. reported "A Matching Pursuit Algorithm for Backtracking Regularization Based on Energy Sorting" [6]. This paper proposes a matching pursuit algorithm for backtracking regularization based on energy sorting. This algorithm uses energy sorting for secondary atom screening to delete individual wrong atoms through the regularized orthogonal matching pursuit (ROMP) algorithm backtracking. The support set is continuously updated and expanded during each iteration. While the signal energy distribution is not uniform, or the energy distribution is in an extreme state, the reconstructive performance of the ROMP algorithm becomes unstable if the maximum energy is still taken as the selection criterion. The proposed method for the regularized orthogonal matching pursuit algorithm can be adopted to improve those drawbacks in signal reconstruction due to its high reconstruction efficiency. The experimental results show that the algorithm has a proper reconstruction.

Ye et al. reported "Incorporating Particle Swarm Optimization into Improved Bacterial Foraging Optimization Algorithm Applied to Classify Imbalanced Data" [7]. In this paper, particle swarm optimization is incorporated into an improved bacterial foraging optimization algorithm, which is applied to classify imbalanced data to solve the problem of how original bacterial foraging optimization easily falls into local optimization. In this study, the borderline synthetic minority oversampling technique (Borderline-SMOTE) and Tomek link are used to pre-process imbalanced data. Then, the proposed algorithm is used to classify the imbalanced data. In the proposed algorithm, the chemotaxis process is first improved. The particle swarm optimization (PSO) algorithm is used to search first and then treat the result as bacteria, improving the global searching ability of bacterial foraging optimization (BFO). Secondly, the reproduction operation is improved and the selection standard of survival of the cost is improved. Finally, the authors improve elimination and dispersal operation, and the population evolution factor is introduced to prevent the population from stagnating and falling into a local optimum. In this paper, three data sets are used to test the performance of the proposed algorithm. The simulation results show that the classification accuracy of the proposed algorithm is better than the existing approaches.

Lin et al. reported "Application of Gray Relational Analysis and Computational Fluid Dynamics to the Statistical Techniques of Product Designs" [8]. During the development of fan products, designers often encounter gray areas when creating new designs. Without clear design goals, development efficiency is usually reduced, and fans are the best solution for studying symmetry or asymmetry. Therefore, fan designers need to figure out an optimization approach that can simplify the fan development process and reduce associated costs. This study provides a new statistical approach using gray relational analysis (GRA) to analyze and optimize the parameters of a particular fan design. During the research, it was found that the single fan uses an asymmetry concept with a single blade as the design, while the operation of double fans is a symmetry concept. The results indicated that the proposed mechanical operations could enhance the variety of product designs and reduce costs. Moreover, this approach can relieve designers from unnecessary effort during the development process and also effectively reduce the product development time.

Hung et al. reported "Applying Educational Data Mining to Explore Students' Learning Patterns in the Flipped Learning Approach for Coding Education" [9]. In this study, the authors applied educational data mining to explore the learning behaviors in data generated by students in a blended learning course. The experimental data were collected from two classes of Python programming-related courses for first-year students in a university in northern Taiwan. During the semester, high-risk learners could be predicted accurately by data generated from the blended educational environment. The f1-score of the random forest model was 0.83, which was higher than the f1-score of logistic regression and decision tree. The model built in this study could be extrapolated to other courses to predict students' learning performance, where the F1-score was 0.77. Furthermore, the authors used machine learning and symmetry-based learning algorithms to explore learning behaviors. By using the hierarchical clustering heat map, this study could define the students' learning patterns including the positive interactive group, stable learning group, positive teaching material group, and negative learning group. These groups also corresponded to the student conscious questionnaire. With the results of this research, teachers can use the mid-term forecasting system to find high-risk groups during the semester and remedy their learning behaviors in the future.

Bartonek et al. reported "Problems of Creation and Usage of 3D Model of Structures and Theirs Possible Solution" [10]. This paper describes problems that occur when creating three-dimensional (3D) building models. The first problem is geometric accuracy; the next is the quality of visualization of the resulting model. The main cause of this situation is that current computer-aided design (CAD) software does not have the sufficient means to precision mapping the measured data of a given object in the field. Therefore, the process of 3D model creation is mainly a relatively high proportion of manual work when connecting individual points, approximating curves and surfaces, or laying textures on surfaces. In some cases, it is necessary to generalize the model in the CAD system, which degrades the accuracy and quality of field data. The paper analyzes these problems and then recommends several variants for their solution. There are two basic methods described: Using topological codes in the list of coordinate points and creating new special CAD features while using Python scripts. These problems are demonstrated on examples of 3D models in practice. These are mainly historical buildings in different locations and different designs (brick or wooden structures). These are four sacral buildings in the Czech Republic (CR): The church of saints Johns of Brno-Bystrc, the Church of St. Paraskiva in Blansko, the Strejc's Church in Židlochovice, and the Church of St. Peter in Alcantara in Karviná city. All of the buildings were geodetically surveyed by the terrestrial method while using the total station. The 3D model was created in both cases in the program AUTOCAD v. 18 and MicroStation.

Chen et al. reported "A Balance Interface Design and Instant Image-based Traffic Assistant Agent Based on GPS and Linked Open Data Technology" [11]. This paper aims to integrate government open data and global positioning system (GPS) technology to build an instant image-based traffic assistant agent with user-friendly interfaces, thus providing more convenient real-time traffic information for users and relevant government units. The proposed system is expected to overcome the difficulty of accurately distinguishing traffic information and to solve the problem of some road sections not providing instant information. Taking the New Taipei City Government traffic open data as an example, the proposed system can display information pages at an optimal size on smartphones and other computer devices, and integrate database analysis to instantly view traffic information. Users can enter the system without downloading the application and can access the cross-platform services using device browsers. The proposed system also provides a user reporting mechanism, which informs vehicle drivers on congested road sections about road conditions. Comparison and analysis of the system with similar applications show that although they have similar functions, the proposed system offers more practicability, better information accessibility, excellent user experience, and an approximately optimal balance (a kind of symmetry) of the important items of the interface design.

Fan et al. reported "Investigation of High-Efficiency Iterative ILU Preconditioner Algorithm for Partial-Differential Equation Systems" [12]. In this paper, the authors investigate an iterative incomplete lower and upper (ILU) factorization preconditioner for partial-differential equation systems. The authors discretize the partial-differential equations into linear equation systems. An iterative scheme of linear systems is used. ILU preconditioners of linear systems are performed on the different computation nodes of multi-central processing unit (CPU) cores. First, the preconditioner of general tridiagonal matrix equations is tested on supercomputers. Then, the effects of partial-differential equation systems on the speedup of parallel multiprocessors are examined. The numerical results estimate that the parallel efficiency is higher than in other algorithms.

Chuang et al. reported "Parameter Optimization for Computer Numerical Controlled Machining Using Fuzzy and Game Theory" [13]. In this study, the precision computerized numerically controlled (CNC) cutting process was chosen as an example, while tool wear and cutting noise were chosen as the research objectives of CNC cutting quality. The effects of quality optimization were verified using the depth of cut, cutting speed, feed rate, and tool nose runoff as control parameters, and actual cutting on a CNC lathe was performed. Further, the relationships between Fuzzy theory and control parameters, as well as quality objectives, were used to define semantic rules to perform fuzzy quantification. The quantified output value was introduced into game theory to carry out the multi-quality bargaining game. Through the statistics of strategic probability, the strategy with the highest total probability was selected to obtain the optimum plan of multi-quality and multi-strategy. Under the multi-quality optimum parameter combination, the tool wear and cutting noise, compared to the parameter combination recommended by the cutting manual, was reduced by 23% and 1%, respectively. This research can indeed ameliorate the multi-quality cutting problem. The results of the research provided the technicians with a set of all-purpose economic prospective parameter analysis methods in the manufacturing process to enhance the international competitiveness of the automated CNC industry.

Bao et al. reported "An Efficient Data Transmission with GSM-MPAPM Modulation for an Indoor VLC System" [14]. The objective of this study was to put forward an efficient and theoretical scheme that is based on generalized spatial modulation to reduce the bit error ratio in indoor short-distance visible light communication. The scheme was implemented while using two steps in parallel: (1) The multi-pulse amplitude and the position modulation signal were generated by combining multi-pulse amplitude modulation with multi-pulse position modulation using transmitted information, and (2) certain light-emitting diodes were activated by employing the idea of generalized spatial modulation to convey the generated multi-pulse amplitude and position modulation optical signals. Furthermore, pulse width modulation was introduced to achieve dimming control in order to improve the anti-interference ability to the ambient light of the system. The two steps above involved the information theory of communication. An embedded hardware system, which was based on the C8051F330 microcomputer and included a transmitter and a receiver, was designed to verify the performance of this new scheme. Subsequently, the verifiability experiment was carried out. The results of this experiment demonstrated that the proposed theoretical scheme of transmission was feasible and could lower the bit error ratio (BER) in indoor short-distance visible light communication while guaranteeing indoor light quality.

Hsueh et al. reported "Fault Diagnosis System for Induction Motors by CNN Using Empirical Wavelet Transform" [15]. In this paper, a novel methodology is demonstrated to detect the working condition of a three-phase induction motor and classify it as a faulty or healthy motor. The electrical current signal data are collected for five different types of fault and one normal operating condition of the induction motors. The first part of the methodology illustrates a pattern recognition technique based on the empirical wavelet transform, to transform the raw current signal into two-dimensional (2-D) grayscale images comprising the information related to the faults. Second, a deep convolutional neural network (CNN) model is proposed to automatically extract robust features from the grayscale images to diagnose the faults in the induction motors. The experimental results show that the proposed methodology achieves a competitive accuracy in the fault diagnosis of the induction motors and that it outperformed the traditional statistical and other deep learning methods.

Shieh et al. reported "Forecasting for Ultra-Short-Term Electric Power Load Based on Integrated Artificial Neural Networks" [16]. In order to improve power production efficiency, an integrated solution regarding the issue of electric power load forecasting was proposed in this study. The solution proposed was to, in combination with persistence and search algorithms, establish a new integrated ultra-short-term electric power load forecasting method based on the adaptive-network-based fuzzy inference system (ANFIS) and back-propagation neural network (BPN), which can be applied in forecasting electric power load in Taiwan. The research methodology used in this paper was mainly to acquire and process the all-day electric power load data of Taiwan Power and execute preliminary forecasting values of the electric power load by applying ANFIS, BPN, and persistence. The preliminary forecasting values of the electric power load obtained therefrom were called suboptimal solutions and the optimal weighted value was finally determined by applying a search algorithm through integrating the above three methods by weighting. In this paper, the optimal electric power load value was forecasted based on the weighted value obtained therefrom. It was proven through experimental results that the solution proposed in this paper can be used to accurately forecast electric power load, with minimal error.

Wang et al. reported "Behavior Modality of Internet Technology on Reliability Analysis and Trust Perception for International Purchase Behavior" [17]. The main research question that this study intends to answer is, "What do users do when a YouTube advertisement appears? Do they avoid or confront them?" The aim of this study is to explore the perceptions and related behaviors of international purchasing and consumers' trust of YouTube advertisements. Statistical analyses focus on the demographics of a sample population in Thailand. The findings are based on data obtained by a questionnaire, the results of which were analyzed by t-test and multiple regression. The results indicate that YouTube advertising has a significant effect on behavioral trends. Moreover, the subjects in the sample reported that they are more likely to avoid YouTube ads than confront them. The study subjects have a low satisfaction with YouTube advertising, and males have a significantly lower satisfaction than females. This study also analyzes the reliability of trust perception toward purchasing. The results indicate that the reliability is greater than 90% at an α level of 5% and a 95% confidence interval.

Lin et al. reported "Application of the Symmetric Model to the Design Optimization of Fan Outlet Grills" [18]. In this study, different designs of the opening pattern of computer fan grills were investigated. The objective of this study was to propose a simulation analysis and compare it to the experimental results for a set of optimized fan designs. The FLUENT computational fluid dynamics (CFD) simulation software was used to analyze the fan blade flow. The experimental results obtained by the simulation analysis of the optimized fan designs were analyzed and compared. The effect of different opening pattern designs on the resulting airflow rate was investigated. Six types of fans with different grills were analyzed. The airflow velocity distribution in the simulated flow channel indicated that the wind speed efficiency of the fan and its influence were comparable to the experimental model. The air was forced by the fan into the air duct. The flow path was separately measured by analog instruments. The three-dimensional flow field was determined by performing a wind speed comparison on nine planes containing the mainstream velocity vector. Moreover, the three-dimensional curved surface flow field at the outlet position and the highest fan rotation speed were investigated. The air velocity distribution at the inlet and the outlet of the fan indicated that among the air outlet opening designs, the honeycomb-shaped air outlet displayed the optimal performance by investigating the fan characteristics and the estimated wind speed efficiency. These optimized designs were the most ideal configurations to compare these results. The air flow rate was evenly distributed at the fan inlet.

Chen et al. reported "Energy Consumption Load Forecasting Using a Level-Based Random Forest Classifier" [19]. In this study, a conventional method of level prediction with a pattern recognition approach was performed by first predicting the actual numerical values using typical pattern-based regression models, and then classifying them into pattern levels (e.g., low, average, and high). A proposed prediction with a pattern recognition scheme was developed to directly predict the desired levels using simpler classifier models without undergoing regression. The proposed pattern recognition classifier was compared to its regression method using a similar algorithm applied to a real-world energy dataset. A random forest (RF) algorithm, which outperformed other widely used machine learning (ML) techniques in previous research, was used in both methods. Both schemes used similar parameters for training and testing simulations. After 10 cross-training validations and five averaged repeated runs with random permutations per data splitting, the proposed classifier shows better computation speed and higher classification accuracy than the conventional method. However, when the number of its desired levels increases, its prediction accuracy seems to decrease and approaches the accuracy of the conventional method. The developed energy level prediction, which is computationally inexpensive and has a good classification performance, can serve as an alternative forecasting scheme.

Chen et al. reported "The Computer Course Correlation between Learning Satisfaction and Learning Effectiveness of Vocational College in Taiwan" [20]. In this paper, the authors surveyed the influence of learning effectiveness in a computer course under the factors of learning attitude and learning problems for students in senior-high school. The authors followed the formula for a regression line as $R = A + BX + \varepsilon$ and simulated it on a Statistical Product and Service Solutions (SPSS) platform with symmetry to obtain the results as follows: (1) In learning attitude, both the cognitive-level and behavior-level are positively correlated with satisfaction. This means the students have a cognitive-level and behavior-level more positively correlated with satisfaction in computer subjects and have a high degree of self-learning effectiveness. (2) In learning problems, the female students had a higher learning effectiveness than male students, and the students who practiced on the computer on their own initiative long-term each week had a higher learning effectiveness.

Gao et al. reported "Locality Sensitive Discriminative Unsupervised Dimensionality Reduction" [21]. Graph-based embedding methods receive much attention due to the use of graph and manifold information. However, conventional graph-based embedding methods may not always be effective if the data have high dimensions and have complex distributions. First, the similarity matrix only considers local distance measurement in the original space, which cannot reflect a wide variety of data structures. Second, the separation of graph construction and dimensionality reduction leads to the similarity matrix not being fully relied on because the original data usually contain lots of noise samples and features. In this paper, the authors address these problems by constructing two adjacency graphs to stand for the original structure featuring similarity and diversity of the data, and then impose a rank constraint on the corresponding Laplacian matrix to build a novel adaptive graph learning method, namely locality sensitive discriminative unsupervised dimensionality reduction (LSDUDR). As a result, the learned graph shows a clear block diagonal structure so that the clustering structure of data can be preserved. Experimental results on synthetic datasets and real-world benchmark data sets demonstrate the effectiveness of our approach.

Symmetry 2020, *12*, 684

Author Contributions: Writing and reviewing all papers, T.-H.M.; English editing, C.T.; Checking and correcting manuscript, J.-F.T. All authors have read and agreed to the published version of the manuscript.

Funding: This research received no external funding.

Acknowledgments: The guest editors would like to thank the authors for their contributions to this special issue and all the reviewers for their constructive reviews. We are also grateful to Dr. Dalia Su, the Managing Editor of Symmetry, for her time and efforts on the publication of this special issue for Symmetry.

Conflicts of Interest: The authors declare no conflicts of interest.

References

1. Kubíček, P.; Bartoněk, D.; Bureš, J.; Švábenský, O. Proposal of Technological GIS Support as Part of Resident Parking in Large Cities–Case Study, City of Brno. *Symmetry* **2020**, *12*, 542, doi:10.3390/sym12040542.
2. Lee, H.-H.; Kwon, J.-H.; Kim, E.-J. Design and Implementation of Virtual Private Storage Framework Using Internet of Things Local Networks. *Symmetry* **2020**, *12*, 489, doi:10.3390/sym12030489.
3. Kwon, J.-H.; Kim, E.-J. Failure Prediction Model Using Iterative Feature Selection for Industrial Internet of Things. *Symmetry* **2020**, *12*, 454, doi:10.3390/sym12030454.
4. Lan, T.-S.; Chuang, K.-C.; Li, H.-X.; Tu, J.-F.; Huang, H.-S. Symmetric Modeling of Communication Effectiveness and Satisfaction for Communication Software on Job Performance. *Symmetry* **2020**, *12*, 418, doi:10.3390/sym12030418.
5. Li, L.; Wang, S.; Zhang, S.; Luo, T.; Chang, C.-C. Homomorphic Encryption-Based Robust Reversible Watermarking for 3D Model. *Symmetry* **2020**, *12*, 347, doi:10.3390/sym12030347.
6. Zhang, H.; Xiao, S.; Zhou, P. A Matching Pursuit Algorithm for Backtracking Regularization Based on Energy Sorting. *Symmetry* **2020**, *12*, 231, doi:10.3390/sym12020231.
7. Ye, F.-L.; Lee, C.-Y.; Lee, Z.-J.; Huang, J.-Q.; Tu, J.-F. Incorporating Particle Swarm Optimization into Improved Bacterial Foraging Optimization Algorithm Applied to Classify Imbalanced Data. *Symmetry* **2020**, *12*, 229, doi:10.3390/sym12020229.
8. Lin, H.-H.; Cheng, J.-H.; Chen, C.-H. Application of Gray Relational Analysis and Computational Fluid Dynamics to the Statistical Techniques of Product Designs. *Symmetry* **2020**, *12*, 227, doi:10.3390/sym12020227.
9. Hung, H.-C.; Liu, I.-F.; Liang, C.-T.; Su, Y.-S. Applying Educational Data Mining to Explore Students' Learning Patterns in the Flipped Learning Approach for Coding Education. *Symmetry* **2020**, *12*, 213, doi:10.3390/sym12020213.
10. Bartoněk, D.; Buday, M. Problems of Creation and Usage of 3D Model of Structures and Theirs Possible Solution. *Symmetry* **2020**, *12*, 181, doi:10.3390/sym12010181.
11. Chen, F.-H.; Yang, S.-Y. A Balance Interface Design and Instant Image-based Traffic Assistant Agent Based on GPS and Linked Open Data Technology. *Symmetry* **2019**, *12*, 1, doi:10.3390/sym12010001.
12. Fan, Y.-H.; Wang, L.-H.; Jia, Y.; Li, X.-G.; Yang, X.-X.; Chen, C.-C. Investigation of High-Efficiency Iterative ILU Preconditioner Algorithm for Partial-Differential Equation Systems. *Symmetry* **2019**, *11*, 1461, doi:10.3390/sym11121461.
13. Chuang, K.-C.; Lan, T.-S.; Zhang, L.; Chen, Y.-M.; Dai, X.-J. Parameter Optimization for Computer Numerical Controlled Machining Using Fuzzy and Game Theory. *Symmetry* **2019**, *11*, 1450, doi:10.3390/sym11121450.
14. Bao, J.-J.; Hsu, C.-L.; Tu, J.-F. An Efficient Data Transmission with GSM-MPAPM Modulation for an Indoor VLC System. *Symmetry* **2019**, *11*, 1232, doi:10.3390/sym11101232.
15. Hsueh, Y.-M.; Ittangihal, V.R.; Wu, W.-B.; Chang, H.-C.; Kuo, C.-C. Fault Diagnosis System for Induction Motors by CNN Using Empirical Wavelet Transform. *Symmetry* **2019**, *11*, 1212, doi:10.3390/sym11101212.
16. Shieh, H.-L.; Chen, F.-H. Forecasting for Ultra-Short-Term Electric Power Load Based on Integrated Artificial Neural Networks. *Symmetry* **2019**, *11*, 1063, doi:10.3390/sym11081063.
17. Wang, S.-L.; Hou, Y.-T.; Kankham, S. Behavior Modality of Internet Technology on Reliability Analysis and Trust Perception for International Purchase Behavior. *Symmetry* **2019**, *11*, 989, doi:10.3390/sym11080989.
18. Lin, H.-H.; Cheng, J.-H. Application of the Symmetric Model to the Design Optimization of Fan Outlet Grills. *Symmetry* **2019**, *11*, 959, doi:10.3390/sym11080959.

Symmetry **2020**, *12*, 684

19. Chen, Y.-T.; Piedad, E., Jr.; Kuo, C.-C. Energy Consumption Load Forecasting Using a Level-Based Random Forest Classifier. *Symmetry* **2019**, *11*, 956, doi:10.3390/sym11080956.
20. Chen, R.-Y.; Tu, J.-F. The Computer Course Correlation between Learning Satisfaction and Learning Effectiveness of Vocational College in Taiwan. *Symmetry* **2019**, *11*, 822, doi:10.3390/sym11060822.
21. Gao, Y.-L.; Luo, S.-Z.; Wang, Z.-H.; Chen, C.-C.; Pan, J.-Y. Locality Sensitive Discriminative Unsupervised Dimensionality Reduction. *Symmetry* **2019**, *11*, 1036, doi:10.3390/sym11081036.

Article

The Effects of Computer-Assisted Learning Based on Dual Coding Theory

Xianghu Liu [1], Chia-Hui Liu [2,*] and Yang Li [3]

[1] College of Foreign Languages, Bohai University, Jinzhou City 121013, China; liuxianghuinchina66@hotmail.com
[2] Department of Industrial Management and Business Administration, St. John's University, New Taipei City 25135, Taiwan
[3] Qingyang No. 1 Primary School, Wensheng District, Liaoyang City 111000, China; Veronica314159@outlook.com
* Correspondence: nancy@mail.sju.edu.tw

Received: 30 January 2020; Accepted: 13 April 2020; Published: 1 May 2020

Abstract: This research explored the integration of dual coding theory and modern computer technology with symmetry into a vocabulary class to improve students' learning attitude and effectiveness. Three research questions are addressed in this research on the effects of computer-assisted learning based on dual coding theory (DCT). This experimental research was carried out in a high school in a remote rural area in China. The study was conducted in two parallel classes (the experimental and the control) in Grade 8 with a total of 88 students. Our research methods included pre- and post-test, questionnaires, and an interview with symmetry as the focus to obtain the results as follows: (1) Using the integration of computer assisted language learning (CALL) and DCT to effectively improve students' learning attitude, (2) transforming students' traditional learning methods into the dual coding method, and (3) enhancing students' vocabulary learning effectiveness.

Keywords: dual coding theory; computer assisted language learning; learning attitude; learning effectiveness

---

1. Introduction

With the rapid development and popularization of network computer technology, how to apply the new technology to education has become a hot topic of general concern to educational scholars. There have been many studies using CALL (Computer Assisted Language Learning) in teaching, and in the process of computer-assisted teaching, many pictures and visual effects are bound to be used to present the language teaching content. As one of the three major elements of language, vocabulary is the material that constitutes the heart of language. Without adequate vocabulary, language loses its meaning [1]. Therefore, vocabulary acquisition is one of the most important parts of language learning. However, teachers often adopt traditional vocabulary teaching methods due to factors such as tight schedules and heavy tasks. Students recite words by themselves, frequently using the rehearsal strategy to memorize words [2]. In order to improve students' vocabulary learning ability, more attention should be paid to exploring correct vocabulary teaching methods. The dual coding theory (DCT), proposed by Paivio in the 1970s, was one of the most important principles at that time [3]. The DCT was first put forward by Paivio [3]. The theory is a description of human cognitive process, including two distinct but interconnected input channels: verbal and non-verbal systems. During the cognitive process, both language generators—logogen—and image generators—imagens—(visual, auditory) were used to activate stimuli. Compared with unitary coding, Paivio strongly believes that using both two systems is more effective than one. The theory attempts to put visual and verbal cognition in equally important positions. By using the visual and verbal system with symmetry in both the left and right hemispheres of the brain, students' learning situations can be improved. This paper aims to investigate the effect of applying computer-assisted dual coding theory (DCT) (a theory about processing by the human cognitive system) to vocabulary teaching, especially in a high school. This research is conducted on the symmetry subject of vocabulary involved in two parallel classes in Grade 8 in a high school in China. The number of research participants was 88 students around 15 years old who had studied English for at least five years. Class one was the control class (CC), which included 43 students (21 girls and 22 boys), and class two was the experiment class (EC), which included 45 students (22 girls and 23 boys). In this research, we investigated the effectiveness of applying the integration of computer assisted language learning (CALL) and DCT. Moreover, we aimed to explore students' attitudes towards visual assisted vocabulary learning based on dual coding theory. This paper is also devoted to researching the changes in the learning method of high school students using dual coding theory and computer-assisted instruction. Finally, this paper discusses to what extent the computer-assisted dual coding theory improved student's vocabulary learning in a high school.

2. Related Work

2.1. Dual Coding Theory

The dual coding theory (DCT) was first put forward by Allan Urho Paivio, a Canadian psychologist from the University of Western Ontario, in 1970s. The theory is a description of the human cognitive process, including two distinct but interconnected input channels: the imagery system and the verbal system. The verbal system deals with modality-specific verbal codes, which are visual, auditory, etc. (e.g., words and book, teacher and study). Verbal system is specialized for processing verbal information (language); it deals with linguistic input and stores linguistic information. The nonverbal system is specialized for the processing of nonverbal objects and events like mental imagery; it deals with visual images and emotional reactions. It has been found that the left hemisphere of the human brain is good at processing verbal information, while the right hemisphere is good at processing representation information [4], which is in line with the DCT's belief that the human cognitive system is composed of two coding systems. Figure 1 is a schematic diagram of the main elements of dual coding theory; it clearly explains the process of the human cognitive system. The model includes the internal organization and connections of the two systems: verbal and nonverbal, and the three levels of processing: representational processing, referential processing, and associative processing.

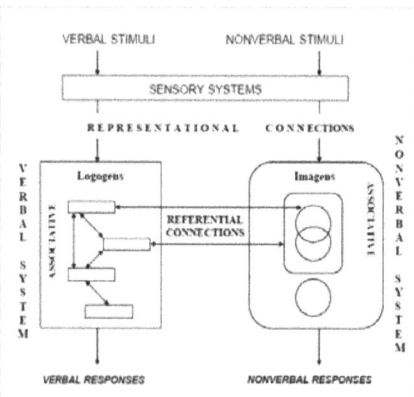

Figure 1. General model of the dual coding theory (DCT) [5].

Figure 1 explains the processing of our cognitive system, which involves the organization of the two coding systems mentioned above: the verbal and nonverbal systems, and the three levels of processing: representational processing, referential processing, and associative processing. The top of the model shows that the cognitive process begins with the sensory system's initial detection of verbal and nonverbal stimuli from the real environment. As vividly shown in Figure 1, the organization of the verbal system is sequential and hierarchical, which indicates that it is modeled like a network. On the other hand, the imagens in the nonverbal system are constructed in an overlapping and nested way. Representation processing is the direct activation of sensory systems and the activation of logogens in a verbal system and imagens in a nonverbal system. For example, when we see a picture of a monkey, the image stimuli will trigger our visual system, while on seeing or hearing the word panda, the verbal stimuli will also activate our verbal system and form logogens. There are two factors to decide which representations are to be activated: the stimulus situation and the individual differences. When "l" occurs in the context of the word "love", it means the letter "l". However, if "l" is put in a series of numbers, it will more likely to be considered as the number "one" [6]. With the application of multimedia in the field of education, the traditional teaching mode can be updated and reformed. Thus, it enhances students' interest and helps the understanding and memory of language. Computer assisted language learning (CALL) refers to the use of computers as the main media to help foreign language teachers in education activities. More specifically, teachers use computer screen-displayed text, pictures, sound, calculations, control, storage, and other functions to improve the quality of their teaching. Levy defines CALL as finding and studying how to apply computers in language teaching [7].

CALL has a history of more than 40 years. The development of CALL can be divided into three stages: behaviorist CALL, communicative CALL, and integrative CALL. Behaviorist CALL started in the late 1950s, and was based on the behaviorist learning model and consisted of drill-and-practice materials. Based on the increasingly prominent communicative approach, communicative CALL became popular in the 1980s. From the early 1990s to the present, integrative CALL has been very popular.

2.2. Vocabulary Learning and Dual Code

In contrast to dual coding theory, the context-availability method denies that the faster identification of concrete and abstract nouns is determined by different types of information processing systems; this theory explains that specific nouns have greater context support [8]. Compared with abstract words, concrete words have stronger or broader associations with context materials. Similarly, Schwanenflugel and Stowe also agree with this explanation and believe that concrete nouns activate more associative information, thereby hastening the process of recognition [9]. However, if the context of the abstract word is meaningful and there is enough verbal information to support it, the abstract word will be recognized as quickly as the concrete word. The difference between context-availability and dual-coding theory lies in the process and place where the information is stored and processed [10]. Many studies have proven and compared the two rival theories' effectiveness in vocabulary learning. By comparing this two theories, Sadoski, Goetz, and Avila concluded that their results were more consistent with the dual coding theory. The results of previous research [11] also showed that pre-teaching with visual aids has a positive effect on vocabulary acquisition compared with pre-teaching with only written context. They believe that this multi-modal approach improves learners' ability to pay attention to vocabulary items and thus increase their vocabulary learning. However, one study [12] concluded that both dual-coding theory and the situational availability method are effective for vocabulary learning, and no one effective vocabulary teaching method is superior to the other. Therefore, it is thought that the dual-coding theory and the context-availability method can be combined or used independently, depending on the subject-matter; it is suggested that interesting pictures should be carefully chosen and used for word recall, and that various techniques should be used to avoid boredom. Research proves there is a link between rote memory and dual coding theory [13]. The authors studied the effects of rote memorization, background, keywords, and keyword methods on the long-term retention of vocabulary by studying 160 ninth-grade students from two schools in Trujillo, Venezuela [14]. The results show that in both long-term and short-term memory, the effects of other methods are lower than that of the context method. Rodríguez and Sadoski claim this result can be explained by DCT [5], and that the information processed through both the verbal system and image system will obtain stronger memory traces and more retrieval paths, thus enhancing vocabulary memory. Context method or rote method primarily activates the verbal system, whose effect is lower than the context method, which is activated in both the verbal and the image system. As more elaboration is offered in the context approach, it is also superior to the keyword approach by which both systems are also activated.

2.3. The Application of Dual Coding Theory to Vocabulary Teaching

The application of dual coding theory in vocabulary teaching in a multimedia situation offers many benefits. In the first place, multimodal input like text, graphics, sound, animation, video, etc., can be provided by computer technology. According to dual coding theory, visual, verbal, and sound sensory stimuli carried out at the same time maximally help foreign language learners to understand the learning materials and master the language forms [15]. Effective vocabulary teaching should be a combination of pronunciation, spelling, word meaning, grammar rules, collocation, internal relations, external relations, and pragmatic rules of words. In this way, with the use of multimedia courseware, students are able to imagine imagens to link the new words with existing knowledge, emotional experience, or real life experience to help them understand and enjoy longer retention of new vocabulary [16]. Mayer explained the concept of cognitive overload in multimedia learning theory [17] and implied that learners should not process too much information which exceeds their available cognitive capacity. Too many pictures can attract students' attention to them but not to the words. The dual coding theory can be used to shape the diversified education samples. The combination of specificity, image, and language has a profound influence in different fields of education: the characterization and understanding of knowledge, the retained memory and learning of school textbooks, effective guidance, individual differences, and the motivation to realize achievements, overcome test anxiety, and master motor skills. Dual coding also has an impact on educational psychology, especially educational research and teacher education [18]. Additionally, one study [19] maintains that the theory of dual coding not only provides a unified interpretation of different topics in education, but its framework can also be applied to other high-level psychological processes. The theory of dual coding provides a concrete model for the behavior and experience of students, teachers, and educational psychologists, and can strengthen the understanding of educational phenomena and teaching practice. Other research [20] investigated the aspect of computer-assisted learning more specifically. The participants were Japanese college science freshmen. The study showed that with online learning, those learning English phrasal verbs with pictures processed information faster and associated non-verbal codes with concepts better. However, the study also found that only relying on picture media is insufficient; other media should also be put into use, reminding us to carefully select pictures, phrasal verbs, and problem formats.

3. Research Method

According to the theory presented in a literature review, compared with the traditional approach, students who accepted dual coding and image creation interventions attained a higher level of vocabulary acquisition. This paper investigates the effectiveness of computer assisted learning based on DCT as the novel teaching method. The proposed research architecture is shown in Figure 2. A framework is used to analyze the influence of computer assisted learning based on DCT teaching effectiveness on the students' studying vocabulary. Furthermore, a questionnaire was designed to investigate the symmetric relationship between variables and statistical methods for analyzing empirical data and verification for answering the research questions. Both quantitative and qualitative research approaches were applied in this study to analyze the data more effectively and reliably. The main research instruments (methods) included: the same pre- and post-questionnaire on students' attitudes, a pretest and a post-test, an interview, and SPSS program version 19.0.

These research questions are stated below:

1. What are students' attitudes towards visual assisted vocabulary learning based on dual coding theory?
2. What are students' opinions of computer-assisted dual coding theory instruction?
3. To what extent does computer-assisted dual coding theory improve student's vocabulary learning in high school?

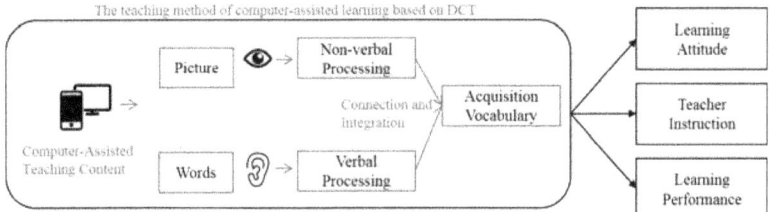

Figure 2. The research architecture.

3.1. Research Participants

The research participants were 88 students around 15 years old selected randomly from two parallel Grade 8 classes in a high school who had studied English for at least five years. During the four months of vocabulary instruction, the teaching method of dual coding theory was consciously applied to teach vocabulary in the experimental class (EC) with the aid of multimedia. Table 1 shows the background information of the participants.

Table 1. Background information of the participants.

Class	Experimental Class		Control Class	
Student Number	Boys	23	Boys	22
	Girls	22	Girls	21
Total Students	45		43	
Teaching Method	DCT-based Instruction		Traditional Method	

The choice of these students was reasonable because the number of samples was consistent with the results of Gay's research [21]. They claimed that when performing correlation analysis, the scale of the sample should exceed 30 in a group. Additionally, the students were all teenagers whose learning methods were easy to form. They did not previously develop a stable learning habit, even though they had an English learning experience for nearly five years [22]. The specific data analysis is conducted in Section 4. Two tests were given to the participants. One was the pre-test which was completed by all the participants before the experiment to examine their level of vocabulary. The other, a post-test, was administrated after the experiment to verify their achievements. The pre-test was a vocabulary test conducted in the first semester in the second grade of junior school (i.e., Grade 8). All the students, including 45 students in the experimental class (EC) and 43 students in the control class (CC), participated in the pre-test. The vocabulary covered in the test was selected from key words in the word list. The structure of the test included English–Chinese translations and Chinese–English translations, each accounting for 20 points, including both concrete and abstract words [22]. In order to increase the validity of the test, the third type of questions was derived from the city high school entrance examination test from recent years. The aim was to test students' mastery of spelling and meaning of words.

3.2. Questionnaires and Interview

Students were reassured the questionnaires were collected anonymously in order to ensure that the data were true and reliable to garner first-hand information about the effect of vocabulary teaching. The questionnaire was distributed to 88 Grade 8 high school students. After a four- month experiment, two questionnaires (Questionnaires I and II) were designed to answer Research Questions 1 and 2. The questionnaires were designed to elicit: students' basic views on learning vocabulary; their own main use of the word memory method normally used; and their views of the teaching methods used by teachers in the classroom. The questionnaire investigated the main attitudes and means of students when encountering difficulties in memorizing words; the last part investigated the teaching methods that students hope to see in the classroom. In order to understand their changes in terms of vocabulary learning methods after the experiment, the researcher handed out the same questionnaire again to all the students in the EC.

The questionnaires in this study were designed based on the research architecture (see Figure 2). Meanwhile, the design of the questionnaires referred to the references related to this research, whose questionnaires have higher reliability and validity (see Table 3). Additionally, the adaption and revision of these questionnaires matched the aims of this study, mentioned above. Furthermore, all questionnaire items were designed by the use of multiple-choice questions because they can be rapidly coded and speedily accumulated to present frequencies of response (Cohen, Manion, and Morrison, 2007) [23]. Such a kind of questionnaires is very easy and convenient to analyze for researchers. In this study, Likert's five-scale is also adopted in this questionnaire. Regarding five degrees from "strongly disagree" to "strongly agree", "1" stands for "strongly disagree", "2" means "disagree", "3" represents "neutral", "4" refers to "agree", and 5 shows "strongly agree". Because there was only one choice for each question, answers to the questions are easier and more convenient for calculation and statistical analysis so that attitudes and opinions of participants can be tested. Finally, the questionnaires (Questionnaires I and II) were to address Research Questions 1 and 2 respectively, along with some interview questions. The statistical results from both indicated the findings or the conclusion of the two research questions. On the whole, the design of the questionnaires was closely related to the research design so that the objectives of this study could be reached smoothly. In order to collect the attitudes and opinions of students in terms of vocabulary learning methods more directly, the researchers surveyed 10 interviewees in the experimental class. The interview questions were divided into eight questions to investigate respondents' views on the teaching methods. Throughout the interview, the interviewees could express their opinions and share their experiences to ensure that the results of the interview would be meaningful. The interview questions were conducted in Chinese throughout, ensuring that every issue could be accurately communicated. The interviewees' answers underwent a truthful translation and analysis in English.

3.3. Research Procedures

The study was carried out in 2018. Eighty-eight high school students participated in the experiment altogether. One of the authors used computer-assisted dual coding theory to improve students' attitude and memory in the EC, while using traditional vocabulary teaching method in the CC. The research procedures included a pilot study, pre-questionnaire, pre-test, vocabulary teaching, post-questionnaire, post-test, and interview. In brief, tests and questionnaires were used to determine the extent to which the computer-assisted dual coding theory could improve students' vocabulary learning. The detailed research procedures have shown in Table 2.

Table 2. Detailed research procedures.

Steps	Procedures	Participants
1	Pilot study	30 selected randomly
2	Pre-questionnaire	EC and CC
3	Pre-test	EC and CC
4	Vocabulary teaching	EC and CC
5	Post-questionnaire	EC and CC
6	Post-test	EC and CC

7	Interview	EC

Cohen [23] pointed out that a pilot study is needed in order to prove the reliability of questionnaires. The questionnaires used in the experiment were administered as a pilot study among 30 students in Grade 8 but not in the EC and CC. Questionnaire I and Questionnaire II were adapted from Zhang [24] and Gu and Johnson [2]. According to the reliability analysis of SPSS shown in Table 3, the reliability coefficients Cronbach's values were 0.80 and 0.76, respectively, which was relatively high, showing that the questionnaires were reliable enough.

Table 3. Reliability statistics of questionnaires.

Questionnaire	Cronbach's Alpha	N of Items
I	0.80	20
II	0.76	20

4. Results and Discussion

In this research, both qualitative and quantitative analyses were used to address the three research questions. The data from vocabulary tests, questionnaires, and interviews were arranged for analysis and discussion.

4.1. Data Analysis and Discussion of Research Question 1 (RQ1)

Research Question 1 is about students' attitude towards visually assisted vocabulary learning based on dual coding theory (DCT). The first questionnaire included two parts: the students' attitude regarding vocabulary learning and their attitude regarding the teaching method. It was distributed to the EC students before and after the experiment in order to find out if the students' attitude changed with the help of DCT teaching. The first and second questionnaires (Questionnaires I and II) consisted of 20 questions each. There are five scales of choice to reveal the degree of affective responses. Before the experiment, the participants' total average values of attitude on vocabulary learning between EC and CC were similar, indicating similar initial attitude levels regarding vocabulary learning. After the experiment, the total average of EC students reached 3.76, while the total average of CC was 2.99. With the help of dual coding theory, the students' attitude towards vocabulary improved significantly in EC. In addition, the independent sample t-test between EC and CC was used to test whether the attitude towards English vocabulary had changed. T-test scores of independent samples from pre-questionnaires in EC and CC on vocabulary attitudes showed that students from EC did not significantly differ from the students from CC about vocabulary attitudes (t (86) = 0.211, p > 0.05). Inspections of both groups' means indicated that the average vocabulary attitudes of EC were similar to CC. The difference between the means was 0.21395 points. The results are shown in Table 4.

Table 4. Results of Research Question 1.

	Class	N	Mean	Std. Deviation	Std. Error Mean
Pre-questionnaire	EC	45	59.4000	3.99659	0.59578
	CC	43	59.1860	5.43908	0.82945
Post-questionnaire	EC	45	75.2444	3.16340	0.47157
	CC	43	59.7907	4.38373	0.66851

The two groups had the same level before the experiment. With the help of DCT in the experiment, students from EC significantly differed from to those in CC in vocabulary learning attitudes (t (86) = 19.027, p < 0.05). Inspections of the two groups' means indicate that the average of EC (75.2444) was significantly higher than the score of CC (59.7907). The difference between the means was 15.45375 points. Therefore, the students in EC significantly improved in their vocabulary learning attitude with the help of DCT. To sum up, the analysis of Questionnaire I and the follow-up interview addressed Research Question 1. EC students' attitudes improved after 4 months of teaching based on computer-assisted dual coding theory, which was in line with many other researchers' claims. They held positive attitudes to pictures, which make the learning process more pleasant and memorable.

4.2. Data Analysis and Discussion of Research Question 2 (RQ2)

In order to verify the change of students' learning method after the research, the questionnaire on students' English vocabulary learning and the interview were designed. The mean values of the EC and CC students' use of traditional learning methods were 3.36 and 3.35, respectively, with only 0.01 difference. Nevertheless, change resulted after the use of the DCT. In the post-questionnaire, the total mean value of the students in EC was 2.37, while the total mean value of the CC students on use of traditional method was 3.34. In the post-questionnaire, the mean difference of EC was 0.08 points lower than that of CC. Through these results, we can see that this experiment made the EC students use the traditional vocabulary learning method less compared to the CC students. Students from EC were not significantly different from CC students regarding the use of traditional learning methods (t (86) = 0.171, $p > 0.05$). Inspections of the two groups' means indicated that the average score for students' use of traditional learning method in EC (33.6222) was similar in that in CC (33.4651). It was obvious that the two groups had the same level before the experiment. The difference between the means was 0.15711 points.

The data statistics show that students from EC significantly improved compared to those in CC regarding vocabulary learning attitudes (t (86) = −11.462, $p < 0.05$). Inspections of the two groups' means indicated that the average score on students' use of traditional learning methods in EC (23.6889) was significantly lower than that in CC (4.18628). It was obvious that the two groups had the same level before the experiment. The difference between the means was −9.68320 points. Therefore, the students in EC used traditional methods less after the vocabulary teaching based on DCT. The mean values of the students in EC and CC on the application of traditional learning method only differed by 0.01. It can be deduced that before this experiment, the students in EC and CC were at similar levels in using the DCT method. The total mean value of EC students in the post-questionnaire increased to 3.24 through the experiment, while the total mean value of the CC students in the post-questionnaire was 2.25, nearly the same as in the pre-questionnaire. The mean difference of the EC in the post-questionnaire was 0.09 points higher than that of the CC in post-questionnaire. These results showed that through this experiment, the EC students used the DCT vocabulary learning method more than the CC students did. The data statistics indicate that students from EC did not significantly differ from CC students on the use of the DCT learning method (t (86) = 0.160, $p > 0.05$). Inspections of the two groups' means indicated that the average score of EC students using the DCT learning method was similar to the score of the CC students. The difference between the means was 0.13798 points. The two groups were at the same level before the experiment. This showed that students from EC differed significantly from those in CC regarding vocabulary learning attitudes (t (86) = 11.076, $p < 0.05$). Inspections of the two groups' means indicated that the average score of students' use of DCT learning method in the EC (32.3778) was significantly higher than that in the CC (22.5349). The difference between the means was 9.84289 points. Table 5 shows that the students in EC used the DCT learning method more with the training of DCT.

Table 5. Results of the students' use of the DCT learning method.

	Class	N	Mean	Std. Deviation	Std. Error Mean
Pre-questionnaire	EC	45	22.5333	4.12090	0.61431
	CC	43	22.3953	3.94691	0.60190
Post-questionnaire	EC	45	32.3778	4.63332	0.69069
	CC	43	22.5349	3.61445	0.55120

It can be concluded from the results of Questionnaire II and the follow-up interview that the students were fully aware of the benefits of DCT and that most of them used this strategy to learn vocabulary. After a period, the participants improved their traditional learning methods including transcription and mnemonics. They began to learn words through visual, auditory, tactile, or emotional means, for example, watching English movies and animations. In this way, they were more interested in vocabulary learning and became more confident regarding English study, which had a positive and deep influence on English vocabulary learning.

4.3. Data Analysis and Discussion of Research Question 3 (RQ3)

In the research procedures, a pre-test and a post-test were conducted before and after implementing the DCT-based instruction. The pre-test was designed to find out whether the EC and CC students were at the same English level, while the post-test aimed to verify whether the application of DCT in junior high school was beneficial in improving students' achievement in terms of their English vocabulary level. The pre-test in the EC and CC on vocabulary was designed to find out whether the students were at the same vocabulary learning level. The mean values of the students in the EC and CC were 25.9333 and 25.9302, respectively, with the mean difference of 0.0031 between them from the students' vocabulary achievements in the pre-test result. Furthermore, the standard deviations of the EC and CC were 1.59716 and 1.67563, respectively. Therefore, the students' scores in these two classes were similar. It seems that the students in EC and CC were generally at the same level in terms of vocabulary learning level. From the results of the independent samples T-test for scores of the pre-test in EC and CC on vocabulary, the EC and CC students' vocabulary achievements in the pre-test were almost at the same level. The significance value was 0.999, which was higher than 0.05, indicating no significant difference between the EC and CC subjects' vocabulary achievements. The post-test was designed to investigate the difference between the EC and CC students after the research. The data show that the mean value of the students in EC on vocabulary was 31.6667, while the mean value of the students in CC was 25.8837 in the post-test. From the mean value, the EC students got higher scores than the CC students did after this experiment; the difference of scores was 5.78295. Regarding the independent samples t-test of students' vocabulary achievements between the students in EC and CC, the significance (2-tailed) value was 0.015, which was lower than 0.05. Thus, we can conclude that through the experiment, the EC and CC students significantly differed in vocabulary achievements. In order to verify that the EC students made significant improvement in vocabulary learning, a paired samples t-test was used. The data reveal that the mean value of EC students' vocabulary score increased from 25.9333 in the pre-test to 31.6667 in the post-test. The students' mean score increased by 8.2245, indicating that by applying dual coding theory to vocabulary teaching in a high school, students' vocabulary achievements can be greatly improved. We also compared the EC students' vocabulary achievements in the pre-test and post-test. It showed t (44) = −5.387, $p < 0.05$. The output of the paired sample t-test showed that EC students' vocabulary achievement in the post-test significantly differed from that in the pre-test. Therefore, we can say that to some extent, applying computer assisted DCT to vocabulary learning can greatly improve students' vocabulary learning. The CC students' vocabulary scores in the pre-test and post-test were also analyzed in this part to further verify the effectiveness of applying DCT to vocabulary learning. We showed that the CC students' mean value of pre-test on vocabulary was 25.9302. Comparatively, in the post-test, the mean value was 25.8837. The CC students before and after the research were approximately at the same level in terms of vocabulary achievement. The results of the pre- and post-tests for the CC students on vocabulary learning can be seen in Table 6, the t (42) score was 0.61 and $p > 0.05$; they did not make significant progress in vocabulary learning, which was different from the result of students in EC.

Table 6. Result of the students' use of the DCT learning method.

	Class	N	Mean	Std. Deviation	Std. Error Mean
Pre-questionnaire	EC	45	25.9333	10.71405	1.59716
	CC	43	25.9302	10.98786	1.67563
Post-questionnaire	EC	45	31.6667	10.83345	1.61496
	CC	43	25.8837	10.91790	1.66496

To sum up, through qualitative and quantitative analysis, it can be concluded that appropriate visual materials have a positive effect on vocabulary teaching. When students learn vocabulary through text such as sentences and translation, compared with through visual materials, they may forget them more easily. In addition, it should be noted that the pictures are more attractive, and the students will be more interested in vocabulary learning; therefore, this method improves the learning effect of students, which is consistent with the results of Hashemi and Pourgharib [25].

4.4. Summary

In this paper, quantitative and qualitative data including a questionnaire, interview, and test were collected, analyzed, and discussed based on the three research questions; Table 7 summarizes the research results related to the three research questions.

Table 7. Summary of the results for the three research questions.

Research Question (RQ)	Research Instrument	Major Finding
RQ 1: What are students' attitudes towards visual assisted vocabulary learning based on dual coding theory?	Questionnaire I (Q1–Q20) and interview (Q1–Q3)	After receiving vocabulary teaching based on DCT, the EC students have a more positive attitude towards vocabulary learning, which is conducive to vocabulary learning.
RQ 2: What are students' opinions of computer-assisted dual coding theory instruction?	Questionnaire II (Q1–Q20) and interview (Q4–Q6)	The EC students tend to use traditional rote learning less but more DCT strategies to remember words. They also offered new suggestions on new vocabulary teaching method.
RQ 3: To what extent does the computer-assisted dual coding theory improve students' vocabulary learning in junior high school?	Pre-test and post-test and interview (Q7–Q8)	With the help of DCT training, EC students' performance in the post-test of vocabulary was significantly better than that in the pre-test.

This paper shows that EC students made great progress in this experiment. First of all, the computer assisted DCT vocabulary learning strategy training has a positive effect on constructing positive vocabulary learning attitudes. Secondly, through training, students use the DCT learning method more than the traditional method in vocabulary learning. They offer more new suggestions about vocabulary learning methods. Finally, applying DCT vocabulary learning has greatly enhanced students' vocabulary achievement. By combining both qualitative and quantitative analysis, some findings of the experiment are concluded:

- **Major findings on Research Question 1:** The results of Questionnaire I and the follow-up interview indicate that the computer-assisted dual coding teaching method can improve participants' attitudes towards vocabulary learning. Through the experimental process, students can acquire visual and audio information via pictures, video, audio, and multimedia, so that they experience higher interest in learning. This finding supports the idea of Cohen and Johnson [26] that students were more interested and paid more attention to vocabulary teaching. In addition, through the process of group discussion and image formation, participants were more willing to engage in learning activities, which increased their learning enthusiasm. To recap the findings of Research Question 1, EC students' attitudes were improved after four months of teaching based on computer-assisted dual coding theory. In other words, they held a positive attitude to pictures, which made the learning process more pleasant, meaningful, and memorable. This finding is in line with Kim and Gilman [20]. Compared with their study, a more detailed analysis process was conducted by the researcher; for example, the study adopted both quantitative and qualitative analysis methods.
- **Major findings on Research Question 2:** In this study, the results of Questionnaire II and the interview show that the interviewees are fully aware of the benefits of DCT and used this strategy more frequently to learn vocabulary, which is consistent with the study of Yanasugondha [18]. After the experiment, the participants improved their traditional learning methods including transcription and mnemonics. Through the training of image formation, students can understand and memorize words with the help of visual, auditory, haptics, feelings, and other aspects, combined with the student's own experience in daily life. The researcher introduced the DCT method at the beginning of the experiment, and then gradually

combined this method with technology to present pictures or videos related to new words in each class. After that, some practice instructions were given to help consolidate students' memory. After a period of time, the participants improved the original method of rote learning. They liked to use visual, auditory, tactile, or emotional tools to learn more words, for example, by watching English movies and animations. In this way, they became more interested in vocabulary learning and more confident in English study, which has a positive and deep influence on vocabulary learning. Furthermore, they can think of other new methods for learning vocabulary, which is in line with Cohen and Johnson [26].

- Major findings on Research Question 3: From the pre- and post-test results, it can be proven that the computer-assisted dual coding theory is beneficial to vocabulary teaching in school. It improved students' vocabulary scores to a large extent. In the comparison with pre- and post-test results, EC students' scores greatly improved, and it is not difficult to imagine that students will improve their vocabulary level in their future studies by using this method effectively. Compared with traditional teaching, the students can shape their own images in the memory process. Additionally, students can not only improve their memory accuracy, but also reduce the burden of learning as well as avoid bad learning habits such as rote learning. They feel that vocabulary learning has become easier for them. After using the vocabulary teaching method, these students have formed an effective and self-disciplined new vocabulary learning method. Therefore, the vocabulary teaching method based on DCT plays a very vital role in improving students' vocabulary proficiency, which is consistent with the results of Hashemi and Pourgharib [25].

5. Conclusions

Based on the symmetry approach, this research proposed the integration of dual coding theory and modern computer technology into vocabulary classes to improve students' learning attitudes and effectiveness. According to the research results summarized above, the implications are as follows: firstly, visual aids like pictures or videos should be presented to help students learn vocabulary. It is proven that teaching vocabulary with vivid pictures and images and can be more meaningful, while also attracting the attention of students. Through the integration of CALL and DCT, students will be more positive concerning learning words, which will deeply influence their future studies. It was shown herein that the integrated method can effectively improve the learning attitude of students. Moreover, during the DCT vocabulary teaching process, instead of being forced to accept this learning method, the students in the experimental class were guided to develop the habit of forming images gradually and naturally. They found this method more effective and interesting, which makes them prefer using this new method. This study helps prove that dual code theory enhances students' learning methods. Outside of using image formation, teachers should also give feedback and evaluation of a student's or group's answers in a timely manner. In order to help students have a deeper understanding of vocabulary, the teacher's own image examples can also be presented to them when necessary. Finally, students will improve their attitude towards learning, consciously associate words with images to remember words, and effectively improve their academic performance. The most significant implication in this research is applying modern technology into the field of education with the advanced DCT concepts to maximally improve the quality of education. To conclude, there is no doubt that the innovative vocabulary teaching approach based on DCT plays a very significant part in enhancing EFL (English as a foreign language) students' vocabulary achievements and language teachers' teaching quality. Meanwhile, the present research enlightens educators regarding future language teaching and research.

Author Contributions: X.L. and Y.L. contributed equally to the conception of the idea, implementing and analyzing the experimental results, writing the manuscript and so on. C.-H.L. contributed to the research design, revision and proofreading of the manuscript and so on. X.L. plays a leading role in completing the article. All authors have read and agreed to the published version of the manuscript.

Symmetry **2020**, *12*, 701

Funding: This research was partly funded by the Scientific Research Funding Programme for PhD-Staff of Bohai University, China. Grant Number: 0518bs005.

Conflicts of Interest: The authors declare no conflict of interest.

References

1. Wilkins, D.A. *Linguistics in Language Teaching*; Edward Arnold: London, UK, 1972.
2. Gu, Y.Q.; Johnson, R.K. Vocabulary Learning Strategies and Language Learning Outcomes. *Lang. Learn.* **1996**, *4*, 643–679.
3. Paivio, A. *Imagery and Verbal Processes*; Holt, Rinehart and Winston (HRW): New York, NY, USA, 1971.
4. Rizzolatti, G.; Umilta, C.; Berlucchi, G. Opposite Superiorities of the Right and Left Cerebral Hemispheres in Discriminative Reaction Time to Physiognomical and Alphabetical Material. *Brain: A J. Neurol.* **1971**, *94*, 431–442.
5. Paivio, A. *Mental Representations: A Dual Coding Approach*; Oxford Psychology Series: New York, NY, USA, 1986.
6. Nuramah, H. The Effects of Using Dual-Coding Theory in Teaching English Reading Comprehension among Vocational Students at Narathiwat Technical College. *Int. J. Arts Humanit. Soc. Sci. Stud.* **2019**, *4*, 1–6.
7. Levy, M. *CALL: Context and Conceptualization*; Oxford University Press: New Yourk, USA, 1997.
8. Schwanenflugel, P.J.; Shoben, E.J. Differential Context Effects in the Comprehension of Abstract and Concrete Verbal Materials. *J. Exp. Psychol. Learn. Mem. Cogn.* **1983**, *9*, 82–102.
9. Schwanenflugel, P.; Stowe, R.W. Context Availability and the Processing of Concrete and Abstract Words in Sentences. *Read. Res. Q.* **1989**, *24*, 114–126.
10. Alam, Y.S.; Oe, Y. E-Learning of English Phrasal Verbs via Pictorial Elucidation and L1 Glosses. In Proceedings of the 12th IEEE International Conference on Advanced Learning Technologies, Rome, Italy, 4–6 July 2012.
11. Alamri, K.; Rogers, V. The Effectiveness of Different Explicit Vocabulary-Teaching Strategies on Learners' Retention of Technical and Academic Words. *Lang. Learn. J.* **2018**, *46*, 622–633.
12. Soylu, B.A.; Yelken, T.Y. Dual-coding Versus Context-availability: Quantitative and Qualitative Dimensions of Concreteness Effect. *Procedia Soc. Behav. Sci.* **2014**, *116*, 4814–4818.
13. Sadoski, M.; Goetz, E.T.; Avila, E. Concreteness Effects in Text Recall: Dual Coding or Context Availability? *Read. Res. Q.* **1995**, *30*, 278–88.
14. Rodriguez, M.; Sadoski, M. Effects of rote, Context, keyword, and Context/keyword Methods on Retention of Vocabulary in EFL Classrooms. *Lang. Learn.* **2000**, *50*, 385–412.
15. Kanellopoulou, C.; Kermanidis, K.L.; Giannakoulopoulos, A. The Dual-Coding and Multimedia Learning Theories: Film Subtitles as a Vocabulary Teaching Tool. *Educ. Sci.* **2019**, *9*, 1–13. https://doi.org/10.3390/educsci9030210
16. Moody, S.; Hu, X.; Kuo, L.-J.; Jouhar, M.; Xu, Z.; Lee, S. Vocabulary Instruction: A Critical Analysis of Theories, Research, and Practice. *Educ. Sci.* **2018**, *8*, 180, 1–22. https://doi.org/10.3390/educsci8040180
17. Mayer, R.E. *Multimedia Learning*; Cambridge University: Cambridge, UK, 2003.
18. Yanasugondha, V. *A Study of English Vocabulary Learning Using the Dual Codling Learning Theory*; Thammasat University: Bangkok, Thailand, 2016.
19. Clark, J.M.; Paivio, A. A Dual Coding Theory and Education. *Educ. Psychol. Rev.* **1991**, *33*, 149–210.
20. Kim, D.; Gilman, D. A. Effects of Text, Audio, and Graphic Aids in Multimedia Instruction for Vocabulary Learning. *Educ. Technol. Soc.* **2008**, *11*, 114–126.
21. Gay, L.R. Educational Research: Competences for Analysis and Application; Merrill: New York, NY, USA, 1992.
22. Paivio, A. Dual Coding Theory, Word Abstractness, and Emotion: A Critical Review of Kousta et al. *J. Exp. Psychol. Gen.* **2013**, *142*, 282–287.
23. Cohen, L.; Manion, L.; Morrison, K. *Research Methods in Education*; Routledge: London, UK, 2007.
24. Zhang, L. *A Study of the Application of Dual Coding Theory to English Vocabulary Instruction in Junior Middle School*; Shenyang Normal University: Shenyang, China, 2016.

25. Hashemi, M.; Pourgharib, B. The Effect of Visual Instruction on New Vocabularies Learning. *Int. J. Basic Sci. Appl. Res.* **2013**, 2, 623–627.
26. Cohen, M.T.; Johnson, H.L. Improving the Acquisition and Retention of Science Material by Fifth Grade Students through the Use of Imagery Interventions. *Instr. Sci.* **2012**, *40*, 925–955.

Article

Proposal of Technological GIS Support as Part of Resident Parking in Large Cities–Case Study, City of Brno

Pavel Kubíček [1], Dalibor Bartoněk [2,*], Jiří Bureš [2] and Otakar Švábenský [2]

[1]   Bc. Brněnské komunikace a.s., Renneska třída 787/1a, 639 00 Brno, Czech Republic; kubicek@bkom.cz

[2]   Civil Engineering, Brno University of Technology, Veveří 331/95, 602 00 Brno, Czech Republic; bures.j@fce.vutbr.cz (J.B.); svabensky.o@fce.vutbr.cz (O.Š.)

*   Correspondence: bartonek.d@fce.vutbr.cz

Received: 2 February 2020; Accepted: 31 March 2020; Published: 3 April 2020

Abstract: Over the last few years, there has been a significant increase in people's dependence on passenger and freight transport. As a result, traffic infrastructure is congested, especially in big city centers and, at critical times, this is to the point of traffic collapse. This has led to the need to address this situation by the progressive deployment of Intelligent Transport Systems (ITS), which are used to optimize traffic, to increase traffic flow, and to improve transport safety, including reduction of adverse environmental impacts. In 2018, the first results of the C-Roads Platform which is a joint initiative of European Member States and road operators for testing and implementing C-ITS services in light of cross-border harmonization and interoperability (C-ROADS) Czech Republic project were put into operation in Brno, closely related to the international initiative to support the data structure for future communication between vehicles and intelligent transport infrastructure. A system of transport organization and safety was introduced in the city of Brno, which manages key information and ensures central management of partial systems of transport organization and safety. The most important part of this system is the parking organization system discussed in this article. The main objective was to streamline the parking system in the city center of Brno and in the immediate vicinity by preventing unauthorized long-term parking, ensuring an increased number of parking places for residents and visitors by increasing the turnover of parking. The aim of the research was to investigate (i) the possibility and optimal use of Geographic Information System (GIS) technology for resident parking system solutions, (ii) the integration of Global Satellite Navigation Systems (GNSS) satellite data and image data collected by cameras on the move and (iii) the possibility of using network algorithms to optimize mobile data collection planning. The aim of our study is to design and optimize the integrated collection of image data localized by satellite GNSS technologies in the GIS environment to support the resident parking system, including an evaluation of its effectiveness. To achieve this goal, a residential parking monitoring system was designed and implemented, based on dynamic monitoring of the parking state using a vehicle equipped with a digital camera system and Global Satellite Navigation Systems (GNSS) technology for measuring the vehicle position, controlled by spatial and attribute data flow from static and dynamic spatial databases in the Geographic Information System (GIS), which integrate the whole monitoring system. The control algorithm of a vehicle passing through the street network works on the basis of graph theory with a defined recurrence interval for the same route, taking into account other parameters such as the throughput of the street network at a given time, its traffic signs and the usual level of traffic density. Statistics after one year of operation show that the proposed system significantly increased the economic yield from parking areas from the original 30% to 90%, and reduced the overall violation of parking rules to only 10%. It further increased turnover and thus the possibility of short-term parking for visitors and also ensured availability of parking for residents in the historical center of Brno and surrounding monitored areas.

Keywords: GIS; monitoring; resident parking; transport

---

1. Introduction

Brno is the second largest city in the Czech Republic and has about 500,000 inhabitants. Every major metropolis faces the challenges of security, housing, cleanliness and transport. Modern information technology, smartphones and other devices are already an essential part of the daily life of citizens and can contribute to the optimization and streamlining of many processes affecting urban life. Brno is also a university city with a strong support for science, research and development, represented by more than 100 scientific research institutions. The most important are, for example, medical and research centers covered by Masaryk University or technical research centers under the auspices of the Brno University of Technology. Brno has the ambition to be a modern city that uses modern technology and thus applies information technology to many areas of social life. The map portal [1] provides an on-line basic map of the city, as well as many thematically oriented map sets, such as, in the area of spatial planning, price maps, 3D building models, temperature maps, maps of water resources, georisk maps, etc. The basic maps relating to the city of Brno can also be found at the geoportal [2].

In the last few years, there has been a significant increase in people's dependence on passenger and freight transport. Increasing traffic has a negative impact on the environment and is also energy intensive. Due to the increase in traffic intensity and density, communications are often very close to capacity. The mobility of people and things is now in the process of fundamental changes, triggered by the rapid development of information and communication technologies and services supporting mobile connectivity. The rapid deployment of mobile digital technologies is also changing the way we provide traffic information or current changes that affect traffic. Intelligent Transport Systems (ITS) are used to optimize transport, increase traffic flow and improve transport safety and reduce its impacts. [3]. Transport in Brno is provided by trams, buses, trolleybuses and trains. Brno has a public international civil airport providing flights to 25 countries across the world, capable of handling up to 557,000 passengers per year.

The proposed system of parking regulation in the center and adjacent parts of the city of Brno is designed to be conceptually consistent with a higher project Transport Organization and Security System (SOBD) supported by a project of the European Union [4]. The subject matter of the information system is the management of key information and central management of future sub-systems of transport organization and safety. Such sub-systems include, e.g., parking regulation systems in the historical center of Brno, parking (for residents, visitor parking, parking lot on the outskirts of the city with good connection to public transport, parking houses, barrier systems), speed measurement, junction control, vehicle weighing and parking systems with payment terminals. The software part of the information system supports activities and manages data in the areas of information management and rules related to transport organization and security, authorization management (e.g., parking privileges), management of information provided by sub-systems (e.g., measurement results), compliance assessment, detection and reporting suspected violations, billing and statistics.

The sharp increase in traffic density, especially in large cities, causes problems with parking of vehicles both in historical centers and in adjacent areas. This situation can no longer be solved extensively, i.e., by increasing the number of parking spaces, or even in a classic way, e.g., by regulating parking via members of the municipal police. Therefore, methods were sought to make the parking system in Brno more efficient using modern technologies.

The main objective of this project is:

1. To prevent long-term (in many cases unauthorized) parking of vehicles in the city center, thus blocking parking capacity;
2. To ensure a sufficient number of parking places for residents and visitors of Brno by increasing the turnover of parking.

The result of the project is a system that automatically monitors the situation of parking in Brno based on a suitable planning algorithm, and automatically evaluates data from the field based on parking rules, and creates and sends offense documentation to the police for further investigation.

Prior to the launch of the parking monitoring system addressed in this article, no similar automated or semi-automated system was in operation to meet the above objectives.

The aim of the research was to investigate (i) the possibility and optimal use of Geographic Information System (GIS) technology for resident parking system solution, (ii) the integration of Global Satellite Navigation Systems (GNSS) satellite data and image data collected by cameras on the move and (iii) the possibility of using network algorithms to optimize mobile data collection planning. The aim of our study is to design and optimize the integrated collection of image data localized by satellite GNSS technologies in the GIS environment to support the resident parking system, including an evaluation of its effectiveness.

Integrated collection of satellite positioning data and image data to deal with parking regulation in large cities due to high traffic density is a topical issue. It is common to use information technology and to collect and evaluate image data acquired mainly by static cameras. The use of satellite-based collection of photographic data in larger areas by mobile cameras, their integration with GIS technology and subsequent evaluation of data as a basis for regulation of parking by authorities is not yet a commonly used technology. Moreover, it has its pitfalls, which were the subjects of our solution.

The system was expected to significantly increase the economic yield from parking areas, reduce the overall rate of parking violations, increase turnover and thus the availability of short-term parking for visitors and residents in the historical center of Brno and the monitored surrounding areas.

The introduction of the parking monitoring system has completed the construction of the Static Object Base Data (SOBD) system as a partial important data structure contributing to the future possibility of using smart communication between vehicles, which is the subject of the transnational project C-ROADS.

2. Related Concepts and Works

The massive development of the automotive industry causes complications not only in transport but also in the possibilities of vehicle parking. The situation is particularly critical in large cities and densely populated conurbations. Extensive solutions to this problem, i.e., by extending parking areas, have been exhausted in many cases. Therefore, an intensive approach to vehicle parking is currently being promoted, which is based on the establishment of appropriate rules for the efficient use of existing parking capacities in a given location. This method belongs to the wider issue of developed cities (Smart Cities). It is a highly topical subject concerning all cities in developed countries around the world.

There are a number of publications in world databases dealing with similar issues and using mathematical, GIS and other tools that inspired the authors to generate ideas for solving the given issue. The topics of improving the organization of transport and increasing the safety and fluidity of transport in the central parts of various cities were monitored. Another monitored topic was the respect of already implemented regulatory measures in transport related to the issue of illegally parked vehicles, reducing congestion, searching for free parking spaces and also the issue of reducing energy consumption and emissions. In terms of technical tools working with spatial location data, the use of GIS tools proved to be the most suitable.

The publications were selected according to the objective defined in the previous chapter, and according to criteria that specifies the wider aspects of the objective. The categories of topics are:

1. Improving the organization of transport, i.e., increasing the safety and flow of traffic in the central part of the city;
2. Increasing respect for regulatory measures in transport and reducing the number of illegally parked vehicles;
3. Reducing congestion by reducing parking space and thereby reducing energy consumption and emissions;
4. Increasing safety and ensuring sufficient parking space for residents and visitors;

5. Use of GIS technology to organize parking in the city.

For each of these categories, the methods used were evaluated and the possibilities of their use in our project were analyzed.

In [5], the authors provide a search character and deal with works aimed at optimizing traffic routes, parking problems and the detection and prevention of traffic accidents. The most widely used methods for the solution were machine learning (ML) and the Internet of Things (IoT). In paper [6], the problem of parking has been addressed by proposing an architecture to automate the parking process using the internet of things, artificial intelligence and multi agent systems. The authors in [7] offer a review of the literature that deals with a wider field of transport problems and its results are useful for solving general transport issues. The study in [8] deals with traffic management based on fog computing. The method is based on cloud computing and is used to control smart city traffic in real time. The paper in [9] presents a machine learning method for traffic management. These include parking monitoring, 5G communications and more. The study in [10] focuses on the design of new road signs for the use of wireless communication technologies. The principle of the proposed solution is the digitization of road signs and their display on the driver's desk of vehicles.

In our project, we used the principles of the machine learning method in the design of the control algorithm for the calculation of the optimized route of the monitoring vehicle and for the search for free parking places. The use of IoT is foreseen only in the C-ROADS project, when the vehicles will be equipped with the appropriate technology to participate in solving situations in an intelligent transport system.

A new parking model for cost and time optimization is described in [11]. Performance is measured by a special parameter. The work in [12] proposes a distributed system that informs drivers about free parking spaces in real time. The detection of vacant places works on the principles of computer vision and machine learning. The study in [13] proposes an intelligent, fully automated parking management system. It informs the driver about the number of free parking spaces in the immediate vicinity and saves fuel.

From these schematics, it was possible to use common computer vision algorithms to recognize images of the license plates of parked vehicles. Localization of a visitor's vehicle is realized by the GNSS method or by their location in the mobile operator's network using the on-board computer in the car or the driver's smartphone. Based on its current location, the web application or the smartphone application of the Brno parking system [14] searches for free parking places in the immediate vicinity, including the possibility of vehicle registration into the system. An evolutionary algorithm was used for a similar function in [15]. Another localization task based on the intelligent algorithm solved in [16] was to find the nearest place for charging electric vehicles. In our project, it was necessary to solve the integration of static and dynamic data from multiple sources. The use of data from static cameras in combination with IoT has been dealt with in [17–19].

The authors in [15] deal with a system for the allocation of free parking spaces—a system of available mathematical models that works on the basis of a genetic algorithm. The advantage of the system is the speed and efficiency of searching.

In [17], the authors present a system for data mining from independent sources, which are stored in city computers. Their system is designed to obtain more information for Smart Cities.

A new Smart City Network Design Tool is described in [18]. The system is based on human machine communication, monitors network traffic and optimizes network services.

In [19] there is an intelligent management system for managing a large number of IoT devices. The system is based on cloud computing and aims to optimize services within a Smart City.

Article [16] shows a case study for charging electric vehicles. The principle is an intelligent concept with the possibility of an ecosystem with a user interface for mobile applications. The project is part of the EU Horizon 2020.

One possible way of relieving traffic and reducing the number of vehicles is their sharing, as solved in [20]. In a study [21], the authors try to solve the problem of parking by introducing a bicycle sharing system where supply and demand are realized using smartphones. An important part of the actual parking space capacity monitoring system is the interoperability addressed in [22] by IoT. In terms of enforceability of offenses when parking vehicles, legal conditions are also important. The work in [23] addresses an analysis of the legal framework for smart city services including transport and parking.

The study in [20] discusses a car sharing system based on the use of cars in 10 different European cities. The knowledge can be used to design a system that allows car sharing based on prediction, usage and other parameters.

The work in [22] describes the Global IoT Services system, which enables interoperability within IoT between cities. The application has been verified in Smart Cities in Spain and South Korea.

The authors in [23] deal with IoT classification within a Smart City. It discusses four IoT cases: an intelligent parking system, intelligent street lighting, intelligent monitoring system and intelligent sensors. Each of these systems is analyzed in this work and their optimization is recommended.

The work in [21] discusses a bicycle sharing system in China. The system was introduced in 2008 and is based on empirical analysis. The result is the design of a new bicycle sharing system called Dockless bikeshare. This system has proved its worth and supports the Chinese Republic.

Most of the above work solves the problem of parking using modern methods from the field of information technology, web applications and multi-criteria decision making. Surprisingly, very few projects use GIS as an integrating element between static and dynamic spatial data. It is this principle that the authors chose as the basic element of the concept presented in the next chapter.

The authors in [24] use a web GIS in combination with multicriterial analysis (MCDA) to find parking spaces. The principle of the solution is group decision-making, which leads to a decrease in the share of information retrieval, average time spent gathering individual information and variability of information retrieval per attribute in the context of parking space selection.

The work in [25] uses GIS technology to select the optimal places for parking and charging electric cars in Germany. The system evaluates demand near points of interest.

The study in [26] presents the use of GIS to monitor traffic density and free parking spaces in Vilnius (Lithuania).

The issue of changing the standards of parking space dimensions in the context of the trend for purchasing and manufacturing larger vehicles that provide the best possible comfort is addressed in [27]. Parking space projects tend to design parking spaces for a representative set of vehicle types, but the effort is to minimize the dimensions of parking spaces. The article evaluates an analysis of vehicles from different countries and with different dimensions that are comparable to the dimensions of parking spaces.

3. The Concept of Transport Organization and Security System in the Historical Center of Brno

The Residential Parking Project in Brno is part of the Transport Organization and Security System (SOBD) [4], which contains:

- A parking regulation system in the historical center of Brno;
- Speed measurement;
- Checking drivers who drive through red traffic lights;
- Vehicle weighing;
- Parking organization (resident parking (R), visitor parking (P), R and P parking, parking houses, barrier systems);
- Parking systems with payment terminals.

In 2018, the first results of the C-ROADS Czech Republic project [28], which is closely linked to the C-ROADS international initiative, which was the result of joint activities of the Czech Republic, Austria and Germany, were put into operation in Brno. The C-ROADS Czech Republic project is co-funded by the EU under the Connecting Europe Facility program (CEF). The main objective of the project was, in cooperation with other European countries, to harmonize the provision of data communication services between vehicles and to allow communication between vehicles and intelligent transport infrastructure, thus creating an environment for the emergence of cooperative intelligent transport systems (C-ITS) including the possibility of using autonomous vehicles. The use of new technologies will contribute to greater safety for road users and smoother and more efficient transport, including achieving the effect of reducing emissions in the atmosphere. C-ITS systems inform drivers in a timely and accurate way about traffic conditions and warn of dangerous locations and other problems around them. In addition, traffic control and information centers receive accurate and comprehensive information on the current traffic situation directly from vehicles. As a result, it is possible to efficiently influence traffic flow and thereby increase traffic flow and safety and reduce its negative environmental impact [28]. The most important part of the traffic management in Brno is the Central Technical Control Center (CTD), which provides dispatching activity, remote supervision of traffic lights, evaluates information from the city surveillance system, monitors public transport preferences, operates a traffic information center [29], manages the parking system, home and paid parking in the city of Brno and cooperates with authorities, police and integrated rescue systems [30].

Citizens can also contribute to improving the quality of road maintenance through a mobile application to report defects on urban roads under the management of the Brno Communication Company [31]. In order to reduce the number of towing vehicles in the block cleaning of public roads, the application is in operation with their clear terms and locations [32].

Partial goals of optimization of the parking system:

1. Improving the organization of transport, i.e., increasing the safety and fluidity of transport in the central part of the city;
2. Increasing respect for traffic-regulatory measures and reducing the number of illegally parked vehicles;
3. Reducing congestion by reducing parking space and thus reducing energy consumption and emissions;
4. Increasing security and ensuring adequate parking space for residents and visitors.

Transport organization is generally a complex issue that cannot be solved without the support of new information technologies. Since it is a method that requires spatial information (position-related information in the reference coordinate system), it is possible to use GIS technologies.

There are four areas currently operating in the system. The central area—Brno center—which has specific rules and entrance to the historical city center is guided by the entry permissions. The other three resident parking areas are linked to the Central District in the northern part and are defined by specific streets. The resident parking system has been phased in. The first stage was launched in 2018. In March 2019, a new concept for residential parking in the city of Brno was approved. More information can be found in [4].

The principle of the proposed solution is based on the use of static (parking areas) and dynamic (monitoring data) spatial data processed in the geographical information system (GIS).

The parking monitoring system is based on the use of a mobile camera system for a large number of parking spaces located in the city of Brno. This technology is original in the Czech Republic. The mobile system is more efficient than the system of a large number of static cameras due to the limited possibilities for their installation, including the necessity of solving the problem of the necessary image quality.

The resident parking areas are shown in Figure 1, together with the types of traffic signs displayed. The red zone A (center of the city) is the entrance restricted zone (resident mode), the green zone B (central area of the city) is without entry restriction (subscription and visitor mode), blue zone C (edge zone) is not controlled during the day. The zones differ in the price of parking, the most expensive being the central zone. Examples of traffic signs illustrate how a car parking area in a given street zone is marked.

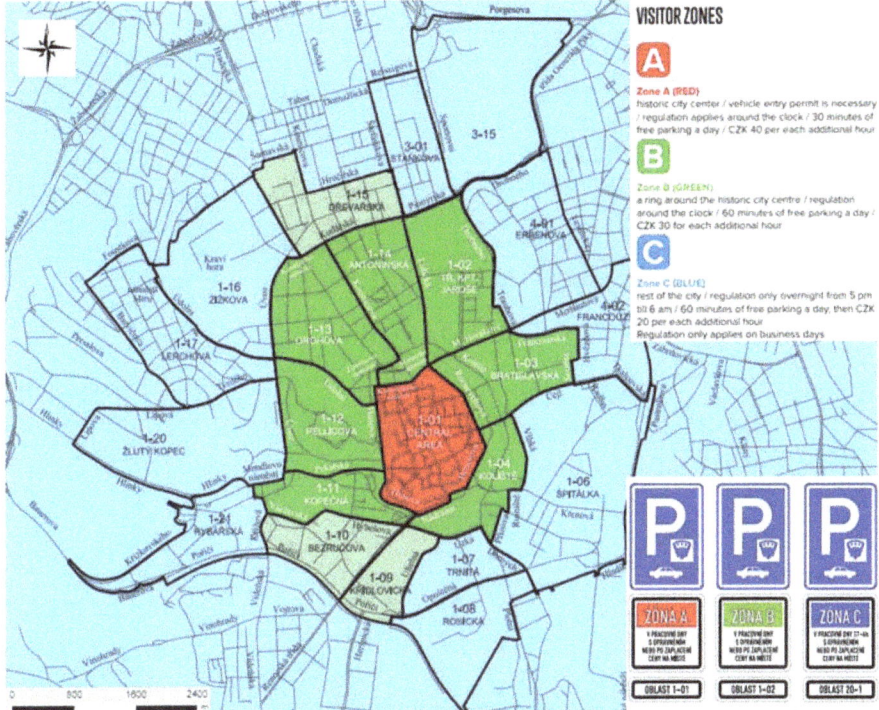

Figure 1. Residential parking areas map [14].

In Figure 1, areas with authorized parking are enclosed in red. Area 1-01 is the central area; areas 1-02, 1-13 and 1-14 are named after the area's most important street. Road signs in partial areas always include a supplementary table that identifies the area, e.g., 1-01, and specifies its restriction.

Resident parking space (Residents) it is intended for a natural resident in a demarcated area or for a property owner in a demarcated area.

Subscription parking place (Subscribers) it is intended for an entrepreneurial natural or legal person established in the demarcated area or for an entrepreneurial natural or legal person with an establishment in the demarcated area.

Visiting persons (Visitors) are neither subscribers nor residents. In the areas of the new parking system, which are marked with an orange band mark, short-term parking is free (60 or 30 min by price band, once every 24 h), then paid via parking meters or mobile app. Another solution is the use of parking houses or parking places. Persons going to visit a particular resident can ask for a temporary parking permit. Each resident (including children and non-residents) can split up to 150 parking hours a year.

Parking houses are buildings for toll parking in the center. Currently, there are eight parking houses with a capacity of 1310 parking spaces.

4. Materials and Methods

The parking system model is based on control data, which consists of three parts.

1. **Static spatial data** locate parking spaces, arising from the project or geodetic survey, and are formed by:

- Parking area polygons;

- The position of vertical traffic signs;
- Residential parking area boundaries.

Horizontal road markings are represented by polygons, and vertical markings are represented by points. In addition to the position, the data includes other attributes such as identification codes, types of stall, Zones of Parking Stages (ZPS) capacity, operating hours, etc.

2. **Dynamical spatial data** consist of data generated by mobile integrated data collection using camera recordings of localized Global Navigation Satellite Systems (GNSS) technologies. Data are captured on the basis of attributes, which are used to start the camera in place of triggers. These data are comprised of:

- Monitor triggers;
- Links;
- Routes and waypoints;
- The street network.

3. **Attribute data** are used for the parametric settings of the camera system based on the orientation of the parking lot to the monitoring vehicle (perpendicular or inclined position of the vehicles, left, right). The data contain these items:

- Vehicle identifiers;
- Driver identifiers;
- Track identifiers.

Figure 2 is a diagram of the data flow between components of the parking system.

The basic pillar of control data is the existence of the polygons of pay stall areas and vertical traffic signs in vector form. The basis for creating polygons is a geodetic survey or project. The result is then a digital map with the polygons of parking areas within the paid parking zone. The Brno Communication Center uses GIS software from City Data Software, Ltd. (CDSw), in which it records the passport data of roads and their features in the city of Brno. One of the applications of this GIS software is Transport, where both horizontal and vertical traffic signs are recorded.

These polygons represent surfaces in individual layers:

- Residential parking areas;
- License plate parking for disabled people;
- Parking for disabled people;
- Prohibited areas—entrances, green areas, obstacles;
- Supply (with authorization);
- Reserved parking spaces—doctors, restaurants, shops, etc.;
- Reserved parking space for a contract/state administration institutions—courts, police, etc.;
- Turn parking;
- Turn parking with time limitation and authorization (visits);
- For motorcycles;
- Closure to residential parking–block cleaning, closing off, etc.

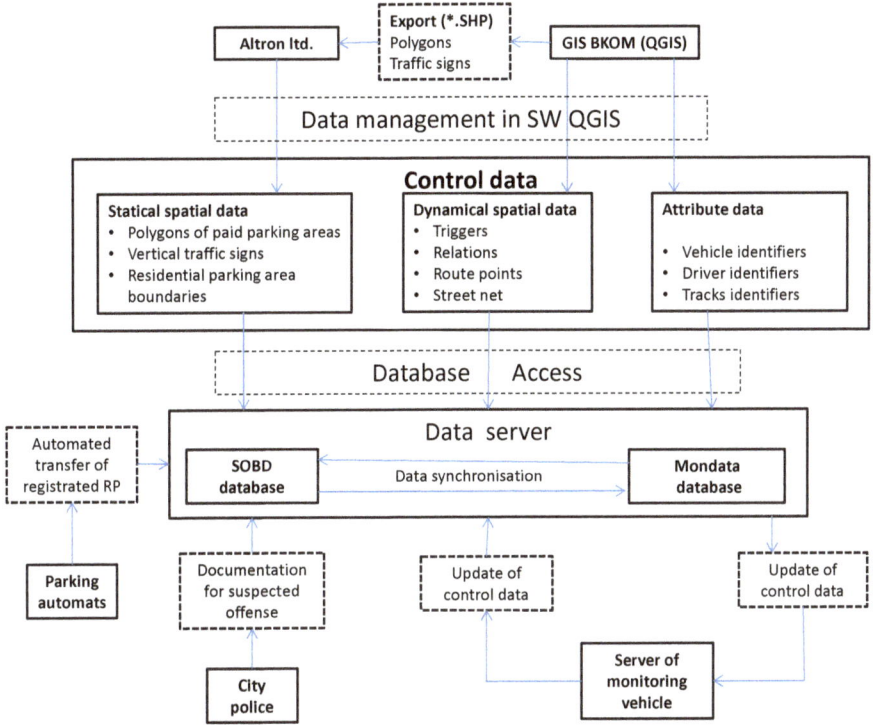

Figure 2. Scheme of data flow between parking system components.

Each base layer polygon has its attribute table, which contains the object identification, time validity, city area, zone type, standing type (longitudinal, perpendicular, oblique), paid parking area designation, street network location, operating time according to the vertical road sign, parking spaces, editing history, etc. These polygons represent the entire parking space, which is usually a few parking spaces. Figure 3 shows an excerpt from a GIS Transport section designed by CDSw in the GISServer on the NexusDB platform.

There are sections of residential parking (fully green), forbidden areas, greenery, etc. (filler—brown grid), disabled places for a specific license plate (green grid), reserved parking places for doctors, restaurants, shops, etc. (purple hatch).

Figure 4 shows a view of the GIS Transport section with additional vertical traffic signs (traffic sign symbols) and a photograph of the parking meter and a vertical traffic sign indicating the parking area in the terrain that is part of the data stored in the database. There is also an example of a photograph of a border of a parking area in the terrain with blue lines. The vertical traffic sign is registered in the Transport application by a point element, with the cell displayed according to the created style in its actual form. An important item of traffic sign attributes is the validity that uniquely links the sign with the resident parking system. Sign attributes, like parking meters, contain a link to photo documentation of the sign in the field. Taking photo documentation, an essential part of the control data, is facilitated by the mobile application from CDSw company (San Diego, CA, USA), Praha, CR. This allows you to take photos of traffic signs through smart phones and assign them to the appropriate point in the Traffic app, so the operator will record the change and can respond to the change. If a new sign is taken, its photo can be assigned to a photo-point with coordinates found by the mobile, and the operator at the PC then assigns photo documentation to point attributes after creating the tag's data point.

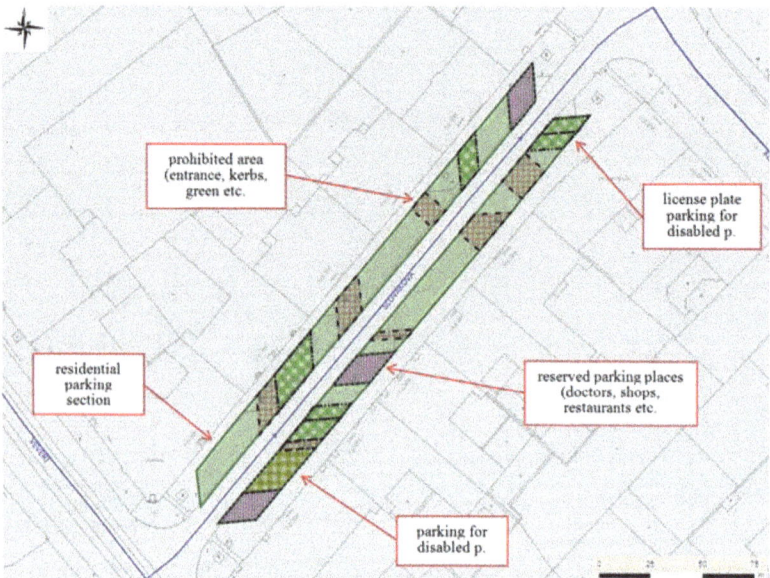

Figure 3. Sections of residential parking in the Transport application.

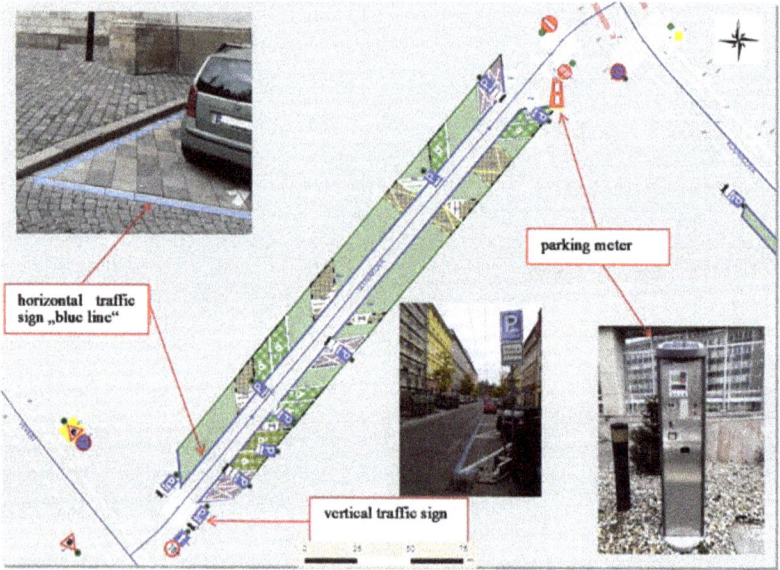

Figure 4. Sections of resident parking with traffic signs.

An example showing how to locate the control data is shown in Figure 5. This illustration shows street polygons, control points, vertical traffic signs, and triggers. Paid parking areas are highlighted in pink.

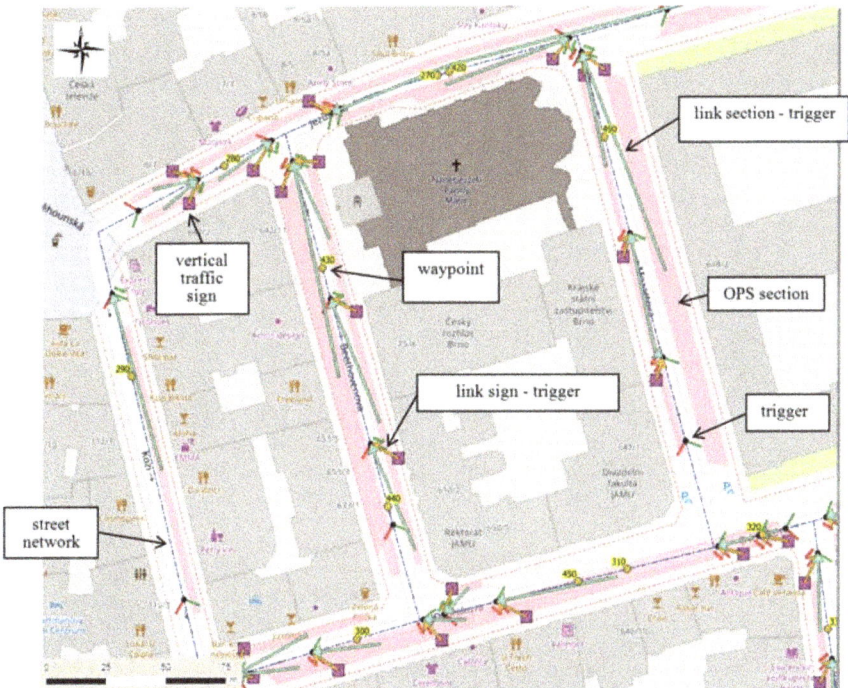

Figure 5. Monitoring control data—sample from the Monitoring Data (MONDATA) database.

Monitoring control data includes so-called monitoring triggers, links between triggers and traffic signs, routes and points, including a street network. Monitoring control data is an important part of the entire system. Monitoring triggers are points with certain coordinates that trigger and stop monitoring when the monitoring car is in the same position. The trigger also contains parameters for a particular section of the pay-per-area section, for example, the type of parking. This allows the recognition process to estimate the angle of rotation of the registration number of the parked cars relative to the monitoring vehicle, thereby increasing the chance of it being correctly recognized from the camera image taken by the camera. The monitoring trigger also includes a preset azimuth under which the vehicle must be moved to trigger the monitoring. Triggers also stop monitoring at the end of the street to avoid recording vehicles outside the pay zone. Links between triggers and traffic signs give a more accurate pairing of road marking documentation and parked vehicle documentation. These are the lines between the trigger and the polygon, representing the parking space in the paid parking area or between the trigger and the vertical traffic sign.

The background data (signs and sections of paid parking) prepared in this way enter the GIS system SOBD, which uses two PostgreSQL™ databases. This is the SOBD database and the Monitoring Data (MONDATA) database. Data management is then performed through the open source geographic information system QGIS™. Data embedded in the SOBD database is synchronized with the MONDATA database data and vice versa, with set rules for editing. The MONDATA database is designed to create and edit triggers, links, waypoints, and street network lines. Some data, such as the attributes of monitoring cars, the drivers of these cars and routes, are without geometry. All this data is necessary and after connecting the monitoring car in the garage to the computer network, this data is inserted into the server in the car.

4.1. Field Data Collection

Figure 6 shows a monitoring vehicle equipped with a camera system. It is a hybrid car equipped with a camera system with control and storage equipment. Control monitoring is performed over defined routes. The number of vehicle plates in residential parking spaces are also recorded, as are traffic signs defining or specifying parking in these parking areas. The result of this recording system is documentation, i.e., photographs in digital form with a mark containing the date and time of the acquisition. Photos are taken five times per second. Stop and start monitoring is automated thanks to the parameters specified for each route.

Figure 6. Monitoring car.

The data is generated as the output from the GIS data base of the Brněnské komunikace (BKOM) data manager. All control data is managed in the QGIS geographic information system. Spatial static data is stored in the SOBD database; spatial dynamic and attribute data is stored in the MONDATA database to monitor parking rules. A field monitoring unit is a monitoring car equipped with cameras and a software server. Field data and documentation is sent to the server and evaluated. In addition to the monitoring vehicle, data from parking meters are transferred to the database. If a violation of the parking rules is found in the evaluation, the documentation together with the data is sent to the database of potential offenses and the municipal police then deal with the offense. Acquired data are subject to regulation by the EU General Data Protection Regulations (GDPR).

The data server is accessed by data from parking meters (number plates of parking cars and paid parking time) and is used to update the control data of the monitoring vehicle. The SOBD database generates potential offenses data, which is evaluated by the Municipal Police. Police evaluate offenses on the basis of registration plate number (RZ) and other parameters (color, etc.) obtained from image and text recognition and evaluate the quality of the recording material in terms of the legal enforceability of the offense.

An example for displaying the saved track of the monitoring vehicle in the form of TrackPoints in the GIS is shown in Figure 7; an example for displaying TrackPoint attributes from a database is shown in Figure 8.

Figure 7. View of saved car positions as TrackPoints—sample from the MONDATA database.

Figure 8. TrackPoints attribute listing from MONDATA database via QGIS.

Routes are used to ensure complete coverage of the site. Each route consists of waypoints whose serial numbers provide navigation to the driver as they drive through the streets. The waypoints, as well as triggers, are registered by the system using GNSS positioning. Occasionally, it is necessary to travel several times so that the monitoring vehicle passes through all the streets with parking areas within the paid parking area. This is especially true for one-way streets. The data also includes a street network, which is represented by line elements that contain topographic information. Monitoring triggers and waypoints are designed to lie directly on the line.

TrackPoints are snapped onto lines representing a street network using an algorithm to obtain information from a monitoring trigger or a waypoint. Therefore, if there is a trigger or a waypoint at the origin of the TrackPoint, this information is also written to the TrackPoint. Once identified in the GIS software, you can view the point attributes. It is essential to have an existing unique attribute item within the point (record) identifier.

Data generated during operation of the monitoring vehicle are generated in addition to the camera records and so-called "Tracks" and "TrackPoints", vehicle position records and other data. Individual tracks (rides) contain a set of TrackPoints, i.e., records of the monitoring car's GNSS position. Furthermore, the track stores information about the driver, the monitoring car, the defined route, the time of the journey, the number of registered registration plates of parked vehicles and more. TrackPoints—that is, individual points—store a specific driving identifier and obtain detailed information that is stored five times per second. This information includes the position, direction of movement, speed, time, and more.

The monitoring vehicle carries out monitoring and documentation based on control data. The data is stored in a database on a server located inside the car. The database can be managed directly on the car's server via remote access, but a database located on the local server of Brno's communication network is used by default and data synchronization of both servers is started after the monitoring car is connected in the garage.

The functionality of the parking system is ensured by checking its compliance with the set rules. These are based on traffic signs and the public decree of the city of Brno. Random checks are carried out by the Municipal Police on the basis of patrols or on the basis of a notice. The systematic technological control is then carried out by the monitoring car, which passes through the streets, falling into the areas of resident parking several times a day.

4.2. Analysis and Methods of Data Evaluation

The passage of the monitoring vehicle is controlled by an algorithm based on graph theory. It is a modified Chinese postman algorithm (The Chinese Postman Problem) with restrictive conditions resulting from traffic signs, with a required time interval of 30 min to capture repeated data, and normal traffic density parameters.

The output of data evaluation is the identification of an offense against parking rules and generation of the necessary documentation on the offense for administrative proceedings via the municipal police. A general mathematical model for evaluating the parking rule is applied for evaluating the data and identifying the offense. The last two items are the subject of this article. The model concept is based on a set of rules on parking that have been approved by the competent administrative authority. Mathematically, we can express the model by function P. It is a Boolean function that depends on the specific area (grounds) A and looks like this:

$$P_A(r, l, t) = R \ x \ (L \cap T) \tag{1}$$

where

$r \in R$ is the vehicle registration plate number,
$l \in L$ is the parking area (parking place),
$t \in T$ is the date and time of parking.

Index A of function P in Equation (1) means that there are different parking rules in each area. If the P_A function is True, then the vehicle with the registration plate r is parked at l at a given time t in accordance with the rules, If the P_A is False, then a rule violation is suspected. The aim is to address the control of the established parking rules in the given areas A and their localities L. The checks will be in two ways:

1. Systematically (by automation), using a monitoring system whose data and documents will be evaluated in appropriate software;
2. Randomly, by the city police lineman.

The monitoring is then carried out on the basis of this data. They determine at which point which camera starts recording and navigate the driver's route. The monitoring results (tracking and TrackPoint records) are transferred to the MONDATA database after connecting the control car in the garage to the computer network. At the same time, the process of automated evaluation of scanned photographic documentation is performed by means of scripts, which ensure the comparison between the individual parking places on the traversed route and attributes of polygons from the SOBD database at the time of scanning. During this process, the photographic documentation showing parking spaces that were not regulated by the resident parking system at the time the monitoring car passed by, i.e., outside the time stamp on the traffic sign indicating the residential parking area, will be deleted. These are mainly parking places where the reserved parking regime is valid for a part of the day, and parking places for disabled people, places of entry, sections of block cleaning, etc. The evaluation uses the position and time of the control vehicle at the time of recording. This information is part of the scanned data (TrackPoints).

5. Results

After completing this data evaluation process, the next phase of comparison is followed, where the remaining photo documentation is subjected to the automated license plate recognition of the parked vehicles, and then these license plates are compared with the SOBD database, which includes information on vehicle registrations parked in regulated resident parking lots. The insertion of their car's license plate into the system is done by drivers when parking in a parking meter or mobile application.

The documentation (in which the registration number of the parked car is not stored) in the SOBD database, together with the identification data of the record and photographic documentation of the traffic sign to which the potential offense was related, is submitted to the Municipal Police for review. If it is found to be a misdemeanor, it is sent to the Department of Transport Administration Activities of the City of Brno (MMB). The time between two consecutive passes of the monitoring vehicle is a minimum of 30 min. If the same parked vehicle is identified on both records, it is subject to analysis on the legitimacy of its parking.

Technical support of the system (service of monitoring vehicles and data server administration) is provided by Eltodo, Praha, CR, the stock company.

In Figure 9a–e is an example of the output of the web documentation of suspected offenses against residential parking generated from the SOBD system for the municipal police. Some data are anonymized for privacy reasons (GDPR). The key part is the photo documentation (Figure 9d) and text with image analysis of the identified license plate (Figure 9b) of the vehicle during the first and second passage. The Municipal Police can choose from more photographs than the specified ones to be included in the offense report. It has a similar structure to this document, but is supplemented by vehicle data obtained from the vehicle register and its owner and specification of the legal provision of the offense. This protocol shall be sent to the owner of the vehicle through the competent authority.

For undeniable image documentation, it is important that in both series of photographs taken after a minimum of 30 min of each pass, a traffic sign indicating the parking area (Figure 9c) is identified based on the satellite location of the route with the marker kept in that place in the SOBD database. In terms of GNSS localization requiring a free horizon for satellite signal reception, it has often been the case in dense urban areas that localization accuracy has been reduced or even GNSS signal reception has been interrupted, causing problems with mobile camera recording. The incompleteness of the image data requires a repeated pass of the same street. In Figure 9d is an example of an incomplete image sequence at the first pass (lack of a traffic sign image) that has necessitated a subsequent second pass, and a subsequent third pass after an offense was identified 30 min later.

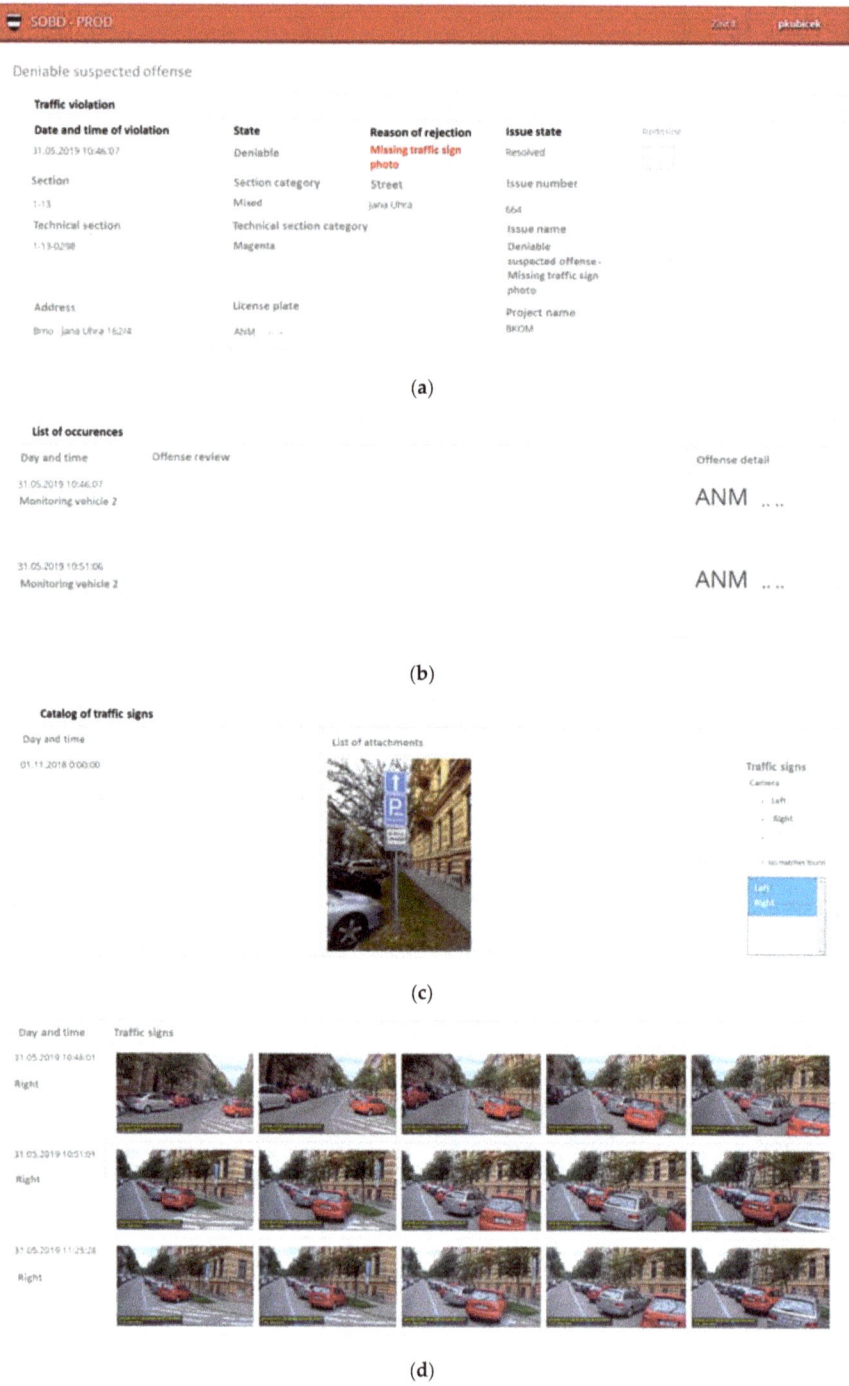

(a)

(b)

(c)

(d)

Figure 9. *Cont.*

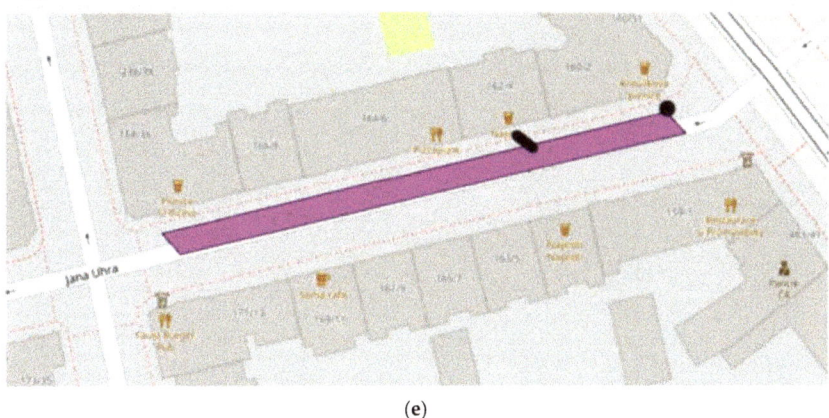

(e)

Figure 9. (**a**) Documentation of suspected offense (header). (**b**) Documentation of suspected offense—identification of vehicle registration number (anonymized). (**c**) Documentation of suspected offense—identification of the parking area traffic sign from the traffic sign catalog in SOBD. (**d**) Documentation of suspected violation—video sequences of transit records. (**e**) Documentation of suspected violation—localization of parking area (purple), parking area traffic signs (black dot) and vehicles suspected of offense (short black line) from QGIS.

At the same time, the parking area is identified, and in the map section (Figure 9e), the specific location of the standing vehicle suspected of a parking violation (short black line) and the corresponding traffic sign indicating the parking area (black dot) is marked.

In the initial phase of trial operation, the system showed up to 60%, in terms of legal enforceability, unrecognizable offenses. This high percentage was subjected to reverse analysis and was found to be due to several factors:

- The need to calibrate the system;
- The illegibility of the registration plate;
- Reduced visibility (snow, rain, fog, night);
- As a result of temporary traffic signs which are not part of the control data (temporary mobile road signs).

The need for system calibration was triggered by the uncertainty of the starting trigger location resulting from the uncertainty of vehicle positioning by GNSS due to the considerable obscurity of the horizon in dense buildings. This resulted in a shift in the image sequence in which not all the necessary objects were recorded for correct image evaluation (e.g., missing an appropriate vertical traffic sign or inappropriate camera angle). The problem was solved by changing the position of the trigger points in the control data by adjusting the orientation of the cameras on the vehicle, and refining the position was solved by using an Inertial Measurement Unit (IMU).

In Table 1, the statistics document the benefits of system calibration by comparing the percentage of qualitative data improvement by the offense demonstrability parameter. After the calibration of the system, the quality of the monitored data was improved by 25% in terms of demonstrable offenses.

Table 1. Statistics of evaluation of qualitative improvement in the provability of the offense.

Period	Provability of Offense [%]	Range of the Sample
Before calibration—from Apr. to Aug. 2019	28	13,251
After calibration—Oct. 2019	35	1695

Table 2 shows in detail the structure of the percentage of the individual parameters in the proportion of data excluded due to the non-verifiability of the offense, which are still subject to improvement.

Table 2. Structure of data discarded by the system due to the non-verifiability of the offense.

Reason for Non-Demonstrable Offense	Before Calibration [%]	After Calibration [%]
Poorly captured vertical road sign (from wrong position)	2	2
Hidden vehicle registration number	3	1
Bad vehicle photo	3	6
Missing street name	18	3
Wrong type of parking	3	2
Bad section type	3	7
Wrong section code	0	0
Wrong time interval	2	1
Incorrectly recognized vehicle registration plate	14	14
Missing catalog photo of road sign	3	4
Hidden road sign	2	1
Missing road sign	11	12
Night shot	7	10
Other	14	19
Temporary traffic sign	7	4
Unreadable road sign	2	1
Vehicle out of section	5	9
ZTP card holder	1	4
Duplicate offense	0	0
Non-sendability of data	0	0

The seemingly high percentage of inadequate documentation for offense proceedings is due to a large number of limiting factors regarding the legal unquestionability of the offense. Failure to comply with any of them makes the offense ineligible.

6. Discussion

The main objectives of this project were to prevent long-term (in many cases unauthorized) parking of vehicles in the center of Brno city, thus blocking the parking capacity, and ensuring a sufficient number of parking spaces for Brno residents and visitors by increasing the turnover of parking, which was accomplished by the introduction of the parking system.

The whole technology is fully automated; the human factor enters the process in the initial phase of setting and modifying the parameters and then at the end in evaluating offenses in administrative proceedings. The benefit of the technology lies in the automated processing of a large number of records that could not be evaluated manually. Another benefit is to free the city center from long-term parked vehicles and increase parking availability.

Our solution utilizes the integration of image data referencing position and time by GNSS satellite technology, while utilizing database resources and Open Source GIS. The methods of graph theory and image recognition were used in algorithmization. In related works (Chapter 2), the methods of the Internet of Things (IoT) [5,22], were used in particular, while Machine Learning (ML) [9,12], and image processing were used [11,12], in other contexts than in our solution. The use of a modified Chinese postman algorithm (The Chinese Postman Problem) has proved to work best for our project.

From the perspective of those interested in parking, information is available on the website or in the mobile application about the availability of parking spaces at the place of need. The availability of parking information is now a frequently used IoT-based application.

The system of static cameras is used for the evaluation of vehicle entrances to the center through a mechanical barrier. For monitoring barrier-free parking spaces in urban conditions, the method of mobile monitoring has clearly proved its worth. The use of static cameras requires multi-angle shots, generating increased demands on the number and installation of cameras, but the success rate of image recognition is high and is up to 89% [33–35].

The illegibility of the license plate is due to multiple factors, e.g., a dirty or missing license plate or one scanned at an unfavorable angle. The technology has only been operational for one year and is constantly being evaluated and improved. The main aim of the system is to force drivers to respect the parking rules.

Through the calibration process of the mobile parking monitoring system, the quality of data was improved from 28% to 35% in terms of provable offenses. We note that this percentage is made up of excellent data in terms of meeting all the necessary conditions for successful enforcement of offenses. In [11], a similar improved success rate from 24% to 43% is reported. The error rate is very individual and depends on local conditions.

Brněnské Komunikace company, Brno, CR is the operator of the resident parking system, but the evaluation is automated and camera records are only available to the municipal police. This complicates the correct setting of control data, as GIS operators cannot access the monitoring car documentation for privacy reasons (according to GDPR), and the municipal police initiative is the only response for correcting settings.

The electronic availability of parking rule information via smart communication devices, which most people use, contributes to the acceptance of the whole system and its increasing popularity. A similar system of regulated parking also operates in the capital of the Czech Republic, Prague.

7. Conclusions

The article introduced a system for the intensive use of parking spaces. It is an automated system based on monitoring the situation in parking places using a mobile camera set equipped with data storage and means for data transfer to the central database. The data is evaluated in a suitable GIS-based software according to parking rules. It is this evaluation that is a critical part of the whole system. In practice, there have been cases where a complaint was sent to municipal police on parking rules violation, which was not fully justified. Therefore, the further development of the described system will be aimed at improving the evaluation process in order to minimize unauthorized cases. Detailed information about the parking system is very well presented on the web or in the form of a smartphone application [36], which lists all the possible parking situations in Brno.

Prior to the introduction of the system, the parking yield was around 30% compared to the current situation after the introduction of the parking system.

In the city center, where the entrances are regulated by a bolt system, the use of static cameras is effective, while for the control of parking in the street network it is cheaper and more flexible in terms of video recording to use a mobile camera system on a vehicle.

The main benefit of the technical solution was the integration of image, location and time data acquired by a moving vehicle equipped with a camera system in combination with GNSS navigation technology.

After the introduction of a monitored parking system, the overall rate of violations of parking rules assessed from the records of offenses in the monitored areas decreased. The statistics of the municipal police show that, in 2019, parking was unjustified, on average, 10% of the time, with a monthly variability of 17% to 7%. Mobile data collection has a number of disturbances affecting the

Symmetry **2020**, *12*, 542

quality of data, and for the correct evaluation of offenses it is necessary to have flawless data. All formulated research goals were fulfilled and the GIS environment proved to be usable both for data integration and for their subsequent analysis and efficient distribution of purpose outputs.

The continuation of the research will consist of improving the data quality in statistically identified problematic parameters, especially license plate recognition, the problem of changes resulting from temporary traffic signs, and traffic in reduced visibility.

The introduction of a regulated parking system in the center of Brno and in the surrounding areas contributed significantly to improving the availability of parking for residents, subscribers and visitors. The system of parking houses and car parks will be gradually extended in the city of Brno, along with the modernization of the transport network. The new parking system offers many advantages. The goal is not to collect money purposefully, but to achieve effective renewal of parking, thus allowing more people to park and deal with the necessary issues. The limitation makes sense where residents have a problem with parking or where there is a day-to-day overpressure. In most of Brno, where a new system was introduced, regulation is only introduced at night. In the most critical places (the first two zones), there is regulation even during the day.

The City of Brno is working to make the most of the use of information technology in conjunction with smart communication devices and smart applications to improve city life.

With the introduction of the automated parking system in the city of Brno as the last part of the SOBD system, one of the important databases for the future possibility of intelligent communication between vehicles, solved in the project C-ROADS Czech Republic, was created.

Author Contributions: P.K. provided support materials and is the administrator of the BKOM company parking system, D.B. elaborated the literature review and system model, J.B. wrote the introductory section, a chapter on experimental results, discussion and conclusion, O.Š. conducted the overall editing of the article the a professional translation. The roles of D.B. and J.B. are in the professional consultancy of GIS technology in support of BKOM. All authors have read and agreed to the published version of the manuscript.

Funding: This paper was elaborated with the support of European project C-ROADS and with the theoretical cooperation of Brno University of Technology, Specific Research Project FAST-S-18-5324 and FAST-J-20-6374.

Acknowledgments: Thanks to all reviewers for their suggestive comments.

Conflicts of Interest: The authors declare no conflict of interest.

References

1. Map Portal of the City of Brno. Available online: https://gis.brno.cz/portal/ (accessed on 31 March 2020). (In Czech).
2. Geoportal State Administration of Land Surveying and Cadastre. Available online: https://geoportal.cuzk.cz (accessed on 31 March 2020).
3. Czech Space Portal. Available online: http://www.czechspaceportal.cz/3-sekce/its-inteligentni-dopravni-systemy/ (accessed on 31 March 2020). (In Czech).
4. Project Parking System Operational Program Transport, 2.3—Improving Traffic Management and Increasing Traffic Safety. Call Number 27. European Union, European Structural and Investment Funds. 2015–2020. Available online: https://ec.europa.eu/transport/sites/transport/files/2019-transport-in-the-eu-current-trends-and-issues.pdf (accessed on 1 April 2020).
5. Zantalis, F.; Koulouras, G.; Karabetsos, S.; Kandris, D. A Review of Machine Learning and IoT in Smart Transportation. *Future Internet* **2019**, *11*, 94. [CrossRef]
6. Belkhala, S.; Benhadou, S.; Boukhdir, K.; Medromi, H. Smart Parking Architecture based on Multi Agent System. *Int. J. Adv. Comput. Sci. Appl.* **2019**, *10*, 378–382. [CrossRef]
7. Fiore, U.; Florea, A.; Lechuga, G.P. An Interdisciplinary Review of Smart Vehicular Traffic and Its Applications and Challenges. *J. Sens. Actuator Netw.* **2019**, *8*, 13. [CrossRef]
8. Ning, Z.L.; Huang, J.; Wang, X.J. Vehicular Fog Computing: Enabling Real-Time Traffic Management for Smart Cities. *IEEE Wirel. Commun.* **2019**, *26*, 87–93. [CrossRef]

9. Pawlowicz, B.; Salach, M.; Trybus, B. Smart City Traffic Monitoring System Based on 5G Cellular Network, RFID and Machine Learning. In *Engineering Software Systems: Research and Praxis*; Kosiuczenko, P., Zielinski, Z., Eds.; Advances in Intelligent Systems and Computing; Springer International Publishing: Basel, Switzerland, 2018; Volume 830, pp. 151–165. [CrossRef]

10. Toh, C.K.; Cano, J.C.; Fernandez-Laguia, C.; Manzoni, P.; Calafate, C.T. Wireless digital traffic signs of the future. *IET Netw.* **2019**, *8*, 74–78. [CrossRef]

11. Hung, F.H.; Tsang, K.F.; Wu, C.K.; Wei, Y.; Liu, Y.C.; Hao, W. Cost and Time-Integrated Road-to-Park Cruising Prevention Scheme in Smart Transportation. *IEEE Access* **2019**, *7*, 54497–54507. [CrossRef]

12. Avalos, H.; Gomez, E.; Guzman, D.; Ordonez-Camachol, D.; Roman, J.; Taipe, O. Where to park? Architecture and implementation of an empty parking lot, automatic recognition system. *Enfoque UTE* **2019**, *10*, 54–64. [CrossRef]

13. Al Maruf, M.A.; Ahmed, S.; Ahmed, M.T.; Roy, A.; Nitu, Z.F. A Proposed Model of Integrated Smart Parking Solution for a city. In Proceedings of the 1st International Conference on Robotics, Electrical and Signal Processing Techniques (ICREST), Dhaka, Bangladesh, 10–12 January 2019; pp. 340–345.

14. Parking System in Brno. Available online: https://www.parkovanivbrne.cz/ (accessed on 31 March 2020). (In Czech).

15. Arellano-Verdejo, J.; Alonso-Pecina, F.; Alba, E.; Arenas, A.G. Optimal allocation of public parking spots in a smart city: Problem characterisation and first algorithms. *J. Exp. Theor. Artif. Intell.* **2019**. [CrossRef]

16. Karpenko, A.; Kinnunen, T.; Madhikermi, M.; Robert, J.; Framling, K.; Dave, B.; Nurminen, A. Data Exchange Interoperability in IoT Ecosystem for Smart Parking and EV Charging. *Sensors* **2018**, *18*, 4404. [CrossRef]

17. Honarvar, A.R.; Sami, A. Multi-source dataset for urban computing in a Smart City. *Data Brief* **2019**, *22*, 222–226. [CrossRef]

18. Rodriguez-Hernandez, M.A.; Gomez-Sacristan, A.; Gomez-Cuadrado, D. SimulCity: Planning Communications in Smart Cities. *IEEE Access* **2019**, *7*, 46870–46884. [CrossRef]

19. Al-Hamadi, H.; Chen, I.R.; Cho, J.H. Trust Management of Smart Service Communities. *IEEE Access* **2019**, *7*, 26362–26378. [CrossRef]

20. Boldrini, C.; Bruno, R.; Laarabi, M.H. Weak signals in the mobility landscape: Car sharing in ten European cities. *EPJ Data Sci.* **2019**, *8*, 7. [CrossRef]

21. Gu, T.Q.; Kim, I.; Currie, G. To be or not to be dockless: Empirical analysis of dockless bikeshare development in China. *Transp. Res. Part A-Policy Pract.* **2019**, *119*, 122–147. [CrossRef]

22. Sotres, P.; Lanza, J.; Sanchez, L.; Santana, J.R.; Lopez, C.; Munoz, L. Breaking Vendors and City Locks through a Semantic-enabled Global Interoperable Internet-of-Things System: A Smart Parking Case. *Sensors* **2019**, *19*, 229. [CrossRef] [PubMed]

23. Weber, M.; Zarko, I.P. A Regulatory View on Smart City Services. *Sensors* **2019**, *19*, 415. [CrossRef] [PubMed]

24. Mohammadreza, J.-N. Exploring the effect of group decision on information search behaviour in web-based collaborative GIS-MCDA. *J. Decis. Syst.* **2019**, *28*, 261–285. [CrossRef]

25. Pagany, R.; Marquardt, A.; Zink, R. Electric Charging Demand Location ModelA User- and Destination-Based Locating Approach for Electric Vehicle Charging Stations. *Sustainability* **2019**, *11*, 2301. [CrossRef]

26. Jakimavicius, M.; Burinskiene, M. A Gis and Multi-Criteria-Based Analysis and Ranking of Transportation Zones of Vilnius City. *Technol. Econ. Dev. Econ.* **2009**, *15*, 39–48. [CrossRef]

27. Matuszková, R.; Heczko, M.; Radimský, M.; Kozák, P. The Comparison of the Parking space dimensions to the modern car fleet of the selected european countries. In *International Conference on Traffic and Transport Engineering*; City Net Scientific Research Center Ltd.: Belgrade, Serbia, 2016; pp. 924–928. ISBN 978-86-916153-3-8.

28. Portal C-Roads. Available online: https://c-roads.cz/ (accessed on 31 March 2020). (In Czech).

29. Traffic Information Center. Available online: https://www.doprava-brno.cz/ (accessed on 31 March 2020). (In Czech).

30. Brněnské Komunikace Portal. Available online: https://www.bkom.cz (accessed on 31 March 2020). (In Czech).

31. Application for Traffic Error Reporting. Available online: https://www.brnaciprobrno.cz (accessed on 31 March 2020). (In Czech).

32. Application of Block Cleaning of Public Roads. Available online: https://cisteni.bkom.cz/ (accessed on 31 March 2020). (In Czech).

33. State Administration of Land Surveying and Cadastre. Available online: www.cuzk.cz (accessed on 31 March 2020).
34. Kubicek, P. GSI support for resident parking monitoring in Brno. In Proceedings of the International Seminar XX. Transportation-Engineering Days-Mobility and Logistics in Cities and Regions of the Future, Mikulov, Czech Republic, 20 March 2019; pp. 58–67, ISBN 978-80-270-5551-7. (In Czech).
35. De Almeida, P.R.; Oliveira, L.S.; Britto, A.S., Jr.; Silva, E.J., Jr.; Koerich, A.L. PKLot—A robust dataset for parking lot classification. *Expert Syst. Appl.* **2015**, *42*, 4937–4949. [CrossRef]
36. Eltodo, A.S. Mobile Control Mechanism—Control Data. Ver. 1.0/2018. (In Czech). Available online: www.eltodo.cz (accessed on 1 April 2020).

Article

Design and Implementation of Virtual Private Storage Framework Using Internet of Things Local Networks

Hwi-Ho Lee [1], Jung-Hyok Kwon [2,*] and Eui-Jik Kim [1,*]

1   School of Software, Hallym University, 1 Hallymdaehak-gil, Chuncheon, Gangwon-do 24252, Korea;
    leehh7028@hallym.ac.kr
2   Smart Computing Laboratory, Hallym University, 1 Hallymdaehak-gil, Chuncheon,
    Gangwon-do 24252, Korea
*   Correspondence: jhkwon@hallym.ac.kr (J.-H.K.); ejkim32@hallym.ac.kr (E.-J.K.); Tel.: +82-33-248-2333;
    Fax: +82-33-242-2524

Received: 30 January 2020; Accepted: 18 March 2020; Published: 24 March 2020

Abstract: This paper presents a virtual private storage framework (VPSF) using Internet of Things (IoT) local networks. The VPSF uses the extra storage space of sensor devices in an IoT local network to store users' private data, while guaranteeing expected network lifetime, by partitioning the storage space of a sensor device into data and system volumes and, if necessary, logically integrating the extra data volumes of the multiple sensor devices to virtually build a single storage space. When user data need to be stored, the VPSF gateway divides the original data into several blocks and selects the sensor devices in which the blocks will be stored based on their residual energy. The blocks are transmitted to the selected devices using the modified speedy block-wise transfer (BlockS) option of the constrained application protocol (CoAP), which reduces communication overhead by retransmitting lost blocks without a retransmission request message. To verify the feasibility of the VPSF, an experimental implementation was conducted using the open-source software *libcoap*. The results demonstrate that the VPSF is an energy-efficient solution for virtual private storage because it averages the residual energy amounts for sensor devices within an IoT local network and reduces their communication overhead.

Keywords: constrained application protocol; Internet of Things local network; sensor device; speedy block-wise transfer; virtual private storage

---

1. Introduction

As the amount of data produced by individuals has increased explosively in recent years, so has the demand for storage solutions for efficiently storing, accessing, and managing user data [1–4]. Two representative storage solutions are typically considered for storing user data: cloud storage and on-premises storage. Cloud storage stores user data on remote servers maintained by third-party service providers such as Google Drive, Apple iCloud, and Dropbox [5–7]. On-premises storage stores user data on local servers dedicated to specific users such as network-attached storage (NAS) [8]. Cloud storage does not require hardware installation at a personal site but is vulnerable to data leakage because it can be accessed by anyone over the Internet [9–11]. Consequently, users are reluctant to store critical data using cloud storage. In contrast, on-premises storage can help to prevent data leakage because the user can set the security policy of the local server to control how the data are stored and who has access [12,13]. However, local storage demands a high cost to install and maintain the hardware, with the risk that all data can become inaccessible due to a local server failure [14].

Meanwhile, with the spread of the Internet of Things (IoT), embedded sensor devices interacting each other over an IoT local network have been deployed in real-world applications [15–18]. As the requirements for IoT applications become more complex, sensor devices tend to be designed with sufficient storage space (e.g., on-board memory and external memory) to perform specific operations reliably [19]. Accordingly, to solve the problems of existing storage solutions, the extra storage space of sensor devices can be merged to build local storage virtually (virtual storage) that does not require additional hardware installation costs and effectively prevents data leakage. Moreover, even when some sensor devices fail, user data stored in the other sensor devices are accessible. Even though the virtual storage is often too small due to the limited number of sensor devices or extra storage size, it can be very useful for storing various kinds of data that require privacy, but are only up to kilobyte in size, such as personal health and financial data. However, sensor devices have physical resources that are inherently limited, such as batteries and microcontroller units (MCUs) [20]. Consequently, the following must be considered when designing virtual storage using such sensor devices.

- Long lifetime: The most important role of storage solutions is to guarantee seamless and reliable data access. However, with virtual storage, some data may not be accessible due to the energy depletion of sensor devices. Each sensor device may have different residual energies—a sensor with low residual energy is discharged before other sensor devices if it is frequently used for storing data. Therefore, before storing data, the virtual storage should be able to select the sensor device based on its residual energy.
- Lightweight data transfer: For storing a large amount of data, the virtual storage must divide the data into multiple blocks and transmit the data to sensor devices in a constrained IoT environment. This causes the sensor devices to suffer from high communication overhead, which leads to a long delay and high energy consumption.

In order to build the storage using sensor devices within an IoT local network, the distributed file system (DFS) solutions such as Ceph, Lustre, Hadoop Distributed File System (HDFS), and Google File System (GFS) can be used [21–24]. The DFS makes it possible to divide a data file into several parts and store them in multiple different devices. To this end, the DFS maintains the list and information of all stored files as metadata (i.e., type, size, attributes, etc.) via name nodes or meta-servers. However, the existing DFS solutions do not consider the limited physical resources of sensor devices composing the IoT local network. Thus, DFS-based storage is highly likely to suffer from unreliable data access due to the energy-depleted sensor devices and the high communication overhead. The recent paradigm of fog computing (a.k.a., edge computing) seems to be an ideal solution to build virtual storage using sensor devices in the IoT local network, because it is expected to benefit an IoT local network by reducing energy consumption, increasing bandwidth utilization, and enhancing security/privacy compared to cloud computing [25–27]. However, when it comes to implementing virtual data storage based on the fog computing paradigm, the research on its implementation technology is still in its infancy. In order to implement virtual data storage based on fog computing, the technologies capable of logically integrating and managing the storage space of edge devices are required, and in addition, various aspects of the data storage system, such as energy consumption, bandwidth, fault tolerance, scalability, and security, should be considered. However, only a few research efforts have been put into fog computing-based virtual data storage [28,29].

This paper presents a virtual private storage framework (VPSF) using Internet of Things (IoT) local networks. The VPSF uses the extra storage space of sensor devices in an IoT local network to store users' private data, while guaranteeing expected network lifetime, by partitioning the storage space of a sensor device into data and system volumes, and, if necessary, logically integrating the extra data volumes of the multiple sensor devices to virtually build a single storage space. When user data need to be stored, the VPSF gateway divides the original data into several blocks and selects the sensor devices in which the blocks will be stored based on their residual energy. The blocks are transmitted to the selected devices using the modified speedy block-wise transfer (BlockS) option of the constrained application protocol (CoAP), which reduces communication overhead by retransmitting lost blocks without a retransmission request message [30–32]. To verify the feasibility of the VPSF, an experimental implementation was conducted using the open-source software *libcoap* [33,34]. The results demonstrate that the VPSF is an energy-efficient solution for virtual private storage because it averages the residual energy amounts for sensor devices within an IoT local network and reduces their communication overhead.

The rest of this paper is organized as follows. Section 2 describes the VPSF design in detail. Section 3 presents the results of the implementation and performance evaluation. Finally, Section 4 concludes the paper.

2. VPSF Design

The VPSF stores user data in the extra storage space of sensor devices while guaranteeing the expected network lifetime of the IoT local network. Accordingly, the VPSF averages the remaining energy of the sensor devices across the IoT local network by building virtual private storage consisting of a specific group of sensor devices, and reduces communication overhead by using the modified BlockS of the CoAP. In this section, the VPSF design is described in detail.

2.1. VPSF Architecture

Figure 1 illustrates the VPSF architecture consisting of the user device, VPSF gateway, and sensor devices. It is assumed that VPSF gateway and sensor devices are static and all sensor devices are within the transmission range of the VPSF gateway. The user device is responsible for the request for data storing and retrieval. In the figure, the user device requests to store the user data. If the user device has the data that need to be stored, it transmits both the storing request message and the data to the VPSF gateway. In contrast, it transmits only the retrieval request message to the VPSF gateway when data retrieval is needed. The storing and retrieval request messages are identified by the VPSF gateway through a message type field in the header of each message.

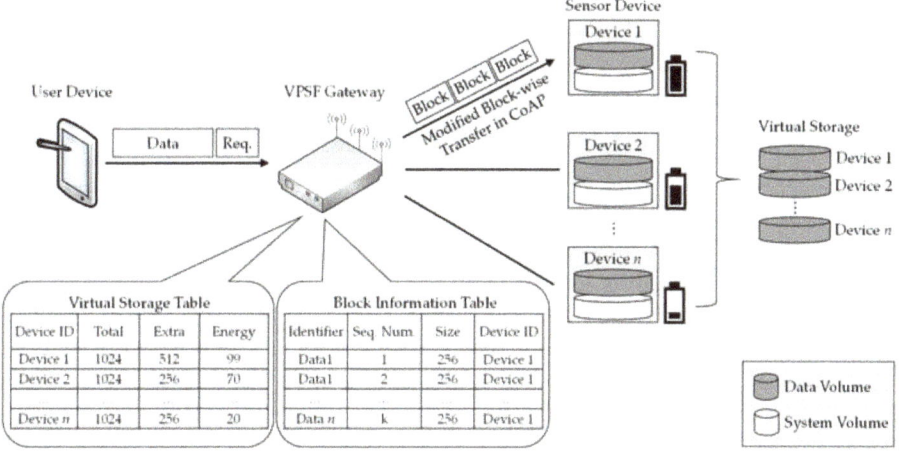

Figure 1. Virtual private storage framework (VPSF) architecture.

The VPSF gateway builds virtual storage by logically integrating the extra data volumes of the multiple sensor devices and manages it by maintaining a virtual storage table (VST) containing the total amount of extra data volumes of all sensor devices in the network and the information of each sensor device (device ID, extra data volume size, and residual energy). For storing data, the VPSF gateway divides the user data into several blocks and selects the sensor devices in which the blocks will be stored based on their residual energy. Then, the VPSF gateway forwards the blocks to the selected sensor devices and maintains a block information table (BIT) containing the information of the stored blocks (original user data identifier, block sequence number, block size, and device ID of the sensor device where the block is stored). The selected sensor device stores only the information contained in the payload of the blocks in binary format. For data retrieval, upon receiving the retrieval request message, the VPSF gateway checks the identifier of the requested data and then searches the BIT to find the device ID of sensor devices where the blocks for the requested data are stored. It then requests the blocks of the corresponding sensor devices in order of sequence number. After the VPSF gateway obtains all blocks, it integrates them into a single data set and transmits the integrated data to the user device.

Each sensor device serves as storage for the blocks. The storage space of a sensor device is divided into data and system volumes by the VPSF gateway—each sensor device separately maintains two volumes. The data volume is a part of the sensor device's storage space that is used to build virtual storage, and the system volume is a storage space required to perform the sensor device's unique operations (e.g., sensor and actuator control). Therefore, upon receiving the blocks from the VPSF gateway, the sensor device stores them in the data volume within its storage space. For data retrieval, it transmits the requested blocks to the VPSF gateway.

The VPSF can have a variety of use cases, such as healthcare, financial, and multimedia data storage, in which it is necessary to consider the physical resources of sensor devices running VPSF such as battery capacity and storage space. One of the potential use cases for VPSF is the personal healthcare data storage. In this use case, the wearable or mobile user devices periodically collect users' health information and generate the healthcare data files. The generated files are transmitted to the VPSF gateway. Then, the VPSF gateway divides the files into multiple blocks and store them in the extra storage space of the sensor devices. In the use case, users maintain personal health information via their own local storage, which is logically created by merging the extra storage space of sensor devices. Therefore, VPSF eliminates the need for users to store sensitive health information in storage managed by third-party service providers, and allows users to maintain it by themselves without additional hardware installation.

2.2. Operations of VPSF Gateway

The VPSF gateway performs three operations: (1) virtual storage building, (2) sensor device selection, and (3) consecutive block retransmission. In the following subsections, the operations of the VPSF gateway are described in detail.

2.2.1. Virtual Storage Building

In a virtual storage building, the VPSF gateway first partitions the storage space of a sensor device into two volumes: system and data. The storage space of the system volume should be sufficient to store the system data needed for installing and running the operating system (OS), applications, and task data generated by processing specific tasks. It is difficult to determine the storage space for the system volume because each sensor device may use a different OS and applications. Consequently, we consider two assumptions to determine the storage space for the system volume. First, the storage space required to store system data is determined by examining the filled storage space when the sensor device does not perform any task, provided as a constant value in VPSF. Second, the sensor device generates task data of the same size at periodic intervals (task data generation interval) and temporally accumulates the data in the storage space. The sensor device transmits the accumulated task data to the intended destination and removes it from the storage space at specific intervals (task data removal interval). The storage space for the data volume (extra storage space) for the i-th sensor device (DV_i) is given by

$$DV_i = T_i - Sys_i - Task_i = T_i - Sys_i - SizeTask_i \frac{RemTask_i}{GenTask_i} \qquad (1)$$

where T_i, Sys_i, $Task_i$, $SizeTask_i$, $GenTask_i$, and $RemTask_i$ are the total storage space, the storage space required to store system data, the storage space required to store task data, the task data size, the task data generation interval, and the task data removal interval for the i-th sensor device, respectively.

The VPSF gateway then logically integrates the data volume of all sensor devices to build virtual private storage. The total storage space of virtual storage is equal to the sum of the storage space of the data volumes. The total storage space of the virtual private storage and the storage space of the data volume for each sensor device are listed in the VST. The VST is updated when a new sensor device is added or the existing sensor device is depleted.

2.2.2. Sensor Device Selection

The purpose of sensor device selection is to guarantee the expected lifetime of the IoT local network by averaging the residual energy of sensor devices, which may differ for each device. Accordingly, when building virtual private storage, the VPSF gateway examines the residual energy of each sensor device and maintains it using the VST. The residual energy values of all sensor devices are represented by $\mathbf{E} = [e_0, e_2, \ldots, e_{n-1}]$, where n is the number of sensor devices. Sensor device selection is initiated when the VPSF gateway receives the storing request message, which contains the size of the user data, from the user device. Algorithm 1 defines the sensor device selection procedure. In the algorithm, the VPSF gateway iteratively performs sensor device selection. Before initiating sensor device selection, the VPSF gateway initializes variables to zero: the number of iterations (k), the device ID of the sensor device selected in the k-th iteration (SSD_k), and the sum of the data volume of the selected sensor devices (SDS). It then calls the list of residual energy for each sensor device from the VST and initiates sensor device selection. During sensor device selection, the VPSF gateway selects a sensor device with the highest residual energy (based on the VST) and verifies whether the data volume of the selected sensor device is large enough to store the user data; if it is not, the sensor device with the next highest residual energy is selected from the remaining list. The VPSF gateway then sums the data volumes of all selected sensor devices and compares the result with the size of the user data. This operation repeats until the sum of the data volumes of the selected sensor devices becomes larger than the size of the user data ($SizeUsr$). After terminating sensor device selection, the VPSF gateway divides the user data into multiple identically-sized blocks and sequentially transmits the blocks to the selected sensor devices.

Algorithm 1. Sensor device selection procedure

1:	**INITIALIZE** k to 0, SSD_k to 0, SDS to 0 // Initialize variables
2:	Call **E** from VST
3:	**REPEAT** // Initiate sensor device selection
4:	$maxE \leftarrow$ max(**E**) // Find the highest residual energy
5:	**FOR** each sensor device, i, $i \in [0, n-1]$
6:	**IF** $E[i] == maxE$ // Find the device ID of the sensor device with the highest residual energy
7:	$SSD_k \leftarrow i$ // Select a sensor device
8:	**ENDIF**
9:	**ENDFOR**
10:	**SSD**$[k] \leftarrow SSD_k$ // List the selected sensor devices
11:	$E[SSD_k] \leftarrow 0$ // Remove the residual energy of the selected sensor device from **E**
12:	$SDS \leftarrow SDS + DV_i$ // Sum the data volumes of the selected sensor devices
13:	$k \leftarrow k + 1$ // Increment the number of iterations by one
14:	**UNTIL** $SizeUsr \leq SDS$ // Terminate sensor device selection
15:	**RETURN SED** // Return the device ID of the selected sensor devices

---

2.2.3. Consecutive Block Retransmission

Consecutive block retransmission aims to reduce communication overhead in the VPSF. Accordingly, the BlockS option of the CoAP is modified to retransmit lost blocks without any request message and then applied to the VPSF. The existing BlockS option of the CoAP uses the non-confirmable (NON) message, which does not require acknowledgment when transmitting the blocks. This allows the server to transmit multiple blocks consecutively. For the existing BlockS option, the confirmable (CON) message that requires acknowledgment is regularly transmitted to verify the client's state. The server transmits the CON message when the number of consecutive transmissions reaches the maximum window size (SPDYWND), which is predefined based on the client's capacity limit. The number of blocks transmitted consecutively is equal to the value of SPDYWND. However, for retransmission, the existing BlockS option only uses the CON message, causing unnecessary retransmission request messages to be transmitted repeatedly.

To solve this problem, in VPSF, the existing protocol is modified to inform the server of the sequence number of the lost blocks. Upon receiving the blocks, the client identifies the lost blocks by examining the gap between their sequence numbers and maintains the sequence number of the lost blocks as a list. The client then piggybacks the list to the acknowledgment and transmits it to the server, thereby providing the server with the sequence number of the lost blocks. Consequently, the server consecutively transmits the lost blocks as NON messages without any retransmission request messages.

Figure 2 illustrates an operational example of the existing and modified BlockS option of the CoAP. MID is the message ID, T is the token, and S is the value of the BlockS option (the sequence number of the blocks, whether more blocks are following, the size of the block, and SPDYWND). In the example, the size of the block is set to 64 B, and SPDYWND is set to 4. The example demonstrates that the modified protocol reduces communication overhead by eliminating retransmission request messages.

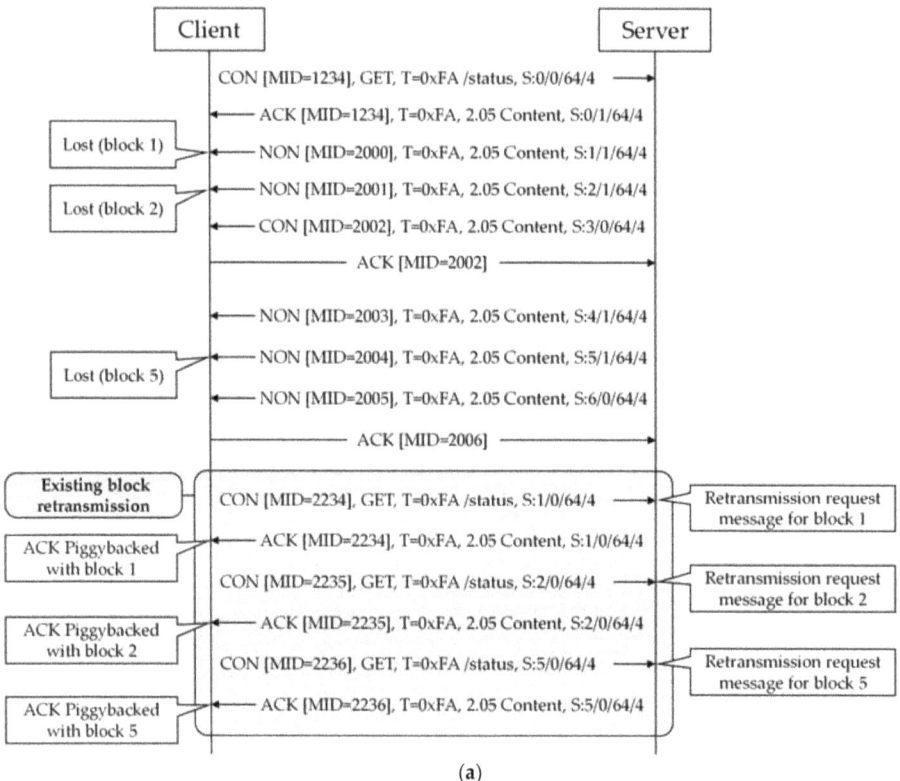

(a)

Figure 2. *Cont.*

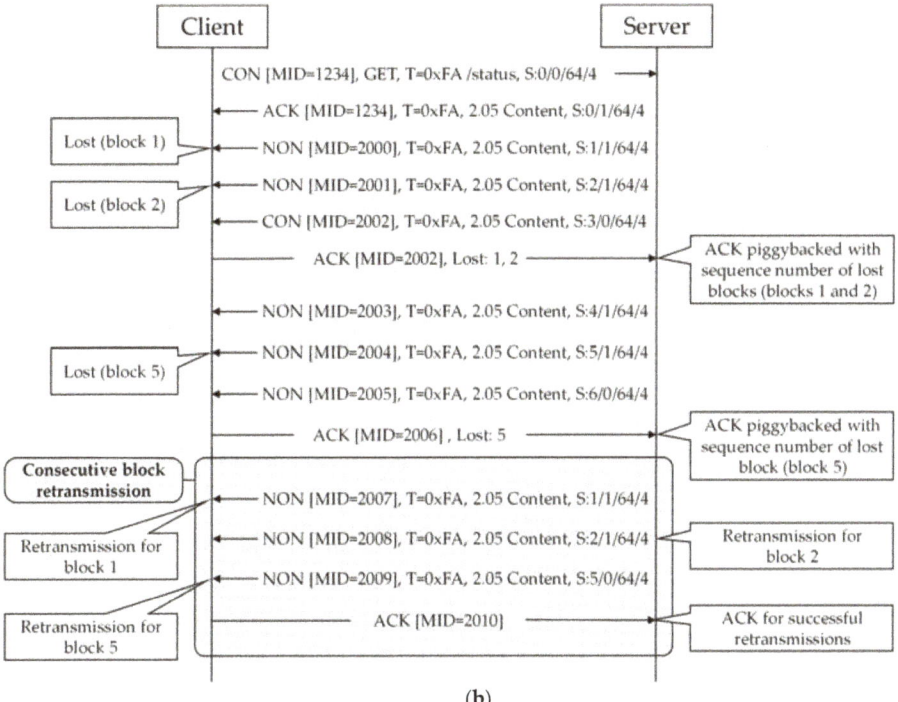

Figure 2. Comparison of two operational examples: (**a**) existing speedy block-wise transfer in CoAP; (**b**) modified speedy block-wise transfer in CoAP.

3. Implementation and Performance Evaluation

The VPSF performance was evaluated through implementation and extensive experiments. The user device was implemented using a personal computer running the Windows 10 OS, and the VPSF gateway and sensor devices were implemented using the open-source hardware Raspberry Pi 3 Model B+ running the Linux-based Raspbian OS.

Figure 3 depicts the VPSF implementation structure. The VPSF gateway was placed between the user device and multiple sensor devices. It communicates with the user device using the hypertext transfer protocol (HTTP) and communicates with the sensor devices using the CoAP. To implement the CoAP for the VPSF, we used the open-source software *libcoap*, a C-implementation of CoAP. We developed a monitoring application running on the VPSF gateway to examine the residual energy of each sensor device and the delay required to successfully store blocks. Each device was equipped with a wireless interface supporting IEEE 802.11 b/g/n and used a 100 Mbps data rate for data transmission.

For the experiment, we implemented four sensor devices, each initially set to have a different residual energy. Table 1 lists the initial settings for the residual energy of each sensor device. In the experiment, the size of user data varied from 100 MB to 1 GB. The block size was fixed as 1000 B, so the number of blocks per user data varied from 10^5 to 10^6. Each sensor device was equipped with a 32 GB SD card, and the storage space of each data volume was set equally to 20 GB. SPDYWND was set to 100 to consecutively transmit 100 blocks. To evaluate the VPSF performance, the VPSF experiment results were compared to legacy virtual private storage (legacy VPS), in which the gateway transmits the same number of blocks to each sensor device using the existing BlockS option of the CoAP.

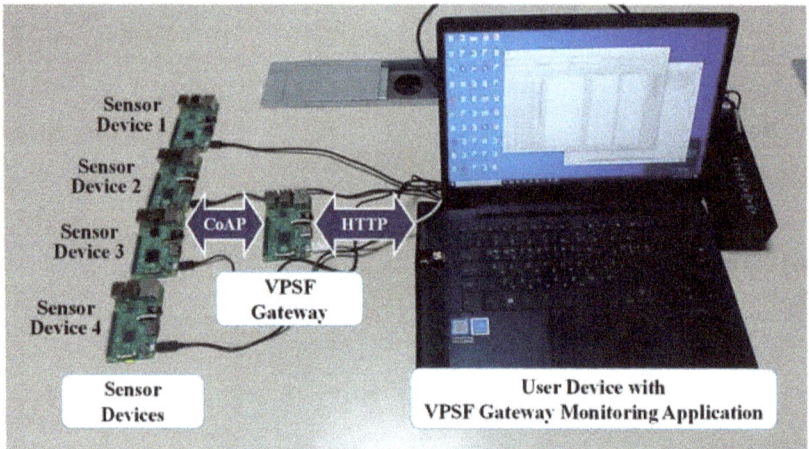

Figure 3. VPSF implementation structure.

Table 1. Initial setting for residual energy of each sensor device.

	Sensor Device 1	Sensor Device 2	Sensor Device 3	Sensor Device 4
Residual energy	8.14 Wh	16.28 Wh	24.42 Wh	32.56 Wh

Figure 4 depicts the changes in the lifetime of the sensor device with the lowest residual energy (Sensor Device 1) when the size of user data increases. The lifetime of the IoT local network is determined by the lifetime of the sensor device with the lowest residual energy. In this experiment, the user device was set to transmit user data continuously until the energy of Sensor Device 1 was depleted. Moreover, the sensor devices were set to overwrite user data to prevent the storage space of its data volume from filling up completely.

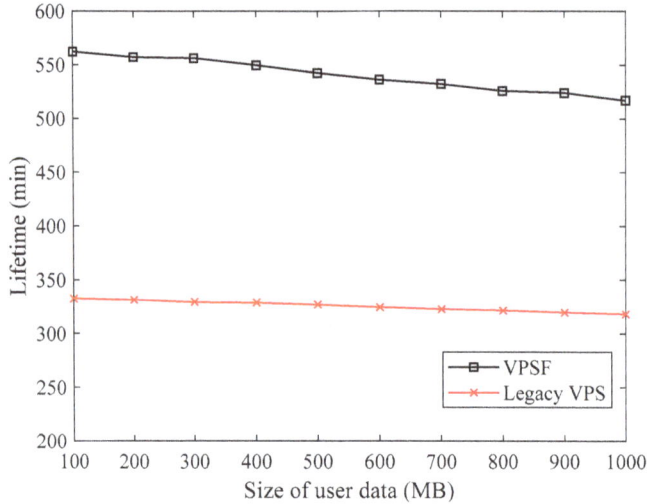

Figure 4. Lifetime of sensor device with the lowest residual energy.

The VPSF exhibited a longer lifetime than legacy VPS, owing to the VPSF's support of sensor device selection and the modified BlockS option. Specifically, Sensor Device 1 was used to store the blocks only after the residual energy of the other sensor devices reached that of Sensor Device 1. Furthermore, the lost blocks were retransmitted without any retransmission request messages in the VPSF. In contrast, in legacy VPS, the same number of blocks were stored in each sensor device whenever the user data were transmitted from the user device, and the retransmission was conducted with the retransmission request messages. In both cases, as the size of user data increased, the lifetime decreased; if the size of user data increased, the sensor device received and stored more blocks because the communication overhead caused by the transmission interval and the user data header decreased. On average, the VPSF exhibited a 65.85% longer lifetime than that of legacy VPS.

Figure 5 depicts the residual energy of each sensor device. In the experiment, the user device was configured to transmit 100 times to the VPSF gateway. Two sizes of user data (500 MB and 1 GB) were considered. In both cases, VPSF had a smaller difference in residual energy among sensor devices compared to legacy VPS because, unlike legacy VPS, the VPSF selected and used some of the sensor devices to store the blocks. Only Sensor Device 4 was used in the 500 MB case, and Sensor Devices 3 and 4 were used in the 1 GB case. The difference in residual energy among sensor devices can be mathematically represented by the fairness index F [35]. The fairness index is calculated as

$$F = \frac{\left(\sum_{i=1}^{n} e_i\right)^2}{n \sum_{i=1}^{n} e_i^2} \tag{2}$$

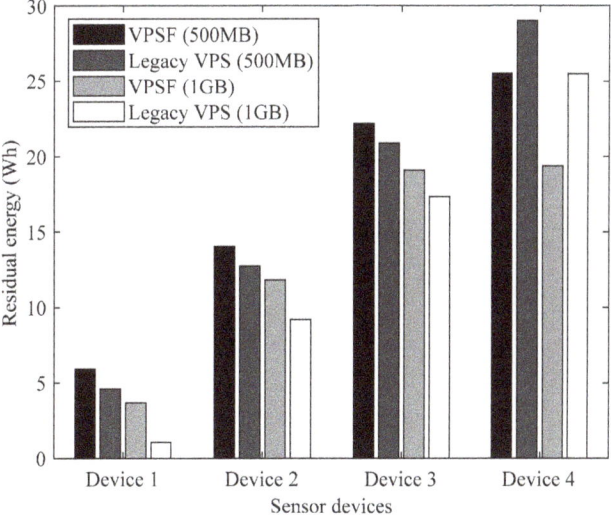

Figure 5. Residual energy of each sensor device.

The higher the fairness index, the smaller the difference in residual energy. Table 2 lists the fairness index of the VPSF and legacy VPS for each case. For 500 MB and 1 GB, respectively, the VPSF exhibited 7.58% and 19.77% higher fairness indexes than legacy VPS.

Table 2. Fairness index.

	VPSF (500 MB)	Legacy VPS (500 MB)	VPSF (1 GB)	Legacy VPS (1 GB)
Fairness index	0.83	0.77	0.82	0.68

Figure 6 depicts the variance in retransmission delay per user data when the size of user data changes. During the experiment, the block delivery ratio was 92.1%, on average, and 7.9% of blocks were lost and retransmitted. The VPSF employed the modified BlockS option, which enabled the server (VPSF gateway) to consecutively retransmit the lost blocks. The retransmission delay of the VPSF was 25.6% shorter than that of legacy VPS, on average. During the experiment, the retransmission delay increased as the size of user data increased—the larger the size of the user data, the higher the number of lost blocks.

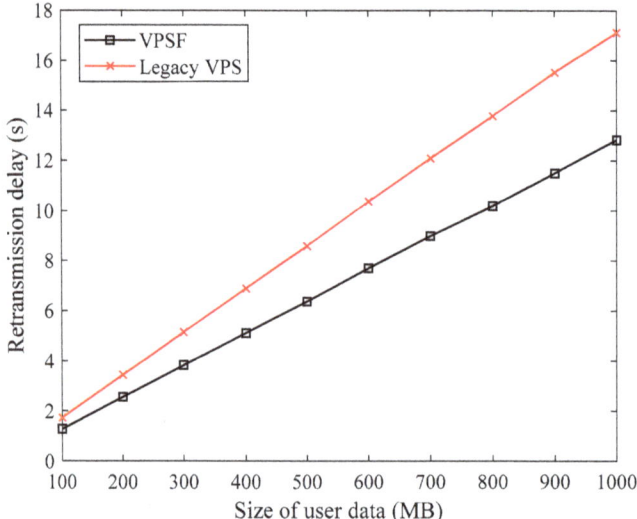

Figure 6. Retransmission delay based on size of user data.

4. Conclusions

In this paper, we presented a VPSF that uses the extra storage space of sensor devices in an IoT local network to store user data while guaranteeing the expected network lifetime. The VPSF gateway performs three operations to build and maintain virtual private storage: (1) virtual storage building, (2) sensor device selection, and (3) consecutive block retransmission. The first operation builds virtual private storage by integrating the extra storage space of sensor devices. The second operation selects a sensor device to store blocks based on the device's residual energy to guarantee the expected lifetime of the IoT local network. The third operation enables the VPSF gateway to retransmit the lost blocks without retransmission request messages to reduce communication overhead. To verify the feasibility of the VPSF, an experiment was performed using a specific implementation and the results were compared with legacy VPS. The results demonstrate that the VPSF exhibited a 65.85% longer lifetime and 25.6% shorter retransmission delay than those of legacy VPS.

Author Contributions: E.-J.K. conceived and designed the overall framework; H.-H.L. performed the open source-based implementation and experiment; J.-H.K. interpreted and analyzed the data; H.-H.L., J.-H.K., and E.-J.K. wrote the paper; E.-J.K. guided the research direction and supervised the entire research process. All authors have read and agreed to the published version of the manuscript.

Funding: This research was supported by Basic Science Research Program through the National Research Foundation of Korea (NRF) funded by the Ministry of Education (NRF-2019R1I1A1A01059787).

Conflicts of Interest: The authors declare no conflict of interest.

References

1. Siddiqa, A.; Karim, A.; Gani, A. Big data storage technologies: A survey. *Front. Inf. Techol. Electron.* **2017**, *18*, 1040–1070. [CrossRef]
2. Dai, H.-N.; Wang, H.; Xu, G.; Wan, J.; Imran, M. Big data analytics for manufacturing internet of things: Opportunities, challenges and enabling technologies. *Enterp. Inf. Syst.* **2019**, 1–25. [CrossRef]
3. Kobusińska, A.; Leung, C.; Hsu, C.-H.; Raghavendra, S.; Chang, V. Emerging trends, issues and challenges in Internet of Things, Big Data and cloud computing. *Future Gener. Compt. Syst.* **2018**, *87*, 416–419. [CrossRef]
4. Yang, C.-T.; Chen, S.-T.; Cheng, W.-H.; Chan, Y.-W.; Kristiani, E. A Heterogeneous Cloud Storage Platform With Uniform Data Distribution by Software-Defined Storage Technologies. *IEEE Access* **2019**, *7*, 147672–147682. [CrossRef]
5. Google Drive. Available online: https://www.google.com/drive/ (accessed on 28 January 2020).
6. iCloud. Available online: https://www.icloud.com (accessed on 28 January 2020).
7. Dropbox. Available online: https://www.dropbox.com (accessed on 28 January 2020).
8. Park, J.K.; Kim, J. Big data storage configuration and performance evaluation utilizing NDAS storage systems. *AKCE Int. J. Graphs Comb.* **2018**, *15*, 197–201. [CrossRef]
9. Nachiappan, R.; Javadi, B.; Calheiros, R.N.; Matawie, K.M. Cloud storage reliability for big data applications: A state of the art survey. *J. Netw. Comput. Appl.* **2017**, *97*, 35–47. [CrossRef]
10. Singh, A.; Chatterjee, K. Cloud security issues and challenges: A survey. *J. Netw. Comput. Appl.* **2017**, *79*, 88–115. [CrossRef]
11. Mansouri, Y.; Toosi, A.N.; Buyya, R. Data storage management in cloud environments: Taxonomy, survey, and future directions. *ACM Comput. Surv.* **2017**, *50*, 1–51. [CrossRef]
12. Kishani, M.; Tahoori, M.; Asadi, H. Dependability analysis of data storage systems in presence of soft errors. *IEEE Trans. Reliab.* **2019**, *68*, 201–215. [CrossRef]
13. Xing, L.; Tannous, M.; Vokkarane, V.M.; Wang, H.; Guo, J. Reliability modeling of mesh storage area networks for Internet of Things. *IEEE Internet Things* **2017**, *4*, 2047–2057. [CrossRef]
14. Wu, Y.; Wang, F.; Hua, Y.; Feng, D.; Hu, Y.; Tong, W.; Liu, J.; He, D. I/O Stack Optimization for Efficient and Scalable Access in FCoE-Based SAN Storage. *IEEE Trans. Parallel Distrib.* **2017**, *28*, 2514–2526. [CrossRef]
15. Ammar, M.; Russello, G.; Crispo, B. Internet of Things: A survey on the security of IoT frameworks. *J. Inf. Secur. Appl.* **2018**, *38*, 8–27. [CrossRef]
16. Sheng, Z.; Tian, D.; Leung, V.C.M. Toward an energy and resource efficient internet of things: A design principle combining computation, communications, and protocols. *IEEE Commun. Mag.* **2018**, *56*, 89–95. [CrossRef]
17. Musaddiq, A.; Zikria, Y.B.; Hahm, O.; Yu, H.; Bashir, A.K.; Kim, S.W. A survey on resource management in IoT operating systems. *IEEE Access* **2018**, *6*, 8459–8482. [CrossRef]
18. Tang, J.; Sun, D.; Liu, S.; Gaudiot, J.-L. Enabling deep learning on IoT devices. *Computer* **2017**, *50*, 92–96. [CrossRef]
19. Ren, J.; Guo, H.; Xu, C.; Zhang, Y. Serving at the edge: A scalable IoT architecture based on transparent computing. *IEEE Netw.* **2017**, *31*, 96–105. [CrossRef]
20. Kurunathan, H.; Severino, R.; Koubaa, A.; Tovar, E. IEEE 802.15. 4e in a nutshell: Survey and performance evaluation. *IEEE Commun. Surv. Tutor.* **2018**, *20*, 1989–2010. [CrossRef]
21. Ceph. Available online: https://ceph.io/ (accessed on 28 January 2020).
22. Lustre. Available online: http://lustre.org/ (accessed on 28 January 2020).
23. Apache Hadoop. Available online: https://hadoop.apache.org/ (accessed on 28 January 2020).
24. Ghemawat, S.; Gobioff, H.; Leung, S.-T. The Google file system. In Proceedings of the 19th ACM Symposium on Operating Systems Principles, Bolton Landing, NY, USA, 19–22 October 2003; pp. 29–43.
25. Hu, P.; Dhelim, S.; Ning, H.; Qiu, T. Survey on fog computing: Architecture, key technologies, applications and open issues. *J. Netw. Comput. Appl.* **2017**, *98*, 27–42. [CrossRef]

26. Yousefpour, A.; Fung, C.; Nguyen, T.; Kadiyala, K.; Jalali, F.; Niakanlahiji, A.; Kong, J.; Jue, J.P. All one needs to know about fog computing and related edge computing paradigms: A complete survey. *J. Syst. Archit.* **2019**, *98*. [CrossRef]
27. Wang, T.; Zhou, J.; Chen, X.; Wang, G.; Liu, A.; Liu, Y. A three-layer privacy preserving cloud storage scheme based on computational intelligence in fog computing. *IEEE Trans. Emerg.* **2018**, *2*, 3–12. [CrossRef]
28. Moysiadis, V.; Sarigiannidis, P.; Moscholios, I. Towards distributed data management in fog computing. *Wirel. Commun. Mob. Comput.* **2018**, *2018*. [CrossRef]
29. Hao, Z.; Novak, E.; Yi, S.; Li, Q. Challenges and software architecture for fog computing. *IEEE Internet Comput.* **2017**, *21*, 44–53. [CrossRef]
30. Shelby, Z.; Hartke, K.; Bormann, C.; Frank, B. The Constrained Application Protocol (CoAP) (RFC 7252). Available online: https://tools.ietf.org/html/rfc7252 (accessed on 28 January 2020).
31. Bormann, C.; Shelby, Z. Block-Wise Transfers in the Constrained Application Protocol (CoAP) (RFC 7959). Available online: https://tools.ietf.org/html/rfc7959 (accessed on 28 January 2020).
32. Cao, Z.; Jin, K.; Fu, B.; Zhang, D. Speeding Up CoAP Block-wise Transfer. Available online: https://tools.ietf.org/id/draft-zcao-core-speedy-blocktran-00.html (accessed on 28 January 2020).
33. Libcoap. Available online: https://libcoap.net (accessed on 28 January 2020).
34. Libcoap Open Source. Available online: https://github.com/obgm/libcoap (accessed on 28 January 2020).
35. Kwon, J.-H.; Kim, E.-J. Asymmetric Directional Multicast for Capillary Machine-to-Machine Using mmWave Communications. *Sensors* **2016**, *16*, 515. [CrossRef] [PubMed]

Article

Failure Prediction Model Using Iterative Feature Selection for Industrial Internet of Things

Jung-Hyok Kwon [1] and **Eui-Jik Kim [2],***

[1] Smart Computing Laboratory, Hallym University, 1 Hallymdaehak-gil, Chuncheon, Gangwon-do 24252, Korea; jhkwon@hallym.ac.kr
[2] School of Software, Hallym University, 1 Hallymdaehak-gil, Chuncheon, Gangwon-do 24252, Korea
* Correspondence: ejkim32@hallym.ac.kr; Tel.: +82-33-248-2333; Fax: +82-33-242-2524

Received: 30 January 2020; Accepted: 8 March 2020; Published: 12 March 2020

Abstract: This paper presents a failure prediction model using iterative feature selection, which aims to accurately predict the failure occurrences in industrial Internet of Things (IIoT) environments. In general, vast amounts of data are collected from various sensors in an IIoT environment, and they are analyzed to prevent failures by predicting their occurrence. However, the collected data may include data irrelevant to failures and thereby decrease the prediction accuracy. To address this problem, we propose a failure prediction model using iterative feature selection. To build the model, the relevancy between each feature (i.e., each sensor) and the failure was analyzed using the random forest algorithm, to obtain the importance of the features. Then, feature selection and model building were conducted iteratively. In each iteration, a new feature was selected considering the importance and added to the selected feature set. The failure prediction model was built for each iteration via the support vector machine (SVM). Finally, the failure prediction model having the highest prediction accuracy was selected. The experimental implementation was conducted using open-source R. The results showed that the proposed failure prediction model achieved high prediction accuracy.

Keywords: failure prediction model; industrial Internet of Things; iterative feature selection; machine learning; manufacturing

---

1. Introduction

Recently, the industrial Internet of Things (IIoT) has been widely adopted by various companies, as it provides connectivity and analysis capabilities, which are the key technologies for advanced manufacturing [1–4]. In IIoT, a large number of sensors are used to periodically detect changes in machine health, manufacturing process, industrial environment, etc. [5–10]. Hence, a huge amount of data is collected from the sensors used in IIoT. In general, the collected data are analyzed to provide useful information for productivity improvement [11]. In particular, failure prediction through data analysis is considered one of the most important issues in IIoT [12]. The failure, such as execution error, long delay, and defective product, leads to a fatal system malfunction and huge maintenance cost, resulting in productivity degradation of enterprises in industrial fields [13]. Therefore, many enterprises have endeavored to predict when and why failures occur for improving their capabilities of failure prevention and recovery, also known as resilience capacity [14]. Especially, the accuracy of failure prediction is an important factor that determines the resilience capacity of enterprises; thus, its improvement is crucial in the IIoT.

To predict failures, we need to build a failure prediction model that determines whether or not a failure occurs [15–17]. In order to build such a failure prediction model, most of the existing studies have used machine learning techniques. Abu-Samah et al. built the failure prediction model on the basis of a Bayesian network, and Kwon and Kim (i.e., our previous work) used the nearest centroid classification to predict machine failures [18,19]. The results of both the studies showed that the built failure prediction model achieved approximately 80% prediction accuracy. However, they did not conduct the feature selection. In other words, they used all the data in the dataset when building the model. Accordingly, in a real IIoT context where a very large number of sensors are used, the prediction accuracy might be significantly degraded because of the impact of the data irrelevant to the failures.

Therefore, feature selection has been considered one of the most important steps in building a failure prediction model. There have been many studies to build the failure prediction model using feature selection, most of which have selected features considering the importance of each feature [20–22]. Moldovan et al. built a failure prediction model using the selected features to improve prediction accuracy and performed feature selection using three algorithms (i.e., random forest, regression analysis, and orthogonal linear transformation) to compare the prediction accuracy of each for the comparative study [20]. Mahadevan and Shah used a support vector machine recursive feature elimination (SVM-RFE) algorithm to rank the features by their importance [21]. The SVM-RFE algorithm was used to compute the norm of the weights of each feature (i.e., the importance of each feature). For feature selection, the authors determined a threshold value specified by the limit of the number of selected features, and then selected the features considering their importance until the number of selected features reached the threshold. The selected features were used for failure detection and diagnosis. Hasan et al. focused on a two-step approach of random forest based-feature selection, which consists of the feature importance measurement and feature selection [22]. In the first step, the importance of each feature was measured using average permutation importance (APIM) score. APIM score is obtained by calculating the mean decreases in the forest's classification accuracy when a specific feature is not available. In the second step, the features with an APIM score greater than a threshold were selected. However, in the existing feature selection methods, the optimal prediction accuracy cannot be obtained because the number of selected features is fixed or all the features in which the importance is greater than a predefined threshold are selected. Specifically, the number of selected features might be too large to obtain optimal prediction accuracy or vice versa.

In this paper, we propose a novel failure prediction model using iterative feature selection, which aims to accurately predict the occurrence of failures. To build the model, the collected data are processed and analyzed using the following steps: (1) preprocessing, (2) importance measurement, (3) feature selection, (4) model building, and (5) model selection. The preprocessing step includes feature elimination, missing data imputation, normalization, and data division. In the importance measurement step, to measure the importance of each feature, the relevancy between each feature and the failure is analyzed using the random forest algorithm [23–26]. Then, the feature selection and model building steps are conducted iteratively. In particular, in the feature selection step, a new feature is selected considering its importance in each iteration and is added to the selected feature set. In the model building step, the failure prediction model is built via the SVM on the basis of the selected feature set updated in each iteration [27–30]. Finally, one of the failure prediction models is selected considering the prediction accuracy in the model selection step. To evaluate the performance of the proposed model, we conducted an experimental implementation using open-source R. We used the semiconductor manufacturing (SECOM) dataset provided by the University of California Irvine (UCI) repository [31]. The results showed that the proposed failure prediction model achieved high prediction accuracy.

The rest of this paper is organized as follows. Section 2 describes the proposed failure prediction model in detail. In Section 3, the results of the implementation and performance evaluation are presented. Finally, Section 4 concludes this paper.

2. Failure Prediction Model Using Iterative Feature Selection

In this section, we present the design of the failure prediction model using iterative feature selection in detail. It is assumed that sensors with unique identification (ID) employed in an IIoT environment periodically generate data, and the data is collected by a data analysis server. In this paper, the sensor device is represented as a feature, and its ID is represented by a feature index. The data analysis server collects data from sensor devices and builds and evaluates the failure prediction model. The data is represented in the form of a matrix format (i.e., dataset) in which each column and row means the feature and data collection time, respectively.

We built the failure prediction model on the basis of the features selected to maximize the prediction accuracy. To this end, feature selection was iteratively conducted, and multiple failure prediction models were built on the basis of the features selected in each iteration. Then, the failure prediction model with the highest prediction accuracy was selected. To build the failure prediction model, we considered five steps, namely preprocessing, importance measurement, feature selection, model building, and model selection. In this section, each step for building the failure prediction model is described in detail.

Figure 1 shows the overall procedure for building the failure prediction model. In the figure, the white square indicates each step, and the grey square indicates the input or output of each step. In the preprocessing step, feature elimination, missing data imputation, normalization, and data division are sequentially conducted. To eliminate invalid features from the input dataset (i.e., collected data), the not applicable (NA) data of each feature are searched, and the variance of each feature is calculated. If the ratio of the NA data of a specific feature is greater than the predefined validity factor determined in the range [0, 1], the corresponding feature is eliminated from the input dataset. Moreover, if the variance of a certain feature is close to zero, the corresponding feature is eliminated from the input dataset. This is because the closer the variance of a certain feature is to zero, the more similar is the value of all the data. In particular, if the variance of a certain feature is zero, all the data of this feature have the same value, which means that the feature is meaningless for the data analysis. To replace the remaining missing data with the appropriate data, an average of the non-missing data of each feature is calculated and imputed. Then, normalization is conducted to match the data scale of each feature. To this end, the standard score is used, which calculates the normalized data of each feature according to Equation (1):

$$x_i' = (x_i - \mu)/\sigma, \tag{1}$$

where x_i' is the $(i+1)$-th normalized data of the feature, x_i is the $(i+1)$-th data of the feature, μ is the average of the feature, and σ is the standard deviation of the feature [32]. The μ and σ values are calculated using Equations (2) and (3), respectively [33].

$$\mu = \frac{1}{m}\sum_{i=0}^{m-1} x_i \tag{2}$$

$$\sigma = \sqrt{\frac{1}{m}\sum_{i=0}^{m-1}(x_i - \mu)^2} \tag{3}$$

where m is the amount of data of the feature. Finally, the normalized dataset is divided into training and test datasets, which are used to build the failure prediction model and to evaluate the performance of the prediction model, respectively.

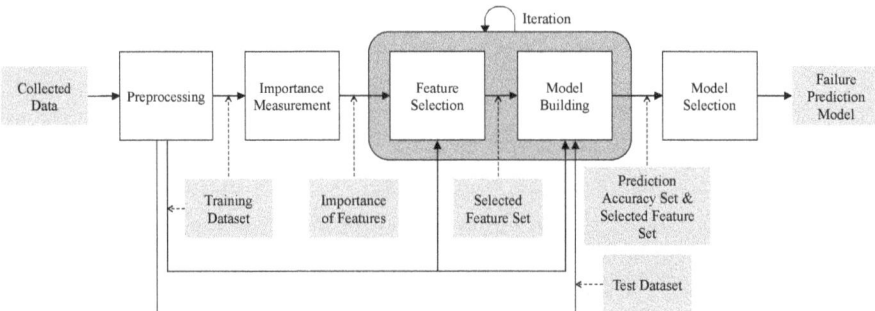

Figure 1. Overall procedure for building failure prediction model.

To measure the importance of each feature, the training dataset (i.e., a part of the preprocessed dataset) is analyzed using the random forest algorithm, which is one of the machine learning techniques for importance measurement. In particular, via the random forest, multiple decision trees are built, and the relevancy between each feature and the failure is analyzed considering the built decision trees comprehensively. In the random forest, multiple subsets of the training dataset are created to build each decision tree differently. Note that each subset consists of different data and features. For this, *n* data (i.e., *n* rows) and *mtry* features (i.e., *mtry* columns) are randomly selected from the training dataset. This operation repeats until the number of created subsets reaches the number of decision trees predefined as *ntree*. Then, each decision tree is built separately using one of the created subsets. After building *ntree* decision trees, the importance of each feature is measured through the mean decrease Gini, which indicates the extent to which each feature affects the correct prediction results. More specifically, in each decision tree, the sum of the difference for Gini impurities between the parent nodes using a particular feature and their child nodes is calculated. Then, the average of the results of all the decision trees is calculated. Note that the decision tree consists of multiple nodes that make decisions using the threshold of a specific feature. In addition, each node has a Gini impurity that is a measurement of the likelihood of an incorrect decision.

Then, the feature selection and model building steps are conducted iteratively. Algorithm 1 shows the overall operation of both the steps. In the algorithm, the set of the importance of features and the set of features are represented by Equations (4) and (5), respectively.

$$\mathbf{IMP} = [imp_0, \cdots, imp_j, \cdots, imp_{nf-1}] \tag{4}$$

$$\mathbf{F} = [f_0, \cdots, f_j, \cdots, f_{nf-1}] \tag{5}$$

where *nf* is the number of features and imp_j is the importance of the $(j+1)$-th feature. The index of each element in **IMP** and **F** (i.e., index of each importance and feature) denotes the sequence number of each feature. At the beginning of the algorithm, the variables are initialized: the selected feature at the $(cnt+1)$-th iteration (sf_{cnt}), the prediction accuracy for the built model at the $(cnt+1)$-th iteration (pa_{cnt}), and the iteration counter (cnt). Then, the feature selection and model building steps are repeated until the importance of a newly selected feature is smaller than the importance threshold. In the algorithm, the maximum number of iterations (*max*) is equal to the number of features whose importance is greater than the importance threshold. Therefore, the iteration terminates when *cnt* reaches *max*. In each iteration, the feature selection step selects a new feature with the highest importance and adds it to the selected feature set represented by Equation (6).

$$\mathbf{SF} = [sf_0, sf_1, \cdots, sf_{cnt}], cnt \in [0, max - 1] \tag{6}$$

Consequently, the number of features included in is incremented by one as the number of iteration increases. Once **SF** is updated, the model building step builds the failure prediction model through SVM on the basis of **SF**. Then, it evaluates the prediction accuracy to update the prediction accuracy set that is expressed by Equation (7).

$$\mathbf{PA} = [pa_0, pa_1, \cdots, pa_{cnt}], cnt \in [0, max - 1] \tag{7}$$

Note that the index of each element in **SF** and **PA** (i.e., index of each selected feature and prediction accuracy) refers to the number of iterations. In the algorithm, modelBuildEvalFunction() is a function to build and evaluate the prediction model. It derives the prediction accuracy by taking **SF** he prediction model. It derives the prediction accuracy by taking **SF** as the input. For example, if $cnt = 2$ (i.e., the third iteration) and the selected feature set is equal to the **SF** $= [f_{39}, f_{12}, f_{19}]$ (i.e., $sf_0 = f_{39}, sf_1 = f_{12},$ and $sf_2 = f_{19}$), modelBuildEvalFunction() first builds the failure prediction model using the training data of the features in the **SF**. Then, it derives pa_2 by evaluating the built model using the test data of the features and updates the prediction accuracy set from **PA** $= [pa_0, pa_1]$ to **PA** $= [pa_0, pa_1, pa_2]$. In the example, the number of elements in **PA** increases from two to three. When cnt reaches max, the operation is terminated. The outputs of this operation are shown in Equations (8) and (9).

$$\mathbf{SF} = [sf_0, sf_1, \cdots, sf_{max-1}] \tag{8}$$

$$\mathbf{PA} = [pa_0, pa_1, \cdots, pa_{max-1}] \tag{9}$$

Algorithm 1: Operation of feature selection and model building steps.

1:       INITIALIZE sf_{cnt} to NULL, pa_{cnt} to 0, and cnt to 0 //Initialize variables
2:       **REPEAT** //Start iteration
3:         /* ========== Feature Selection Step ========== */
4:         $maxImp \leftarrow$ max(**IMP**) //Find the highest importance from **IMP**
5:         **FOR** each feature index, j, $j \in [0, nf - 1]$
6:           **IF IMP**$[j]$ == $maxImp$ //Find feature index with the highest importance
7:           $sf_{cnt} \leftarrow$ **F**$[j]$ //Select a new feature
8:           **F**$[j]$, **IMP**$[j] \leftarrow 0$ //Remove the selected feature from **IMP** and **F**
9:         **ENDIF**
10:        **ENDFOR**
11:      **SF**$[cnt] \leftarrow sf_{cnt}$ //Update the selected feature set
12:      /* ========== Model Building Step ========== */
13:      $pa_{cnt} \leftarrow$ modelBuildEvalFunction(**SF**) //Build and evaluate model
14:      **PA**$[cnt] \leftarrow pa_{cnt}$ //Update the prediction accuracy set
15:      $cnt \leftarrow cnt + 1$ //Increment iteration counter by one
16:      **UNTIL** $cnt \geq max$ //Terminate iteration if condition becomes TRUE
17:      **RETURN SF** and **PA** //Return selected feature set and prediction accuracy set

---

The model selection step selects the failure prediction model having the highest prediction accuracy, referring to **PA** and **SF**. In particular, this step first searches for the highest prediction accuracy in **PA** by using max(**PA**), where max() is the function that searches the element having the maximum value in a given set. Then, it derives the index of max(**PA**) from **PA**. For this, each element in **PA** is compared with max(**PA**), and the index of the element that is equal to max(**PA**) is derived from **PA**. Finally, a failure prediction model is selected, taking into account the index of max(**PA**) and **SF**. More specifically, the elements (i.e., selected features) in which the index is less than or equal to the index of max(**PA**) are extracted from **SF**. Then, one of the failure prediction models built in the model building step is selected by comparing the features used in the model building step and the extracted features from **SF**.

3. Implementation and Performance Evaluation

An experimental implementation was conducted to verify the feasibility of the proposed failure prediction model by using the open-source R version 3.4.3. For this, the SECOM dataset provided by the UCI repository was used. This dataset consists of 1567 data elements and 591 features, and the data were collected from a semiconductor manufacturing process by monitoring the sensors and the process measurement point. The data of 590 features were measured from different sensors, and the data of the remaining feature were the results of the house line test represented by Pass and Fail.

For feature elimination, we set the validity factor to 0.1, which was empirically determined to maximize the prediction accuracy. Thus, features with more than 10% NA data and features having zero variance were eliminated from the dataset. Through feature elimination, the number of features reduced from 591 to 393. The ratio of the training dataset and the test dataset was set to 7:3. To measure the importance of each feature, we used the randomForest and caret packages. We set n, $mtry$, and $ntree$ to 1000, 19, and 500, respectively. Through this setting, 500 decision trees were built using the randomly created 1000×19 matrix. The importance of each feature was measured through the mean decrease Gini. Figure 2 shows the importance of the top 30 features. The x-axis and the y-axis denote the mean decrease Gini and the feature, respectively. In the figure, F60 has the highest mean decrease Gini (i.e., 1.52) among all the features.

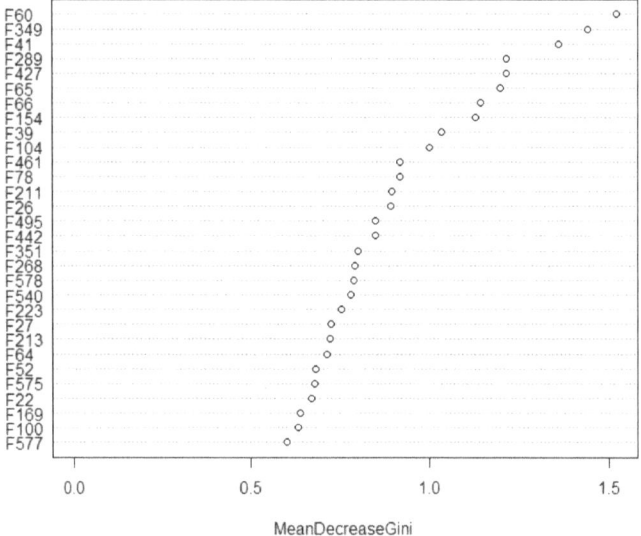

Figure 2. Importance of each feature.

For iterative feature selection, we set the importance threshold to 0.7, which was chosen in the range of [0.1, 1], taking into account the feature selection results of the existing research for the comparative study. Therefore, max was determined as 24. This implied that the number of iterations was 24. The training dataset contained 70 fails and 1038 passes. This imbalance of the training dataset made it difficult for the failure prediction model to predict a fail case. To address this problem, the sampling was conducted before building the failure prediction model. In particular, some of the pass data were removed to have less effect on the model building. With the results of feature selection and the sampled training dataset, the failure prediction model was built using the SVM. To this end, we used the e1071 package in R. Table 1 lists the obtained **SF** and **PA**. In the table, max(**PA**) is 0.72, and its index is 7. As a result, the failure prediction model that was built using eight features (i.e., F60, F349, F41, F289, F427, F65, F66, and F154) was selected.

Table 1. Selected feature set and prediction accuracy set.

Index	0	1	2	3	4	5	6	7	8	9	10	11
SF	F60	F349	F41	F289	F427	F65	F66	F154	F39	F104	F461	F461
PA	0.69	0.64	0.66	0.7	0.7	0.69	0.71	0.72	0.71	0.7	0.67	0.71

Index	12	11	13	14	15	16	17	18	19	20	21	22
SF	F211	F26	F495	F442	F351	F268	F578	F540	F223	F27	F213	F64
PA	0.66	0.7	0.69	0.64	0.68	0.65	0.64	0.6	0.56	0.64	0.65	0.63

For the performance evaluation, the prediction accuracy of the proposed model was compared to that of the existing models. We considered three existing models, which were built on the basis of a fixed number of features (i.e., 12 and 24 features) and all the features, respectively. Figures 3–6 show the receiver operating characteristic (ROC) curve for the three failure prediction models that used different numbers of features. The ROC curve is a performance measure of the prediction model and presents the relationship between the true positive rate (TPR) and the false positive rate (FPR) [34]. The TPR and FPR values were calculated using Equations (10) and (11), respectively.

$$TPR = \frac{TP}{TP + FN} \tag{10}$$

$$FPR = \frac{FP}{FP + TN} \tag{11}$$

where TP, FN, FP, and TN are the true positive, false negative, false positive, and true negative, respectively. With the ROC curve, the area under the curve (AUC) was used to evaluate the prediction accuracy of the model. In particular, the greater the AUC was, the higher was the prediction accuracy. In Figure 3, the failure prediction model that was built on the basis of iterative feature selection has the greatest AUC among the considered models. This was because the model was iteratively built using a different number of features; one of the models, i.e., the one with the highest prediction accuracy, was selected. As shown in Figures 4 and 5, if a fixed number of features are used for building the model, the prediction accuracy might be relatively degraded. The reason was that irrelevant features made it difficult to build accurate prediction models. If more features irrelevant to the failure were used to build the model, the prediction accuracy of the model decreased. Therefore, in the case that all the features in the dataset were used as shown in Figure 6, the AUC decreased significantly. Quantitatively, the proposed model achieved 14.3% and 22.0% higher AUC than in the fixed number of features and the all features cases, respectively.

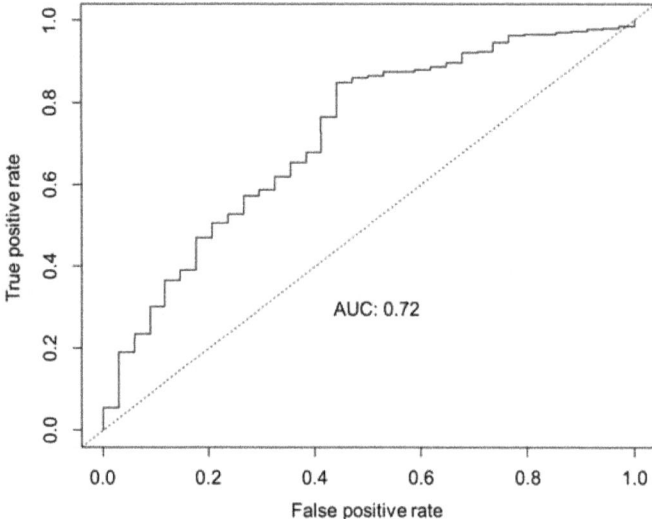

Figure 3. Receiver operating characteristic (ROC) curve for failure prediction model based on iterative feature selection.

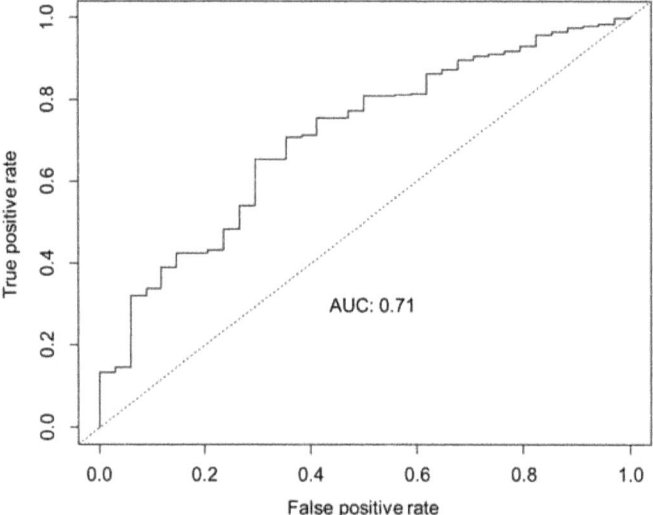

Figure 4. ROC curve for failure prediction model based on a fixed number of features (12 features).

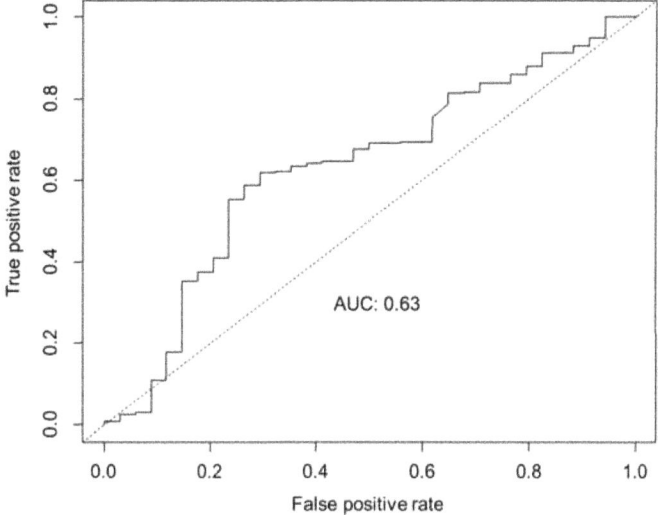

Figure 5. ROC curve for failure prediction model based on a fixed number of features (24 features).

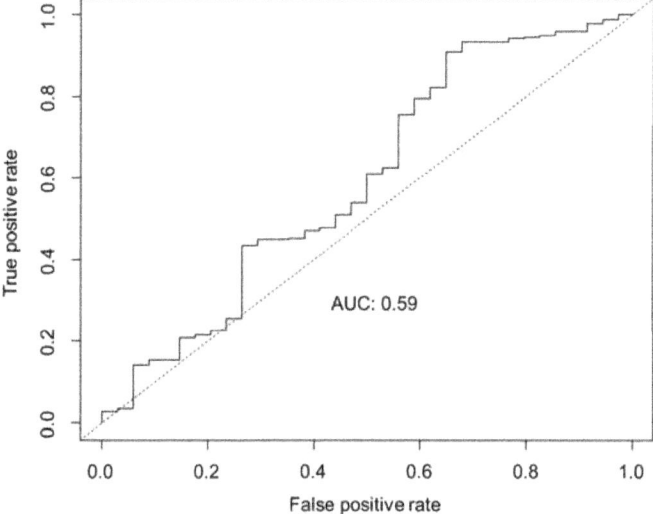

Figure 6. ROC curve for failure prediction model based on all features.

4. Conclusions

In this paper, we proposed a failure prediction model using iterative feature selection with the aim of predicting the failure occurrences. The procedure for building the failure prediction model consisted of the following five steps: (1) preprocessing, (2) importance measurement, (3) feature selection, (4) model building, and (5) model selection. In the first step, feature elimination, missing data imputation, normalization, and data division are sequentially conducted. The importance measurement step is to measure the importance of each feature by using the random forest. The third and the fourth steps were iteratively performed to build the failure prediction models using the various selected feature sets and to obtain the prediction accuracy of the built models. In the last step, the failure prediction model with the highest prediction accuracy was selected. The experimental implementation

Symmetry **2020**, *12*, 454

was conducted to evaluate the performance of the proposed model using the open-source R and the SECOM dataset given by the UCI repository. Through the experimental implementation, we obtained the importance of features representing the relevancy between each feature and failure. Moreover, we obtained the selected feature set and prediction accuracy set, each of which contains twenty-four features and prediction accuracy measurements. In the experiments, the proposed failure prediction model was built with eight features, and we compared the prediction accuracy of the proposed model with that of the failure prediction model built based on 12 features, 24 features, and all features. The results showed that the proposed model achieved 1.4%, 14.3%, and 22.0% higher AUC than that of other models. Comprehensively, the contributions of this paper are as follows. (1) We presented and discussed the problems of the existing failure prediction models for IIoT. (2) Through the importance measurement and iterative feature selection, we derived the feature index and the number of features that maximize the prediction accuracy of the failure prediction model. (3) We verified the feasibility of our work by conducting open-source-based implementation and extensive experiments. In our work, the proposed failure prediction model was implemented using only a limited dataset. Therefore, future work involves performing additional experiments with extended datasets to assess whether the proposed model is useful for various IIoT applications.

Author Contributions: E.-J.K. conceived and designed the failure prediction model; J.-H.K. performed the implementation and experiment; E.-J.K. interpreted and analyzed the data; J.-H.K. and E.-J.K. wrote the paper; E.-J.K. guided the research direction and supervised the entire research process. All authors have read and agree to the published version of the manuscript.

Funding: This research was supported by Hallym University Research Fund, 2019 (HRF-201911-010).

Conflicts of Interest: The authors declare no conflict of interest.

References

1. Cheng, J.; Chen, W.; Tao, F.; Lin, C.-L. Industrial IoT in 5G environment towards smart manufacturing. *J. Ind. Inf. Integr.* **2018**, *10*, 10–19. [CrossRef]
2. Kaur, K.; Garg, S.; Aujla, G.S.; Kumar, N.; Rodrigues, J.J.; Guizani, M. Edge Computing in the Industrial Internet of Things Environment: Software-Defined-Networks-Based Edge-Cloud Interplay. *IEEE Commun. Mag.* **2018**, *56*, 44–51. [CrossRef]
3. Zhu, C.; Rodrigues, J.J.; Leung, V.C.; Shu, L.; Yang, L.T. Trust-based Communication for the Industrial Internet of Things. *IEEE Commun. Mag.* **2018**, *56*, 16–22. [CrossRef]
4. Talavera, J.M.; Tobon, L.E.; Gomez, J.A.; Culman, M.A.; Aranda, J.M.; Parra, D.T.; Quiroz, L.A.; Hoyos, A.; Garreta, L.E. Review of IoT applications in agro-industrial and environmental fields. *Comput. Electron. Agric.* **2017**, *42*, 283–297. [CrossRef]
5. Lee, H.-H.; Kwon, J.-H.; Kim, E.-J. FS-IIoTSim: A flexible and scalable simulation framework for performance evaluation of industrial Internet of things systems. *J. Supercomput.* **2018**, *74*, 4385–4402. [CrossRef]
6. Liao, Y.; Loures, E.D.F.R.; Deschamps, F. Industrial Internet of Things: A Systematic Literature Review and Insights. *IEEE Internet Things J.* **2018**, *5*, 4515–4525. [CrossRef]
7. Ometov, A.; Bezzateev, S.; Voloshina, N.; Masek, P.; Komarov, M. Environmental Monitoring with Distributed Mesh Networks: An Overview and Practical Implementation Perspective for Urban Scenario. *Sensors* **2019**, *19*, 5548. [CrossRef]
8. Kontogiannis, S. An Internet of Things-Based Low-Power Integrated Beekeeping Safety and Conditions Monitoring System. *Inventions* **2019**, *4*, 52. [CrossRef]
9. Haseeb, K.; Almogren, A.; Islam, N.; Din, I.U.; Jan, Z. An Energy-Efficient and Secure Routing Protocol for Intrusion Avoidance in IoT-Based WSN. *Energies* **2019**, *12*, 4174. [CrossRef]
10. Nguyen, D.T.; Le, D.-T.; Kim, M.; Choo, H. Delay-Aware Reverse Approach for Data Aggregation Scheduling in Wireless Sensor Networks. *Sensors* **2019**, *19*, 4511. [CrossRef]

11. Lade, P.; Ghosh, R.; Srinivasan, S. Manufacturing Analytics and Industrial Internet of Things. *IEEE Intell. Syst.* **2017**, *32*, 74–79. [CrossRef]

12. Kammerer, K.; Hoppenstedt, B.; Pryss, R.; Stökler, S.; Allgaier, J.; Reichert, M. Anomaly Detections for Manufacturing Systems Based on Sensor Data—Insights into Two Challenging Real-World Production Settings. *Sensors* **2019**, *19*, 5370. [CrossRef] [PubMed]

13. Lim, H.K.; Kim, Y.; Kim, M.-K. Failure Prediction Using Sequential Pattern Mining in the Wire Bonding Process. *IEEE Trans. Semicond. Manuf.* **2017**, *30*, 285–292. [CrossRef]

14. Sanchis, R.; Canetta, L.; Poler, R. A Conceptual Reference Framework for Enterprise Resilience Enhancement. *Sustainability* **2020**, *12*, 1464. [CrossRef]

15. Wang, J.; Yao, Y. An Entropy-Based Failure Prediction Model for the Creep and Fatigue of Metallic Materials. *Entropy* **2019**, *21*, 1104. [CrossRef]

16. Hamadache, M.; Dutta, S.; Olaby, O.; Ambur, R.; Stewart, E.; Dixon, R. On the Fault Detection and Diagnosis of Railway Switch and Crossing Systems: An Overview. *Appl. Sci.* **2019**, *9*, 5129. [CrossRef]

17. Nam, K.; Ifaei, P.; Heo, S.; Rhee, G.; Lee, S.; Yoo, C. An Efficient Burst Detection and Isolation Monitoring System for Water Distribution Networks Using Multivariate Statistical Techniques. *Sustainability* **2019**, *11*, 2970. [CrossRef]

18. Abu-Samah, A.; Shahzad, M.K.; Zamai, E.; Said, A.B. Failure Prediction Methodology for Improved Proactive Maintenance using Bayesian Approach. *IFAC Pap. Online* **2015**, *48*, 844–851. [CrossRef]

19. Kwon, J.-H.; Kim, E.-J. Machine Failure Analysis Using Nearest Centroid Classification for Industrial Internet of Things. *Sens. Mater.* **2019**, *31*, 1751–1757. [CrossRef]

20. Moldovan, D.; Cioara, T.; Anghel, I.; Salomie, I. Machine learning for sensor-based manufacturing processes. In Proceedings of the 2017 IEEE 13th International Conference on Intelligent Computer Communication and Processing (ICCP), Cluj-Napoca, Romania, 7–9 September 2017; pp. 147–154. [CrossRef]

21. Mahadevan, S.; Shah, S.L. Fault detection and diagnosis in process data using one-class support vector machines. *J. Process Control* **2009**, *19*, 1627–1639. [CrossRef]

22. Hasan, M.A.M.; Nasser, M.; Ahmad, S.; Molla, K.I. Feature Selection for Intrusion Detection Using Random Forest. *J. Inf. Secur.* **2016**, *7*, 129–140. [CrossRef]

23. Wu, D.; Jennings, C.; Terpenny, J.; Gao, R.X.; Kumara, S. A Comparative Study on Machine Learning Algorithms for Smart Manufacturing: Tool Wear Prediction Using Random Forests. *J. Manuf. Sci. Eng.* **2017**, *139*, 071018-1–071018-2. [CrossRef]

24. Awan, F.M.; Saleem, Y.; Minerva, R.; Crespi, N. A Comparative Analysis of Machine/Deep Learning Models for Parking Space Availability Prediction. *Sensors* **2020**, *20*, 322. [CrossRef] [PubMed]

25. Wu, D.J.; Feng, T.; Naehrig, M.; Lauter, K. Privately evaluating decision trees and random forests. *Proc. Priv. Enhancing Technol.* **2016**, *2016*, 335–355. [CrossRef]

26. Garg, R.; Aggarwal, H.; Centobelli, P.; Cerchione, R. Extracting Knowledge from Big Data for Sustainability: A Comparison of Machine Learning Techniques. *Sustainability* **2019**, *11*, 6669. [CrossRef]

27. Lin, X.; Li, C.; Zhang, Y.; Su, B.; Fan, M.; Wei, H. Selecting Feature Subsets Based on SVM-RFE and the Overlapping Ratio with Applications in Bioinformatics. *Molecules* **2018**, *23*, 52. [CrossRef]

28. Liao, C.H.; Wen, C.H.-P. SVM-Based Dynamic Voltage Prediction for Online Thermally Constrained Task Scheduling in 3-D Multicore Processors. *IEEE Embed. Syst. Lett.* **2017**, *10*, 49–52. [CrossRef]

29. Shen, Y.; Wu, C.; Liu, C.; Wu, Y.; Xiong, N. Oriented Feature Selection SVM Applied to Cancer Prediction in Precision Medicine. *IEEE Access* **2018**, *6*, 48510–48521. [CrossRef]

30. Li, Z.; Wang, N.; Li, Y.; Sun, X.; Huo, M.; Zhang, H. Collective Efficacy of Support Vector Regression with Smoothness Priority in Marine Sensor Data Prediction. *IEEE Access* **2019**, *7*, 10308–10317. [CrossRef]

31. SECOM Data Set in University of California Irvine (UCI) Machine Learning Repository. Available online: https://archive.ics.uci.edu/ml/datasets/secom (accessed on 30 January 2020).

32. Schenatto, K.; de Souza, E.G.; Bazzi, C.L.; Gavioli, A.; Betzek, N.M.; Beneduzzi, H.M. Normalization of data for delineating management zones. *Comput. Electron. Agric.* **2017**, *143*, 238–248. [CrossRef]

33. Singh, D.; Singh, B. Investigating the impact of data normalization on classification performance. *Appl. Soft Comput.* **2019**, in press. [CrossRef]

34. Ali, L.; Niamat, A.; Khan, J.A.; Golilarz, N.A.; Xingzhong, X.; Noor, A.; Nour, R. An Optimized Stacked Support Vector Machines Based Expert System for the Effective Prediction of Heart Failure. *IEEE Access* **2019**, *7*, 54007–54014. [CrossRef]

Article

Symmetric Modeling of Communication Effectiveness and Satisfaction for Communication Software on Job Performance

Tian-Syung Lan [1,2], Kai-Chi Chuang [1], Hai-Xia Li [1,*], Jih-Fu Tu [3] and Huei-Sheng Huang [2]

1   College of Mechanical and Control Engineering, Guilin University of Technology, Guilin 541004, China; tslan888@yahoo.com.tw (T.-S.L.); s1038901@gmail.com (K.-C.C.)
2   Department of Information Management, Yu Da University, Miaoli County 36143, Taiwan; 99310024@ydu.edu.tw
3   Department of Industrial Management and Business Administration, St. John's University, Tamsui District, New Taipei City 25135, Taiwan; tu@mail.sju.edu.tw
*   Correspondence: 13737381010@163.com; Tel.: +86-1373-738-1010

Received: 14 February 2020; Accepted: 3 March 2020; Published: 5 March 2020

Abstract: Job performance is an issue highly related to the repetition of one enterprise. Because of the popularity of the Internet, consumer electronics have boomed rapidly and remove the space limitation stems. Users in the Taiwanese community send messages or share information through communication software that leads to more dependence from business. Various business problems have been solved and job performance has increased through the diversified functions on communication software. Thus, this research supposed that staff are willing to continuously use communication software LINE(a new communication app which allows one to make FREE voice calls and send FREE messages), and they agree that the varied functions of communication software would mean that information delivery more symmetrically affects their job performance. According to the research outcomes, communication effectiveness significantly influenced communication satisfaction and job performance, and communication satisfaction significantly influenced job performance. As organizational communication must be conducted through media that disseminate information, and different media have different communication effects, the relationship between communication effectiveness and job performance was completely mediated by communication satisfaction. The research suggested companies or organizations use LINE as a symmetric communication method to not only help employees improve their job performance, but also help enterprises achieve their goals or raise the profit, or even steady development for enterprises.

Keywords: symmetric model; communication effectiveness; communication satisfaction; job performance

---

1. Introduction

Exchanging information by multiple roles connects the social activities in our daily life, such as commercial activities, academic activities, and personal activities. This information or message exchanges often relay thoughts and emotions, and this kind of behavior is called communication. The way of information exchange is quite diverse; formal ways of exchanging information include through writing or language. Since the Internet network has been developed, electronic communication began to prevail, breaking the limitation of time and distance, which means both sides can perform information exchange without time and place limitations. Mobile devices have become more and more popular and broken through the limitation of space. With the use of the network, users can transmit instant messages anywhere using applications. The transmission methods of information have also become more diverse, such as multi-person consultations or discussions, instant image or video transmissions, and even paying function that has been extra added on some communication software. As the public displays increasing reliance on mobile devices, using instant communication applications as the pipeline in commercial applications can not only solve many business problems in an instant, but also help enterprises improve their business performance.

A team of Taiwan Institute for Information Industry (FIND) surveyed and analyzed the usage behavior on a Taiwanese community website. The survey found that the top two most frequently used by Taiwan people were LINE and Facebook [1], as shown in Figure 1.

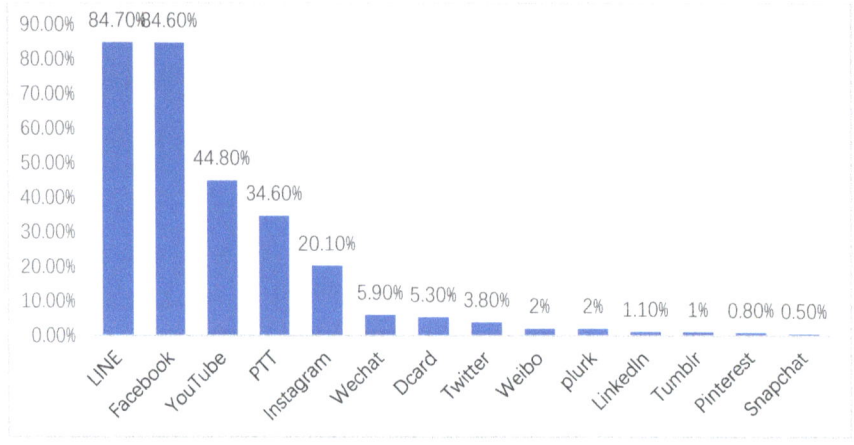

Figure 1. Frequency of using communication application more than three times a week [1].

As the most frequently used communication application, another survey about LINE, which was done by The Nielsen Company, found that LINE is a combination of communications, payments, entertainment, and other features [2]. Making calls or sending instant messages is the users' primary behavior, as shown in Figure 2.

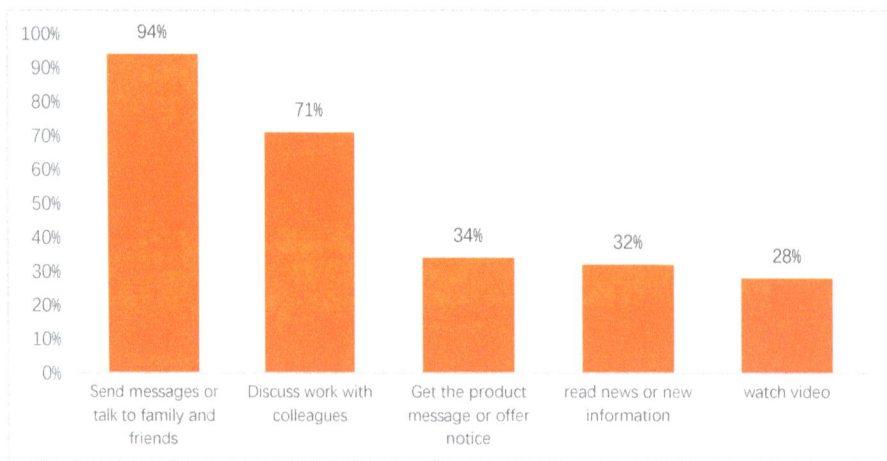

Figure 2. Most frequently used features of LINE [2].

LINE is a mobile application that can be used in mobile devices like smart phone, tablet, or even computer. LINE was first developed for individual users, which means individuals can send texts, stickers, or pictures to each other. With the development of the program, users can also make real-time video and set up a group. Group members can send messages, stickers, pictures, videos, audios, or even files; make audio or video calls with other members; and create albums or notes within the group. Many companies or organizations, thus, set up groups for different departments or units. They even set up groups for business partners or cooperators. The main intention is to facilitate communication and to reduce misunderstanding or mistakes.

According to the 2016 survey from Taiwan Institute for Information Industry, the application most people visit weekly was LINE, whereas Nielsen's 2016 survey also found that 91% of Taiwan people had used LINE in the last seven days, which means LINE was the most used communication application in Taiwan. At the same time, according to the same Nielsen's 2016 survey mentioned above, two of the most engaged behavior on LINE was to communicate with family and friends (94%) and to discuss work (71%). Unlike the traditional phone or application with text-only messaging, LINE has become the first choice to make contact with people in Taiwan. Therefore, it is worth exploring whether LINE's competence in communication effectiveness and for commercial communication can positively affect job performance.

As LINE is a new and recent software that has just appeared in Taiwan since 2012 [3], there is no other research discuss about the communication effectiveness, communication satisfaction, or job performance about LINE. Due to the lack of previous research, this research held that the more willing the workers from different levels of enterprises of different occupations, such as education organization, vehicle industry, and financial industry are to continuously use LINE as communication, the more they can use LINE's diversified function of information transmission, and certainly they are satisfied with the results of the information transmission, thereby continuously improving their job performance. Symmetrically, if the communication effectiveness is high enough to accurately and smoothly transmit messages, people within the company will be satisfied with the results of communication, and ultimately lead to improved job performance.

2. Literature Review

2.1. Communication Effectiveness

Carl Rogers, an American psychologist, said that active listeners not only listened to what the other person was saying, they also tried to understand and respond to the feelings behind those words, and would place themselves in the speaker's position to understand the message deeply [4]. Whether expressing one's opinions or ideas, or consulting or discussing with others at work, effective communication is essential, and effective communication can not only bring the membership closer, but also help members build a commitment to the organization, willing to contribute to the organization [5]. According to the Sproull and Kiesler's study of the situation about users self-exposing on the Internet, the Internet did help users expose themselves faster and deeper [6]. Yang believed that language cues can compensate for the lack of non-verbal cues, and that both parties could use online communication to subtly arrange the meaning of language and words. As people interact on the Internet over time, intimacy can also be created between the parties and help people expose themselves [7]. Levinger suggested that the transition in self-disclosure relationships begins with a shift from superficial revelations to intimate revelations, which means that the longer people communicate with friends on the Internet, the more intimate they feel [8].

Effective organizational communication depends on appropriate communication mechanisms and media [9]. As organizational communication must be conducted through media that disseminate information, and different media have different communication effects, the communication media chosen to be used by the organization affects their ability to process information, to learn, and determines the results of communication. In addition, the medium and methods of communication used by organizations also influence their behavior patterns [10]. Comparing to an environment with a high social presence, there is usually more information hiding and untrustworthy behavior in a low social presence environment [11].

In addition to exploring people's feelings of the existence of other participating individuals through digital media, in recent years, social presence has also begun to study the social cues contained by the digital media itself through interactive forms, so that people can produce a presence of a social actor. This social perception does not come from others involved in the media, but from the social cues provided by the media itself [12]. Instant messaging tools are mostly used in text-based communication, but there are also emoticons (emojis) available to users. Social context cues can be divided into two categories: dynamic and static. Emoticons (emojis) are dynamic and a substitute for facial expressions, it should be able to achieve more communication help at the social context level. Ambler believed that communication effectiveness could be assessed from the degree to which the sender can select and match the appropriate communication medium according to the situation and the extent to which the recipient correctly understands the message [13]. Gray believed that only by mastering the goal of communication can it be able to further measure the effectiveness of communication, and if the sender is not clear enough about the communication objectives, then there is no basis for measurement of communication effectiveness [14].

The key to successful communication is the correctness of information transmission, which was not easy to occur if there are different perceptions between the sender and the recipient. Successful communication can also be called effective communication. Effective communication can help deliver the correct message between employees, solve problems, and successfully achieve organizational goals.

2.2. Communication Satisfaction

Hargie and Dickson clearly pointed out that communication was the process of conveying a message that allowed two parties to understand and influence each other through sharing and building common perspectives [15]. Li and Tsai defined communication satisfaction as the different communication levels and forms in an organization, and the satisfaction levels of the individual members with their perception of quality and quantity [16]. Communication satisfaction relays a level of recognition. Communication was also a marketing process, marketing itself in the workplace to give cognitive subjective feelings. Communication helped employees to improve their productivity, the main purpose not only was meant to increase mutual understanding, avoiding misunderstanding between superiors and colleagues but also let one's professional ability to be found and one's job performance was able to be seen. Communication in the modern workplace symbolized as a positive image. Hecht defined communication satisfaction as an emotional response to the achievement of communication goals and expectations [17,18].

Krone, Jablin, and Putnam believed that there are four characteristics of communication: communication frequency, communication mode, communication content, and communication direction [19]. Communication frequency refers to the number of contacts between members of the organization. The more frequent communication involves between supervisors, volunteers, and full-time personnel; the closer the relationship between the members of the organization and the organization will perform. Communication mode refers to the channel used to convey information, that is, how the supervisor delivers the information to the members of the organization. According to the organizational structure, organizational communication can be divided into two categories, formal communication and informal communication. Formal mode refers to non-private pipelines, such as memos, group meetings, and other communication pipeline and methods that the supervisors trust or depend on. In contrast, the informal mode is very private, and it is often like a verbal, face-to-face individual teaching, and a temporary and improvisational communication pipeline [20]. The communication content is not referring to actual messages conveyed, but is related to the extent of direct communication to which the sender used in order to change the recipient's actions. Further, command-style information is direct communication, and vice versa. The communication direction refers to the change or flow of information horizontally and vertically within an organization. The communication flow between supervisor and subordinate initially occurs in a single direction, but the current human management environment has changed a lot. Schmuck and Runkel believed that the direction of communication has changed to be a loop with feedback, and communication is rarely one-way, it should be two-way.

Robertson pointed out that ideal communication can establish a good connection between managers and employees, but not only for sharing information and ideas [21]. Therefore, the level of leadership communication behavior was different comparing with other organizations. If the leader could understand the leadership's organizational communication characteristics, along with the communication technology, communication would be easier to succeed naturally.

After the employees started to use the communication software as the communication channel, they were able to produce a sense of satisfaction in communication when they achieved a communication smoothly and completed the tasks delivered by the organization, which was the main purpose of communication as they were able to gain satisfaction through the effectiveness of message transmission in communication.

2.3. Job Performance

Job performance results from the actions of the members in the organization and is expected by the organization and its members to be achieved with minimal resources [22]. According to Campbell's research structure, Borman and Motowidlo proposed that job performance is divided into task performance and contextual performance [23]. Task performance refers to the outcome of an individual's work that directly relates to the organization expects, and is judged by whether it meets the requirements of a formal role [24]. Contextual performance refers to the actions of an individual who voluntarily performs informal activities; adheres to complete the task; is willing to cooperate with others; follows rules and procedures; and is able to endorse, support, and defend the organization's goals.

Wright and Boswell measured job performance in five perspectives, which are support, goal Emphasis, team building, workshop facilitation, and global rating [25]. Castro, Dounlas, Hochwasser, Ferris, and Frink represented employees' performance with eight characteristics: work habits, planning and analytical skills, job knowledge, management skills, communication skills, developments in other aspects, interpersonal relationships, and overall assessment [26]. If a performance assessment could be performed properly, it would not only enable the employees to understand the benefit of executing it, but also affect the work efficiency and the mission direction in the future [27].

The pursuit of an enterprise is to maximize job performance with limited resources. Under such conditions, the joint efforts of all members are necessary. Therefore, it is necessary for all members to stay together and achieve their goals.

3. Research Design

3.1. Research Assumptions

As discussed in the previous section, Levinger [8] suggested that the longer people communicate with friends on the Internet, the more intimate they feel. Krone, Jablin, and Putnam [19] believed that the more frequent communication occurs between supervisors, volunteers, and full-time personnel, the closer the relationship between the members of the organization and the organization will perform, meaning higher communication satisfaction, and both Borman and Motowidlo [23] and Wright and Boswell [25] pointed out that job performance was mostly resulted from different communication form.

Based on the research purposes and the framework presented in Figure 3, the following hypotheses are developed.

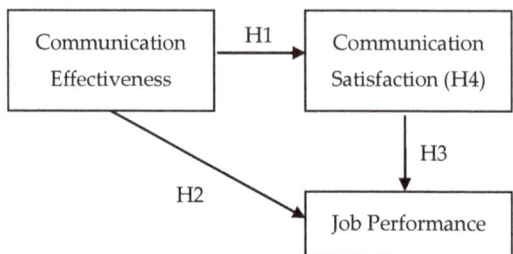

Figure 3. Research framework.

Hypothesis 1 (H1). *Communication effectiveness has significant effects on communication satisfaction.*

Hypothesis 2 (H2). *Communication effectiveness has significant effects on job performance.*

Hypothesis 3 (H3). *Communication satisfaction has significant effects on job performance.*

Hypothesis 4 (H4). *Communication satisfaction has a mediation effect between communication effectiveness and job performance.*

3.2. Sample of Research

The subjects of this research are the members who have used communication software to explain, distribute, and communicate at work in different levels of enterprises from all occupations. The convenience sampling method was used and conducted for the qualified anonymous samples. The questionnaires were used to collect data, and to understand the respondents' views on the research issues and to analyze their behaviors. The questionnaires were distributed in paper and the number of distributed questionnaires in detail as shown in Table 1.

Table 1. Detail of distributed questionnaires.

	Distributed	Response	Valid	Invalid	Valid Response Rate
Paper Questionnaires	480	455	445	10	92.7%

3.3. Instrument and Procedures

Based on the aforementioned research purpose and structure, the research questionnaire was divided into four parts: basic data, using behavior, communication satisfaction, and job performance. A 5-point Likert Scale was used to score and quantify, on a scale of 1–5, respectively, in five categories: strongly disagree, disagree, no opinion, agree, and strongly agree. The first part of the questionnaire was basic data, including four basic questions: age, gender, the amount of salary, and work area. The second part was the using communication effectiveness scale, the third part was the communication satisfaction scale, and the fourth part was the job performance scale.

Yan suggested that effective communication in an organization refers to the ability of communication showed within an organization, and that communication effectiveness can be explained from two aspects, "people" and "things", respectively. "People" refers to communication satisfaction and communication harmony, and "things" refers to the effectiveness and correctness of the messages. The purpose of communication within an organization is to enhance understanding, build consensus, coordinate actions, brainstorm ideas, meet the needs of members, and thus achieve predetermined objectives. The achievement of the above objectives can be used as a judgment on effective communication [28].

There were many pieces of research on communication effectiveness, and the factors of it in this research were based on the findings of Yan [29], which measured communication effectiveness in four factors, communication satisfaction, communication harmony, messages effectiveness, and messages timeliness (correctness). The findings had been revised to develop suitable dimensions and questions for the purposes and subjects of this research, as shown in Table 2.

Table 2. Dimensions and questions of communication effectiveness.

Dimensions	Questions
Harmony	1. I am satisfied with the overall harmonious communication in my department.
Effectiveness	2. I am satisfied with the overall effective communication in my organization.
Timeliness	3. I am satisfied with the overall timely communication in my organization.
Satisfaction	4. I am satisfied with the overall communication in my organization.

Communication satisfaction is a degree of awareness, defined by Price as a kind of emotional orientation that members of society produced towards their group [30]. Crino and White defined communication satisfaction as "one individual's perception of different forms of communication and its respective satisfactions in the organization" [31]. According to the previous literature of various studies, the communication satisfaction scale had been revised and developed as shown in Table 3.

Table 3. Dimensions and questions of communication satisfaction.

Dimensions		Questions
	1.	I am satisfied with the information from LINE about my job compared to my colleagues' job.
	2.	I am satisfied with the information from LINE about the judgment of me from my supervisors.
	3.	My supervisors know about my effort from LINE.
	4.	I am satisfied with the report on how my work problems were handled.
	5.	The extent to which my supervisors know and understand the problems the staff is facing through LINE.
	6.	The extent to which my supervisors can listen and pay attention to me
Communication	7.	My supervisors are able to provide guidance on solving work-related issues.
Satisfaction	8.	The extent to which my supervisors are willing to listen with an open mind.
	9.	The extent of correct and fluent communication with my colleagues though LINE.
	10.	Using communication software (LINE) can achieve smooth communication.
	11.	Using communication software (LINE) can clearly deliver information about work.
	12.	Using communication software (LINE) can clearly deliver company policies and goals.
	13.	Because of organizational communication, I feel that I am a very important part of the organization.
	14.	Conflict can usually be resolved through an appropriate communication pipe (LINE).
	15.	Communication through LINE is very important for productivity.

Edwards integrated the views of several scholars and argued that individual job performance can be studied in terms of demand and ability, and demand and supply [32]. Demand and ability refer to the fit between the abilities the work demanded and an individual's abilities, whereas demand and supply refer to the fit between an individual's needs and the job attributes. Based on the above studies, this research defined job performance as individual job performance, which measures the behavior of members and the results of their work. According to the relevant researches and previous literature, this research revised Robbins [33], Liu, and Liao's questionnaires [34] to measure job performance, and developed the questions of the job performance scale, as shown in Table 4.

Table 4. Factors and questions of job performance.

Dimensions	Questions
Job Performance	1. I appreciate the results of the members' work. 2. I can continuously improve my work quality and efficiency by using LINE. 3. I am willing to assist my colleagues in the extra work and strive for organizational performance. 4. I can comply with the organization's rules through LINE. 5. I can carry out orders or works through LINE. 6. I have many skills to handle works. 7. I can offer specific proposals for my job through LINE. 8. I can solve work problems alone. 9. I can handle emergent works through LINE. 10. I can participate in organizing affairs or meetings with a positive attitude. 11. I understand my job and responsibility through LINE. 12. I value my job. 13. I offer my help to my colleagues through LINE.

To increase the stability and reliability of the questionnaire, a reliability analysis was used to identify the Cronbach's α coefficient in each dimension. The Cronbach's α of the variable communication effectiveness was 0.791. The Cronbach's α of the variable communication satisfaction was 0.815. The Cronbach's α of the variable job performance as 0.758. The reliability of each variable was passed.

4. Results and Discussion

This research used regression analysis to explore the explanation and relationship among the variables. One independent variable was used to explain or predict another variable by building and testing the regression equation. Therefore, this research used regression analysis to examine whether communication effectiveness influenced communication satisfaction and job performance, to examine whether communication satisfaction influenced job performance, and to examine whether communication satisfaction had an intermediary effect in communication effectiveness and job performance.

To explore whether there was a correlation between two variables—communication effectiveness and communication satisfaction—regression analysis was carried out with communication effectiveness as independent variable and communication satisfaction as the dependent variable. Through regression analysis, it could be found that the coefficient of the independent variable was 0.711. The p-value was 0.021, which was less than 0.05, so the result was significant, indicating that communication effectiveness had a positive effect on communication satisfaction. The coefficient of determination was 0.601, which meant 60.1% of communication satisfaction could be explained by communication effectiveness. The date was shown in Table 5.

Table 5. Regression analysis of communication satisfaction on communication effectiveness.

Mode		Unstandardized Coefficients		Standardized Coefficients	t	Sig.
		B	Std. Error	Beta		
1.	(Constant)	1.011	0.093		10.881	0.000
	CE	0.711	0.027	0.775	25.853	0.021
		R	R^2	Adjusted R^2	Std. Error of the Estimate	
		0.775 [a]	0.601	0.600	0.32125	

Coefficients (a) a. Dependent variable: CS

Next, we discuss whether communication effectiveness is related to job performance by, first, taking communication effectiveness as the independent variable and job performance as the dependent variable for regression analysis. The coefficient of the independent variable was 0.753. Its p-value was 0.005, which was less than 0.05, so the result was significant, indicating that communication effectiveness had a positive effect on job performance. The coefficient of determination was 0.665, which meant 66.5% of job performance could be explained by communication effectiveness. The date was shown in Table 6.

Table 6. Regression analysis of job performance on communication effectiveness.

Mode		Unstandardized Coefficients		Standardized Coefficients	t	Sig.
		B	Std. Error	Beta		
1.	(Constant)	0.815	0.086		9.504	0.000
	CE	0.753	0.025	0.816	29.660	0.005
		R	R^2	Adjusted R^2	Std. Error of the Estimate	
		0.816 [a]	0.665	0.664	0.29670	

Coefficients (a) a. Predictors: (Constant), CE.

According to Baron and Kenny's view of defining the mediator [35], as shown in Figure 4, there are three conditions for meeting the mediator. The first is that the variation of the independent variable can significantly explain the variation of the mediator, which means path a is significant. The second one is that the variation of the mediator can significantly explain the variation of the dependent variable, which means path b is significant. The last one is when controlling independent variables and mediators are being considered at the same time, the relationship effect of the independent variable for the dependent variable (path c) is less significant, and in the case, the mediation effect of the mediator becomes strongest which c = 0. In other words, the independent variable can significantly explain the variation of the dependent variable, which means path c is significant. However, when both the independent variable and mediator are considered, the effect of the previous argument is reduced. As a result, the mediation effect of the mediator can be examined by four regression patterns.

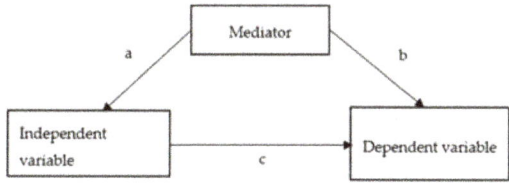

Figure 4. Defining the mediator [35].

Model 1: The effect of the independent variable on the mediator is significant.
Model 2: The effect of the independent variable on the dependent variable is significant.
Model 3: The effect of the mediator on the dependent variable is significant.
Model 4: Examine the effect of the mediator and the independent variable on the dependent variable at the same time, and compare the result of model 2, at which point the independent variable of model 4 should have less effect on the dependent. There should be two scenarios:

1. Complete mediation: If only the effect of the mediator is still significant, the effect of the independent variable is no longer significant, and then the mediator has a complete mediation effect between the relationship of the independent variable and the dependent variable.

2. Partial mediation: If the effect of the mediator and the independent variable are both still significant, then the mediator has a partial mediation effect between the relationship of the independent variable and the dependent variable.

In this section, to explore the mediation effect of communication satisfaction on communication effectiveness and job performance, regression analysis was taken to verify the hypotheses, and the communication satisfaction was used as a mediator to analyze to understand whether communication satisfaction has an impact on communication effectiveness and job performance.

To examine the mediation effect of communication satisfaction between communication effectiveness and job performance, four steps were carried out. The first step was the regression analysis of communication effectiveness on communication satisfaction. The p-value was 0.021, and the coefficient was 0.711, which had reached a significant level. The second step was the regression analysis of communication effectiveness on job performance. The p-value was 0.005, and the coefficient was 0.753, which had also reached a significant level. The third step was the regression analysis of communication satisfaction on job performance. The p-value was 0.000, and the coefficient was 0.922, which had reached a significant level. The last step was the regression analysis of communication effectiveness and communication satisfaction on job performance. The p-values were 0.182 and 0.002, and the coefficient was 0.245 and 0.715, which meant only communication satisfaction had reached a significant level. Based on the above four-step analysis results, the research found that communication satisfaction had a complete mediation effect between communication effectiveness and job performance. The analysis result can be found in Table 7.

Table 7. Mediation effect analysis.

Dependent Variable	Communication Satisfaction		Job Performance		Job Performance		Job Performance	
	Model 1		Model 2		Model 3		Model 4	
Dependent Variable	Coe.	Sig.	Coe.	Sig.	Coe.	Sig.	Coe.	Sig.
Communication Effectiveness	0.711	0.021	0.753	0.005	—	—	0.245	0.182
Communication Satisfaction	—	—	—	—	0.922	0.000	0.715	0.002

According to the above-mentioned analysis, the analysis results were used to review the research hypotheses and were explained. This research contained four research hypotheses, which were summarized and verified, as shown in Table 8.

Table 8. The result of hypotheses verification.

Hypothesis	Result
H1: Communication effectiveness has significant effects on communication satisfaction.	valid
H2: Communication effectiveness has significant effects on job performance.	valid
H3: Communication satisfaction has significant effects on job performance.	valid
H4: Communication satisfaction has a mediation effect between communication effectiveness and job performance.	Complete mediation

According to the research result, three hypotheses developed by this research were valid. As pointed out by Levinger [8], communication effectiveness can make people feel more or less intimate with each other and lead to a level of communication satisfaction. As job performance was mostly affected by different kind of communication, as mentioned by Borman and Motowidlo [23] and Wright and Boswell [25], this research assumed that both communication effectiveness and communication satisfaction would affect job performance, which was later proved by the research result as valid.

5. Conclusions and Discussion

Because of the popularity of the Internet, consumer electronics have boomed rapidly and break the space limitation stems. Users send messages or share information by communication software that leads to more dependence from businessmen. Various business problems have been solved and job performance has been risen up through the diversified functions on communication software.

In this study, the communication effectiveness is validated by the results of communication and led to the improvement of one's job performance by our proposed symmetric model. This research held that the more willing the workers are to use LINE as communication, the more they can use LINE's diversified function of information transmission, and certainly, they are satisfied with the results of the information transmission, thereby improving their job performance. Relatively, if the communication effectiveness is high enough to accurately and smoothly transmit messages, people within the company will be satisfied with the results of communication, which ultimately leads to an improvement in job performance. Therefore, the researchers believed that communication effectiveness has significant effects on communication satisfaction and job performance.

5.1. Conclusions

The results showed that communication effectiveness had a positive effect on communication satisfaction, and its explanatory power was 60.1%. Communication effectiveness had a positive effect on job performance, and its explanatory power was 66.5%. As communication effectiveness had significant positive effects on both communication satisfaction and job performance, and the explanatory power was both higher than 60%, when organizations or enterprises use communication software to communicate, they should pay attention to and strengthen the effectiveness of using LINE or other communication software to strengthen workers' communication satisfaction and job performance.

As there was only the variable communication satisfaction remained significant after the four-step regression analysis for the mediation effect, it was proved that communication satisfaction had a complete mediation effect between communication effectiveness and job performance, which meant the satisfaction of using LINE as a communication software would affect communication effectiveness and job performance. Therefore, organizations or companies should pay more attention of the usage of LINE, or even consider the utilization of communication software as the primary communication pipe, such as how their employees think about using LINE as the major communication pipe. As mentioned in the above chapters, LINE users can send texts, stickers, or pictures to each other. They can also make real-time video and set up a group. Group members can send messages, stickers, pictures, videos, audios, or even files; make audio or video calls with other members; and create albums or notes within the group. Many companies or organizations, thus, set up groups for different departments or units. They even set up groups for business partners or cooperators. The main intention is to facilitate communication and to reduce misunderstanding or mistakes. As the results of this research showed that employees were satisfied with using LINE as a major communication pipe in work place, it was suggested that employers can make good use of LINE continuously to improve employees' communication effectiveness and job performance. Communication software often improve their usefulness by updating their software functions, employers or supervisors can always make good use of the updated functions to assign tasks, establish to-do lists, establish personal memo, or any others ways to improve communication satisfaction, communication effectiveness, and job performance. As every company or organization must have the needs of communication to keep their work ongoing or keep their business developing, the results of this research (using communication software like LINE could ultimately improve job performance) could even help companies or organizations for their steady development.

5.2. Discussions

According to the findings of the research, future researchers could further the research by conducting follow-up discussions on the above phenomena and exploring other variables that can affect job performance. In addition, this research suggested that future researchers could continue to expand the sample size or add some new influence variables, or adjust the causal sequence between the variables in the framework or re-establish the research structure for re-discussion. It is also possible to lock in the comparison of industry differences, such as high-tech and traditional industries, it might result in giving some professional advice on work communication. Finally, future researchers can introduce more updated variables into the research structure, such as salary, benefits, and so on. It is believed that these future research results could provide some valuable information and the strategic direction of implementation.

Future research is suggested to explore the negative effect on employees of continuously using communication software as the major communication methodology in the work place, such as privacy issues, messiness of information, taking up non-working hours, etc.

5.3. Limitation

Although the results of the research were satisfied and the findings could contribute to help steady development of companies, there were two major limitations should be noted: First, the number of samples could be larger to strengthen the persuasiveness of the research or be more widespread to different levels of position or areas of industry to ensure the credibility of the results. Second, the research also limited to the users of communication software in Taiwan. The results of other countries might different from Taiwan's.

Author Contributions: Conceptualization, T.-S.L.; Data curation, K.-C.C. and H.-X.L.; Methodology, T.-S.L. and J.-F.T.; Project administration, T.-S.L.; Software, H.-X.L.; Validation, H.-S.H.; Visualization, K.-C.C. and H.-S.H.ng; Writing—original draft, T.-S.L. and J.-F.T. All authors have read and agreed to the published version of the manuscript.

Funding: This research received no external funding.

Conflicts of Interest: The authors declare no conflicts of interest.

References

1. Institute for Information Industry. Service System Drives Research and Development Plans for Emerging Businesses. 2017. Available online: https://www.iii.org.tw/ (accessed on 20 September 2019).
2. LINE. LINE Annual Seminar. 2016. Available online: https://meet.bnext.com.tw/articles/view/38786 (accessed on 20 September 2019).
3. Watchinese. App "LINE" is Booming! 2012. Available online: https://www.watchinese.com/article/2012/4402 (accessed on 25 November 2019).
4. Dessler, G. *Human Resource Management*; Prentic-Hall: Upper Saddle River, NJ, USA, 2005.
5. Rollinson, D. *Organizational Behavior and Analysis*, 2nd ed.; Prentice-Hall: New Upper Saddle River, NJ, USA, 2002; pp. 627–631.
6. Sproull, L.; Kiesler, S. Computers, Networks and Work. *Sci. Am.* **1991**, *265*, 116–127. [CrossRef]
7. Yang, C.C. Developing intimacy in computer-mediated communication: A study of the functions of verbal and non-verbal cues. *J. World Press Coll.* **1995**, *5*, 19–38.
8. Levinger, G. Toward the analysis of close relationships. *J. Exp. Soc. Psychol.* **1980**, *16*, 510–544. [CrossRef]
9. Shiang, J. Preconditions of digital democracy. *RDEC Bimon.* **2004**, *28*, 52–66.
10. Lu, W.S. The principle and practicality of organizational learning: The viewpoint of organization communication. *Pers. Mon.* **1996**, *23*, 24–43.
11. Hassanein, K.; Head, M. Manipulating social presence through the web interface and its impact on attitude towards online shopping. *Int. J. Hum. Comp. Stu.* **2007**, *65*, 689–708. [CrossRef]
12. Tung, F.W.; Deng, Y.S. Social cues in human-computer interaction-speech and interactivity effects on children in E-learning. *J. Des.* **2006**, *11*, 81–97.

13. Ambler, S. *Agile Modeling: Effective Practices for Extreme Programming and the Unified Process*, 1st ed.; John Wiley & Sons, Inc.: New York, NY, USA, 2002.
14. Gray, D. How to Measure Your Communication Effectiveness. Unpublished Work. 2005. Available online: http://communicationnation.blogspot.com/ (accessed on 14 October 2019).
15. Hargie, O.; Dickson, D. *Skilled Interpersonal Communication: Research, Theory, and Practice*; Routledge: London, UK, 2004.
16. Lee, Y.D.; Tsai, W.Y. A study on the construction of communication satisfaction inventory for the business employees in Taiwan. *J. Natl. Cheng-Kung Univ.* **1998**, *33*, 257–283.
17. Hecht, M.L. Measures of communication satisfaction. *Hum. Commun. Res.* **1978**, *4*, 350–368. [CrossRef]
18. Hsu, L.L.; Lai, S.Q.; Hsu, C.S.; Li, C.Y. The effect of blog and MSN usage on the interpersonal communication: A social presence theory perspective. *Chaoyang Bus. Manag. Rev.* **2011**, *10*, 19–48.
19. Krone, K.J.; Jablin, F.M.; Putnam, L.L. Communication theory and organizational communication: Multiple perspectives. In *Handbook of Organizational Communication*; Sage Publications: Los Angeles, CA, USA, 1987; pp. 18–40.
20. Schmuck, R.A.; Runkel, P.J. *Handbook of Organization Development in Schools*; National Press Books: Los Angeles, CA, USA, 1972.
21. Robertson, E. Using leadership to improve the communication climate. *Strateg. Commun. Manag.* **2002**, *7*, 24–27.
22. Liang, H.Y.; Wu, C.F.; Chang, S.C. Apply the balanced scorecard to measure strategic performance in healthcare system. *Tzu Chi Nurs. J.* **2005**, *4*, 26–31.
23. Borman, W.C.; Motowidlo, S.J. *Expanding the Criterion Domain to Include Elements of Contextual Performance*; Jossey-Bass: Los Angeles, CA, USA, 1993; pp. 71–98.
24. Motowidlo, S.J.; Van Scotter, J.R. Evidence that task performance should be distinguished from contextual performance. *J. Appl. Psychol.* **1994**, *79*, 475–480. [CrossRef]
25. Wright, P.M.; Boswell, W.R. Desegregating HRM: A review and synthesis of micro and macro human resource management research. *J. Manag.* **2002**, *28*, 247–276. [CrossRef]
26. Castro, S.L.; Douglas, C.; Hochwarter, W.A.; Ferris, G.R.; Frink, D.D. The effects of positive affect and gender on the influence tactics-job performance relationship. *J. Leadersh. Organ.* **2003**, *10*, 1–18. [CrossRef]
27. Lee, M.L.; Chou, P.F. Discussion on employee's cognitive attitude and job performance. *J. Far East Coll.* **2003**, *20*, 869–882.
28. Yan, C.H. A Study on Relations of Communication Capability, Attitude and Communication Effectiveness. *Master's Thesis.* 1999. Available online: https://ndltd.ncl.edu.tw/ (accessed on 10 October 2019).
29. Yen, C.C.; Wei, W.L. The effects of the key factors of organizational communication on organizational communication effectiveness and design performance. *J. Des.* **2011**, *16*, 61–85.
30. Price, J.H. *Handbook of Organization Measurement*; Health & Company: Washington, DC, USA, 1972; pp. 156–157.
31. Crino, M.D.; White, M.C. Satisfaction in communication: an examination of the Downs-Hazen measure. *Psychol. Rep.* **1981**, *49*, 831–838. [CrossRef]
32. Edwards, J.R. Person-job fit: A conceptual integration, literature review, and methodological critique. In *International Review of Industrial and Organizational Psychology*; Cooper, C.L., Robertson, I., Eds.; John Wiley & Sons, Inc.: New York, NY, USA, 1991; pp. 283–357.
33. Robbins, S.P. *Organizational Behavior*, 11th ed.; Prentice-Hall: New Upper Saddle River, NJ, USA, 2006.
34. Liou, J.C.; Liao, W.T. The Relationships among Knowledge Continuity Management, Job capability and Personal Work Performance. In Proceedings of the 2010 The 13th Conference on Interdisciplinary and Multifunctional Business Management, Taipei, Taiwan, 26 June 2010; pp. 189–202.
35. Baron, R.; Kenny, D. The moderator-mediator variable distinction in social psychological research: Conceptual, strategic, and statistical considerations. *J. Personal. Soc. Psychol.* **1986**, *51*, 1173–1182. [CrossRef]

Article

Homomorphic Encryption-Based Robust Reversible Watermarking for 3D Model

Li Li [1], Shengxian Wang [1], Shanqing Zhang [1,*], Ting Luo [2] and Ching-Chun Chang [3]

[1] Department of Computer Science, Hangzhou Dianzi University, Hangzhou 330018, China; lili2008@hdu.edu.cn (L.L.); wsx1131@163.com (S.W.)

[2] Collage of Science and Technology, Ningbo University, Ningbo 315000, China; luoting@nbu.edu.cn

[3] Department of Computer Science, University of Warwick, Coventry CV47AL, UK; ching-chun.chang@warwick.ac.uk

* Correspondence: sqzhang@hdu.edu.cn; Tel.: +86-130-7360-1029

Received: 31 January 2020; Accepted: 21 February 2020; Published: 1 March 2020

Abstract: Robust reversible watermarking in an encrypted domain is a technique that preserves privacy and protects copyright for multimedia transmission in the cloud. In general, most models of buildings and medical organs are constructed by three-dimensional (3D) models. A 3D model shared through the internet can be easily modified by an unauthorized user, and in order to protect the security of 3D models, a robust reversible 3D models watermarking method based on homomorphic encryption is necessary. In the proposed method, a 3D model is divided into non-overlapping patches, and the vertex in each patch is encrypted by using the Paillier cryptosystem. On the cloud side, in order to utilize addition and multiplication homomorphism of the Paillier cryptosystem, three direction values of each patch are computed for constructing the corresponding histogram, which is shifted to embed watermark. For obtaining watermarking robustness, the robust interval is designed in the process of histogram shifting. The watermark can be extracted from the symmetrical direction histogram, and the original encrypted model can be restored by histogram shifting. Moreover, the process of watermark embedding and extraction are symmetric. Experimental results show that compared with the existing watermarking methods in encrypted 3D models, the quality of the decrypted model is improved. Moreover, the proposed method is robust to common attacks, such as translation, scaling, and Gaussian noise.

Keywords: three-dimensional models; cloud computing; histogram shifting; encrypted model; decrypted model

---

1. Introduction

Due to the development of outsourced storage in the cloud, reversible watermarking in an encrypted domain has been developed for security in the cloud [1–4]. However, the cloud cannot introduce distortion of original content during watermark embedding. Therefore, the reversible watermarking method is required [5,6]. In addition, the watermark carrier is vulnerable during transmission, and the embedded watermark is expected to resist common attacks [7,8]. Therefore, robust reversible watermarking in an encrypted domain has greatly attracted researchers for potential applications.

In general, watermarking can be divided into robust and fragile watermarking methods in terms of their robustness. Robust watermarking [9] is used to protect security and resist attacks, while fragile watermarking [10,11] is used to provide integrity authentication. For the occasions with high data security requirements, such as judicial authentication, medical images, etc., more researchers focus on fragile watermarking in the encrypted domain.

Reversible watermarking in the encrypted domain can be divided into reserving room before encryption (RRBE) and vacating room after encryption (VRAE). The RRBE method reserves embedding room before encrypting the original image [12–15]. For example, the vacated bits, which are reserved by self-embedding before encryption, can be substituted by the watermark in the encrypted domain [4]. With the development of reversible watermarking [16,17], the original image can be restored absolutely after extracting the watermark. The second type directly implements watermark embedding by modified the encrypted image [18,19] after encryption. For instance, Xiang divided the original image into patches to be encrypted, and then the histogram of statistical values was calculated in the encrypted domain for shifting to embed watermark [20].

However, these methods are only applied to images, and cannot be used in 3D models directly due to different structures between images and 3D models. Ke et al. proposed a robust watermarking method on the basis of self-similarity [21]. In that method, a 3D model is divided into patches, and watermark bits were embedded by changing the local vector length of a point in each patch. Feng et al. divided a 3D model into patches, then embedded a watermark into each patch by modulating angle quantization [22]. However, those methods are not reversible. Jiang et al. proposed a 3D model watermarking method on the basis of stream cipher encryption [1]. The watermark was embedded by flipping the least significant bits (LSBs) of the vertex coordinates. Since the original 3D models have high spatial correlation, the watermark can be extracted successfully. Shah proposed a watermarking method based on the homomorphic Paillier cryptosystem, which used VRAE framework to vacate space before encryption [2]. However, those methods are fragile to attacks and cannot protect their copyrights.

To our best of knowledge, although the aforementioned watermarking methods on encrypted 3D models have been developed, the research on robustness for encrypted 3D models is rarely reported. In this paper, in order to protect the security of a 3D model in the cloud, we proposed a homomorphic encryption-based robust reversible watermarking method. In this method, the original model is first divided into patches to facilitate patch encryption using the Paillier cryptosystem. Then, the watermark is embedded by constructing the symmetrical direction histogram and shifting histogram in the encrypted domain, and the robust interval is reserved during the histogram shifting. Last, the receiver extracts the watermark in the encrypted model or the decrypted model by constructing a direction histogram of patches, and restores the original model through the method of histogram shifting which is the opposite to the embedding process. The contributions of the paper are organized as follows.

(1) The proposed method can directly construct direction histogram in the encrypted model so that the watermark can be extracted and the original encrypted model can be restored in the encrypted domain.

(2) The proposed method is robust to several common attacks by reserving the robust interval during the histogram shifting for watermark embedding.

(3) The proposed method not only has higher security and capacity, but also has less distortion compared with the original model.

The rest of this paper is organized as follows. In the second part, the Paillier cryptosystem is briefly introduced. In the third part, the related robust reversible watermarking method flow is proposed. The experimental results are shown in Section 4. The conclusions of the thesis are discussed in Section 5.

2. Paillier Cryptosystem

The Paillier cryptosystem [23], which was proposed by Paillier Pascal in 1999, has homomorphism and probability. Homomorphism means that one arithmetic operation of two ciphertexts are equal to another arithmetic operation of two corresponding plaintext. Moreover, homomorphism includes addition and multiplication homomorphism. Probability means that different ciphertexts, which are obtained by encrypting the same plaintext with different parameters, can be decrypted to the same plaintext. The following describes the processes of key generation, encryption, and decryption, two properties, and the application of modular multiplication inverse (MMI) [24] in the Paillier cryptosystem.

- Key Generation

Randomly pick up two large primes numbers p and q. Calculate $N = pq$ and $\lambda = lcm(p - 1, q - 1)$, where $lcm(\cdot)$ stand for the lowest common multiple. Afterwards, select $g \in Z_{N^2}^*$ randomly, which satisfies

$$gcd(L(g^\lambda mod N^2), N) = 1 \tag{1}$$

where $L(u) = (u - 1)/N$, and $gcd(\cdot)$ means the greatest common divisor of two inputs. $Z_{N^2} = \{0, 1, 2, \ldots, N^2 - 1\}$ and $Z_{N^2}^*$ are the numbers in Z_{N^2} which prime with N^2. Finally, we get the public key (N, g) and corresponding private key λ.

- Encryption

Select a parameter $r \in Z_{N^2}^*$ randomly. The plaintext $m \in Z_N$ can be encrypted to the corresponding ciphertext c by

$$c = E[m, r] = g^m \cdot r^N mod N^2 \tag{2}$$

where $E[\cdot]$ denotes the encryption function. Due to the nature of the Paillier cryptosystem, for the same plaintext m, different ciphertexts c can be obtained by choosing different r. After decryption, different ciphertexts can be restored to the same plaintext m, which ensures the security of the ciphertext.

- Decryption

The original plaintext m can be obtained by

$$m = D[c] = \frac{L(c^\lambda mod N^2)}{L(g^\lambda mod N^2)} mod N \tag{3}$$

Moreover, two important characteristics are described as follows (which has been applied in the proposed method).

- Lemma One

For two plaintexts $m_1, m_2 \in Z_N$, compute corresponding ciphertexts c_1, c_2 with r_1, r_2 according to Equation (1), respectively. The Equation $c_1 = c_2$ holds if and only if $m_1 = m_2$ and $r_1 = r_2$.

- Homomorphic Multiplication

For $\forall r_1, r_2 \in Z_N^*$, two plaintexts $m_1, m_2 \in Z_N$ and corresponding ciphertexts $E[m_1, r_1], E[m_2, r_2] \in Z_{N^2}^*$ satisfy

$$c_1 \cdot c_2 = E[m_1, r_1] \cdot E[m_2, r_2] = g^{m_1 + m_2} \cdot (r_1 \cdot r_2)^N mod N^2 \tag{4}$$

$$D[c_1 \cdot c_2] = D[E[m_1, r_1] \cdot E[m_2, r_2] mod N^2] = m_1 + m_2 mod N \tag{5}$$

The original Paillier cryptosystem only has addition homomorphism and multiplication homomorphism. The subtraction homomorphism can be achieved through modular multiplication inverse (MMI).

- Modular Multiplication Inverse (MMI)

For two coprime integers y and z, the existence of an integer θ satisfies

$$\theta \cdot y = 1 mod z \tag{6}$$

where θ is called the modular multiplicative inverse of y, and θ can be obtained according to the extended Euclidean method [25].

3. The Proposed Method

In order to protect the security of 3D model in the cloud, a homomorphic encryption-based robust reversible watermarking method is proposed. Figure 1 shows the flowchart of the proposed method. Firstly, the original model is divided into patches, and vertices in each patch are encrypted using the Paillier cryptosystem. In the cloud, three direction values of each patch are computed, and the direction histogram is constructed for shifting to embed the watermark. At last, the watermark can be extracted from direction histogram, and the original 3D model can be restored by histogram shifting.

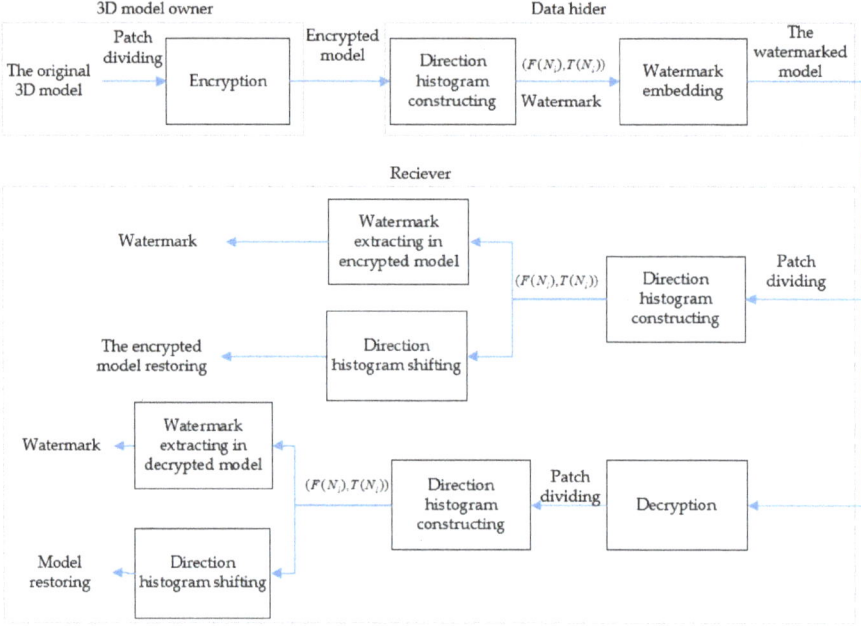

Figure 1. Flowchart of the proposed method.

3.1. Preprocessing

Because the input of the Paillier cryptosystem should be a positive integer, the vertex coordinates firstly are converted from decimal to positive integer.

3D models are consisted of vertex data and connectivity data. The vertex data includes the coordinates of each vertex in the spatial domain. The connectivity data reflects the connection relationship between vertices. A 3D model devil and its local region are illustrated in Figure 2. Each vertex and each face of the 3D model have a corresponding index number, respectively. For a 3D model M, let $\{v_i\}_{i=0}^{N_V}$ represents the sequence of vertices, where $v_i = (v_{i,x}, v_{i,y}, v_{i,z})$ and N_V is the number of vertices. Note that each coordinate $|v_{i,j}| < 1$, $j \in \{x, y, z\}$, and the significant digit of each coordinate is 6.

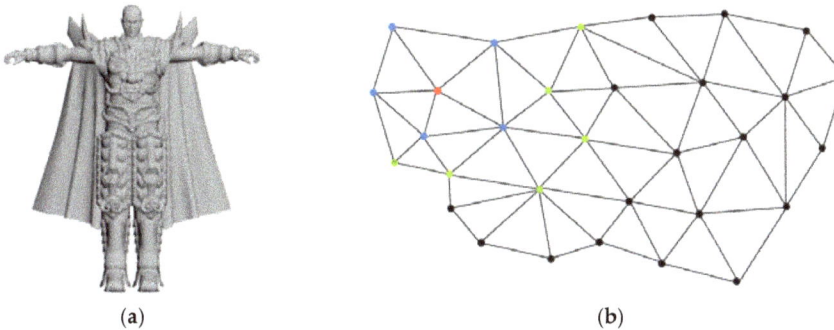

(a)   **(b)**

Figure 2. A 3D model devil; (**a**) original model, (**b**) local region.

Normally, uncompressed vertices are 32-bit floating point numbers with a precision of 6 digits. The first four significant digits of vertex coordinates can accurately display the 3D model. Therefore, the vertex coordinates are converted into an integer with four significant digits by using Equation (7).

$$v'_{i,j} = \lfloor v_{i,j} \cdot 10^4 \rfloor, \quad j \in \{x, y, z\} \tag{7}$$

Moreover, all vertex coordinates should be converted to positive integers for encryption by using Equation (8).

$$v'_{i,j} = v'_{i,j} + 10000, \quad j \in \{x, y, z\} \tag{8}$$

After preprocessing, the pre-processed 3D model is computed, and denoted as M'.

3.2. Patch Dividing and Patch Encryption

The section describes how to divide the model into several non-overlapping patches and perform encryption by using the Paillier cryptosystem.

3.2.1. Patch Dividing

For the vertex of the 3D model, if two vertices v_i and v_k are connected by a edge, v_k is a neighbor of v_i. All neighbors of v_i constitute the 1-ring neighborhood of v_i, and all 1-ring neighborhood of the neighbors of the vertex v_i constitute its 2-ring neighborhood. $N(v_i)$ is the 2-ring neighborhood of the vertex v_i, and $N(v_i)$ is computed by

$$N(v_i) = \{v_k | 0 \leq |v_i v_k| \leq 2, k = 0, 1, \ldots, N_V\} \tag{9}$$

where N_V are the number of the vertices of the 3D model, and $|v_i v_k|$ represents the number of vertices between v_i and v_k. As illustrated in Figure 2, the blue vertices are the 1-ring neighborhood of the red vertex, and the green vertices are the 2-ring neighborhood of the red vertex.

When the 3D model is divided into patches, it is necessary to ensure patches do not overlap each other. Suppose that the unclassified and classified sets are S_Y and S_N, respectively. $S_Y = \{v_i\}_{i=0}^{N_V}$ and S_N are initially empty. Suppose that the lth patch is denoted as $P^{(l)}$. A 3D model is divided into patches by the following rules, and initially $l = 1$.

Step 1: The first vertex v_i is selected according to the order of vertex index, and v_i and its 1-ring neighborhood are used as the $P^{(l)}$. Vertices in $P^{(l)}$ are sorted by

$$P^{(l)}(p) = \begin{cases} v_i, & p = 1 \\ v_k, & p = 2, 3, \ldots N_l \end{cases} \tag{10}$$

where N_l is the number of vertices in $P^{(l)}$.

Step 2: Update the unclassified set and the classified set by using Equation (11).

$$S_N = S_N \cup N(v_i), \quad S_Y = S_Y - N(v_i) \tag{11}$$

where $S_N \cup N(v_i)$ is the union of two sets, and $S_Y - N(v_i)$ is the vertex set that exist in S_Y but not in $N(v_i)$. $N(v_i)$ is put into the classified set for ensuring patches do not overlap each other.

Step 3: Determine whether the unclassified set S_Y is empty. If S_Y is empty, then the division of patches ends. If S_Y is not empty, then continue to select the $(l+1)^{th}$ patch from Step 1, until S_Y is empty.

As illustrated in Figure 3, the local region of 3D model devil can be divided into five patches, and each color in Figure 3 represents a patch.

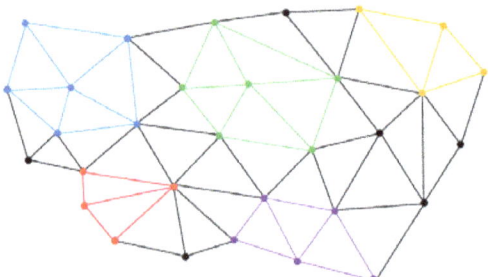

Figure 3. Patch dividing of 3D model devil.

3.2.2. Patch Encryption

Let $P^{(l)}(p, j), j \in \{x, y, z\}$ be the j-axis coordinates of the pth vertex in $P^{(l)}$. Referring to Equation (2), an integer $r_1(l) \in Z_N^*$ can be randomly selected to encrypt $P^{(l)}(p, j)$ with the public key (N, g).

$$C^{(l)}(p, j) = E[P^{(l)}(p, j), r_1(l)] = g^{P^{(l)}(p,j)} \cdot r_1(l)^N \bmod N^2 \tag{12}$$

where $p \in [1, N_l], j \in \{x, y, z\}$, $C^{(l)}$ denotes the encrypted vertex coordinates, and $E[M']$ represents the encrypted model.

3.3. Watermark Embedding

Firstly, three direction values of each patch in ciphertext are computed. Then, according to the possible values of the direction in ciphertext, the mapping table is constructed to map the direction values in ciphertext to the direction values in plaintext. The direction histogram is constructed by counting the direction values of all patches. Lastly, the watermark is embedded by histogram shifting.

3.3.1. Three Direction Values Calculation of Each Patch

In order to calculate three direction values of each patch, a vector $M(p)$ is defined by using Equation (13).

$$M(p) = \begin{cases} 1 & if\, p = 2, 3, \ldots N_l \\ -1 & if\, p = 1 \end{cases} \tag{13}$$

Suppose that $d^{(l)}(j), j \in \{x, y, z\}$ denotes the j-axis direction value of the lth patch $P^{(l)}$, which is calculated by Equation (14).

$$d^{(l)}(j) = \sum_{p=2}^{N_l} [P^{(l)}(p, j) \cdot M(p) + P^{(l)}(1, j) \cdot M(1)] \tag{14}$$

In the encrypted domain, without the private key λ, the encrypted vertex coordinates cannot by decrypted to obtain the vertex coordinate in plaintext, so the direction value $d^{(l)}(j)$ cannot be directly calculated. In the proposed method, the direction value in ciphertext can be calculated using the MMI method. $C^{(l)}$ represents the encrypted patch corresponding to the original patch $P^{(l)}$. In order to calculate the direction value in ciphertext, the modular multiplicative inverse $\theta_{C^{(l)}(p,j)}$ of $C^{(l)}(p,j)$ should be calculated through the extended Euclidean method. $\theta_{C^{(l)}(p,j)}$ satisfies

$$\theta_{C^{(l)}(p,j)} \cdot C^{(l)}(p,j) = 1 \bmod N^2 \tag{15}$$

For the lth patch, the vectors $M_1^{(l)}$ and $M_2^{(l)}$ are defined by using Equations (16) and (17), respectively.

$$M_1^{(l)}(p,j) = \begin{cases} C^{(l)}(p,j) & if\, p = 2,3,\ldots N_l \\ \theta_{C^{(l)}(p,j)} & if\, p = 1 \end{cases} \tag{16}$$

$$M_2^{(l)}(p,j) = \begin{cases} \theta_{C^{(l)}(p,j)} & if\, p = 2,3,\ldots N_l \\ C^{(l)}(p,j) & if\, p = 1 \end{cases} \tag{17}$$

Since the direction value $d^{(l)}(j)$ may be negative, two direction values $c_{d1}^{(l)}(j)$ and $c_{d2}^{(l)}(j)$ are re-defined. If $d^{(l)}(j)$ is positive, $c_{d1}^{(l)}(j)$ is the ciphertext corresponding to $d^{(l)}(j)$. If $d^{(l)}(j)$ is negative, $c_{d2}^{(l)}(j)$ is the ciphertext corresponding to $d^{(l)}(j)$. $c_{d1}^{(l)}(j)$ and $c_{d2}^{(l)}(j)$ can be calculated by Equation (18).

$$\begin{cases} c_{d1}^{(l)}(j) = M_1^{(l)}(1,j)^3 \prod_{p=2}^{N_l} M_1^{(l)}(p,j) \bmod N^2 \\ c_{d2}^{(l)}(j) = M_2^{(l)}(1,j)^3 \prod_{p=2}^{N_l} M_2^{(l)}(p,j) \bmod N^2 \end{cases} \tag{18}$$

After $c_{d1}^{(l)}(j)$ and $c_{d2}^{(l)}(j)$ are calculated, $d^{(l)}(j)$ is obtained by querying the mapping table. The following is the corresponding equation derivation and proof. To facilitate understanding, a patch consisting of four vertices is used as an example. Suppose that P_1, P_2, P_3, P_4 denote the j-axis coordinate of the pth vertex as illustrated in Figure 4, and $\hat{c}_1, \hat{c}_2, \hat{c}_3, \hat{c}_4$ is the ciphertext corresponding to P_1, P_2, P_3, P_4. $\theta_1, \theta_2, \theta_3, \theta_4$ is the modular multiplicative inverses corresponding to $\hat{c}_1, \hat{c}_2, \hat{c}_3, \hat{c}_4$, which satisfies

$$\begin{cases} \theta_1 \cdot c_1 = \theta_1 \cdot g^{P_1} \cdot r_1^{N} = 1 \bmod N^2 \\ \theta_2 \cdot c_2 = \theta_2 \cdot g^{P_2} \cdot r_1^{N} = 1 \bmod N^2 \\ \theta_3 \cdot c_3 = \theta_3 \cdot g^{P_3} \cdot r_1^{N} = 1 \bmod N^2 \\ \theta_4 \cdot c_4 = \theta_4 \cdot g^{P_4} \cdot r_1^{N} = 1 \bmod N^2 \end{cases} \tag{19}$$

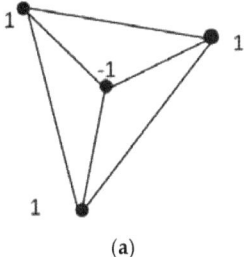

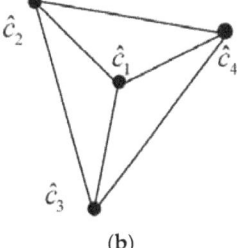

(a)   (b)

Figure 4. The patch with four vertices. (a) $M(p)$ correspond to the vertex. (b) The encrypted coordinate.

Then the direction value in ciphertext can be calculated by using Equation (20).

$$\begin{cases} c_{d1}(j) = M_1(1,j)^3 \prod_{p=2}^{N_l} M_1(p,j) \bmod N^2 = \theta_1{}^3 \cdot \hat{c}_2 \cdot \hat{c}_3 \cdot \hat{c}_4 \\ c_{d2}(j) = M_2(1,j)^3 \prod_{p=2}^{N_l} M_2(p,j) \bmod N^2 = \hat{c}_1{}^3 \cdot \theta_2 \cdot \theta_3 \cdot \theta_4 \end{cases} \tag{20}$$

It can be derived to the following equation.

$$\begin{cases} c_{d1}^{(l)}(j) = g^{P_2+P_3+P_4} \cdot r_1{}^{3N} \cdot \theta_1 \bmod N^2 \\ c_{d2}^{(l)}(j) = g^{3P_1} \cdot r_1{}^{3N} \cdot \theta_1 \cdot \theta_2 \cdot \theta_3 \bmod N^2 \end{cases} \tag{21}$$

According to Carmichael theory, the following equation holds.

$$\begin{cases} g^{N\lambda} = 1 \bmod N^2 \\ r_1{}^{N\lambda} = 1 \bmod N^2 \end{cases} \tag{22}$$

Hence, the following equation holds.

$$g^{N\lambda} \cdot r_1{}^{N\lambda} = 1 \bmod N^2 \tag{23}$$

According to Equations (19) and (23), Equation (24) can be derived.

$$\begin{cases} \theta_1 = g^{N\lambda-P_1} \cdot r_1{}^{N(\lambda-1)} \bmod N^2 \\ \theta_2 = g^{N\lambda-P_2} \cdot r_1{}^{N(\lambda-1)} \bmod N^2 \\ \theta_3 = g^{N\lambda-P_3} \cdot r_1{}^{N(\lambda-1)} \bmod N^2 \\ \theta_4 = g^{N\lambda-P_4} \cdot r_1{}^{N(\lambda-1)} \bmod N^2 \end{cases} \tag{24}$$

According to Equations (22) and (24), Equation (20) can be simplified as

$$\begin{cases} c_{d1}^{(l)}(j) = g^{3N\lambda+P_2+P_3+P_4-3P_1} \bmod N^2 \\ c_{d2}^{(l)}(j) = g^{3N\lambda+3P_1-P_2-P_3-P_4} \bmod N^2 \end{cases} \tag{25}$$

3.3.2. Constructing the Mapping Table

Due to the spatial correlation of the 3D model, the vertex coordinates are relatively close in space. According to the experiments on multiple 3D model, the direction values are usually in a certain range, and the maximum direction value is usually related to the number of vertices in the patch. As illustrated in Figure 5, the blue line shows the change in the maximum direction value when the number of vertices in the patch changes. The red line is the fitted curve of the blue line, and its fitting function $F(N_l)$ satisfies

$$F(N_l) = 1.925 \cdot (N_l - 1)^3 - 60.6 \cdot (N_l - 1)^2 + 528 \cdot (N_l - 1) - 609 \tag{26}$$

Therefore, the direction values are all within a certain range. When the number of vertices of the patch changes, the direction values does not exceed $F(N_l)$. Moreover, in order to obtain robustness, the robust interval $T(N_l)$ is designed in the process of histogram shifting. The robust interval $T(N_l)$ is related to the number of the patch, which is defined by

$$T(N_l) = t \cdot (N_l - 1) \tag{27}$$

where t represents the strength of robustness. Hence, the change of direction values is $F(N_l) + T(N_l)$ at most.

Suppose that d_p denotes the absolute of direction values, then $d_p \in [0, 2F(N_l) + T(N_l)]$. With the public key, and the ciphertext c_{d_p} corresponding to d_p can be calculated by

$$c_{d_p} = g^{d_p} \bmod N^2, \quad d_p = 0, 1, 2, \ldots, 2F(N_l) + T(N_l) \tag{28}$$

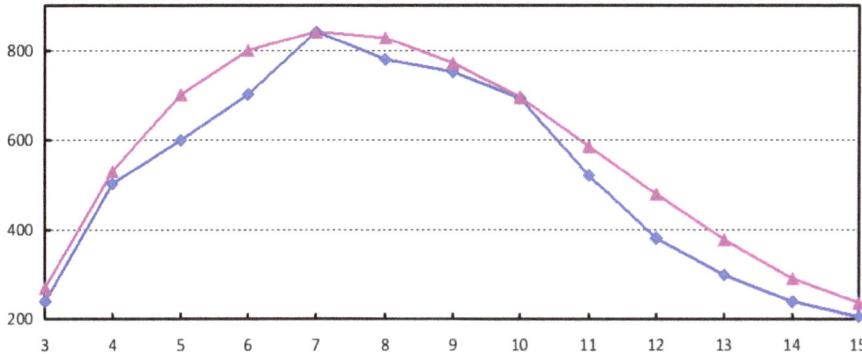

Figure 5. The blue line represents relationship between the maximum direction value and the number of vertices of the patch, and the red line is the fitted curve of the blue line.

Hence, the mapping table can be constructed as illustrated in Figure 6, and the direction values in ciphertext can be mapped to the direction values in plaintext through the mapping table c_{d_p}. The mapping method is described as follows: c_{d_p} is a ciphertext set obtained by encrypting all possible values d_p. When $c_{d1}^{(l)}(j)$ matches the value $c_{d_p}[m]$ in c_{d_p}, it indicates $d^{(l)}(j) \geq 0$, and $d^{(l)}(j) = d_p[m]$. When $c_{d2}^{(l)}(j)$ matches the value $c_{d_p}[m]$ in c_{d_p}, it indicates $d^{(l)}(j) < 0$, and $d^{(l)}(j) = -d_p[m]$, where $m \in [0, 2F(N_l) + T(N_l))$, and $d_p[m]$ represents the mth value in the mapping table. Therefore, without the private key, the direction values in plaintext can be obtained by querying the mapping table.

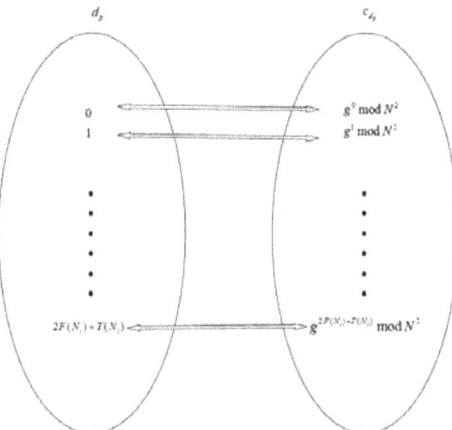

Figure 6. The mapping table.

3.3.3. Constructing the Symmetrical Direction Histogram

In the proposed method, the direction values in ciphertext are first calculated using the MMI method. Then, according to all possible direction values, the mapping table can be constructed, so the direction values in ciphertext can be mapped to direction values in plaintext. Last, the direction histogram can be constructed by counting all direction values. The direction histogram of all patches with six vertices is shown as Figure 7. It is found that most direction values are concentrated in the central area, and only a small part of the direction values are beyond the central area. Moreover, the direction histogram is symmetrical visually.

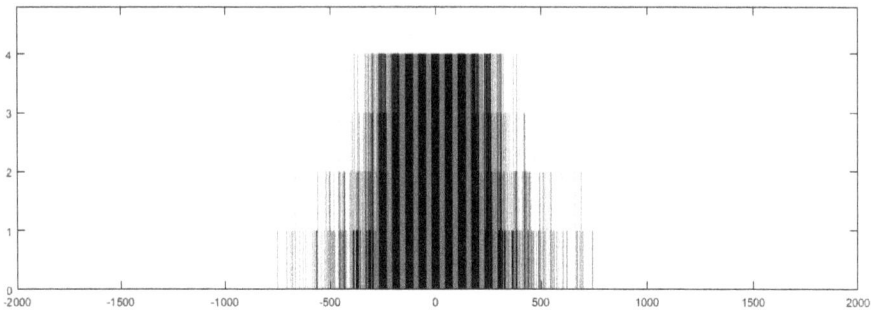

Figure 7. The direction histogram of patches with seven vertices.

3.3.4. Embedding Watermark by Histogram Shifting

In the proposed method, the watermark is embedded by shifting the direction histogram. In order to embed the watermark, the changed direction values should exceed the range of original histogram. Using $F(N_l)$ and $T(N_l)$ as embedding keys, the embedded function $B(N_l)$ is defined by Equation (29) to change the direction values.

$$B(N_l) = \left\lceil \frac{F(N_l) + T(N_l)}{N_l - 1} \right\rceil = t + 528 + \varphi(N_l - 1) \tag{29}$$

where $\varphi(N_l - 1)$ is the function about $N_l - 1$, $\varphi(N_l - 1)$ will not change, and $\lceil \cdot \rceil$ means to round up. Suppose that $\beta = t + 528$, and $B(N_l)$ can be changed by modifying β. Moreover, three bits can be embedded by changing three direction values of a patch. Suppose that $C_w^{(l)}$ denotes the encrypted patch with watermark. If the watermark bit '0' needs to be embedded, the vertex coordinate is not changed, which means $C_w^{(l)} = C^{(l)}$. If the bit '1' needs to be embedded, the ciphertext $C^{(l)}(p, j)$ in the patch $C_w^{(l)}$ is changed by Equation (30) to obtain $C_w^{(l)}(p, j)$.

$$C_w^{(l)}(p, j) = \begin{cases} C^{(l)}(p, j) \cdot g^{B(N_l)} = g^{P^{(l)}(p,j) + B(N_l)} \cdot r_1(k)^N \bmod N, & if\, d^{(l)}(j) \in [0, f(N_l))\, and\, p = 2, 3, \ldots, N \\ C^{(l)}(p, j) \cdot g^{B(N_l)} = g^{P^{(l)}(p,j) + B(N_l)} \cdot r_1(k)^N \bmod N, & if\, d^{(l)}(j) \in (-f(N_l), 0)\, and\, p = 1 \end{cases} \tag{30}$$

where $C_w^{(l)}(p, j)$ is the encrypted patch with watermark. Suppose that $P_w^{(l)}(p, j)$ denotes the decrypted patch with watermark. The operation in ciphertext is equivalent to change the coordinates $P^{(l)}(p, j)$ to $P_w^{(l)}(p, j)$ in plaintext.

$$P_w^{(l)}(p, j) = \begin{cases} P^{(l)}(p, j) + B(N_l), & if\, d^{(l)}(j) \in [0, F(N_l))\, and\, p = 2, 3, \ldots, N \\ P^{(l)}(p, j) + B(N_l), & if\, d^{(l)}(j) \in (-F(N_l), 0)\, and\, p = 1 \end{cases} \tag{31}$$

Suppose that $d_w^{(l)}(j)$ denotes the direction value with watermark. After embedding the bit '0', $d_w^{(l)}(j)$ is still in the range $(-F(N_l), F(N_l))$, so $(-F(N_l), F(N_l))$ is called the 0-bit area. After embedding the bit '1', $d^{(l)}(j)$ will be changed by the size of $F(N_l) + T(N_l)$ for making $d_w^{(l)}(j)$ within the range $(-2F(N_l) - T(N_l), -F(N_l) - T(N_l)]$ or $[F(N_l) + T(N_l), 2F(N_l) + T(N_l))$. $(-2F(N_l) - T(N_l), -F(N_l) - T(N_l)]$ and $[F(N_l) + T(N_l), 2F(N_l) + T(N_l))$ are called the 1-bit area. Finally, the encrypted model with watermark can be obtained. For example, after embedding 1000 bits, the direction histogram is shown in Figure 8. When embedding the bit '0', the direction values are still in the 0-bit area. When embedding the bit '1', the direction values will be shifted into the 1-bit area. Moreover, the 0-bit area and the 1-area are separated by the robust interval of size $T(N_l)$. Hence, the watermark is embedded successfully.

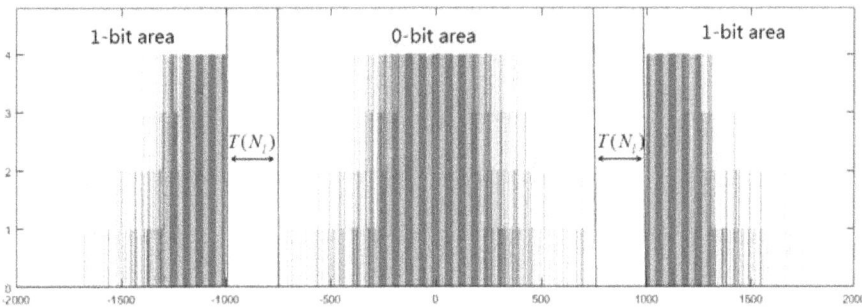

Figure 8. The watermarked histogram. After embedding the watermark, the original direction histogram can be divided into 0-bit area and 1-bit area. The 0-bit area and 1-bit area are separated by the robust interval of size $T(N_l)$.

3.4. Watermark Extraction

Watermark extraction includes extracting the watermark in the encrypted model and extracting the watermark in the decrypted model.

3.4.1. Extracting Watermark in an Encrypted Domain and Restore the Original Encrypted Model

The watermarked model is firstly divided into patches, and the direction values in ciphertext is calculated and mapped to the direction values in plaintext using the MMI method and the mapping table. Then, the direction histogram is constructed, and the watermark is extracted from direction histogram. Finally, with the embedding key $(F(N_l), T(N_l))$, the embedding function $B(N_l)$ can be obtained to restore the original encrypted model. Let $w^{(l)}(j)$ be the watermark embedded in the j-axis of the lth patch, and $w^{(l)}(j)$ is extracted by Equation (32).

$$w^{(l)}(j) = \begin{cases} 0, & if \, d^{(l)}(j) \in (-F(N_l), F(N_l)) \\ 1, & else \end{cases} \tag{32}$$

The original encrypted model can be restored by histogram shifting, which is reverse to the embedding process. In order to restore the original encrypted model, the modular multiplicative inverse $\theta_{g^{B(N_l)}}$ of $g^{B(N_l)}$ need to be calculated through the extended Euclidean method.

$$\theta_{g^{B(N_l)}} \cdot g^{B(N_l)} = 1 \bmod N^2 \tag{33}$$

Therefore, the original encrypted vertex coordinate $C^{(l)}(p, j)$ can be obtained by Equation (34).

$$C^{(l)}(p,j) = \begin{cases} C_w^{(l)}(p,j) \cdot \theta_{g^{B(N_l)}} = C^{(l)}(p,j) \cdot g^{B(N_l)} \cdot \theta_{g^{B(N_l)}} \bmod N^2 \\ \quad if \, d^{(l)}(j) \in [F(N_l) + T(N_l), 2F(N_l) + T(N_l)) \, and \, p = 2,..,N \\ C_w^{(l)}(p,j) \cdot \theta_{g^{B(N_l)}} = C^{(l)}(p,j) \cdot g^{B(N_l)} \cdot \theta_{g^{B(N_l)}} \bmod N^2 \\ \quad if \, d^{(l)}(j) \in (-2F(N_l) - T(N_l), -F(N_l) - T(N_l)] \, and \, p = 1 \\ C_w^{(l)}(p,j) \\ \quad if \, d^{(l)}(j) \in (-F(N_l), F(N_l)) \end{cases} \tag{34}$$

where $C_w^{(l)}(p,j)$ is the vertex coordinate with watermark in the patch $C_w^{(l)}$, and the processing in ciphertext is equivalent to the change in plaintext by using Equation (35).

$$P^{(l)}(p,j) = \begin{cases} P_w^{(l)}(p,j) - B(N_l), & if\, d^{(l)}(j) \in [F(N_l) + T(N_l), 2F(N_l) + T(N_l)) \, and\, p = 2,..,N \\ P_w^{(l)}(p,j) - B(N_l), & if\, d^{(l)}(j) \in (-2F(N_l) - T(N_l), -F(N_l) - T(N_l)] \, and\, p = 1 \\ P_w^{(l)}(p,j), & if\, d^{(l)}(j) \in (-F(N_l), F(N_l)) \end{cases}$$ (35)

After the above process, $d_w^{(l)}$ is restored to $d^{(l)}$, and the original encrypted model can be obtained.

3.4.2. Extracting Watermark in Decrypted Model

With the private key, the watermarked model can be decrypted. With the embedding key, the watermark can be extracted and the original 3D model can be restored. Firstly, the watermarked model is divided into patches and the direction values of each patch are calculated by using Equation (14). Then, the direction histogram is constructed and the watermark is extracted from direction histogram by using Equation (32). Finally, with the embedding key, the original model can be restored by using Equation (35).

The decrypted model with watermark may be vulnerable to some common attacks such as noise interference during transmission. Since the robust interval during histogram shifting is reserved, the proposed method is robust to common attacks, such as Gaussian noise, translation, scaling, etc. As illustrated in Figure 8, the 0-bit area and the 1-bit area are separated by the robust interval of size $T(N_l)$. After the decrypted watermarked model is attacked slightly, it will cause a small range fluctuation of the direction values. However, if the direction values do not enter the error area, the receiver can still correctly extract the watermark. In order to improve the accuracy of watermark extraction after being disturbed, the watermark is extracted by using

$$w^{(l)}(j) = \begin{cases} 0, & if\, d^{(l)}(j) \in (-F(N_l) - T(N_l)/3, F(N_l) + T(N_l)/3) \\ 1, & else \end{cases}$$ (36)

4. Experimental Results and Discussion

The proposed method processed 3D model and implemented the watermark method in MATLAB R2016b under Window 7. We implemented the following experiment on 40 3D models and calculated the average of 40 3D models. Figure 9 shows six models used in the experiment.

The quality of the decrypted watermarked model is evaluated by the signal-to-noise ratio (SNR). The higher the value SNR, the better the imperceptibility after embedding watermark. SNR is computed as

$$SNR = 10\lg \frac{\sum_{i=1}^{N_V} [(v_{i,x} - \overline{v}_x)^2 + (v_{i,y} - \overline{v}_y)^2 + (v_{i,z} - \overline{v}_z)^2]}{\sum_{i=1}^{N_V} [(g_{i,x} - v_{i,x})^2 + (g_{i,y} - v_{i,y})^2 + (g_{i,z} - v_{i,z})^2]}$$ (37)

where $\overline{v}_x, \overline{v}_y, \overline{v}_z$ are the mean of vertex coordinates, $v_i(v_{i,x}, v_{i,y}, v_{i,z})$ are the original coordinates, and $g_i(g_{i,x}, g_{i,y}, g_{i,z})$ are the coordinates of the watermarked model M_w.

In addition, the bit error rate (BER) is used to measure the error rate of the extracted watermark. The lower the value, the higher the accuracy of the extracted watermark.

4.1. The Value of β

According to Equation (29), $B(N_l) = \beta + \varphi(N_l - 1)$, and the embedding function $B(N_l)$ is changed by changing the value of β. According to Equation (31), if the value of β is large, the distortion of the decrypted model is high and the accuracy of watermark extracting is high, and vice versa. In order to observe the effect of β on the quality of decrypted model and the bit error rate of the extracted watermark, we changed the value of β to perform on 40 tested models and calculated their average. The relationship between the value of β and the distortion *SNR* is illustrated in Figure 10a. As β increases, *SNR* gradually decreases. When $\beta = 588$, *SNR* of the decrypted model is slightly greater than 30 dB. Based on imperceptible considerations, in order to obtain better model quality, the value of β cannot exceed 588. The relationship between the value of β and BER is shown in Figure 10b. When $\beta \geq 528$, the watermark was correctly extracted without being attacked. Therefore, the value of b cannot be less than 528.

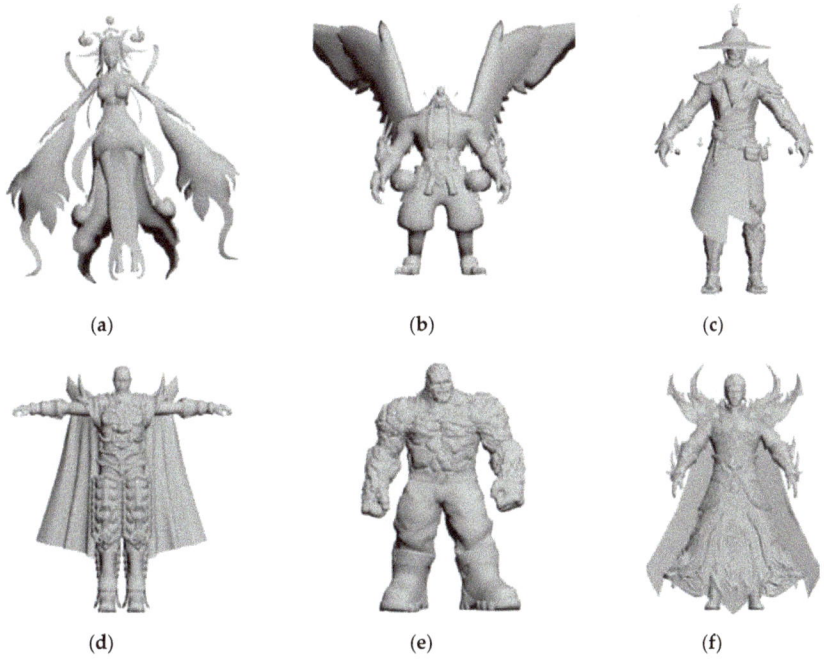

Figure 9. Six tested 3D models. (**a**) Fairy. (**b**) Boss. (**c**) Solider. (**d**) Devil. (**e**) Thing. (**f**) Lord.

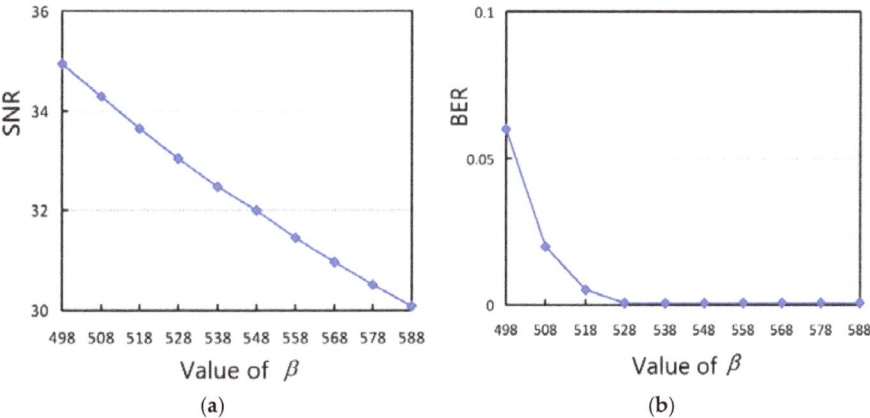

Figure 10. The effect of β on the distortion of decrypted model and the bit error rate of the extracted watermark. (**a**) β is related to signal-to-noise ratio (*SNR*). (**b**) β is related to bit error rate (BER).

4.2. The Value of t

As shown in Figure 7, the 0-bit area and 1-bit area are separated by the robust interval of size $t \cdot (N_l - 1)$. If the robust interval is large, the robustness is high. However, as t increases, the quality of the decrypted model is reduced. Therefore, t needs to be adjusted according to the actual application scenario. If higher robustness is required, a greater value of t can be assigned. If better quality of decrypted model is required, a smaller value of t is set. In order to choose a suitable value, experiments were conducted on 40 models to test the robustness with different values of t.

As illustrated in Figure 11, the BER of watermark extraction is low under Gaussian noise (0.01). By increasing t, the BER could be reduced. When $t = 50$, the watermark could be extracted correctly. Therefore, when higher robustness is required, the value of t can be assigned to be 50.

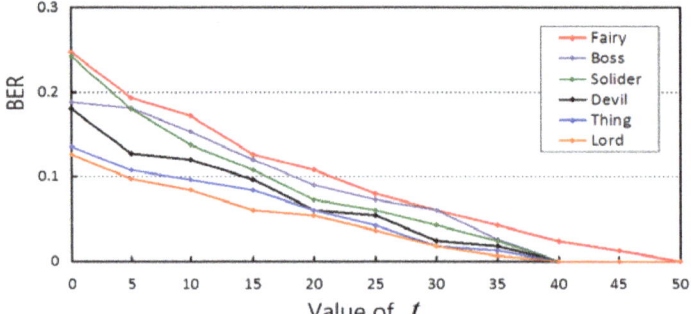

Figure 11. The BER under Gaussian attacks (the strength is 0.01).

4.3. Feasibility of the Watermarking

In order to show the feasibility of the proposed watermarking method, the 3D model devil with 30,000 vertices was tested, and other models had similar results. The watermark was a 1024-bit pseudo-random sequence. Firstly, the original model was divided into patches and the encrypted model was obtained by encrypting the 3D model with the public key as illustrated in Figure 12. Secondly, with the embedding key, the watermark was embedded to obtain the watermarked model as illustrated in Figure 12c. Then, the directly decrypted model (as shown in Figure 12d) is obtained by decrypting the encrypted model; *SNR* of the decrypted model was 30.93. Lastly, the watermark was extracted and the model was restored (as shown in Figure 12e), and the *SNR* of the restored model approaches infinity, which shows that the restored model was exactly the same as the original model. Figure 12f shows that all watermark bits were correctly extracted. The experimental results showed that the proposed method achieved reversibility of embedding and extraction, and the restoration of the original model. Figure 13 shows five decrypted 3D models had less distortion compared to the original 3D model, and Figure 13f shows the *SNR* of the decrypted models were close to 30, which denotes the proposed method can obtain good quality.

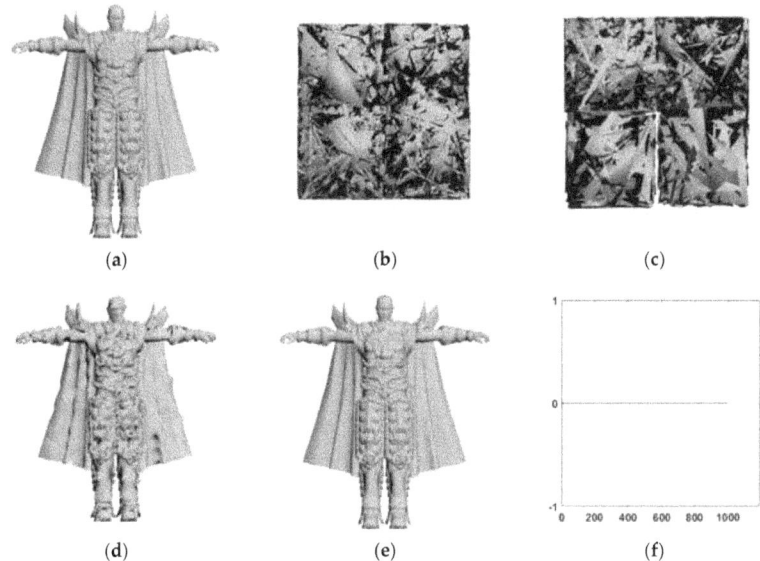

Figure 12. Experiment with 3D model 'devil' (**a**) The original model; (**b**) the encrypted model; (**c**) the watermarked model; (**d**) the decrypted model. After decryption, the *SNR* was 30.93. (**e**) The restored model. After restoration, the *SNR* approached infinity. (**f**) The bit error rate after watermark extraction.

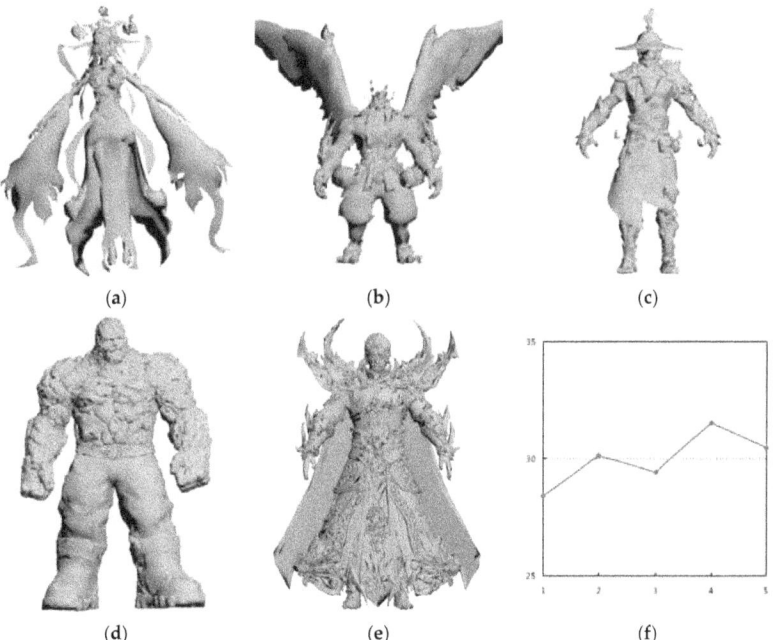

Figure 13. Five watermarked 3D models. (**a**) The watermarked "Fairy"; (**b**) the watermarked "Boss"; (**c**) the watermarked "Solider"; (**d**) the watermarked "Thing"; (**e**) the watermarked "Lord"; (**f**) *SNR* of the five watermarked models.

4.4. Robustness Analysis

In order to compare the robustness under attacks, several attacks were performed on the decrypted 3D model. Table 1 shows the bit error rate of watermark extraction under different attacks.

Table 1. The BER under several common attacks.

Model	t	SNR	Gaussian (0.005)	(0.01)	(0.02)	Translation	Scaling 0.8	1.2	1.5
Fairy	40	30.96	1.75%	2.63%	6.15%	1	0.21	0.073	0.18
Boss	40	30.96	1.24%	1.52%	6.26%	1	0.17	0.064	0.19
Solider	35	31.44	1.98%	2.67%	5.98%	1	0.13	0.053	0.15
Devil	30	31.44	2.06%	2.41%	6.74%	1	0.16	0.069	0.19
Thing	30	31.44	1.68%	2.47%	6.45%	1	0.18	0.057	0.21
Lord	25	32.1	2.28%	3.24%	8.64%	1	0.23	0.071	0.24

4.4.1. Robustness Against Translation Attacks

The robustness against the translation attacks was tested. As shown in Table 1, the method perfectly resisted translation attacks. When the model was subjected to a translation attack, the vertex coordinates of the patch increased by a certain value at the same time. According to Equation (14), when the vertex coordinates in a patch are changed by the same size, it can be known that its direction values will not change. Therefore, the watermark can be extracted correctly.

4.4.2. Robustness Against Scaling Attacks

The robustness against scaling attacks was tested by different levels (0.8, 1.2, 1.5) on the decrypted 3D model. As shown in Table 1, the proposed method was robust to scaling attacks. When the model was attacked, the vertex coordinates of the patch were multiplied by a certain coefficient at the same time. According to Equation (14), its direction values also increased or decreased accordingly. As illustrated in Figure 7, the direction values of most patches were concentrated in the central area. Therefore, when the scaling size was increased, most of the vertices were still in the original area, and only a small number of vertices were offset. On this condition, the robustness was high. When the scaling size was decreased, the 1-bit area was easily shifted to the 0-bit area, which affected the accuracy of extracting the watermark. Therefore, the robustness was much higher when the 3D model was amplified compared with other levels of attacks.

4.4.3. Robustness to Gaussian Noise Attacks

The robustness against Gaussian noise attacks was tested by performing different degrees (0.005, 0.01, 0.02) on the decrypted 3D model. As shown in Table 1, the robustness against Gaussian noise attacks was high. When the model was attacked by Gaussian noise, the vertex coordinates were slightly disturbed. According to Equation (14), its direction values were also slightly modified. As illustrated in Figure 7, the direction values of most patches were concentrated in the central area, and only a few vertices were in the non-central area. Therefore, when the model was slightly disturbed, the direction values of the central area were slightly disturbed, and only a few direction values of non-central area were shifted.

However, the proposed method cannot resist the attacks of cropping and simplification, it is because those attacks will influence the order of the vertices. Moreover, the proposed method cannot resist salt and pepper noise, mainly because the attack obviously changes the relative position between vertices.

4.5. Compared with the Existing Watermark Method in an Encrypted Domain

To our knowledge, few effective robust reversible watermarking methods for 3D model in the encryption domain has been reported in the literature. In order to show the effectiveness of the proposed method, Jiang [1] is extended to the encrypted 3D model. From Table 2, the proposed method has a slightly higher embedding capacity compared with the Jiang [1], and it is mainly because a patch has three coordinate axes and three bits can be embedded. To sum up, the proposed method has good security and robustness, and the decrypted 3D model has low distortion.

Table 2. Compared to the method of Jiang [1].

	Capability	Robustness	Security	SNR of Decrypted Model	SNR of Restored Model	BER
The proposed method	0.396	yes	high	30.08	$+\infty$	0
Method of Jiang [1]	0.365	no	low	5.35	31.97	4.22%

5. Conclusions

In this paper, a robust reversible three-dimensional (3D) model watermarking method based on homomorphic encryption is presented for protecting the copyright of 3D models. The 3D model is divided into non-overlapping patches, and the vertex in each patch is encrypted by using the Paillier cryptosystem. On the cloud side, three direction values of each patch are computed, and the symmetrical direction histogram is constructed for shifting to embed watermark. In order to obtain robustness, the robust interval is designed in the process of histogram shifting. The watermark can be extracted from the direction histogram, and the original encrypted model can be restored by histogram shifting. Experimental results show that the decrypted 3D models have less distortion compared with the existing methods, which denotes the proposed method can embed more secret data without increasing the 3D models distortion. Moreover, the proposed method can resist a series of attacks compared to the existing watermarking methods on encrypted 3D model. Thus, the proposed method is efficient to protect copyright of 3D models in the cloud when the cloud administrator does not know the content of the 3D models, but the existing methods have no ability.

In the future, we will investigate the following two possible research directions. (1) Reduce the distortion of the directly decrypted 3D model. (2) Further improve the robustness against more kinds of attacks, such as cropping and salt and pepper noise.

Author Contributions: Conceptualization and funding acquisition are credited to L.L. Methodology and writing—original draft are due to S.W. Conceptualization and supervision are credited to S.Z. Writing—review & editing is credited to T.L. Formal analysis and investigation are originated by C.-C.C. All authors have read and agreed to the published version of the manuscript.

Funding: This work was partially supported by National Natural Science Foundation of China (No. 61370218, No. 61971247), Public Welfare Technology and Industry Project of Zhejiang Provincial Science Technology Department (No. LGG19F020016), and Ningbo Natural Science Foundation (No. 2019A610100).

Conflicts of Interest: The authors declare no conflict of interest.

References

1. Jiang, R.Q.; Zhou, H.; Zhang, W.M. Reversible Data Hiding in Encrypted Three-Dimensional Mesh Models. *IEEE Trans. Multimed.* **2018**, *20*, 55–67. [CrossRef]
2. Shah, M.; Zhang, W.M.; Hu, H.G. Homomorphic Encryption-Based Reversible Data Hiding for 3D Mesh Models. *Arab. J. Sci. Eng.* **2018**, *43*, 8145–8157. [CrossRef]
3. Wu, H.T.; Cheung, Y.M.; Tang, S.H. A high-capacity reversible data hiding method for homomorphic encrypted images. *J. Vis. Commun. Image Represent.* **2019**, *62*, 87–96. [CrossRef]
4. Xiang, S.J.; Luo, X.R. Reversible Data Hiding in Homomorphic Encrypted Domain by Mirroring Ciphertext Group. *IEEE Trans. Circuits Syst. Video Technol.* **2018**, *28*, 3099–3110. [CrossRef]

5.   Liu, J.; Wang, Y.; Li, Y. A robust and blind 3D watermarking algorithm using multiresolution adaptive parameterization of surface. *Neurocomputing* **2017**, *237*, 304–315. [CrossRef]
6.   Weng, S.; Zhao, Y.; Pan, J.S. Reversible watermarking based on invariability and adjustment on pixel pairs. *IEEE Signal Process. Lett.* **2008**, *15*, 721–724. [CrossRef]
7.   Amini, M.; Ahmad, M.; Swamy, M. A Robust multibit multiplicative watermark decoder using vector-based hidden markov model in wavelet domain. *IEEE Trans. Circuits Syst. Video Technol.* **2016**, *1*, 402–413. [CrossRef]
8.   Benedens, X.; Busch, C. Towards blind detection of robust watermarks in polygonal models. *Comput. Graph. Forum* **2000**, *19*, 199–208. [CrossRef]
9.   Malvar, H.S.; Florencio, D.A. Improved spread spectrum: A new modulation technique for robust watermarking. *IEEE Trans. Signal Process* **2003**, *51*, 898–905. [CrossRef]
10. Coltuc, D. Low distortion transform for reversible watermarking. *IEEE Trans. Image Process.* **2012**, *21*, 412–417. [CrossRef]
11. Wu, Q.L.; Wu, M. Adaptive and blind audio watermarking algorithm based on chaotic encryption in hybrid domain. *Symmetry* **2018**, *10*, 284. [CrossRef]
12. Feng, J.B.; Lin, I.C.; Tasi, C.S. Reversible watermarking: Current status and key issues. *Int. J. Netw. Secur.* **2006**, *2*, 161–171.
13. Ma, K.W.; Zhang, X.; Zhao, N. Reversible data hiding in encrypted images by reserving room before encryption. *IEEE Trans. Inf. Forensics Secur.* **2013**, *8*, 553–562. [CrossRef]
14. Zhang, W.; Ma, K.; Yu, N. Reversibility improved data hiding in encrypted images. *Signal Process.* **2014**, *94*, 118–127. [CrossRef]
15. Shiu, C.-W.; Chen, Y.C.; Hong, W. Encrypted image-based reversible data hiding with public key cryptography from difference expansion. *Signal Process. Image Commun.* **2015**, *39*, 226–233. [CrossRef]
16. Ni, Z.; Shi, Y.Q.; Ansari, N. Reversible data hiding. *IEEE Trans. Circuits Syst. Video Technol.* **2006**, *16*, 354–362.
17. Wu, H.T.; Cheung, Y.-M. Reversible watermarking by modulation and security enhancement. *IEEE Trans. Instrum. Meas.* **2010**, *59*, 221–228.
18. Zhang, X. Separable reversible data hiding in encrypted image. *IEEE Trans. Inf. Forensics Secur.* **2012**, *16*, 826–832. [CrossRef]
19. Zhang, X.; Qian, Z.; Feng, G. Efficient reversible data hiding in encrypted images. *J. Vis. Commun. Image Represent.* **2015**, *25*, 322–328. [CrossRef]
20. Xiang, S.J.; Yang, L. Robust and reversible image watermarking algorithm in homomorphic encrypted domain. *Ruan Jian Xue Bao/J. Softw.* **2018**, *29*, 957–972. (In Chinese)
21. Ke, Q.; Xie, D.Q. A self-similarity based robust watermarking scheme for 3D point cloud models. *Inf. Jpn.* **2010**, *16*, 287–291.
22. Feng, X. A new watermarking algorithm for point model using angle quantization index modulation. In Proceeding of the National Conference on Electrical Electronics and Computer Engineering, Xi'an, China, 12–13 December 2015; pp. 962–968.
23. Paillier, P. Public-key cryptosystems based on composite degree residuosity classes. In *International Conference on the Theory and Applications of Cryptographic Techniques*; Springer: Berlin/Heidelberg, Germany, 1999; pp. 223–238.
24. Zheng, P.J.; Huang, J.W. Discrete wavelet transform and data expansion reduction in homomorphic encrypted domain. *IEEE Trans. Image Process.* **2013**, *22*, 2455–2468. [CrossRef] [PubMed]
25. Donald, K. *The Art of Computer Programming*, 3rd ed.; Addison-Wesley: Boston, MA, USA, 1997; pp. 325–515.

Article

A Matching Pursuit Algorithm for Backtracking Regularization Based on Energy Sorting

Hanfei Zhang [1,*], Shungen Xiao [2,*] and Ping Zhou [1]

[1] Information service and Information Research Center, Huaiyin Normal University, Huai'an 223700, China; zhouping_hytc@163.com
[2] School of Information Mechanical and Electrical Engineering, Ningde Normal University, Ningde 352100, China
* Correspondence: zhanghanfei2016@163.com (H.Z.); xiaoshungen022@163.com (S.X.)

Received: 9 December 2019; Accepted: 29 January 2020; Published: 3 February 2020

Abstract: The signal reconstruction quality has become a critical factor in compressed sensing at present. This paper proposes a matching pursuit algorithm for backtracking regularization based on energy sorting. This algorithm uses energy sorting for secondary atom screening to delete individual wrong atoms through the regularized orthogonal matching pursuit (ROMP) algorithm backtracking. The support set is continuously updated and expanded during each iteration. While the signal energy distribution is not uniform, or the energy distribution is in an extreme state, the reconstructive performance of the ROMP algorithm becomes unstable if the maximum energy is still taken as the selection criterion. The proposed method for the regularized orthogonal matching pursuit algorithm can be adopted to improve those drawbacks in signal reconstruction due to its high reconstruction efficiency. The experimental results show that the algorithm has a proper reconstruction.

Keywords: backtracking; energy sorting; atom screening

1. Introduction

Magnetic resonance (MR) image reconstruction technology has been long-established in clinical medical detection with the rapid development of medical image processing technology. It has become an essential means of medical diagnosis [1–3]. In practical medical applications, the traditional approach is to sample data according to the Shannon–Nyquist sampling technique. The data collected in this way can adequately represent the original signal, but they have massive amounts of redundancy. Therefore, these methods often lead to the overflow of acquisition data and the waste of sensors. It is of considerable significance to reduce the amount of data. The method of extracting a sinusoidal signal from the noise has attracted many scientists and using the compressibility of the signal to sample data is a new subject. It originates from the study of the acquisition of a finite-rate-of-innovation signal. Fixed deterministic sampling kernels are used to double the innovation rate instead of acquiring continuous signals at twice the Nyquist sampling frequency.

The compressed sensing (CS) [4–7] based on sparse representation has attracted significant attention as a new sampling theory in recent years. It breaks the limitation of Nyquist's sampling theorem, compresses signal sampling simultaneously, saves a lot of time and storage space, and has become a new research direction in the field of signal processing [8–10]. CS theory has been widely used in many biomedical imaging systems and physical imaging systems, such as computed tomography, ultrasound medical imaging, and single-pixel camera imaging. Compressed sensing magnetic resonance imaging (CS-MRI) based on CS can reconstruct high-quality MR images through a small amount of sample data, which significantly shortens the scanning time, speeds up the processing of MR images, and improves work efficiency. The compressed sensing mainly includes two aspects: the first is the sampling and compression of the signal, and the second is the reconstruction of the original signal. The former is for sparse or compressible high-dimensional signals to acquire low dimensional measurement values through a measurement matrix. At the same time, the latter uses these low dimensional measurement data to restore the original signal as much as possible. However, how to design a recovery algorithm with fewer observation times, excellent reconstruction performance, and low complexity are essential challenges in the study of CS.

The basic pursuit manner [11–14] has been put forward by some scholars for this problem. The convex optimization process has a good reconstruction effect, but it is often disadvantageous to practical applications because it takes an excessively long time to run. For this reason, the greedy iterative algorithm [15–18] has been favored by the vast majority of researchers because of its low complexity and simple geometric principle. Among all the kinds of reconstruction algorithms studied at present, the greedy algorithm is the most widely used. However, in greedy algorithms, more attention is paid to a sparse unknown reconstruction algorithm, which does not need the precondition of known signal sparseness. The representative algorithms are the sparsity adaptive matching pursuit and the regularized adaptive matching pursuit algorithms. They approximate sparsity by setting an initial step and expanding the support set step by step, while the backtracking adaptive orthogonal matching pursuit uses backtracking detection to reconstruct the unknown sparseness signal. In recent years, a forward–backward pursuit (forward–backward pursuit) algorithm was proposed to estimate sparsity by iteratively accumulating the difference between the front and back steps.

An energy-based adaptive matching pursuit algorithm increases the sparsity level gradually according to the increase of the iteration residual energy. Furthermore, the adaptive matching-pursuit-based difference reconstruction algorithm uses the rate of change between the measurement matrix and the residual inner product elements to approximate the sparsity adaptively. The proposed BRAMP algorithm is also an adaptive algorithm for compressed sensing reconstruction.

The orthogonal matching pursuit algorithm (OMP) [19–21], the regularized orthogonal matching pursuit algorithm (ROMP) [22,23], uses each atom and the residual value of the measurement matrix for the inner product. Then, the atom that is most matched with the residual is placed in the support set using some principles. Once the atom is selected, it will not be deleted until the end of the iteration. The other is a class of compressive sampling matching pursuit algorithm (CoSaMP) [24,25], the subspace tracking algorithm (SP) [26,27]. After selecting the matched atoms, they added a backtracking function to delete unstable atoms to better guarantee the quality of the reconstructed signal. The OMP algorithm continues the principle of atom selection in a matching pursuit algorithm. Although the signal can be accurately reconstructed with only one atom being selected in each iteration, the efficiency of the algorithm is low. The ROMP algorithm, stagewise orthogonal matching pursuit algorithm, and generalized orthogonal matching pursuit algorithm can select more than one atom in each iteration, which speeds up the convergence of the algorithm. However, they cannot guarantee that the selected atoms in each iteration are correct. If the wrong atoms are selected in the previous iteration, the choice of atoms in the next iteration will be affected. The CoSaMP algorithm and the SP algorithm can select more than one atom at each iteration.

Meanwhile, the backtracking procedure is introduced to improve the reconstruction accuracy. The above algorithms increase the number of atoms to candidate sets to improve the performance of the algorithm. Due to the influence of noise observation, the performance of reconstructing signals by the above algorithms is not ideal. A regularized orthogonal matching pursuit algorithm uses regularization criteria as atomic screening rules. It can ensure that the energy of selected atoms is much larger than that of non-selected atoms, and its reconstruction performance is better than other greedy algorithms.

In this research, the regularization method was adopted to select the atomic advantage effectively, and the ROMP and SP algorithms were used to screen the atomic backtracking strategy. Further, a matching pursuit algorithm for regular backtracking based on the energy ranking (ESBRMP) was proposed. The experimental results show that this algorithm had a better reconstruction effect.

2. Compressed Sensing Theory

Let x be the N length of the original signal, y is the M length of the observed signal, $\Phi_{M \times N} (M < N)$ is the measurement matrix, and they meet with $y = \Phi x$. If x includes K sparse signals and $M \geq K \times \lg(N)$ between K, M, and N, x could achieve the accurate reconstruction. The problem to be solved in this paper is how to reconstruct the signal x from the observed signal y, which is usually solved using the following optimization problem:

$$\min \|x\|_0, \ s.t. \ y = \Phi x. \tag{1}$$

In practice, a certain degree of error is allowed. Therefore, the original optimization problem can be transformed into a simpler approximate solution. δ is a minimal constant:

$$\min \|x\|_0, \ s.t. \ \|y - \Phi x\|_2^2 \leq \delta. \tag{2}$$

The minimum norm problem is an NP difficult problem, and it is challenging to solve the problem directly. The matching pursuit algorithm provides a powerful tool for the approximate solution, and Tropp and Gilbert [18] pointed out that the methods for sparse signal reconstruction have a specific stability. Furthermore, the OMP algorithm continues the selection rule of atoms in the matching pursuit algorithm and realizes the orthogonalization of the selected atom set recursively to ensure the optimization of the iteration, thus reducing the number of iterations. Needell and Vershynin [22], based on the OMP algorithm, proposed the ROMP algorithm, where the regularization process is used in the OMP algorithm for a known sparsity. The difference between the ROMP algorithm and the OMP algorithm is that, first, the algorithm selects multiple atoms as a candidate set based on the relevant principles, and second, some atoms are selected by the regularization principle from the candidate set, and then incorporated into the final support set to realize the rapid and effective selection of the atom. The SP and CoSaMP algorithms use the idea of back-stepping filtering. The reconstruction quality and the reconstruction complexity of these algorithms are similar to that of linear programming (LP).

3. Reconstruction Processes

The ROMP algorithm can accurately reconstruct all the matrices and all sparse signals that satisfy the restricted isometry property (RIP) [28], and the reconstruction speed is fast. The ROMP algorithm first selects the atoms according to the correlation principle and calculates the correlation coefficient by calculating the absolute value of the inner product between the residual and each atom in the measurement matrix Φ:

$$u = \left\{ u_j \mid u_j = \left| < r, \varphi_j > \right|, j = 1, 2, 3 \cdots, N \right\}. \tag{3}$$

The ROMP algorithm uses the regularization process to carry out the two filters of the atom. Through Equation (4), the correlation coefficients of the atoms corresponding to the index value set are divided into several groups. That is, the correlation coefficient of the atom corresponding to the medium index is divided into several groups according to Equation (4):

$$\left| u(i) \right| \leq 2 \left| u(j) \right|, \ i, j \in J_0; J_0 \in J \tag{4}$$

The key to the regularization process is to select a set of atomic index values corresponding to the most significant energy correlation coefficients from the perception matrix, store them in the updated support set, and complete the secondary selection. Then, the atomic index value corresponding to a group of correlation coefficients with the maximum energy is deposited into J_0. The regularization process allows the ROMP algorithm to obtain the support set $|\Lambda|$ with a lower atomic number than Φ_Λ to reconstruct the signal accurately for most iterations. For the atoms that have not been selected into the support set, the regularization process can ensure that their energy is much smaller than the energy of the selected atoms, which is a simple and effective way of undertaking atomic screening. It could improve the stability of the signal reconstruction. After a particular iteration to get the support set for the reconstruction of the signal, the least squares method is used for the signal approximation and the remainder update. The flow chart of the ROMP algorithm is shown in Figure 1. It can be expressed as:

$$\hat{x} = \arg\min\|y - \Phi_\Lambda x\|_2, \tag{5}$$

$$r_{new} = y - \Phi_\Lambda \hat{x}. \tag{6}$$

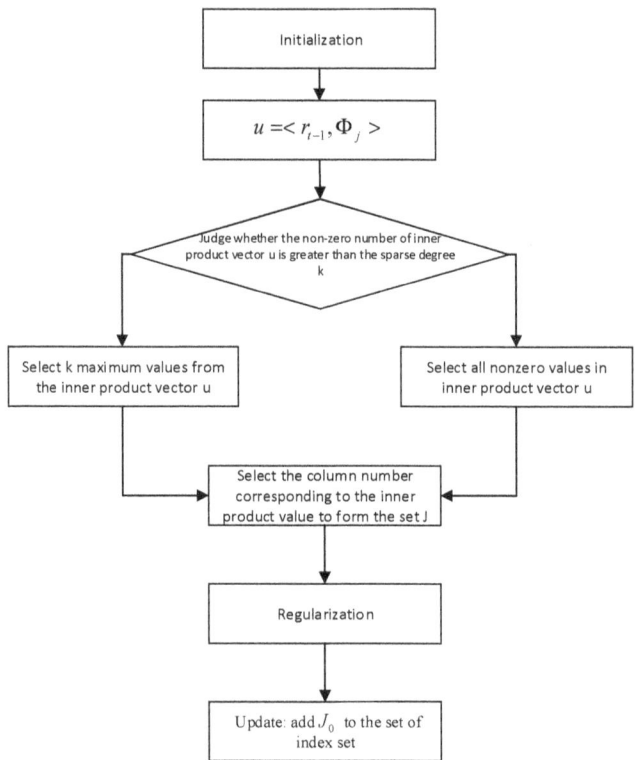

Figure 1. The flow chart of the regularized orthogonal matching pursuit (ROMP) algorithm.

The ROMP algorithm is represented as follows:

(1) Initialization: $r_0 = y$, $\Lambda = \phi$, iterating $t = 1$, repeating the following steps K times or until $|\Lambda| \geq 2K$.

(2) Calculation: $u = < r_{t-1}, \Phi_j >$.

(3) The set of the largest non-zero coordinates of K or all its non-zero coordinates, and the small one is set to J.

(4) Regularization: In all subsets with comparable coordinates $J_0 \in J$, where $|u(i)| \leq 2|u(j)|$, $i, j \in J_0 J_0 \in J$, select the maximum energy for reconstructing the original signal.

(5) Update: Add J_0 to the index set $\Lambda = \Lambda \cup J_0$, $x = \arg\min\|y = \Phi_\Lambda x\|_2$, $r = y - \Phi x$.

The ROMP algorithm selects the atom through a regularization criterion in a reasonable condition. When the signal energy distribution is uniform or showing the distribution of an extreme energy state, i.e., the maximum total energy as the selection criteria, the algorithm may not accurately choose the required columns, and therefore the ROMP algorithm performance becomes unstable. The ROMP algorithm for energy sorting is proposed to solve the unstable problem, which is combined with the advantages of the ROMP algorithm and the SP algorithm. For the selection principle, first, a screening is carried out using the correlation criterion to select the column vector with the maximum inner product of the column and the iterative error vector. Then, the set of column vectors with the energy ratio less than two is selected in the selected column vector based on the regularization criterion. Lastly, the algorithm selects the set of columns that meet the requirements in all the column sets through the energy screening criteria.

The steps of the energy sorting are as follows:

(1) The correlation criterion and regularization standard selects a set of all columns: $\Sigma_i, i = 1, 2, 3, \cdots L$.

(2) For the Σ_i set of all columns, the energy E_i, the number of column vectors Num_i, and the energy average Eve_i are counted, where $i = 1, 2, 3, \cdots L$.

(3) Select the maximum n energy set E_j by setting the energy threshold, $j = 1, 2, \cdots n$.

(4) Select a column E_l from E_j that is lower than the threshold.

(5) Find the descending order of E_l through energy values, and select the set of energy averages not less than k ($0 < k < 1$) times of later from the maximum energy value. This is the set of columns that are screened.

In the above steps, the purpose of step 3 is to ensure that the selected set energy is more significant than most of the sets. The purpose of steps 4 and 5 is to ensure that the selected set energy distribution is more reasonable than others. The set of columns that are filtered can contain more useful signal information. The flow chart of the ESBRMP algorithm is shown in Figure 2.

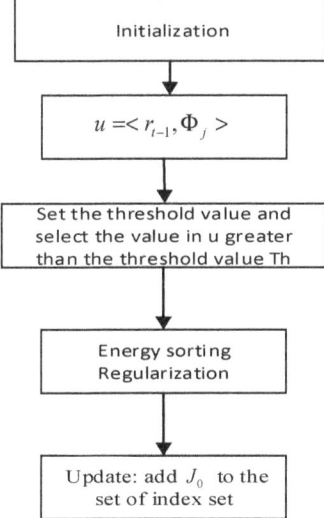

Figure 2. The flow chart of the matching pursuit algorithm for regular backtracking based on the energy ranking (ESBRMP).

The steps of the ESBRMP algorithm are as follows:

(1) Initialization: Set the residual $r_0 = y$, $\Lambda = \phi$.
(2) Calculate the inner product between the residuals r_{i-1} and the atoms of the observation matrix.
(3) Set the threshold value, select the value larger than the threshold value Th from u, and make up the set J of the sequence number j corresponding to these values.
(4) Energy sorting and finding subsets $J_0 \in J$.
(5) Update the index set $\Lambda_i = \Lambda_{i-1} \cup J_0$ and update the support set $\Gamma_i = \Gamma_{i-1} \cup J_0$.
(6) Solve the least squares problem $\hat{\theta} = \text{argmin} \|y - A_t \theta_t\|$.

(7)   Backtracking update support set: Based on the backtracking idea, a new support set is made up of the larger aL elements ($0 < a < 1$, A is the number of B)

(8)   Update the residual $\hat{r}_t = y - A_t\hat{\theta}$.

(9)   Judge whether $\|\hat{r}_t\|_2 \leq \|\hat{r}_{t-1}\|_2$ is established. If it is established, stop iterating; if it is not established, determine whether the number of initial stages s can be reached. If it is reached, the iteration is stopped; if it is not reached, return to the second step and continue to iterate.

4. Experimental Results and Discussion

The one-dimensional Gaussian random signal with an original signal length was reconstructed under different numbers of sparsity and measurement. The measurement matrix was a Gaussian random matrix. The length, sparsity, compression ratio, and the reconstruction performance of the observed signal are shown in Figures 3 and 4. Figure 3 shows the ESBRMP algorithm's reconstruction signal and residual. Figure 4 is the traditional OMP algorithm's reconstruction signal and residual.

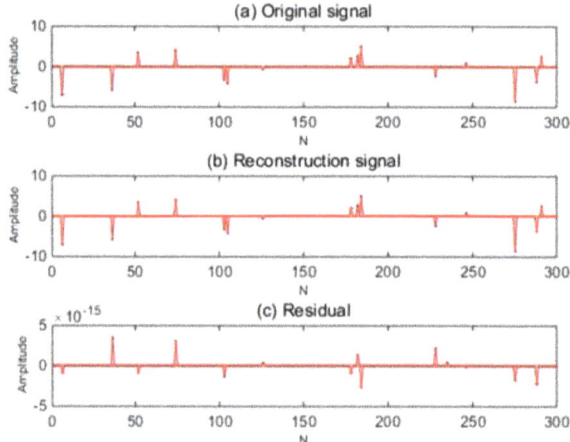

Figure 3. ESBRMP algorithm's reconstruction result.

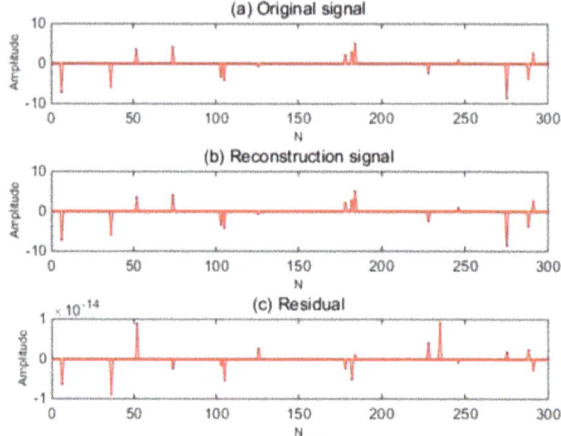

Figure 4. OMP algorithm's reconstruction result.

From the above experiments, it can be seen that the ESBRMP algorithm had a better effect on the reconstruction of the one-dimensional signal, and the residual of reconstruction was small. The related experiments were carried out on the reconfiguration rate, sparsity, and measurement of the signal, as shown in Figures 5 and 6. Under different sparsities, the relationship between the measurement and the signal reconstruction rate is shown in Figure 5. When the sparsity was low, the original signal could be restored with a lower number of measurements, and the lower number of measurements produced a lower signal reconstruction rate when the sparsity was high.

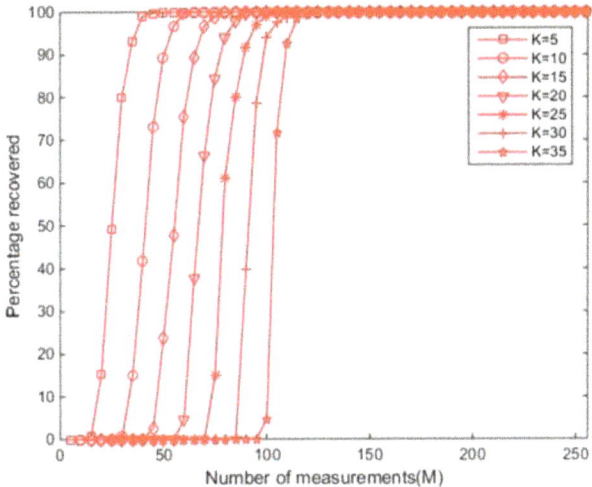

Figure 5. The ESBRMP algorithm's relationship between the signal reconfiguration rate and the number of measurements.

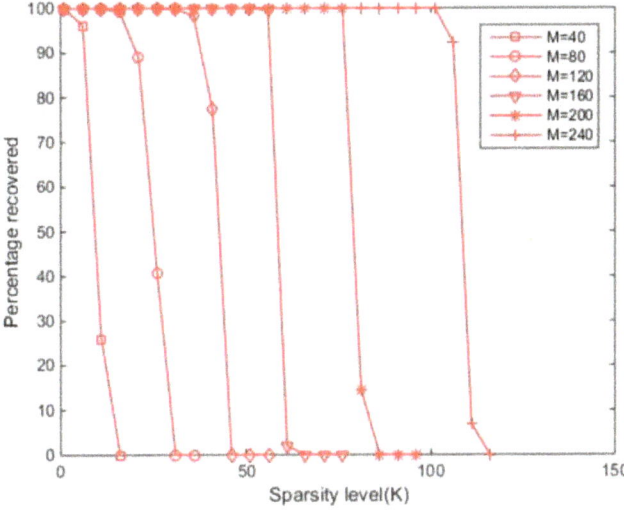

Figure 6. The ESBRMP algorithm's relationship between the signal reconfiguration rate and the sparsity.

Under different numbers of measurements, the relation between the sparsity and the signal reconstruction rate is shown in Figure 6. When the number of measurements was low, the original signal could be restored with a lower sparsity, and the lower sparsity produced a lower signal reconstruction rate when the number of measurements was higher. Overall, regarding the signal reconstruction rate, the size of the sparsity was directly proportional to the number of measurements. The sparsity was more significant than usual, and the more measurements we needed to ensure that the signal had a high reconstruction rate.

The performance of the ESBRMP algorithm was compared to other typical greedy pursuit algorithms, such as the OMP, ROMP, SP, and CoSaMP algorithms. Moreover, the comparison between the exact reconstruction probability and reconstruction accuracy was verified.

The accurate reconstruction probability of the signal was compared with other algorithms. The accurate reconstruction of the signal was defined as the actual signal, which gives the same position of the non-zero elements in the recovery signal in the ideal condition without noise. The accurate reconstruction rate of the signal for different measurements M is given in Figures 7 and 8. From Figure 7, for all reconstruction algorithms, the exact reconstruction probability of the first signal increased with the increase of the number of measurements M. For this algorithm, when the number of measurements was more significant than 35, the reconstruction probability of the ESBRMP algorithm was close to 1. When the number of measurements was greater than 25, the reconstruction probability of the ESBRMP algorithm was more than the OMP, ROMP, and SP algorithms. Overall, for the same signal, the number of measurements required to stabilize the reconstructed signal using the ESBRMP algorithm was less than the OMP, ROMP, and SP algorithms.

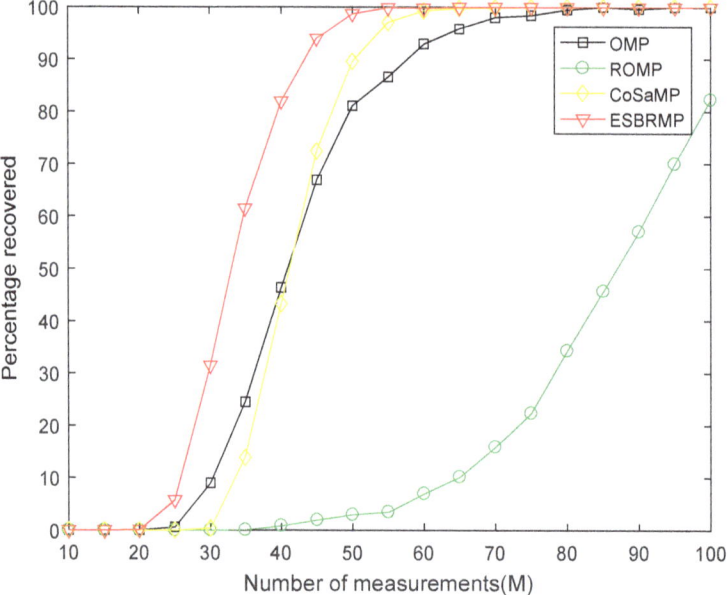

Figure 7. The relationship between the signal recovery rate and measurement.

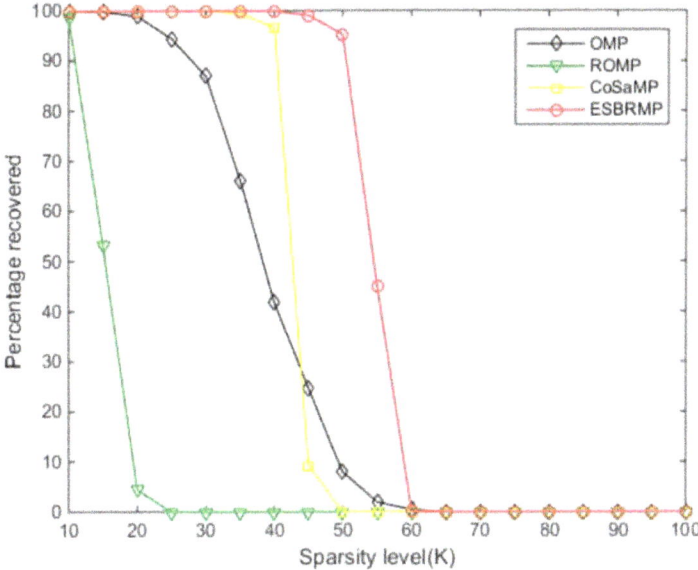

Figure 8. The relationship between the signal recovery rate and sparsity.

When the sparsity is greater than 60 in Figure 8, the reconstruction probability was close to 0. When the number of measurements was between 25 and 60, the reconstruction probability of the ESBRMP algorithm was higher than the other algorithms. Overall, for the same signal, the sparsity required to stabilize the reconstructed signal using the ESBRMP algorithm was higher than the other algorithms. The accurate reconstruction rate of all kinds of algorithms decreased gradually with the increase of sparsity, which was because the amount of information contained in the signal was related to the sparsity K of the signal. The sparsity K was more extensive than others, which meant there was more meaningful information. In the signal reconstruction, the atoms contained in the observation matrix were determined. More atoms were needed for the reconstruction of the signal with a larger sparsity K, while the number of atoms needed to satisfy the dictionary, the possibility of representing the signal, and the precision reconstruction rate was lower than others. On the contrary, for signals with a smaller sparsity K, the number of atoms used to represent the reconstruction was smaller. Moreover, there were many kinds of atom combinations satisfied in the dictionary, which made it possible to represent the signal and it had a higher precision reconstruction rate.

In order to further illustrate the performance of the ESBRMP algorithm, Lena images with the size of 256×256 were selected to compare the peak signal to noise ratio (PSNR) and the reconstruction time of the reconstructed images. First, an orthogonal wavelet transform (coif3) was used for the transform, then each column of the transformed matrix was reconstructed, and finally, the reconstructed image was obtained using the inverse wavelet transform. The measurement matrix was an orthogonal observation matrix. Table 1 compares the average PSNR and reconstruction time of the reconstructed images with different compression ratios. Under the same conditions, the larger the PSNR, the higher the quality of the reconstructed images.

Table 1. Qualities of images reconstructed and running time by different algorithms. PSNR: Peak signal to noise ratio.

Algorithms	M/N = 0.3		M/N = 0.4		M/N = 0.5	
	PSNR (dB)	T (s)	PSNR (dB)	T (s)	PSNR (dB)	T (s)
OMP	23.7848	16.3249	26.3581	35.4372	29.6349	62.9146
ROMP	19.3325	2.8873	22.6383	3.0273	26.8823	3.5338
CoSaMP	22.1336	4.1558	24.0518	6.3511	25.8473	9.2915
ESBRMP	26.2538	3.2903	28.6912	5.1083	31.2703	8.9474

From Table 1, it can be seen that the PSNR value of the reconstructed image of the ESBRMP algorithm was higher than that of other algorithms, and even in the case of a low sampling rate, it still had a better reconstruction effect. The reconstruction time of the ESBRMP algorithm was higher than the ROMP algorithm and less than the other algorithms.

Table 2 shows the reconstruction effects of different images at the same sampling rate. It can still be seen that the ESBRMP algorithm also had a strong reconstruction ability and reasonable reconstruction time for other images, which shows that the ESBRMP algorithm had better applicability than others.

Table 2. PSNR of the different images using different algorithms.

Algorithms	Lena	Fruits	Cameraman	Pepers
OMP	29.6349	30.9803	28.0214	29.1471
ROMP	26.8921	28.8023	24.0257	25.7125
CoSaMP	25.8473	27.2755	24.1903	25.0361
ESBRMP	31.2703	33.4108	29.1827	30.5297

The reconstruction time was related to the number of atoms needed for the signal reconstruction; the more atoms used for reconstruction, the longer the reconstruction time. Through the analysis of the accurate reconstruction rate of signal reconstruction, the results show that the larger the signal sparsity, the more atoms that were needed, and the longer the reconstruction time. On the contrary, the smaller the signal sparsity, the fewer atoms that were needed, and the shorter the reconstruction time. The reconstruction probability of the ESBRMP algorithm in the environment without noise was more than for the OMP, ROMP, and SP algorithms, and had a high probability of signal reconstruction.

5. Conclusions

In this paper, a matching pursuit algorithm for backtracking regularization based on energy sorting (ESBRMP) was proposed. The algorithm uses energy sorting to carry out two atomic screening and uses backtracking to delete individual unreliable atoms. Experimental results showed that the ESBRMP algorithm could reconstruct sparse signals with a high probability and had a high reconstruction accuracy without a noisy environment.

Author Contributions: H.Z. proposed the framework of this work and carried out all of the experiments, and S.X. drafted the manuscript. P.Z. offered useful suggestions and helped to modify the manuscript. All authors have read and agreed to the published version of the manuscript.

Funding: This study was funded by the Huaian Natural Science Research Project (Grant No. HABZ201919), the project of application research and science and technology of Huaian (industrial and agricultural) (Grant Nos. HAGZ2014009), Young excellent talent support program of Huaiyin Normal University (Grant No. 13HSQNZ01), Science and Technology Guiding Project of Fujian Province, China (2019Y0046), Natural Science Foundation of Fujian Province of China (No. 2019J01846, No. 2018J01555, No. 2017J01773), Special subject of Ningde normal university serving local enterprises (Grant Nos. 2019ZX403 and 2018ZX409).

Conflicts of Interest: This study is for academic research and submission purposes only. The authors in this study declare that they have no competing interests.

Abbreviations

ESBRMP	backtracking regularization matching pursuit algorithm based on energy sorting
MR	magnetic resonance
CS	compressed sensing
CS-MRI	compressed sensing magnetic resonance imaging
OMP	orthogonal matching pursuit algorithm
ROMP	regularized orthogonal matching pursuit algorithm
CoSaMP	compressive sampling matching pursuit algorithm
SP	subspace tracking
LP	linear programming
RIP	restricted isometry property
BRAMP	Backtracking Regularized Adaptive Matching Pursuit

References

1. Yahyazadeh, S.; Mehraeen, R. A Comparison of the Diagnostic Value of Magetic Resonance Mammography Versus Ultrasound Mammography in Moderate-and High-risk Breast Cancer Patients. *J. Evolut. Med. Dent. Sci.* **2018**, *7*, 5629–5633. [CrossRef]
2. Afsar, J.; Khazaei, A.; Zolfigol, M.A. A novel nano perfluoro ionic liquid as an efficient catalyst in the synthesis of chromenes under mild and solvent-free conditions. *Iran. J. Catal.* **2018**, *9*, 37–49.
3. Nasiriavanaki, M.; Xia, J.; Wan, H.; Bauer, A.Q.; Culver, J.P.; Wang, L.V. High-resolution photoacoustic tomography of resting-state functional connectivity in the mouse brain. *Proc. Natl. Acad. Sci. USA* **2014**, *111*, 21–26. [CrossRef] [PubMed]
4. Chkifa, A.; Dexter, N.; Hoang, T.; Webster, C.G. Polynmial approximation via compressed sensing of high-dimensional functions on lower sets. *Math. Comput.* **2018**, *87*, 1415–1450. [CrossRef]
5. Donoho, D.L. Compressed Sensing. *IEEE Trans. Inf. Theory* **2006**, *52*, 1289–1306. [CrossRef]
6. Baraniuk, R. Compressive sensing. *IEEE Signal Process. Mag.* **2007**, *24*, 118–121. [CrossRef]
7. Saha, T.; Srivastava, S.; Khare, S.; Stanimirović, P.S.; Petković, M.D. An improved algorithm for basis pursuit problem and its applications. *Appl. Math. Comput.* **2019**, *355*, 385–398. [CrossRef]
8. Wei, Y.; Liu, S. Numerical analysis of the dynamic behavior of a rotor-bearing-brush seal system with bristle interference. *J. Mech. Sci. Tech.* **2019**, *33*, 3895–3903. [CrossRef]
9. Xiao, S.; Liu, S.; Song, M.; Ang, N.; Zhang, H. Coupling rub-impact dynamics of double translational joints with subsidence for time-varying load in a planar mechanical system. *Multibody Syst. Dyn.* **2019**. [CrossRef]
10. Xiao, S.; Liu, S.; Jiang, F.; Song, M.; Cheng, S. Nonlinear dynamic response of reciprocating compressor system with rub-impact fault caused by subsidence. *J. Vib. Control* **2019**, *25*, 1737–1751. [CrossRef]
11. Chen, S.S.; Donoho, D.L.; Saunders, M.A. Atomic decomposition by basis pursuit. *SIAM Rev.* **2001**, *43*, 129–159. [CrossRef]
12. Chi-Duran, R.; Comte, D.; Diaz, M.; Silva, J.F. Automatic detection of P- and S-wave arrival times: new strategies based on the modified fractal method and basic matching pursuit. *J. Seismol.* **2017**, *21*, 1171–1184. [CrossRef]
13. Zhang, F.; Jin, Z.; Sheng, X.; Li, X.; Shi, J.; Liu, X. A direct inversion for brittleness index based on GLI with basic-pursuit decomposition. *Chin. J. Geophys.* **2017**, *60*, 3954–3968.
14. Cui, L.; Yao, T.; Zhang, Y.; Gong, X.; Kang, C. Application of pattern recognition in gear faults based on the matching pursuit of characteristic waveform. *Measurement* **2017**, *104*, 212–222. [CrossRef]
15. Nam, N.; Needell, D.; Woolf, T. Linear Convergence of Stochastic Iterative Greedy Algorithms with Sparse Constraints. *IEEE Trans. Inf. Theory* **2017**, *63*, 6869–6895.
16. Tkacenko, A.; Vaidyanathan, P.P. Iterative greedy algorithm for solving the FIR paraunitary approximation problem. *IEEE Trans. Signal Process.* **2006**, *54*, 146–160. [CrossRef]
17. Mallat, S.G.; Zhang, Z. Matching pursuits with time-frequency dictionaries. *IEEE Trans. Signal Process.* **1993**, *41*, 3397–3415. [CrossRef]
18. Tropp, J.A.; Gilbert, A.C. Signal Recovery from Random Measurements Via Orthogonal Matching Pursuit. *IEEE Trans. Inf. Theory* **2007**, *53*, 4655–4666. [CrossRef]

19. Li, H.; Ma, Y.; Fu, Y. An Improved RIP-Based Performance Guarantee for Sparse Signal Recovery via Simultaneous Orthogonal Matching Pursuit. *Signal Process.* **2017**, *144*, 29–35. [CrossRef]
20. Michel, V.; Telschow, R. The Regularized Orthogonal Functional Matching Pursuit for Ill-Posed Inverse Problems. *Siam J. Numer. Anal.* **2016**, *54*, 262–287. [CrossRef]
21. Needell, D.; Vershynin, R. Greedy signal recovery and uncertainty principles. *Proc. SPIE Int. Soc. Opt. Eng.* **2008**, *6806*, 1–12.
22. Needell, D.; Vershynin, R. Uniform Uncertainty Principle and Signal Recovery via Regularized Orthogonal Matching Pursuit. *Found. Comput. Math.* **2007**, *9*, 317–334. [CrossRef]
23. Needell, D.; Tropp, J.A. CoSaMP: Iterative signal recovery from incomplete and inaccurate samples. *Appl. Comput. Harmon. Anal.* **2008**, *26*, 301–321. [CrossRef]
24. Rao, M.S.; Naik, K.K.; Reddy, K.M. Radar Signal Recovery using Compressive Sampling Matching Pursuit Algorithm. *Def. Sci. J.* **2016**, *67*, 94–99. [CrossRef]
25. Dai, W.; Milenkovic, O. Subspace pursuit for compressive sensing signal reconstruction. *IEEE Trans. Inf. Theory* **2009**, *55*, 2230–2249. [CrossRef]
26. Satpathi, S.; Chakraborty, M. On the Number of Iterations for Convergence of CoSaMP and Subspace Pursuit Algorithms. *Appl. Comput. Harmon. Anal.* **2017**, *43*, 568–576. [CrossRef]
27. Candes, E.; Tao, T. Decoding by linear programming. *IEEE Trans. Inf. Theory* **2005**, *51*, 4203–4215. [CrossRef]
28. Candes, E.; Romberg, J.; Tao, T. Robust uncertainty principles: Exact signal reconstruction from highly incomplete frequency information. *IEEE Trans. Inf. Theory.* **2006**, *52*, 489–509. [CrossRef]

Article

Incorporating Particle Swarm Optimization into Improved Bacterial Foraging Optimization Algorithm Applied to Classify Imbalanced Data

Fu-Lan Ye [1], Chou-Yuan Lee [1,*], Zne-Jung Lee [1], Jian-Qiong Huang [1] and Jih-Fu Tu [2,*]

[1] School of Technology, Fuzhou University of International Studies and Trade, Fuzhou 350202, China;
     yfl@fzfu.edu.cn (F.-L.Y.); lrz@fzfu.edu.cn (Z.-J.L.); hjq@fzfu.edu.cn (J.-Q.H.)
[2] Department of Industrial Engineering and Management, St. John's University,
     New Taipei City 25135, Taiwan
*   Correspondence: lqy@fzfu.edu.cn (C.-Y.L.); tu@mail.sju.edu.tw (J.-F.T.)

Received: 24 December 2019; Accepted: 29 January 2020; Published: 3 February 2020

Abstract: In this paper, particle swarm optimization is incorporated into an improved bacterial foraging optimization algorithm, which is applied to classifying imbalanced data to solve the problem of how original bacterial foraging optimization easily falls into local optimization. In this study, the borderline synthetic minority oversampling technique (Borderline-SMOTE) and Tomek link are used to pre-process imbalanced data. Then, the proposed algorithm is used to classify the imbalanced data. In the proposed algorithm, firstly, the chemotaxis process is improved. The particle swarm optimization (PSO) algorithm is used to search first and then treat the result as bacteria, improving the global searching ability of bacterial foraging optimization (BFO). Secondly, the reproduction operation is improved and the selection standard of survival of the cost is improved. Finally, we improve elimination and dispersal operation, and the population evolution factor is introduced to prevent the population from stagnating and falling into a local optimum. In this paper, three data sets are used to test the performance of the proposed algorithm. The simulation results show that the classification accuracy of the proposed algorithm is better than the existing approaches.

Keywords: particle swarm optimization; improved bacterial foraging optimization; imbalanced data

---

1. Introduction

In machine learning the imbalanced distribution of categories is called an imbalanced problem. When conventional algorithms are directly applied to this problem, the classification results tend to be biased towards most classes, resulting in a few classes not being correctly identified. Moreover, most of the traditional algorithms train classifiers based on the maximization of overall accuracy, meaning they ignore the misclassification of a few samples, thus affecting the classification results of traditional classifiers [1–3]. However, in many practical applications, a few samples are often more valuable than most samples, such as in bank fraud user identification, medical cancer diagnosis, and network hacker intrusion [4–9].

Imbalanced data mining is an important problem in data mining. Various algorithms, including k nearest neighbor (KNN), decision tree (DT), artificial neural network (ANN), and the genetic algorithm (GA), have been recommended for data mining [10–17]. However, these algorithms usually assume that datasets are distributed evenly among different classes and that some classes may be ignored. In the literature, some methods for dealing with imbalanced data have been proposed. These methods include adjusting the size of training datasets, cost-sensitive classifiers, and snowball methods [18–20]. These methods may result in the loss of information in general rules and the incorrect classification of additional classes. Ultimately, they can lead to an over-matching of data and poor performance due to having too many specific rules. Traditional optimization methods can no longer solve the complex problems faced by many datasets. In recent years, people have proposed a hybrid intelligent system to improve the accuracy of data mining rather than use a separate method. The hybrid method combines the best results of various systems to improve the accuracy [21–23].

Particle swarm optimization (PSO) was first invented by Dr. Eberhart and Dr. Kennedy [24,25]. It is a population-based heuristic algorithm used for simulating social behavior, such as birds clustering to promising locations, in order to find accurate targets in multi-dimensional space. PSO uses groups of individuals (called particles) to perform searches as with evolutionary algorithms, and particles can be updated from each iteration to the other [26–30]. In order to find the optimal solution, each particle changes its search direction based on two factors: its best previous location (p_{best}) and all other members' best locations (g_{best}) [31–34]. Shi et al. called p_{best} the cognitive part and g_{best} the social part [35].

The bacterial foraging optimization (BFO) algorithm is a bionic intelligent algorithm which was proposed by Passino in 2002 according to *Escherichia coli* in the human intestine [36,37]. The bacterial foraging chemotaxis process makes its local search ability stronger, but the global search ability of bacteria foraging can only be achieved by elimination and dispersal, and the global search ability is not strong enough to be limited by elimination and dispersal probability; thus it easily to falls into a local search optimal problem. In this paper, the incorporation of particle swarm optimization into an improved bacterial foraging optimization algorithm applied to the classification of imbalanced data is proposed. The borderline synthetic minority oversampling technique (Borderline-SMOTE) and Tomek link are used to pre-process imbalanced data. Thereafter, the proposed algorithm is used to classify imbalanced data.

Because PSO has a strong global search ability, individual effect, and group effect, PSO is incorporated into the improvement of the chemotaxis process of the improved BFO algorithm. The proposed algorithm improves the global searching ability and efficiency through the strong global search ability of PSO. In addition to embedding PSO into the BFO algorithm's chemotaxis process to improve the BFO algorithm's vulnerability to local optimization, in the improved replication operation, the crossover operator is introduced into the replication parent to increase the diversity of the population, while retaining the best individual. In the improved elimination and dispersion operation, the population evolution factor f_{evo} is proposed, and $(1 - f_{evo})$ is introduced to replace the P_{ed} in the original BFO algorithm so as to prevent the population from falling into a local optimum and achieving evolution stagnation. The purpose of this study was to improve the classification accuracy of ovarian cancer microarray data and to improve the practicability and accuracy of doctors' judgment of ovarian cancer microarray data.

This paper is organized as follows: Section 2 reviews PSO and BFO. Section 3 shows the proposed algorithms. Section 4 presents the experimental results and discussion. This section also describes an in-depth comparison of the proposed algorithm with other methods. Finally, a conclusion is given.

2. A Brief Description of Bacterial Foraging Optimization and Particle Swarm Optimization

In this paper, the bacterial foraging optimization algorithm is improved. Firstly, PSO is incorporated into the BFO chemotaxis process to improve the chemotaxis process. For this reason, this section introduces the basic concepts of bacterial foraging optimization and particle swarm optimization.

2.1. Bacterial Foraging Optimization

Passino introduced bacteria foraging optimization as a solution to distributed optimization and control problems. It is an evolutionary algorithm and a global random search algorithm. The BFO algorithm mainly solves the optimization problem by using four process iterative calculations: chemotaxis, swarming, reproduction, elimination, and dispersal [38]. In the chemotaxis process, there are two basic movements of *E. coli* in the process of foraging, namely, swimming and tumbling. Usually, in areas with poor environmental conditions (for example, toxic areas), bacteria may tumble more frequently, and in areas with a good environment, they will swim more often. Let $P(j, k, l) = \{\theta^i(j, k, l) | i = 1, 2, \ldots S\}$ indicate the ith bacterium in the population of the S bacteria at the jth chemotaxis process, kth reproduction process, and lth elimination and dispersal process. Let $L(i, j, k, l)$ be the cost at the location $\theta(j, k, l)$ of the ith bacterium. When the bacterial population size is S and N_c is the length of the bacteria in one direction of the chemotactic operation, the chemotaxis operation of each step of the ith bacterium is expressed as

$$\theta^i(j+1, k, l) = \theta^i(j, k, l) + \alpha(i) \frac{\delta(i)}{\sqrt{\delta^T(i)\delta(i)}} \tag{1}$$

where $\alpha(i) > 0$ represents the step unit of the forward swimming and $\delta(i)$ represents a unit vector in the random direction vector after the tumbling. In the swarming process, in addition to searching for food in their own way, each bacterial individual receives an appeal signal from other individuals in the population; that is, the individual will swim to the center of the population and will also receive a repulsive force signal from nearby individuals to maintain a safe distance between it and other individuals. Hence, the decision-making behavior of each bacterial individual in BFO which finds food is affected by two factors. The first is its own information, that is, the purpose of individual foraging to maximize the energy acquired by the individual in unit time, and the other is information from other individuals, that is, foraging information transmitted by other bacteria in the population. The mathematical expression is described as

$$\begin{aligned} L_{cc}(\theta, P(j, k, l)) &= \sum_{i=1}^{S} L_{cc}^i\left(\theta, \theta^i(j, k, l)\right) \\ &= \sum_{i=1}^{S}\left[-x_{attract} \exp\left(-y_{attract} \sum_{m=1}^{p}\left(\theta_m - \theta_m^i\right)^2\right)\right] \\ &+ \sum_{i=1}^{S}\left[-x_{repellent} \exp\left(-y_{repellent} \sum_{m=1}^{p}\left(\theta_m \theta_m^i\right)^2\right)\right] \end{aligned} \tag{2}$$

where $L_{cc}(\theta, P(j, k, l))$ denotes the penalty for the actual cost function, S is the number of bacteria, θ_m is the location of the fittest bacterium, and $x_{attract}$, $x_{repellent}$, $y_{attract}$, and $y_{repellent}$ are different coefficients. The swarming process is minimized mathematically.

$$L_{sw}(i, j, k, l) = L(i, j, k, l) + L_{cc}(\theta, P(j, k, l)) \tag{3}$$

In the swarming process, the number of biologically-motivated choices is expressed as N_s. In the reproduction process, according to the strength of the foraging ability of the bacteria, the appropriate cost L is selected; that is, L ranks the sum of the cost of all the locations experienced by the ith bacteria in the chemotaxis operation, and the elimination ranks 50% later. The number of bacteria in the population, the reproduction process of the remaining bacteria, and the new individuals generated by themselves which are identical to themselves have the same foraging ability and the same location, and the replication operation maintains the invariance of the population size. After N_{re} reproduction steps the elimination and dispersal process occurs, where N_{ed} is the number of steps of elimination and dispersal. These operations occur with a certain probability P_{ed}. When the individual bacteria meet the probability P_{ed} of elimination and dispersal, the individual dies and randomly generates a new individual at any location in the solution space. These new bacteria may have different bacterial foraging capabilities than the original bacteria, conducive to jumping out of the local optimal solution. A flow diagram of bacteria foraging optimization is presented in Figure 1.

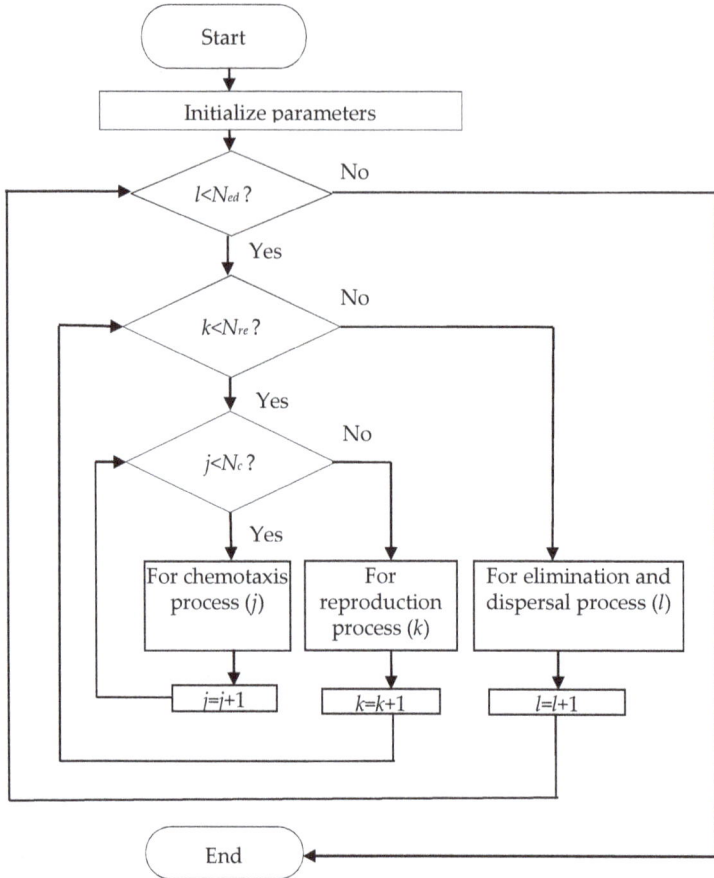

Figure 1. A flow diagram of bacterial foraging optimization (BFO).

2.2. Particle Swarm Optimization

PSO is a bionic algorithm used for the study of birds searching for food in nature. It regards birds as a particle in space, and a bird swarm is subject to PSO [39,40]. A single particle carries corresponding information—i.e., its own velocity and location—and determines the distance and direction of its motion according to the corresponding information of the particle itself. The PSO is used to initialize a group of particles which are randomly distributed into a solution space to be searched and then iterated according to a given equation. The equation of the mature particle swarm optimization algorithm includes two optimum concepts. The first is the local optimum p_{best} and the other is the global optimum g_{best}. The local optimum is the best solution obtained by each particle in the search, and the global optimum is the best solution obtained by this particle swarm. The PSO algorithm has the characteristics of memory, using positive feedback adjustment; the principle of the algorithm is simple, the parameters are few, and the applicability is good. The formulae of PSO are Equations (4) and (5), as described.

$$v_i^{t+1} = wv_i^t + c_1 \times rand_1^t \times \left(x_i^{p_{best}} - x_i^t\right) + c_2 \times rand_2^t \times \left(x^{g_{best}} - x_i^t\right) \tag{4}$$

$$x_i^{t+1} = x_i^t + v_i^{t+1} \tag{5}$$

In Equation (4), v_i^t and v_i^{t+1} denote the velocity of the ith particle in iterations t and $t+1$, w is the inertia weight, c_1 and c_2 are learning factors, $rand_1^t$ and $rand_2^t$ are random numbers between $[0, 1]$ in iteration t, x_i^{pbest} is the best location of the ith particle, and x^{gbest} is the best location of fitness found by all particles in the population. In Equation (5), x_i^t and x_i^{t+1} denote the location of the ith particle in iterations t and $t+1$. A flow chart of PSO is shown in Figure 2.

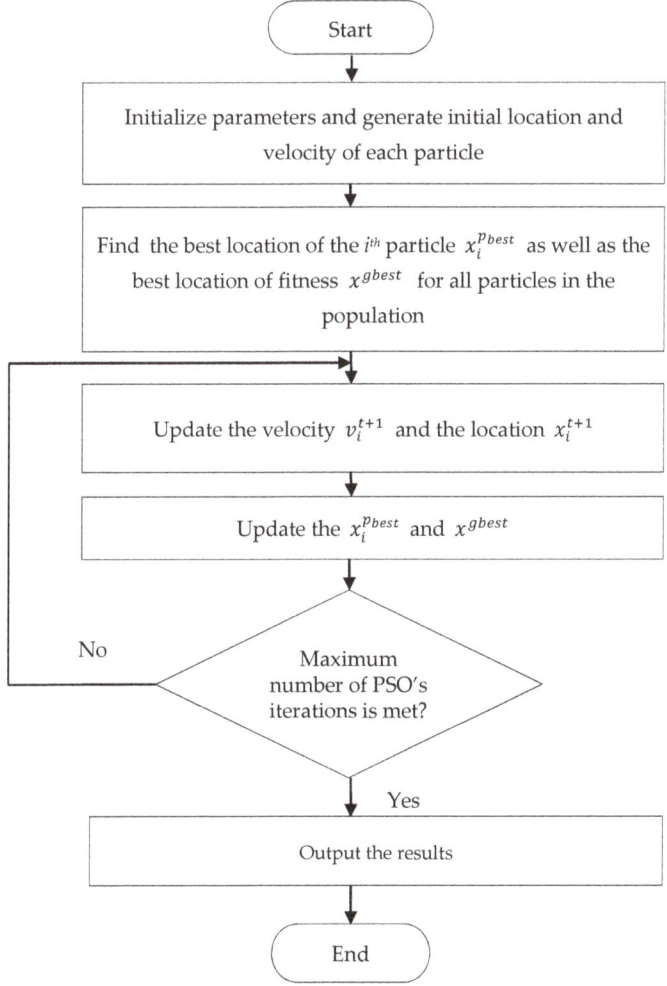

Figure 2. A flow chart of the particle swarm optimization (PSO) algorithm.

3. The Proposed Algorithm

In this paper, the incorporation of particle swarm optimization into an improved bacterial foraging optimization algorithm applied to the classification of imbalanced data is proposed. Three datasets are used for testing the performance of the proposed algorithm. One consists of ovarian cancer microarray data, and the other two, obtained from the UCI repository, are a spam email dataset and zoo dataset. The ovarian cancer microarray data were obtained from Taiwan's university. There are 9600 features in the microarray data of ovarian cancer, which were collected from China Medical University Hospital, with an imbalance ratio of about 1:20 [41,42]. The instances of microarray data we used included

ovarian tissue, vaginal tissue, cervical tissue, and myometrium, including six benign ovarian tumors
(BOT), 10 ovarian tumors (OVT), and 25 ovarian cancers (OVCA). The spam email dataset and zoo
dataset were obtained from the UCI repository [43]. For the spam email dataset, there were 4601 emails
with 58 features, as shown in Table 1, and the imbalance ratio was about 1:1.54. For the zoo dataset,
there were 101 instances with 17 features, as shown in Table 2, and the imbalance ratio was about 1:25.

Table 1. The 58 features of the spam email dataset.

Number	Meaning	Range	Maximum Value
1–48	Frequency of occurrence of a particular word	[0, 100]	<100
49–54	Frequency of occurrence of a particular character	[0, 100]	<100
55	Travel length of capital letters	[1, …]	1102.5
56	Longest capital travel	[1, …]	9989
57	Total travel length of capital letters	[1, …]	15,841
58	Spam ID (1 for spam)	[0, 1]	1

Table 2. The 17 features of the zoo dataset.

Number	Feature Name	Data Type
1	Animal name	Continuous
2	Hair	Nominal
3	Feathers	Continuous
4	Eggs	Nominal
5	Milk	Nominal
6	Airborne	Nominal
7	Aquatic	Nominal
8	Predator	Nominal
9	Toothed	Nominal
10	Backbone	Nominal
11	Breathes	Nominal
12	Venomous	Nominal
13	Fins	Nominal
14	Legs	Nominal
15	Tail	Nominal
16	Domestic	Nominal
17	Catsize	Nominal

Figure 3 shows a flow chart of the proposed algorithm. In Figure 3, the used parameters are
set first. The approaches of the Borderline-SMOTE and Tomek link are used for pre-process data.
Thereafter, the improved BFO algorithm is applied to classify imbalanced data so as to solve the
shortcoming of falling into a local optimum in the original BFO algorithm.

In order to over-sample the minority instances, the Borderline-SMOTE is designed in the proposed
algorithm; the main idea of SMOTE is to balance classes by generating synthetic instances from the
minority class [44]. For the subset of minority instances m_i, k nearest neighbors are obtained by
searching. The k nearest neighbors are defined as the smallest distance between the Euclidean distance
and m_i, and n synthetic instances are randomly selected from them which are recorded as Y_j, $j = 1, 2,$
… , n. This is done to create a new minority instance as in Equation (6) as described, where *rand* is the
random number between [0, 1].

$$m_{new} = m_i + rand * \left(Y_j - m_i \right) \qquad (6)$$

In the proposed algorithm, as a data cleaning technology, the Tomek link is effectively applied to
eliminate the overlap in the sampling method [45]. The Tomek link is used to remove unnecessary
overlaps between classes until the nearest neighbor pairs at the minimum distance belong to the same
class. Suppose that the nearest neighbors (m_i, m_j) of a pair of minimal Euclidean distances belong to

different classes. $d(m_i, m_j)$ represents the Euclidean distance between m_i and m_j. If there is no instance m_l satisfying Equation (7), we call (m_i, m_j) a pair of Tomek link.

$$d(m_i, m_l) < d(m_i, m_j) \text{ or } d(m_j, m_l) < d(m_i, m_j) \tag{7}$$

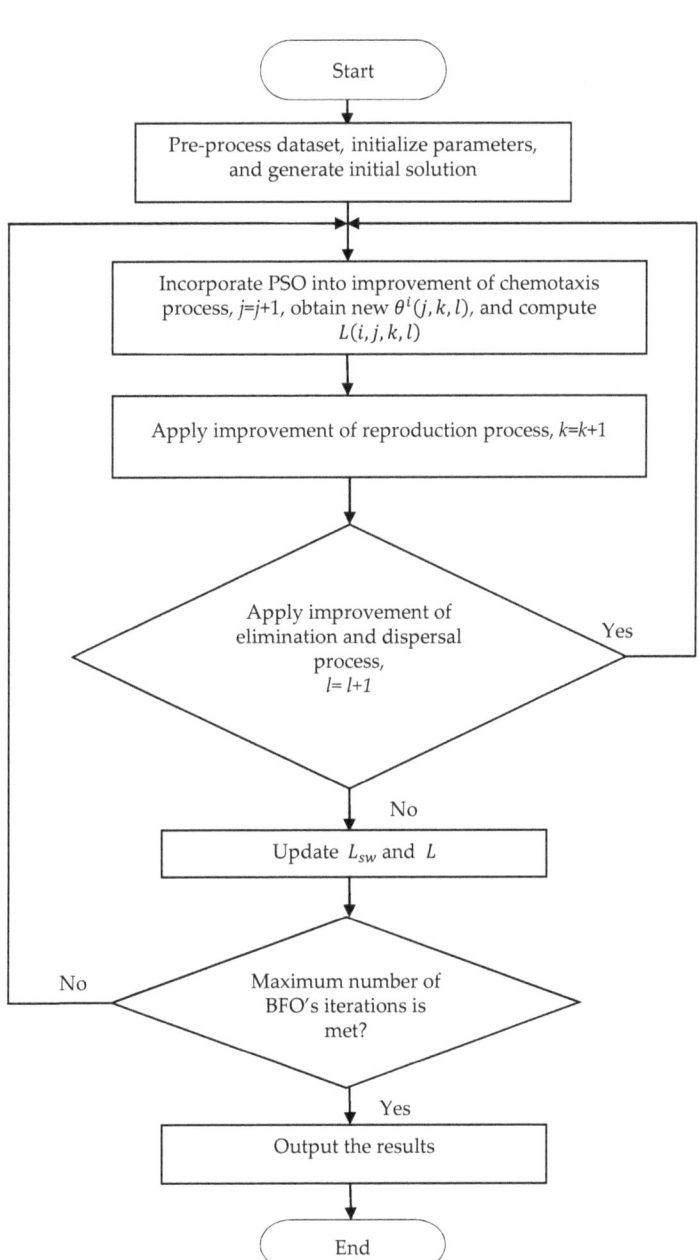

Figure 3. A flow diagram of the proposed algorithm.

In this paper, the parameter k used for SMOTE was set to k = 3. After preprocessing data, the solution of location θ^i was generated. Thereafter, the improved BFO algorithm was performed. Aiming at the BFO algorithm shortcoming of falling into a local optimum, we propose the incorporation of particle swarm optimization into an improved bacterial foraging optimization to solve these problems. An improved BFO proposed algorithm improves the chemotaxis process, reproduction process, and the elimination and dispersal process.

3.1. Improvement of Chemotaxis Process

The original BFO algorithm mainly searches within the process of chemotaxis. When the chemotaxis searches the target area, the swimming and tumbling operation of the chemotaxis process directly affects the effect of the algorithm. While a large swimming step makes the global search ability strong, a small swimming step makes the local search ability strong. Because of the characteristics of chemotaxis, the BFO algorithm has good local search ability because it can change direction in chemotaxis, meaning the local search accuracy is very good. However, the global search ability of bacteria can only rely on the elimination and dispersal operation process, and its global search ability is not good.

Because PSO has strong memory and global search ability, individual effect, and group effect, in this paper, the PSO is incorporated into the chemotaxis process of the original BFO so as to solve the problem of how the original BFO algorithm easily falls into local optimization. By using particles to search first and then treat particles as bacteria, the global search ability of the original BFO algorithm is improved. The purpose of this study is to find an effective algorithm which combines the advantages of PSO, including fast convergence speed, strong search ability, and the good classification effect of the BFO algorithm, to improve the accuracy of imbalanced data.

3.2. Improvement of Reproduction Process

In the reproduction process of the original BFO algorithm, half of the good bacteria (S/2) are replicated using the current bacterial position generation cost L as the basis for good or bad arrangement in the bacterial population with a population size of S, and the sub-population generated by replication replaces the other half of the bad bacteria in the original bacterial population.

Because each parent has one of the same offspring in the bacterial population with size S after replication, the diversity of the population is reduced. In this paper, the cost of the current bacterial location is used to rank the values as good and bad, and half of the excellent bacteria S/2 are reproduced. The reproduced sub-population replaces the worse S/2 bacteria in the original bacterial population. In order to increase the diversity of the population and prevent the loss of the best individual, a hybrid operator is introduced into the parent individual (excluding the best parent individual) to cross with the best individual. The hybrid equation is [46]

$$\sigma = \sigma + rand * (\sigma_{best} - \sigma) \tag{8}$$

where σ is the parent individual (excluding the best parent individual), σ_{best} is the best parent individual, and *rand* is the random number with entries on [0, 1].

3.3. Improvement of Elimination and Dispersal Process

The elimination–dispersal operation helps the BFO algorithm jump out of the local optimal solution and find the global optimal solution. In the elimination–dispersal process of the original BFO, elimination and dispersal is carried out according to the given fixed probability P_{ed} without considering the evolution of the population.

In this paper, the elimination–dispersal operation is improved by introducing the population evolution factor and elimination–dispersal is carried out according to the evolution of the population, which is conducive to the effectiveness of the algorithm and prevents the population from falling into a local optimum due to slow evolution. The formula of the population evolution factor f_{evo} is

$$f_{evo} = \frac{L_{gen} - L_{gen-1}}{L_{gen-1} - L_{gen-2} + rand} \tag{9}$$

where L_{gen} represents the optimal generation cost at the iteration *gen* and *rand* prevents the denominator from being 0. In this paper, $(1 - f_{evo})$ is used to replace P_{ed} as in the original BFO algorithm. When $f_{evo} > 1$, the evolution is accelerated. At this time, the evolution degree of the population is faster and the population is in a fast and effective optimization state. Elimination–dispersal with a lower elimination–dispersal probability $(1 - f_{evo})$ can retain the current favorable location information. When $0 \leq f_{evo} < 1$, the evolution slows down. When the evolution degree of the population is slow, the population falls into a local optimum to a large extent. It is necessary for elimination–dispersal with a high elimination–dispersal probability $(1 - f_{evo})$ to jump out of the local optimum solution so as to prevent the population from not evolving.

In order to overcome the shortcoming of the BFO algorithm easily falling into a local optimum and uncertain orientation during the chemotaxis process, PSO is incorporated into the BFO algorithm in this paper, that is to say, PSO is added to the chemotaxis process of each individual bacterium, which is the cost of each bacterium according to PSO. For the improved chemotaxis process, PSO is performed to obtain the updated location of the θ^i. The procedure of the proposed algorithm is detailed as follows.

(1) The particle swarm population of size S is initialized. Here, PSO is added to the chemotaxis process of each individual bacterium, and the swarm population size S of PSO is the same as that of the BFO algorithm. The initial velocity and position of each particle is randomly generated. The maximum number of PSO iterations is T. The BFO algorithm parameters N_c, N_s, N_{re}, N_{ed}, $x_{attract}$, $x_{repellent}$, $y_{attract}$, and $y_{repellent}$ are set. The number of BFO iterations is $N_c \times N_{re} \times N_{ed}$.

(2) The cost L, defined as the classification accuracy of each particle, is calculated. The best location of the *i*th particle x_i^{Pbest} and the best location of the cost x^{gbest} for all particles in the population are found. x_i^{Pbest} is updated and x^{gbest} if x_i^{Pbest} and x^{gbest} are improved.

(3) Equation (4) is applied to update the velocity v_i^{t+1} and Equation (5) is applied to update the location x_i^{t+1}. In Equation (4), the velocity of each particle must be limited to the range of the set maximum velocity v_{max}. If the velocity of each particle exceeds the limit, the velocity is expressed as v_{max}.

(4) If the set termination condition is met, it will stop; otherwise, the process goes back to step 2. The termination condition is usually to reach the best location x^{gbest} of the cost for all particles in the population, or to exceed the set PSO's maximum number of iterations T. Through Equation (4) and Equation (5) particles treated as bacteria, PSO is completed to obtain the updated position x_i^{t+1}. In other words, the PSO is performed to obtain the updated location of θ^i in the improved chemotaxis process.

(5) In the swarming process, the cost of L_{sw} is evaluated by Equation (3).

(6) In the improved reproduction process, Equation (8) is performed to increase the diversity of the population and avoid losing the best individual; in other words, the parent individual (excluding the best parent individual) crosses the best individual.

(7) In the improved elimination–dispersal process, the population evolution factor f_{evo} is used in Equation (9). The new θ^i by PSO is generated according to $(1 - f_{evo})$. In the improved BFO algorithm, P_{ed} is replaced with $(1 - f_{evo})$.

(8) If the maximum number of BFO iterations is met, the algorithm is over. Finally, we output the classification accuracy results in this implementation.

The proposed algorithm is performed and cost L is defined as the classification accuracy. This experiment used a classification accuracy based on the confusion matrix, which can test the performance of the classification method. The confusion matrix is shown in Table 3.

Table 3. The confusion matrix.

Predicted / Actual	Actual Positive	Active Negative
Predicted positive	TP (true positive)	FP (false positive)
Predicted negative	FN (false negative)	TN (true negative)

TP and FP represent the true positive class and the false positive class, respectively; FN and TN represent the false negative class and the true negative class, respectively. When the predicted value is a positive example, it is recorded as P (positive). When the predicted value is a negative example, it is is recorded as N (negative). When the predicted value is the same as the actual value, it is recorded as T (true). Finally, when the predicted value is opposite to the actual value, it is is recorded as F (false). The four results of defining examples in the data set after model classification are TP: predicted positive, actual positive actual; FP: predicted positive, actual negative; TN: predicted negative, actual negative; and FN: predicted negative, actual positive. The classification accuracy calculation formula is

$$\text{Classification accuracy} = (\text{TP} + \text{TN})/(\text{TP} + \text{FN} + \text{FP} + \text{TN}) \times 100\% \qquad (10)$$

The receiver operating characteristic curve (ROC curve) and area under the curve (AUC) can test the performance of the classification results. This is because the ROC curve has a favorable characteristic: when the distribution of positive and negative instances in the test dataset changes, the ROC curve can remain unchanged. Class imbalance often occurs in the actual data set, i.e., there are many more negative instances than positive instances (or vice versa) and the distribution of positive and negative instances in the test data may change with time. The area under the ROC curve is calculated as the evaluation method of imbalanced data. It can comprehensively describe the performance of classifiers under different decision thresholds. The AUC calculation formula is

$$\text{Area Under the Curve (AUC)} = \frac{1 + \left(\frac{\text{TP}}{\text{FP} + \text{FN}}\right) - \left(\frac{\text{FP}}{\text{TN} + \text{FP}}\right)}{2} \qquad (11)$$

4. Simulation Results and Discussion

In this study, our purpose was to obtain an effective algorithm with which to improve the accuracy of imbalanced data. In order to verify the performance of the proposed algorithm, ovarian cancer microarray data, a spam email dataset and a zoo dataset are used for simulation experiments. The Borderline-SMOTE and Tomek link approaches are used for preprocess data to increase the numbers of minority classes until they are the same number as the majority class. In the simulation experiment, some parameters of the algorithm need to be determined. In this experiment, the BFO algorithm parameters were set as $S = 50$, $N_c = 100$, $N_s = 4$, $N_{re} = 4$, $N_{ed} = 2$, $P_{ed} = 0.25$, $x_{attract} = 0.05$, $x_{repellent} = 0.05$, $y_{attract} = 0.05$, $y_{repellent} = 0.05$, $\alpha(i) = 0.1$, and $i = 1, 2, \ldots S$. The number of BFO iterations was $N_c \times N_{re} \times N_{ed} = 100 \times 4 \times 2 = 800$. This study evaluated the results when adopting 10-fold cross validation with random partitions. The maximum number of PSO iterations was set to 5000 and the other parameters were set as inertia weight $w = 0.6$, learning factors $c_1 = c_2 = 1.5$, and maximum velocity of each particle $v_{max} = 2$ [47].

The parameter value of the algorithm is the key to the performance and efficiency of the algorithm. In evolutionary algorithms there are no general methods for determining the optimal parameters of the algorithm. Most parameters are selected by experience. There are many BFO and PSO parameters. Knowing how to determine the optimal BFO and PSO parameters to optimize the performance of the algorithm is a very complex optimization problem. In the parameter setting of PSO and BFO, in order to jump off the local solution to find the global solution without spending a lot of calculation time, we used empirical values.

4.1. Comparing and Analyzing the Classification Accuracy of the Proposed Algorithm and Other Methods

(1) In addition to the proposed algorithm, we also employ other existing approaches for comparison. The approaches used include the support vector machine (SVM), DT, random forest (RF), KNN, and BFO. The SVM is a learning system that uses a hypothesis space of linear function in a high-dimensional feature space. DT uses partition information entropy minimization to recursively partition the dataset into smaller subdivisions and then generate a tree structure. RF is an ensemble learning method for classification that constructs multiple decision trees during training time and outputs the class that depends on the majority of the classes. KNN is a method used to classify objects based on the closest training examples in an n-dimensional pattern space. The BFO algorithm is described in Section 2.1.

(2) Tables 4–6 list the classification performances of the ovarian cancer microarray data, spam email dataset, and zoo dataset, respectively. From Table 4, the average classification accuracy in the proposed algorithm for the ovarian cancer microarray data can be seen to be 93.47%. From Table 5, the average classification accuracy of the proposed algorithm for the spam email dataset can be seen to be 96.42%. As shown in Table 6, the average classification accuracy for the zoo dataset of the proposed algorithm is 99.54%. From Tables 4–6, it is clearly evident that the proposed approach has the best classification results given a fair comparison for all compared approaches. This is because the performance of the classification for the three tested datasets can be found based on intelligent information. In fact, the proposed approach has similar performance, meaning it performs well in classification accuracy.

(3) In the comparison results it can be found that the classification accuracy of the original BFO method in Table 4 was 89.93%, which is not better than the proposed algorithm classification accuracy of 93.47%. In Table 5, the classification accuracy of the original BFO method can be seen to be 94.27%, which is not better than the proposed algorithm classification accuracy of 96.42%. In Table 6, the classification accuracy of the original BFO method can be seen to be 94.38%, which is not better than the proposed algorithm classification accuracy of 99.54%. Because the original BFO algorithm can change direction in the chemotaxis operation, its local search ability is better; the global search, however, can only rely on elimination and dispersal operation, and the global search ability is not very good. Hence, the classification accuracy is not better than the proposed algorithm.

(4) The proposed algorithm provides a better classification effect because PSO is incorporated into the improved chemotaxis process. PSO has memory and global search abilities, so we first used particles for global search and then treat these particles as bacteria, and the chemotaxis operation improved the global search ability. The PSO algorithm introduced in this paper only uses its global operation and uses the memory of PSO to improve the bacterial search ability. In the improved reproduction operation, the crossover operator is introduced to the replica parent to increase the diversity of the population while the best individual is retained. In the improved elimination and dispersal operation, the $(1 - f_{evo})$ replaces P_{ed} in the original BFO, and is introduced to prevent the population from dying and falling into a local optimum.

Table 4. The classification accuracy for microarray data of ovarian cancer. Legend: RF, random forest; SVM, support vector machine; DT, decision tree; KNN, k nearest neighbor.

Approaches	Classification Accuracy
SVM	88.45%
DT	85.71%
RF	83.66%
KNN	80.88%
BFO	89.93%
The proposed algorithm	93.47%

Table 5. The classification accuracy for the spam email dataset.

Approaches	Classification Accuracy
SVM	93.51%
DT	90.83%
RF	91.68%
KNN	90.64%
BFO	94.27%
The proposed algorithm	96.42%

Table 6. The classification accuracy for the zoo dataset.

Approaches	Classification Accuracy
SVM	93.55%
DT	92.71%
RF	90.32%
KNN	91.46%
BFO	94.38%
The proposed algorithm	99.54%

4.2. Analysis of ROC and AUC

In this experiment, the area below the ROC is also called the AUC and is used to evaluate the performance of the proposed approach. The value of the AUC is from 0 to 1.0, and the closer to 1.0, the better the effect of the model classifier. The value of the AUC is 0.979 for the ovarian cancer microarray data, as shown in Figure 4. The value of the AUC is 0.987 for the spam email dataset, as shown in Figure 5. The value of the AUC is 0.995 for the zoo data, as shown in Figure 6. Hence, the experimental results show that the proposed algorithm has good classification performance.

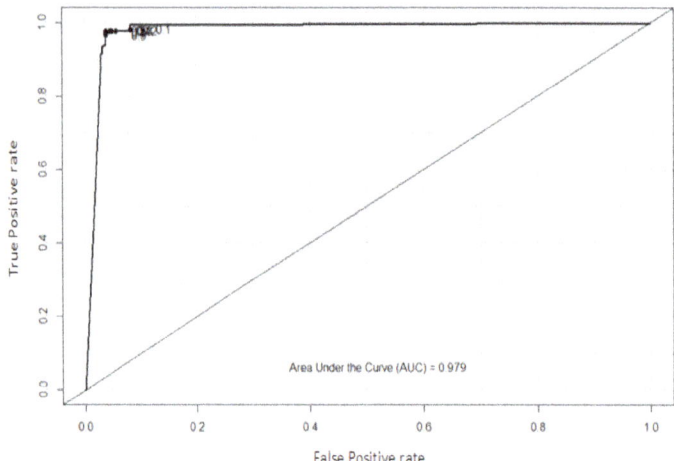

Figure 4. The receiver operating characteristic (ROC) and the area under the curve (AUC) for the microarray data of ovarian cancer.

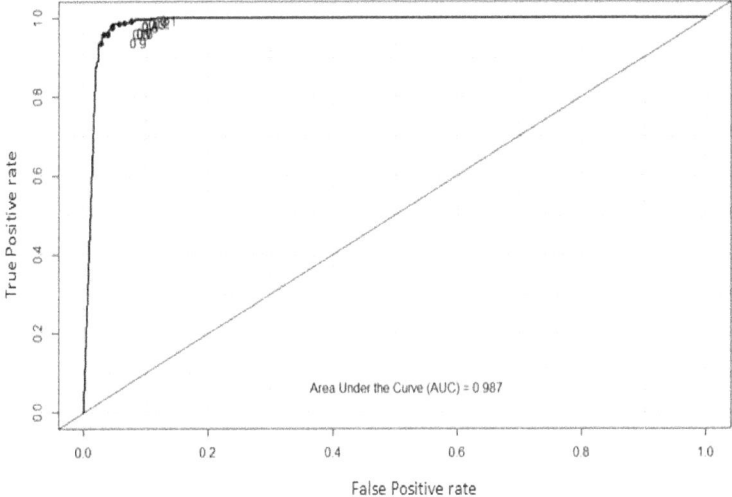

Figure 5. The ROC and AUC for the spam email dataset.

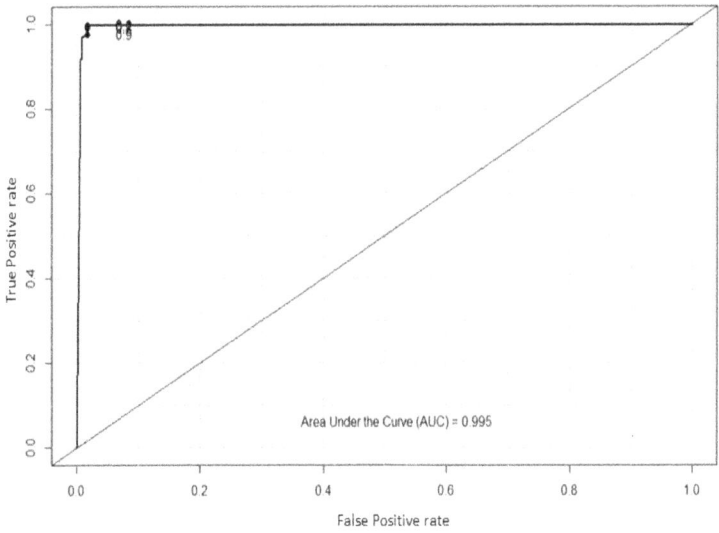

Figure 6. The ROC and AUC for the zoo dataset.

5. Conclusions

This paper has proposed the incorporation of particle swarm optimization into an improved bacterial foraging optimization algorithm applied to the classification of imbalanced data. The Borderline-SMOTE and Tomek link approaches were used to pre-process data. Thereafter, the intelligent improved BFO was applied to the classification of imbalanced data so as to solve the shortcoming of falling into a local optimum in the original BFO algorithm. Three datasets were used for testing the performance of the proposed algorithm. The proposed algorithm includes an improved chemotaxis process, an improved reproduction process, and an improved elimination and dispersal process. In this paper, the global search ability of the BFO was improved by using particles to search and then treating particles as bacteria in the improved chemotaxis process. After the improved chemotaxis, the swarming operations, improved reproduction operations, and improved elimination and dispersal operations were performed. The average classification accuracy of the proposed algorithm for the ovarian cancer microarray data was 93.47%. The average classification accuracies of the spam email dataset and the zoo dataset of the proposed algorithm were 96.42% and 99.54%, respectively. The value of the AUC was 0.979 for the ovarian cancer microarray data, 0.987 for the spam email dataset, and 0.995 for the zoo dataset. The experimental results showed that the proposed algorithm in this research can achieve the best accuracy in the classification of imbalanced data compared with existing approaches.

In this paper, PSO was introduced into an improved bacterial foraging optimization algorithm and applied to the classification of imbalanced data. Based on the research results, we put forward the following suggestions:

(1) Improvement of the algorithm's operation: The key to implementing the optimization is the operation of the algorithm. Designing an excellent operation plays an important role in improving the performance and efficiency of the algorithm. In BFO, this will become a key area of research into BFO to improve chemotaxis and reproduction and the elimination and dispersal operation process, and to coordinate the local mining ability and global exploring ability of the processing algorithm.
(2) Selection of algorithm parameters: The parameter value of the algorithm is key to the performance and efficiency of the algorithm. In evolutionary algorithms, there is no general method to determine the optimal parameters of the algorithm. At present, there are many BFO parameters. Determining the optimal parameters of BFO to optimize the performance of the algorithm itself is a complex optimization problem.
(3) Combining with other algorithms: Combining the advantages of BFO and other algorithms to propose more efficient algorithms is a valuable topic in BFO research.

Author Contributions: Methodology, F.-L.Y. and C.-Y.L.; software, F.-L.Y., J.-Q.H., and J.-F.T.; formal analysis, F.-L.Y., C.-Y.L., and Z.-J.L.; investigation, F.-L.Y., C.-Y.L., and J.-F.T.; resources, C.-Y.L. and Z.-J.L.; data curation, F.-L.Y., J.-Q.H., and J.-F.T.; original draft preparation, F.-L.Y., C.-Y.L., and J.-F.T.; review and editing, F.-L.Y., C.-Y.L., and J.-F.T. All authors have read and agreed to the published version of the manuscript.

Funding: This research received no external funding.

Acknowledgments: This research was supported by the Major Education and Teaching Reform Projects in Fujian Undergraduate Colleges and Universities in 2019 under grant FBJG20190284. This work was also supported by projects under 2019-G-083.

Conflicts of Interest: The authors declare no conflict of interest. The funders had no role in the design of the study; in the collection, analyses, or interpretation of data; in the writing of the manuscript; or in the decision to publish the results.

References

1. Hu, J.J.; Yang, H.Q.; Lyu, M.R.; King, I.R.; So, A.M.C. Online Nonlinear AUC Maximization for Imbalanced Data Sets. *IEEE Trans. Neural Netw. Learn. Syst.* **2018**, *29*, 882–895. [CrossRef]
2. Huang, X.L.; Zou, Y.X.; Wang, Y. Cost-sensitive sparse linear regression for crowd counting with imbalanced training data. In Proceedings of the 2016 IEEE International Conference on Multimedia and Expo (ICME), Seattle, WA, USA, 11–15 July 2016.
3. Padmaja, T.M.; Dhulipalla, N.; Bapi, R.S.; Krishna, P.R. Imbalanced data classification using extreme outlier elimination and sampling techniques for fraud detection. In Proceedings of the International Conference on Advanced Computing and Communications, Guwahati, India, 18–21 December 2007; pp. 511–516.
4. Lin, S.W.; Ying, K.C.; Lee, C.Y.; Lee, Z.J. An intelligent algorithm with feature selection and decision rules applied to anomaly intrusion detection. *Appl. Soft Comput.* **2012**, *12*, 3285–3290. [CrossRef]
5. Lee, C.Y.; Lee, Z.J. A Novel Algorithm Applied to Classify Unbalanced Data. *Appl. Soft Comput.* **2012**, *12*, 2481–2485. [CrossRef]
6. Zhang, Y.; Wu, L. Stock market prediction of S&P 500 via combination of improved BCO approach and BP neural network. *Expert Syst. Appl.* **2009**, *36*, 8849–8854.

7. Xia, C.Q.; Han, K.; Qi, Y.; Zhang, Y.; Yu, D.J. A Self-Training Subspace Clustering Algorithm under Low-Rank Representation for Cancer Classification on Gene Expression Data. *IEEE/ACM Trans. Comput. Biol. Bioinform.* **2018**, *15*, 1315–1324. [CrossRef] [PubMed]

8. Esfahani, M.S.; Dougherty, E.R. Incorporation of Biological Pathway Knowledge in the Construction of Priors for Optimal Bayesian Classification. *IEEE/ACM Trans. Comput. Biol. Bioinform.* **2014**, *11*, 202–218. [CrossRef] [PubMed]

9. Sadreazami, H.; Mohammadi, A.; Asif, A.; Plataniotis, K.N. Distributed-Graph-Based Statistical Approach for Intrusion Detection in Cyber-Physical Systems. *IEEE Trans. Signal Inform. Process. Netw.* **2018**, *4*, 137–147. [CrossRef]

10. Mathew, J.; Pang, C.K.; Luo, M.; Leong, W.H. Classification of imbalanced data by oversampling in kernel space of support vector machines. *IEEE Trans. Neural Netw. Learn. Syst.* **2018**, *29*, 4065–4076. [CrossRef]

11. Zhang, J.; Bloedorn, E.; Rosen, L.; Venese, D. Learning rules from highly imbalanced data sets. In Proceedings of the Fourth IEEE International Conference on Data Mining, ICDM'04, Brighton, UK, 1–4 November 2004; Volume 1, pp. 571–574.

12. Jiang, Y.; Zhou, Z.H. Editing training data for *k*NN classifiers with neural network ensemble. In Proceedings of the International Symposium on Neural Networks, Dalian, China, 19–21 August 2004; Volume 1, pp. 356–361.

13. Tao, Q.; Wu, G.W.; Wang, F.Y.; Wang, J. Posterior probability support vector Machines for imbalanced data. *IEEE Trans. Neural Netw.* **2005**, *16*, 1561–1573. [CrossRef]

14. Zhang, J.; Mani, I. *k*NN approach to imbalanced data distributions: A case study involving information extraction. In Proceedings of the ICML'2003 Workshop on Learning from Imbalanced Datasets, Washington, DC, USA, 21 August 2003.

15. Elaidi, H.; Elhaddar, Y.; Benabbou, Z.; Abbar, H. An idea of a clustering algorithm using support vector machines based on binary decision tree. In Proceedings of the 2018 International Conference on Intelligent Systems and Computer Vision (ISCV), Fez, Morocco, 2–4 April 2018; pp. 1–5.

16. Ye, D.; Chen, Z. A rough set based minority class oriented learning algorithm for highly imbalanced data sets. In Proceedings of the IEEE International Conference on Granular Computing, Hangzhou, China, 26–28 August 2008; pp. 736–739.

17. Yang, X.; Song, Q.; Cao, A. Clustering nonlinearly separable and imbalanced data set. In Proceedings of the 2004 2nd International IEEE Conference on Intelligent Systems, Varna, Bulgaria, 22–24 June 2004; Volume 2, pp. 491–496.

18. Lu, Y.; Guo, H.; Feldkamp, L. Robust neural learning from imbalanced data samples. In Proceedings of the 1998 IEEE International Joint Conference on Neural Networks, Anchorage, AK, USA, 4–9 May 1998; Volume 3, pp. 1816–1821.

19. Wang, J.; Jean, J. Resolve multifont character confusion with neural network. *Pattern Recognit.* **1993**, *26*, 173–187. [CrossRef]

20. Searle, S.R. *Linear Models for Unbalanced Data*; Wiley: New York, NY, USA, 1987.

21. Wang, J.; Miyazaki, M.; Kameda, H.; Li, J. Improving performance of parallel transaction processing systems by balancing data load on line. In Proceedings of the Seventh International Conference on Parallel and Distributed Systems, Taipei, Taiwan, 4–7 December 2000; pp. 331–338.

22. Crepinsek, M.; Liu, S.H.; Mernik, M. Replication and comparison of computational experiments in applied evolutionary computing: Common pitfalls and guidelines to avoid them. *Appl. Soft Comput.* **2014**, *19*, 161–170. [CrossRef]

23. De Corte, A.; Sörensen, K. Optimisation of gravity-fed water distribution network design: A critical review. *Eur. J. Oper. Res.* **2013**, *228*, 1–10. [CrossRef]

24. Kennedy, J.; Eberhart, R.C. A discrete binary version of the particle swarm algorithm. In Proceedings of the 1997 IEEE International Conference on Systems, Man, and Cybernetics. Computational Cybernetics and Simulation, Orlando, FL, USA, 12–15 October 1997; Volume 5.

25. Eberhart, R.; Kennedy, J. A new optimizer using particle swarm theory. In Proceedings of the Sixth International Symposium on Micro Machine and Human Science (MHS' 95), Nagoya, Japan, 4–6 October 1995.

26. Jia, J.Y.; Zhao, A.W.; Guan, S.A. Forecasting Based on High-Order Fuzzy-Fluctuation Trends and Particle Swarm Optimization Machine Learning. *Symmetry* **2017**, *9*, 124. [CrossRef]

27. Xue, H.X.; Bai, Y.P.; Hu, H.P.; Xu, T.; Liang, H.J. A Novel Hybrid Model Based on TVIW-PSO-GSA Algorithm and Support Vector Machine for Classification Problems. *IEEE Access* **2019**, *7*, 27789–27801. [CrossRef]

28. Kim, J.J.; Lee, J.J. Trajectory Optimization with Particle Swarm Optimization for Manipulator Motion Planning. *IEEE Trans. Ind. Inform.* **2015**, *11*, 620–631. [CrossRef]
29. Liu, H.M.; Yan, X.S.; Wu, Q.H. An Improved Pigeon-Inspired Optimisation Algorithm and Its Application in Parameter Inversion. *Symmetry* **2019**, *11*, 1291. [CrossRef]
30. Salleh, I.; Belkourchia, Y.; Azrar, L. Optimization of the shape parameter of RBF based on the PSO algorithm to solve nonlinear stochastic differential equation. In Proceedings of the 2019 5th International Conference on Optimization and Applications (ICOA), Kenitra, Morocco, 25–26 April 2019.
31. Medoued, A.; Lebaroud, A.; Laifa, A.; Sayad, D. Feature form extraction and optimization of induction machine faults using PSO technique. In Proceedings of the 2013 3rd International Conference on Electric Power and Energy Conversion Systems, Istanbul, Turkey, 2–4 October 2013.
32. Yeom, C.U.; Kwak, K.C. Incremental Granular Model Improvement Using Particle Swarm Optimization. *Symmetry* **2019**, *11*, 390. [CrossRef]
33. Lee, J.H.; Kim, J.W.; Song, J.Y. Distance-Based Intelligent Particle Swarm Optimization for Optimal Design of Permanent Magnet Synchronous Machine. *IEEE Trans. Magn.* **2017**, *53*, 1–4. [CrossRef]
34. Yu, X.; Chen, W.N.; Gu, T.L. Set-Based Discrete Particle Swarm Optimization Based on Decomposition for Permutation-Based Multiobjective Combinatorial Optimization Problems. *IEEE Trans. Cybern.* **2018**, *48*, 2139–2153. [CrossRef]
35. Shi, Y.; Eberhart, R. A modified particle swarm optimizer. Proceeding of the 1998 IEEE International Conference on Evolutionary Computation, World Congress on Computational Intelligence, Anchorage, AK, USA, 4–9 May 1998.
36. Passino, K.M. Biomimicry of bacterial foraging for distributed optimization and control. *IEEE Control Syst. Mag.* **2002**, *22*, 52–67.
37. Abraham, A.; Biswas, A.; Dasgupta, S. Analysis of reproduction operator in bacterial foraging optimization algorithm. In Proceedings of the IEEE World Congress on Computational Intelligence, Hong Kong, China, 1–6 June 2008; pp. 1476–1483.
38. Bidyadhar, S.I.; Raseswari, P. Bacterial Foraging Optimization Approach to Parameter Extraction of a Photovoltaic Module. *IEEE Trans. Sustain. Energy* **2018**, *9*, 381–389.
39. Noguchi, T.; Togashi, S.; Nakamoto, R. Based maximum power point tracking method for multiple photovoltaic and converter module system. *IEEE Trans. Ind. Electron.* **2002**, *49*, 217–222. [CrossRef]
40. Raza, A.; Yousaf, Z.; Jamil, M. Multi-Objective Optimization of VSC Stations in Multi-Terminal VSC-HVdc Grids, Based on PSO. *IEEE Access* **2018**, *6*, 62995–63004. [CrossRef]
41. Lu, S.J. Gene Expression Analysis and Regulator Pathway Exploration with the Use of Microarray Data for Ovarian Cancer. Master's Thesis, National Taiwan University of Science and Technology, Taipei, Taiwan, 2006.
42. Lee, Z.J. An integrated algorithm for gene selection and classification applied to microarray data of ovarian cancer. *Int. J. Artif. Intell. Med.* **2008**, *42*, 81–93. [CrossRef] [PubMed]
43. Blake, C.; Keogh, E.; Merz, C.J. *UCI Repository of Machine learning Databases*; Department of Information and Computer Science, University of California: Irvine, CA, USA, 1998; Available online: https://archive.ics.uci.edu/ml/datasets.php (accessed on 24 December 2019).
44. Gosain, A.; Sardana, S. Farthest SMOTE: A modified SMOTE approach. In *Computational Intelligence in Data Mining*; Springer: Singapore, 2019; pp. 309–320.
45. Devi, D.; Purkayastha, B. Redundancy-driven modified Tomek link based undersampling: A solution to class imbalance. *Pattern Recogn. Lett.* **2017**, *93*, 3–12. [CrossRef]
46. Liu, L.; Shan, L.; Yan, J.H. An Improved BFO Algorithm for Optimising the PID Parameters of Servo System. In Proceedings of the IEEE the 30th Chinese Control and Decision Conference (2018 CCDC), Shenyang, China, 9–11 June 2018; pp. 3831–3836.
47. Abd-Elazim, S.M.; Ali, E.S. A hybrid particle swarm optimization and bacterial foraging for power system stability enhancement. *Complexity* **2015**, *21*, 245–255. [CrossRef]

Article

Application of Gray Relational Analysis and Computational Fluid Dynamics to the Statistical Techniques of Product Designs

Hsin-Hung Lin [1,2,*] , Jui-Hung Cheng [3] and Chi-Hsiung Chen [1]

1   Department of Creative Product Design, Asia University, Taichung City 41354, Taiwan
2   Department of Medical Research, China Medical University Hospital, China Medical University, Taichung 404, Taiwan
3   Department of Mold and Die Engineering, National Kaohsiung University of Science and Technology, Kaohsiung 80778, Taiwan
*   Correspondence: hhlin@asia.edu.tw or a123lin0@gmail.com; Tel.: +886-04-2332-3456-1051

Received: 27 September 2019; Accepted: 20 January 2020; Published: 3 February 2020

Abstract: During the development of fan products, designers often encounter gray areas when creating new designs. Without clear design goals, development efficiency is usually reduced, and fans are the best solution for studying symmetry or asymmetry. Therefore, fan designers need to figure out an optimization approach that can simplify the fan development process and reduce associated costs. This study provides a new statistical approach using gray relational analysis (GRA) to analyze and optimize the parameters of a particular fan design. During the research, it was found that the single fan uses an asymmetry concept with a single blade as the design, while the operation of double fans is a symmetry concept. The results indicated that the proposed mechanical operations could enhance the variety of product designs and reduce costs. Moreover, this approach can relieve designers from unnecessary effort during the development process and also effectively reduce the product development time.

Keywords: gray relational analysis; flow-field analysis; fan design; CFD; product design evaluation; symmetry

---

1. Introduction

During the development of new fan products, it is necessary to repeatedly experiment and test to optimize the product. However, the conventional design and development of a fan is usually limited by standard methods, and the fan is the best solution to study symmetry or asymmetry. This method consists of multiple rounds of simulations and experiments. When a designer comes up with a new idea, it takes a long time to test and verify the parameters of the impeller profile. During the research, the single fan uses an asymmetry concept and the singular blade is used as the design, while the double fans constitute an asymmetry concept. In order to optimize the best method of symmetry or asymmetry regarding the mechanical principle of the blades, a new statistical method of gray relational analysis (GRA) analysis and the optimization of specific fan design parameters are required. In 2012, Kim et al. [1] suggested that a detailed blade design and optimized tip clearance is important for performance, and the geometric parameters of a blade were calculated and the results served as the flow criteria. The geometric parameters of a blade were also determined by calculations and verified by the simulation results of Computational Fluid Dynamics (CFD) and experiments. In 2010, Hurault et al. [2] studied the impact of the turbulence in axial-flow fans, and the fans that had been studied were provided with radial, swept-forward, and swept-backward blades. He compared the results of experiments and CFD with those obtained by Rhee et al. [3]. However, there are many gray zones for the blade parameters during the process of development, and therefore most of the parameters are difficult to determine. In 2009, Lai et al. [4] applied the method of gray relational analysis (GRA) to product design evaluation (PDE) models when designing new product models. The final results solved the problem of complicated probabilities in the application of ergonomics to human comfort. In 2011, Wei et al. [5] proposed an optimal alternative solution package with the concept of the largest gray correlation degree, and the package was used to determine the negative ideal solution with a minimal degree of gray relation. The method is simple and effective, and it is also easy to calculate. In 2012, Qiu et al. [6] applied GRA to the verification of simulation models and simulation techniques for modeling, and improved the technique of GRA by considering data curves' geometrical shape. The rationality and effectiveness of GRA have been further verified by case studies. Li et al. [7] proposed that the important achievements by the continuous and diverse values of the gray system theory can be predictable and controllable. Under indeterminate conditions, he applied GRA to typical gray matrix problems and solved the problem of indeterminate and gray zones. The theory of GRA is one of the most mature and most broadly utilized gray system theories. During the analysis, the calculations were conducted on the basis of value comparisons, and in general, the comparison of parameters was also an important index for the analysis. Gray relational analysis supplies a simple way of analyzing a sequence of relationships or behaviors of a system. The analysis has the characteristics of quantitative and sequential analysis, and it can be applied to a random sequence of major and minor factors. This approach can analyze and confirm the factors affecting the target factors or the factors' degrees of improvement. It substantially affects the quantitative analysis of the factors of a system with a trend of dynamic development [8].

Li's work aims to investigate the possibility of using tip nozzles on ducted fans under conditions of large blade pitch angles and high ruggedness. The aerodynamic performance and flow field of the hovering ducted fan are studied numerically at a certain range of blade pitch angles at three operating speeds. Numerical experiments were performed using a shear stress transfer k-ω turbulence model and a fine, high-quality structured grid. The maximum thrust, peak efficiency, and stall margin of a ducted fan with a sharp jet are the main objectives of this study. The results show that under the condition of stall margin, the thrust of the fan with the tip nozzle increased by 30%. The improvement in aerodynamics seems to increase with increasing blade pitch angle because the separation flow at the front of the blade becomes uniform and reattaches to the blade surface due to the entrainment of the tip jet. The nozzles that are angled in the downward flushing direction can increase the nozzle ejection efficiency at larger blade pitch angles. Tip nozzles are suitable for fans with large pitch angles and high ruggedness [9]. Wang research proposed an integrated device called a wind energy fan (WEF), which uses wind energy to directly drive a fan connected to a wind turbine through a drive shaft. This vertical wind turbine can achieve underground ventilation. A test platform was established to test the WEF performance, considering three transmission ratios and two wind turbines with three and five blades. The results show that the transmission ratio has a significant effect on the fan air volume and should be selected to obtain the rated air volume. A wind turbine with three blades is easier to start, and its air volume is 5.43–17.85% higher than a wind turbine with five blades. Based on the aerodynamic characteristics of vertical fans and axial fans, a method of matching power and speed was proposed. This scheme is an effective wind energy technology, which can realize the active utilization of wind energy [10]. Wu used CFD simulations to study the transient characteristics of blade forces in fans with uneven blade spacing. Based on this, a "[T] -h" model for predicting blade forces was developed, and then a prediction based on simulation results and CFD research was developed based on the Lowson model [11]. David evaluated the performance of these underground fan systems in four different deep gold mines in South Africa. Of the six systems, the overall efficiency of the auxiliary fan system was 5%, with an average fan efficiency of 33% of the 33 fans. The results show that these fans deviate significantly from the design operating point. Therefore, current underground fan practices have significant shortcomings. Our detailed studies have concluded that the combination of underground auxiliary fan systems can lead to significant energy inefficiencies. Therefore, maintaining good underground fan operation (such as optimal fan selection, pipe design, and maintenance) is critical to the efficiency of the mine ventilation network [12].

It is clear from the above analysis that no one has yet attempted to apply GRA to fan design. Based on this observation, a new concept of applying GRA to fan design is proposed in this study. After the relationship between parameters of a fan design is determined by GRA, the performance of new fan designs can be improved by the optimization of parameters. To verify the performance improvement, the CFD software, FLUENT, is used to obtain numerical results of the fan's performance, including flow rate and static pressure [1].

2. Model for Investigation

2.1. Development of the Model

Gray relational analysis is utilized in this study to establish the relationship between the indeterminate and gray zones of parameters for fan products. From the results of related methods, the optimal approach for the parameter analysis of a product can be determined by the results obtained. The flow chart of this study is shown in Figure 1. It includes the principles for the calculation of GRA. When using GRA to assess each of the fan parameters, a value is considered valid if it surpasses the threshold value of 0.7, which is recommended.

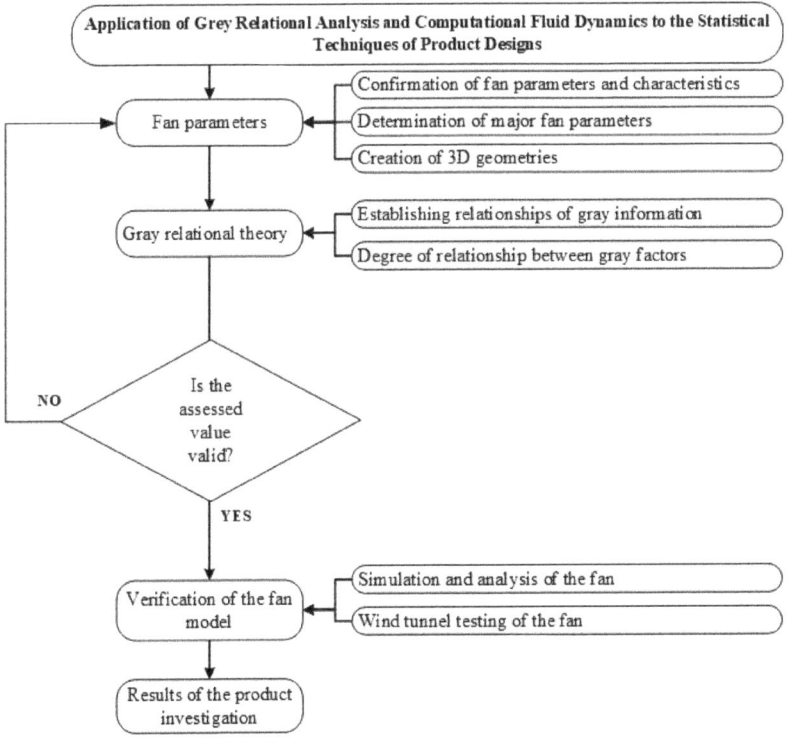

Figure 1. Framework of the development procedure.

For the evaluation of design parameters, it is usually difficult to predict the performance gain due to design optimization without making prototypes for measurement. However, the cost of making prototypes can be huge when the design optimization is based on a large number of design parameters. Therefore, simulation by CFD software is an important tool for a designer to predict the performance indicators of a new fan design, including air-flow rate and static pressure. By comparing these indicators, which are available from CFD simulation, the flow-field characteristics can be captured, and the optimal design can be determined among several candidates. The CFD simulation results are also compared with the experiment results in this study for the validation of this method.

2.2. Fan Model for Investigation

A schematic diagram of a symmetrical dual-impeller fan model in a case study is shown in Figure 2a, in which the initial impeller diameter is 80 mm. The main components, including the impellers, motors, and the base, are shown in Figure 2b, which is an exploded view of the fan model.

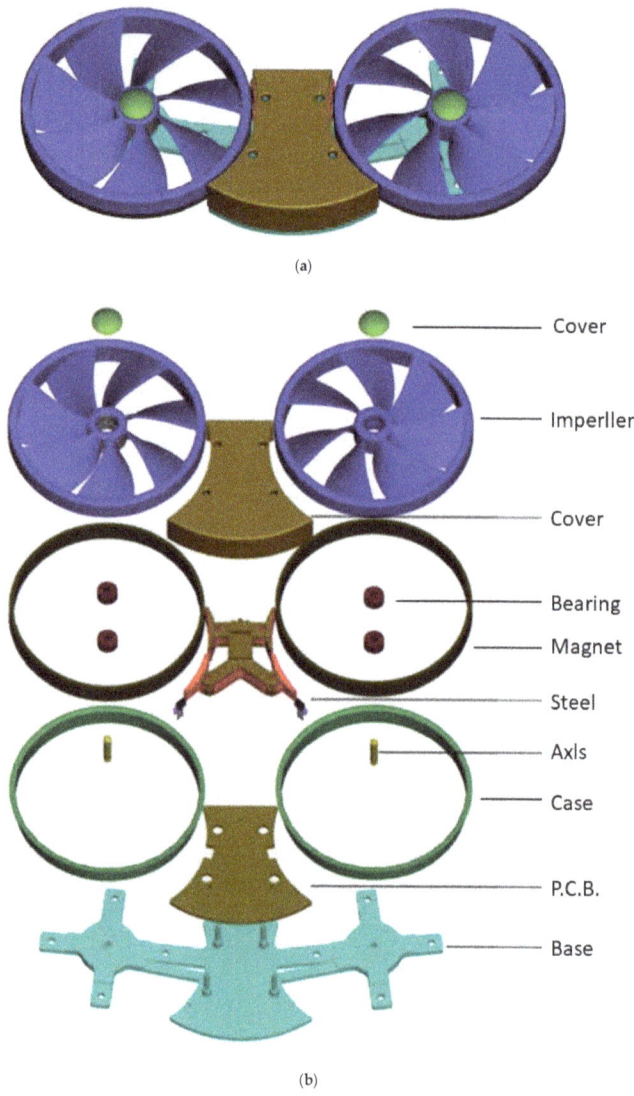

(a)

(b)

Figure 2. Fan parameters: (**a**) assembled dual-impeller fan; (**b**) exploded view of dual-impeller fan.

The operational principle of fans is mostly by means of the rotation of blades causing the pressure difference between the fore and aft ends to happen, driving the rapid flow of the surrounding air. This takes away the heat of the heat-dissipating body and results in a temperature decrease. For a typical design, after the design of a cooling element is shaped, the impedance curve of the element is fixed [13]. Therefore, it is the most often used approach in the typical cooling element design process to change the design of a fan to match the cooling element and enhance the overall cooling efficiency [14,15]. Therefore, it is rather important to find out and know the performance curves of different fans when designing cooling elements [16,17].

2.3. Fan Parameters Affecting the Performance Curve

1. Blade pitch angle: The larger the pitch angle, the larger the pressure difference between the blade's upper and lower surfaces. Under the same rotation speed, the air pressure is also larger with a larger pitch angle. However, when the pressure of the lower surface is too large, the phenomenon of recirculation may occur, and this instead reduces the fan's performance. Therefore, the blade pitch angle should also be increased to a certain extent.
2. Blade spacing: When the distance between the blades is too small, this leads to air-flow disturbance, which increases the friction on the blade surfaces and reduces fan efficiency. When the distance between blades is too large, this leads to an increase of pressure loss and insufficient air pressure [18].
3. The number of blades: This affects other specifications of fan blades, such as the sectional curve and pitch angle. The width of each blade usually depends on its height. To guarantee that blade spacing will not affect the air pressure, the approach of increasing the number of blades is usually adopted as a remedy in comparatively thinner fans.

3. Research Methods

3.1. Gray Relational Theory

Assuming a space in relation to the gray information as

$$\{Q(X), R\} \tag{1}$$

where Q(X) is the factor set in relation to the gray information, and R is the relation of mutual influence. The factor subset $X_0(k)$ is taken as the reference sequence, and X_i (k), $i \neq 0$ is the comparison sequence [8]:

$$X_0 = [x_0(1), x_0(2), \cdots, x_i(k)] \tag{2}$$

$$X_i = [x_i(1), x_i(2), \cdots, x_i(k)], \ i \in I, k \in N \tag{3}$$

The correlation coefficient in relation to the gray information for $X_i(k)$ on $X_0(k)$ is defined as

$$r_i(k) = r[X_0(k), X_i(k)] \tag{4}$$

The correlation degree in relation to the gray information for X_i on X_0 is

$$r(X_0, X_i) = \frac{1}{n}\sum_{k=1}^{n} r[X_0(k), X_i(k)] = \frac{1}{n}\sum_{k=1}^{n} r_i(k) \tag{5}$$

where the quantitative model of the correlation coefficient of gray information relationship for $X_i(k)$ on $X_0(k)$ is defined as

$$r_i(k) = r[X_0(k), X_i(k)] = \frac{\Delta min + \zeta \Delta max}{\Delta_{0,i}(k) + \zeta \Delta max} \tag{6}$$

In the equation, $\Delta_{0,i} = |X_0(k) - X_i(k)|$ is the absolute difference of two comparison sequences, $\Delta\min = \min_{i \in I} \min_k |X_0(k) - X_i(k)|$ is the minimum of the absolute differences of all comparison sequences [19], $\Delta\max = \max_{i \in I} \max_k |X_0(k) - X_i(k)|$ is the maximum of the absolute differences of all comparison sequences, and ζ is the distinguishing coefficient. Its value is adjusted according to the practical demands of the system. Typically, its value is between 0 and 1, and is usually assigned as 0.5.

From the analysis mentioned above, four major equations of GRA and the quantitative model of the correlation degree are employed to establish the analysis model in relation to the gray information. The procedure is as follows.

Step 1: The initialization of the original sequence.

Step 2: Obtain the difference sequence, $\Delta_{0,i} = |X_0(k) - X_i(k)|$.

Step 3: Obtain the minimum of the absolute differences of all comparison sequences $\Delta\min$ and the maximum value $\Delta\max$.

Step 4: Calculate the gray correlation degree $r_i(k)$. The distinguishing coefficient is assigned as 0.5. Substitute the difference sequence, the minimum, and the maximum of the absolute differences into the quantitative model of the correlation degree in relation to the gray information to obtain the gray correlation degree $r_i(k)$.

Step 5: Calculate the correlation degree in relation to the gray information X_i on X_0.

Step 6: Sort the degree of relationship between the major factor and all other factors in the gray system.

3.2. Governing Equations

In three-dimensional Cartesian coordinates, the governing equations are as follows (FLUENT User's Guide) [1,20].

(1) Continuity equation:

$$\frac{\partial u}{\partial x} + \frac{\partial v}{\partial y} + \frac{\partial w}{\partial z} = 0 \tag{7}$$

(2) Momentum equations:

X direction:

$$\frac{\partial u}{\partial t} + \frac{\partial (u^2)}{\partial x} + \frac{\partial (uv)}{\partial y} + \frac{\partial (uw)}{\partial z} = -\frac{1}{\rho}\frac{\partial P}{\partial x} + v\left[\frac{\partial^2 u}{\partial x^2} + \frac{\partial^2 u}{\partial y^2} + \frac{\partial^2 u}{\partial z^2}\right] \tag{8}$$

Y direction:

$$\frac{\partial v}{\partial t} + \frac{\partial (uv)}{\partial x} + \frac{\partial (v^2)}{\partial y} + \frac{\partial (vw)}{\partial z} = -\frac{1}{\rho}\frac{\partial (P - P_0)}{\partial z} + v\left[\frac{\partial^2 v}{\partial x^2} + \frac{\partial^2 v}{\partial y^2} + \frac{\partial^2 v}{\partial z^2}\right] \tag{9}$$

Z direction:

$$\frac{\partial w}{\partial t} + \frac{\partial (uw)}{\partial x} + \frac{\partial (vw)}{\partial y} + \frac{\partial (w^2)}{\partial z} = -\frac{1}{\rho}\frac{\partial P}{\partial z} + v\left[\frac{\partial^2 w}{\partial x^2} + \frac{\partial^2 w}{\partial y^2} + \frac{\partial^2 w}{\partial z^2}\right] \tag{10}$$

(3) Energy equation:

$$\frac{\partial T}{\partial t} + \frac{\partial (uT)}{\partial x} + \frac{\partial (vT)}{\partial y} + \frac{\partial (wT)}{\partial z} = \alpha\left(\frac{\partial^2 T}{\partial x^2} + \frac{\partial^2 T}{\partial y^2} + \frac{\partial^2 T}{\partial z^2}\right) + \frac{q}{\rho C_P} \tag{11}$$

(4) Governing equations can be represented by the general equations as follows:

$$\frac{\partial (\rho\varphi)}{\partial t} + \frac{\partial (\rho\varphi u)}{\partial x} + \frac{\partial (\rho\varphi v)}{\partial y} + \frac{\partial (\rho\varphi w)}{\partial z} = \frac{\partial}{\partial x}\left(\Gamma\frac{\partial\varphi}{\partial x}\right) + \frac{\partial}{\partial y}\left(\Gamma\frac{\partial\varphi}{\partial y}\right) + \frac{\partial}{\partial z}\left(\Gamma\frac{\partial\varphi}{\partial z}\right) + s \tag{12}$$

where $\frac{\partial(\rho\varphi u)}{\partial x} + \frac{\partial(\rho\varphi v)}{\partial y} + \frac{\partial(\rho\varphi w)}{\partial z}$ is the convective term, $\frac{\partial}{\partial x}\left(\Gamma\frac{\partial\varphi}{\partial x}\right) + \frac{\partial}{\partial y}\left(\Gamma\frac{\partial\varphi}{\partial y}\right) + \frac{\partial}{\partial z}\left(\Gamma\frac{\partial\varphi}{\partial z}\right)$ is the diffusive term, S is the source term, and $\frac{\partial(\rho\varphi)}{\partial t}$ is the unsteady term and is not considered when the system is in steady state. Symbol $\varnothing$ represents physical variables such as u, v, w, k, ε, and T (Table 1). The velocity components in the x, y, and z directions are u, v, and w, respectively; Γ is the corresponding diffusivity of each physical variable. Since we are looking for a steady-state solution, the variables are independent of time. Therefore, the partial derivatives of u, v, w, and T with respect to t are equal to zero.

Table 1. Symbols of independent variables.

Continuity	1
X-momentum	u
Y-momentum	v
Z-momentum	w

3.3. Standard k−ε Turbulence Model

Due to its extensive range of applications and reasonable precision, the standard $k-\varepsilon$ model has become one of the main tools that are used for the calculation of turbulent flow fields. The standard $k-\varepsilon$ turbulence model is a type of semi-empirical turbulence mode. Based on the fundamental physical control equations, the model can be used to derive the transport equations for the turbulence kinetic energy (k) and the rate of dissipation of turbulence energy (ε) as follows.

Turbulence kinetic energy equation (k)

$$\frac{\partial}{\partial t}(\rho k) + \frac{\partial}{\partial x_i}(\rho k u_i) = \frac{\partial}{\partial x_j}\left[\left(\mu + \frac{\mu_t}{\sigma_k}\right)\frac{\partial k}{\partial x_j}\right] + G_k + G_b - \rho\varepsilon - Y_M \tag{13}$$

(1) Equation of the rate of dissipation (ε)

$$\frac{\partial}{\partial t}(\rho\varepsilon) + \frac{\partial}{\partial x_i}(\rho\varepsilon u_i) = \frac{\partial}{\partial x_j}\left[\left(\mu + \frac{\mu_t}{\sigma_\varepsilon}\right)\frac{\partial\varepsilon}{\partial x_j}\right] + C_{1s}\frac{\varepsilon}{k}(G_k + C_{3\varepsilon}G_b) - C_{2g}\rho\frac{\varepsilon^2}{k} \tag{14}$$

(2) Coefficient of turbulent viscosity (μ_t)

$$\mu_t = \rho C_\mu \frac{k^2}{\varepsilon} \tag{15}$$

where G_k indicates the turbulence kinetic energy that is generated by the laminar velocity gradient, G_b indicates the turbulence kinetic energy that is generated by buoyancy, Y_M indicates the fluctuation that is generated by the excessive diffusion in compressible turbulent flows, and σ_k and σ_ε are the turbulence Prandtl number of kinetic energy and dissipation, respectively. Further, $C_{1\varepsilon}$, $C_{2\varepsilon}$, and $C_{3\varepsilon}$ are empirical numbers, and their recommended numbers are shown in Table 2.

Table 2. Coefficients of standard k−ε turbulence model.

$C_{1\varepsilon}$	$C_{2\varepsilon}$	C_u	C_k	C_ε
1.44	1.92	0.09	1.0	1.3

The $k-\varepsilon$ model is based on the assumption that the flow field is fully turbulent and the molecular viscosity is negligible. Therefore, better results will be obtained from the calculation of fully turbulent flow fields.

3.4. Performance Testing Equipment for Wind Turbines

The main device of the performance testing equipment for fans is an outlet-chamber wind tunnel that conforms to AMCA 210-99. The principal parts include flow setting means, multiple nuzzles, flow-rate regulating devices, etc. The major function is to supply a good and stable flow field for measurement and acquire the complete performance curves [21].

3.5. Calculation of Flow Rates

Regarding the measured pressure difference between the nozzle outlet and inlet (PL_5 and PL_6), the flow rates on the cross-sections of nozzles shown in Figure 3 can be obtained by the nozzle coefficients. For the calculation of the outlet flow rate of the fan under test, the effect of density variations must be considered.

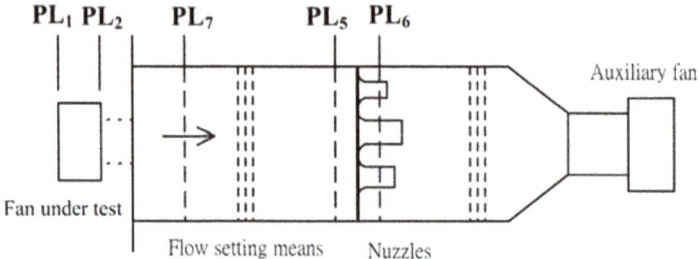

Figure 3. Schematic of measurement planes.

The equation for the calculation of flow rates in a test chamber with multiple nuzzles [22,23] is

$$Q_5 = 265.7Y \sqrt{\Delta P/\rho_5} \sum_n (C_n A_{6n}) \tag{16}$$

where

Q_5	the total flow rate measured by a bank of nozzles, CMM
ΔP	the pressure difference across the nozzles, mm-Aq
ρ_5	the air density upstream of the nozzles, kg/m^3
Y	expansion factor
C_n	the discharge coefficient of the nth nozzle (Nozzle Discharge Coefficient
A_{6n}	the cross-sectional area of the nth nozzle's throat, m^2

3.6. Method of Measurements

(a) Start the measurement from the point of the maximum flow rate (i.e., the point at which the static pressure of a fan is zero). Pay attention to the pressure difference across the nozzles, which should be between 0.5 inch-Aq and 2.5 inch-Aq. If the differential pressure reading is not within this range, this indicates that the flow rate measured for the time being is incorrect. It is required to adjust the nozzle switch to respond to the variations in flow rate accordingly.

(b) After the completion of the data acquisition on the point of maximum flow rate, adjust the pressure to adequate values by means of the shutter of the auxiliary fan and inverter.

(c) Increase the pressure sequentially; the nozzle switch, the shutter of the auxiliary fan, and the inverter must be adjusted during each of the changes. After the system turns stable, then acquire a group of data by the data acquisition system [24,25].

(d) Store 10 sets of data in 10 different files, and use a computer program to calculate the values of air flow rate (Q), pressure (ΔP), and efficiency (η).

(e) Import the calculation results into CAD software to draw the performance curves of the fans

This section expatiates on the procedures of the performance–curve measurement of fans based on the experience acquired after many rounds of measurements.

$$P_s = P_t - P_v \tag{17}$$

$$P_t = P_{t_2} - P_{t_1} \tag{18}$$

where P_s is the static pressure of the fan under test;

P_t is the total pressure of the fan under test;

P_v is the dynamic pressure of the fan under test;

P_{t_2} is the total pressure at the fan's outlet (or plane PL_2);

P_{t_1} is the total pressure at the fan's inlet (or plane PL_1).

Since in this experiment there was no duct at the inlet of the fan under test, therefore $P_{t_1} = 0$ On the other hand, the measured static pressure at the outlet is the same as the static pressures measured at the measuring plane PL_7. Therefore, $P_{s_2} = P_{s_7}$.

$$P_{t_2} = P_{s_7} + P_v \tag{19}$$

$$P_s = P_{s_7} \tag{20}$$

It is concluded from the above equation that the static pressure of the fan under test happens to be equal to the static pressure obtained at the outlet test chamber P_{t_7}. The calculation of dynamic pressures is

$$P_{v_2} = \frac{\rho_2 V_2^2}{19.6} \tag{21}$$

where P_{v_2} is the outlet dynamic pressure of the fan under test, mm-Aq;

V_2 is the outlet air velocity of the fan under test, m/s;

ρ_2 is the outlet air density of the fan under test, kg/m^3;

and $V_2 = \frac{Q_2}{60A_2} = \frac{Q}{60A_2} \cdot \frac{\rho}{\rho_2} = \frac{Q}{50\rho_2A_2}$

where Q_2 is the outlet flow rate of the fan under test, CMM;

Q is the standard flow rate of the fan under test, CMM;

A_2 is the outlet cross-sectional area of the fan under test, m^2;

ρ is the density of air at STP (1.2 kg/m^3).

$$P_t = P_s + P_v = P_s + P_{v_2}.$$

$$P_t = P_s + \frac{\rho_2 V_2^2}{19.6} \tag{22}$$

3.7. Method of Measuring the Performance Curves of Fans

With a fixed amount of power, the flow rate varies inversely proportional to the output air pressure. Since the efficiency of fans changes as the flow rate varies, a non-linear relationship between the flow rate and the air pressure exists, and this forms the performance curve of fans [26]. The measurement process is shown in Figure 4.

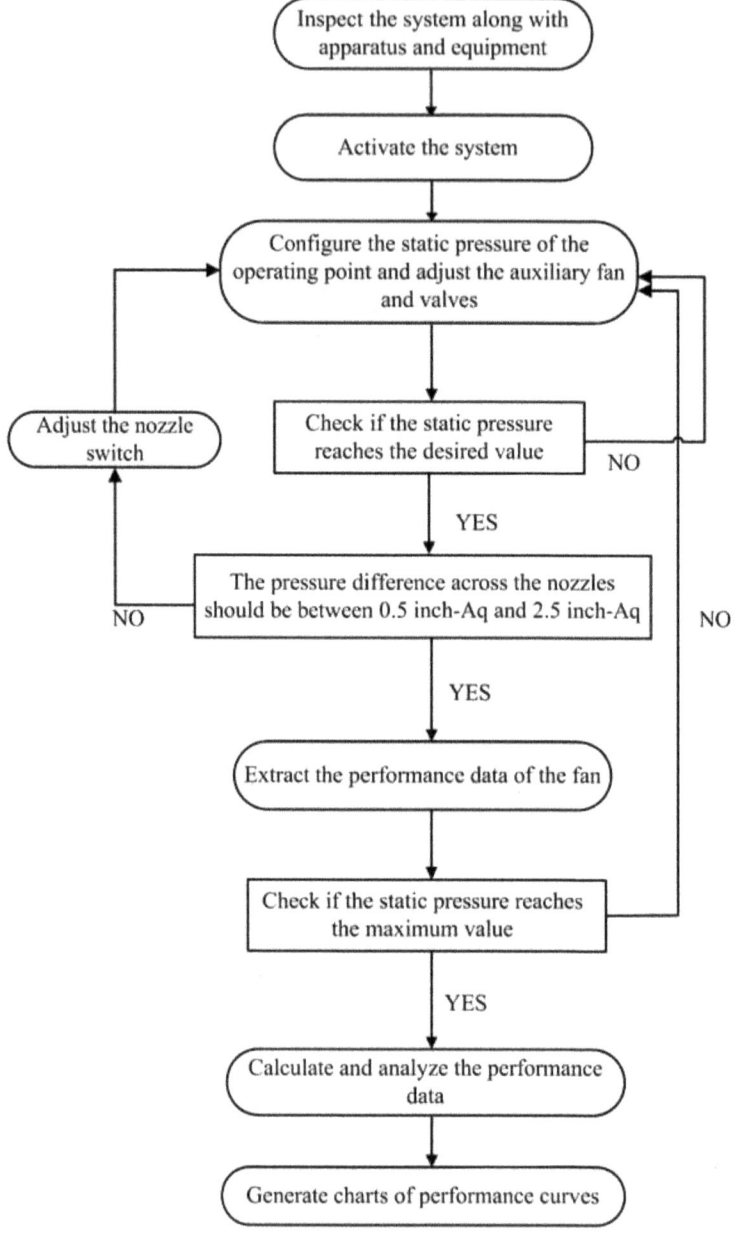

Figure 4. Operational flow chart of fan performance measurements.

3.8. Fan Performance Test Equipment

In terms of performance measurement, the detailed installation and operation of measurement equipment and instruments are described as follows. Regarding the fan performance measurement equipment, the fan performance test body used in this paper uses the AMCA 210-99 standard export wind tunnel, mainly including the main body. The main functions of the rectifier plate, multi-nozzle, and air volume adjustment device are to simulate the air flow conditions downstream of various fans, and to provide a good and stable measurement flow field, so that a complete performance curve can be obtained.

The test platform includes the body, rectifier plate, multiple nozzles, and auxiliary fans (see Figures 4–9) to provide an ideal measurement benchmark; with the air volume adjustment device, it can simulate the outlet of the fan to be tested for various system impedances and even use in free air. The details are as follows:

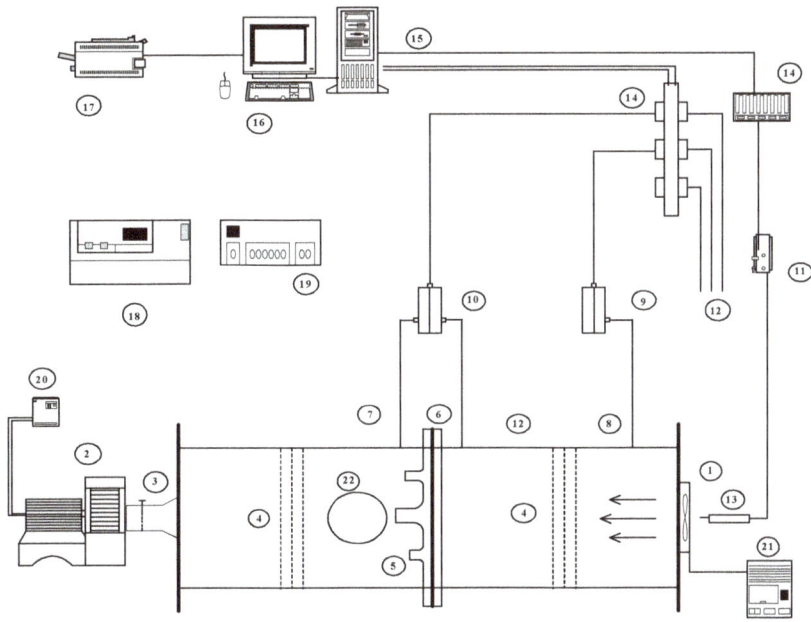

1. Test fan	12. Thermocouple
2. Auxiliary fan	13. Fiber Optic Tachometer
3. Air volume adjustment device	14. Multi-function capture interface card
4. Rectifier	15. Multi-function signal conversion card
5. Multi-nozzle	16. personal computers
6. Static pressure hole in front of nozzle	17. Laser printer
7. Static pressure hole behind nozzle	18. Thermometer and hygrometer
8. Air inlet static pressure hole	19. Barometer
9. No. 1 pressure converter	20. Digital inverter
10. No. 2 pressure converter	21. Power Supplier
11. Optical fiber tachometer adjustment device	22. Hand hole

Figure 5. Main specifications of fan performance test.

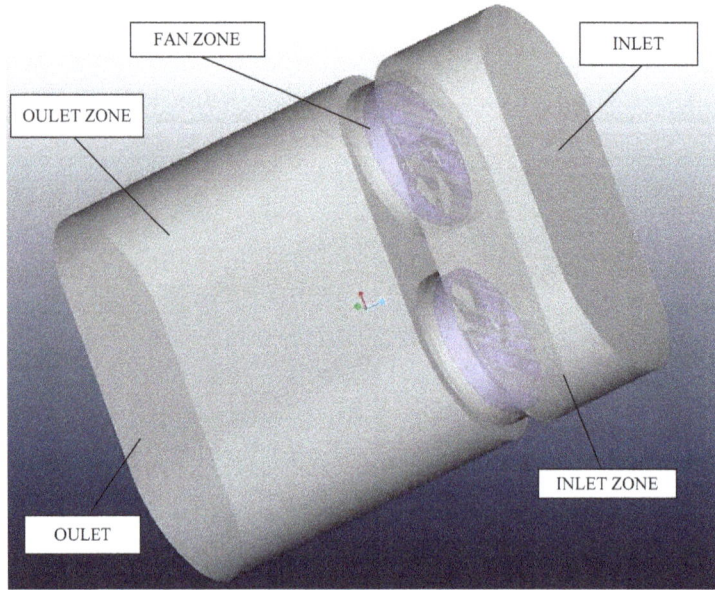

Figure 6. Model of the dual-impeller fan.

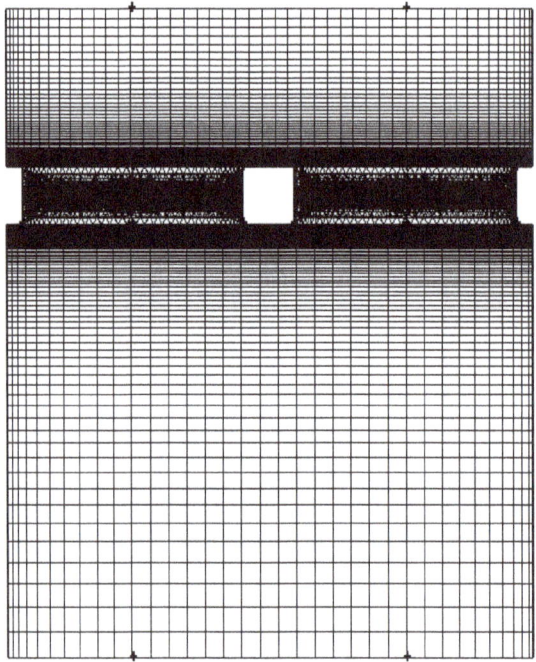

Figure 7. Structure of the numerical model of the dual-impeller fan for the case study.

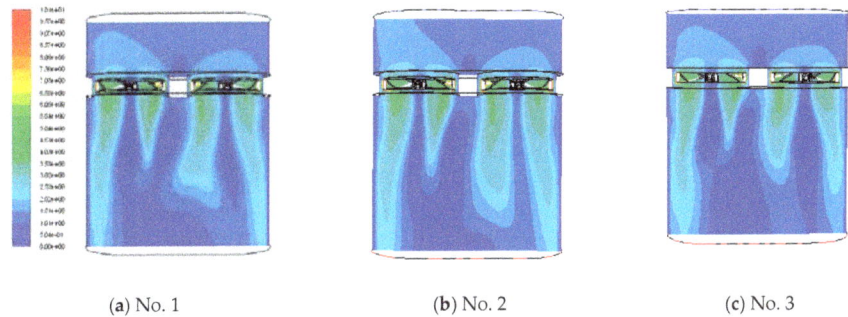

(**a**) No. 1   (**b**) No. 2   (**c**) No. 3

Figure 8. Contours of velocity on the centerline section.

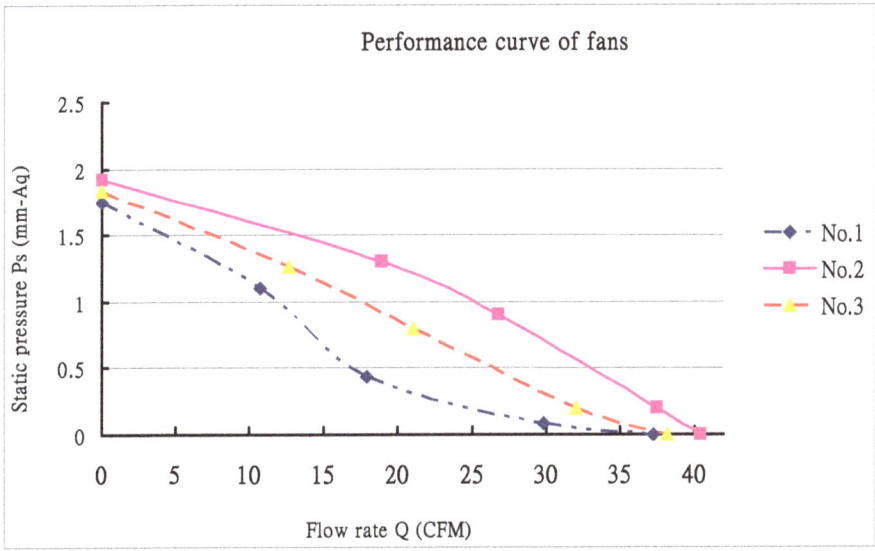

Figure 9. Fan performance curves.

(1) The body

The cross-sectional area of the outlet wind tunnel should be designed to be more than 16 times the maximum measurable area of the air outlet of the axial flow fan (because the test surface required by the axial flow fan is large, the wind tunnel is designed in this way).

(2) Rectifier

There is one set consisting of a front and one back box, with three pieces in each group. The area opening rate should be maintained at 50–60%. It is used to stabilize the fluid flow and ensure the reliability of measurement. Since the measurement of Sections 6 and 7 downstream of the nozzle and the static pressure of the fan are located upstream of the rectifier plate, in order to avoid the design of the rectifier plate affecting the measurement of these two sections, the maximum bounce velocity of the rectifier plate must be maintained at Sections 2 and 6 within 10% of the flow rate. Meanwhile, the measurement section (upstream of the nozzle) is also encountered downstream of the rectifier, so it specifies a local maximum speed of 0.1 M downstream of the rectifier unless the local maximum speed is less than 2 m/s; otherwise, it must not exceed 25% of the average flow rate.

(3) Multi-nozzle

This wind tunnel has seven nozzles with throat diameters of 30, 25, 25, 20, 15, 10, and 5. The nozzles with different diameters can measure different air volumes. The test fans with different nozzles can measure different air volumes. As air flows through the nozzle, a speed boundary layer is formed between the solid surface, and the correction factor is needed when calculating the flow rate. When the fluid velocity is slow, the speed boundary layer is relatively large, and the error is also relatively small when estimating the flow rate. It is large, so the Reynolds number will be set above 12,000 during the measurement; in order to avoid excessive changes in the air properties such as density and temperature, the flow rate will be controlled below Mach number 0.1 during the test. In order to prevent the flow fields between the nozzles from interfering with each other, the position of the nozzles is also clearly specified in the AMCA (Air Movement and Control Association, AMCA) specification, as shown in Figure 5.

4. Case Study

To investigate the influence of various parameters on fan performance, three different fan designs are investigated in this study and their parameters are shown in Table 3.

Table 3. Table of fan parameters.

Fan Design No.	No. 1	No. 2	No. 3
Rotation speed	2000	2000	2000
Leading-edge radius	61	63	65
Blade count	7	9	11
Outside diameter of the fan	70	72	74
Hub incidence angle	1	3	6
Tip incidence angle	0	3	6
Hub stagger angle	1	2	3
Tip stagger angle	0.75	0.85	0.95

The purpose of this step is to find new fan designs with potential performance gains, and those three representative designs as shown in Table 4 are categorized in order to determine the design direction of this study based on the results obtained from GRA.

Table 4. Models of new fan designs.

No. 1	No. 2	No. 3
7 blades	9 blades	11 blades

4.1. Analysis of the Correlation Degree of Gray Information

The procedures of building the analysis model in relation to the gray information are explained sequentially as follows.

Step 1: The initial values of the design parameters for evaluation are shown in Table 5. These values are converted by GRA for initialization, and the results are shown in Table 6.

Table 5. Initial values of design parameters.

Fan Design No.		1	2	3
Rotation speed	X_0	2000	2000	2000
Leading-edge radius	X_1	61	63	65
Blade count	X_2	7	9	11
Outside diameter of the fan	X_3	70	72	74
Hub incidence angle	X_4	1	3	6
Tip incidence angle	X_5	1	3	6
Hub stagger angle	X_6	1	2	3
Tip stagger angle	X_7	0.75	0.85	0.95

Table 6. Initialization of design parameters for gray relational analysis (GRA) $X_0 = [x_0(1), x_0(2), \cdots, x_i(k)]$.

Fan Design No.		1	2	3
Rotation speed	X_0	0.3333	0.3333	0.3333
Leading-edge radius	X_1	0.3228	0.3333	0.3439
Blade count	X_2	0.2593	0.333	0.4074
Outside diameter of the fan	X_3	0.3241	0.333	0.3426
Hub incidence angle	X_4	0.1	0.3	0.6
Tip incidence angle	X_5	0.1	0.3	0.6
Hub stagger angle	X_6	0.1667	0.3333	0.5
Tip stagger angle	X_7	0.2941	0.3333	0.3725

Step 2: Obtain the difference sequence, $\Delta_{0,i} = |X_0(k) - X_i(k)|$, as shown in Table 7.

Table 7. Difference sequence $\Delta_{0,i} = |X_0(k) - X_i(k)|$.

Fan Design No.		1	2	3
Leading-edge radius	X_1	0.0106	0.0000	0.0106
Blade count	X_2	0.0741	0.0000	0.0741
Outside diameter of the fan	X_3	0.0093	0.0000	0.0093
Hub incidence angle	X_4	0.2333	0.0333	0.2667
Tip incidence angle	X_5	0.2333	0.0333	0.2667
Hub stagger angle	X_6	0.1667	0.0000	0.1667
Tip stagger angle	X_7	0.0392	0.0000	0.0392

Step 3: From Table 7, the maximum and minimum values of the difference sequence can be determined as $\Delta\min = \Delta_{1,1}(2) = 0.0000$ and $\Delta\max = \Delta_{4,3}(1) = 0.2667$, respectively.

Step 4: Set the threshold value for gray correlation degrees at 0.5. The gray correlation degrees of various variance factors can be obtained as shown in Table 8.

Table 8. Gray correlation degree $r_i(k)$.

Fan Design No.		1	2	3
Leading-edge radius	X_1	0.9265	1.0000	0.9265
Blade count	X_2	0.6429	1.0000	0.6429
Outside diameter of the fan	X_3	0.9351	1.0000	0.9351
Hub incidence angle	X_4	0.3637	0.8000	0.3334
Tip incidence angle	X_5	0.3637	0.8000	0.3334
Hub stagger angle	X_6	0.4445	1.0000	0.4445
Tip stagger angle	X_7	0.7727	1.0000	0.7727

Step 5: Calculate each variance factor X_i for its average difference in the design parameters X_0 of the correlation degree in relation to the gray information $r(X_0, X_i)$. The resulting correlation degrees in relation to the gray information are shown in Table 9.

Table 9. Correlation degrees in relation to the gray information $r(X_0, X_i)$.

Factor		Correlation Degree
Leading-edge radius	X_1	0.9510
Blade count	X_2	0.7619
Outside diameter of the fan	X_3	0.9567
Hub incidence angle	X_4	0.4990
Tip incidence angle	X_5	0.4990
Hub stagger angle	X_6	0.6297
Tip stagger angle	X_7	0.8485

4.2. Configuration of the Numerical Model

As shown in Figure 6, a numerical model of the dual-impeller fan was built for the case study. The dimensions of the inlet and outlet zones were determined based on the recommended values in order to reflect a real scenario of no impedance to the air flow into the ambient.

4.3. Settings of Model Parameters

A. Settings of boundary conditions

The main consideration of the settings of boundary conditions is to reflect the physical phenomena of the surrounding environment and objects around the target model. It is critical to meet the physical phenomena or else the calculation result of the simulation might be affected. A designer might also be misguided into making a wrong decision. In this case study, the boundary conditions include the inlet boundary condition, outlet boundary condition, and wall boundary condition, which are described as follows.

1. Inlet boundary condition: The inlet condition is for the initial calculation. In order to simulate the condition of a fan in an infinite domain, a normal atmospheric pressure of $P0$ is set at the inlet.

2. Outlet boundary condition: In order to simulation the air flow that is generated by the rotating impellers into the ambient, a normal atmospheric pressure of $P0$ is also set at the outlet.

3. Wall boundary condition: For a fluid flow passing along a wall, it needs to satisfy not only the non-permeable condition but also the no-slip condition.

In addition to the above-mentioned conditions, this case study includes the following assumptions in order to simplify the complexity of the flow field calculation.

1. The flow field is at a steady state and the fluid is non-compressible air.
2. The turbulence model that is used in this case study is k–ε with an eddy correction.
3. The influence of gravitation is neglected.
4. Relevant fluid properties, including the viscosity coefficient, density, and specific heat, are constants.
5. A rotation speed of 2000 RPM is set for the fluid in the rotating zone.
6. The fluid velocity at the surface of a solid is zero, and this is the no-slip condition.
7. The heat radiation term and the buoyancy term are neglected, while physical properties are independent of temperature. This is because when the temperature of fluid is different at different locations, the buoyancy force is generated due to the variation in its density. However, air is driven by fans under forced convection while the natural convective effect is much less effective; therefore, the buoyance term can be neglected. On the other hand, the heat convective term due to the fluid's sensible heat and latent heat is much larger than the heat radiation term, and therefore the radiation term can be neglected.

B. Mesh settings

As shown in Figure 7, the total number of cells is 1,957,013 for the dual impellers and 2,659,498 for the entire system, including the inlet and the outlet. As the mesh for the inlet and the outlet is used for the analysis of the upstream and downstream flow fields and for the boundary conditions, more cells are required at the locations that are closer to the dual impellers in order to simulate the complicated flow field locally. For the domain that is upstream to the dual impellers, the size of a cell is the largest at the inlet. Similarly, the size of a cell is the largest at the outlet for the domain that is downstream to the impellers. This is because no complex geometry exists at either the inlet or the outlet.

4.4. Simulation Results of Fans

The results of numerical simulation make it easy to understand the aerodynamic characteristics and the flow field of fans, which serve as the foundation for further investigation, analysis, and improvement. The contours of pressure, as shown in Figure 8, allow us to better understand the influence of pressure on the entire system in the flow field being analyzed as well as the velocity distribution of the fluid at the centerline section.

Lastly, the one to be compared is the resulting flow rate by numerical calculations. Based on the predicted flow rates of Table 10 by simulation at the outlet, it is known that the flow rate of 40.4 CFM in No. 2 is the maximum, whereas the change of incidence angle still has the effect of increasing the flow rate, but for the phenomenon of recirculation occurring along the upper edge of the impeller and between the blades, no big improvement is observed.

Table 10. Predicted flow rates by simulation.

Fan Design No.	1	2	3
Rotation speed when the maximum flow rate occurs (RPM)	2000	2000	2000
Maximum static pressure Ps (mm-Aq)	1.75	1.92	1.83
Maximum flow rate Q (CFM, Cubic feet per minute)	37.3	40.4	38.2

The weighted averages of the correlation degrees $x_1 \sim x_n$ are determined by the following equation. By applying the weighted averages to the flow rate and the static pressure of each fan design, the resulting values of maximum flow rate and maximum static pressure are shown in Table 11.

Table 11. Weighted averages of the maximum flow rate and the maximum static pressure.

Fan Design No.	1	2	3
Maximum static pressure Ps (mm-Aq)	0.3182	0.3491	0.3327
Maximum flow rate Q (CFM)	0.3218	0.3486	0.3296

In this study, simulation of three kinds of different fan designs designated as No. 1, No. 2, and No. 3 was conducted separately. Verifications of the various results obtained, including flow rates and air pressures, were also conducted by the simulation. With the simulation results obtained, consistency verification was further conducted on these results by the correlation degree of gray information. Observation and comparison were conducted both on the maximum static pressure and the maximum flow rate. It can be found in the simulation results that the maximum flow rate of No. 2 is apparently 9% higher than that of No. 1, whereas the maximum static pressure of No. 2 is also about 8% higher than that of No. 1, as shown in Figure 6.

4.5. Comparison Between the Results of Simulation and Experiment

Method of measuring the performance curve of a fan

The testing of fan characteristics is accomplished on a wind tunnel, as shown in Figure 9. The performance of a fan is usually determined by several operating points instead of a single point of static pressure versus air flow rate, because it is typically not considered as a stable system. Moreover, when a fan operates under a constant input power, the resulting flow rate varies inversely proportional to the output air pressure. In this study, the procedure of measuring fan performance is as follows.

1. Preparatory work for measurements

    A. Turn on the thermometer, hygrometer, barometer, fiber-optic tachometer, and inverter one hour before measurement. Make sure the equipment operates at a stable state. A testing workbench with a wind tunnel is shown in Figure 10a. The fan to be tested is mounted on the front plate of the main chamber. Care should be taken to ensure that the fan is sealed adequately to prevent leakage.

    B. Turn on the test fan and the auxiliary blower for several minutes until both of them run stably. Adjust the blast gate from fully open to fully closed and check the air flow through the chamber. Check the readings of each of the equipment.

    C. Measure the pressure difference between the free-flow condition (free deliver) and the no-flow condition (shut off). Divide the pressure difference into nine segments for determining the pressure increment and the data acquisition points.

(a) Testing workbench with a wind tunnel   (b) Specimen under wind-tunnel testing

Figure 10. Wind-tunnel testing.

2. Measurement procedure

    A. Start the measurement from the free-flow condition with a static pressure of 0. Pay attention to the pressure difference between the nozzle array. The pressure difference needs to be in the range of 0.5–2.5 mm-Aq or the measured air-flow rate could be incorrect. In this case, it is required to select another nozzle from the nozzle array for a different range of air-flow rates.

    B. After the data under the free-flow condition are determined, use the blast gate and the inverter of the auxiliary blower to adjust the pressure to a desired range.

    C. Increase the pressure to the next range of these nine segments by swapping the nozzle and adjusting the blast gate and the inverter. Use the data-acquisition system to take data of system readings after it has been stabilized. Repeat this step for all of these nine segments.

    D. Pull out the data that are recorded in files and calculate the air-flow rate, air pressure, and efficiency of the computer program.

    E. Summarize the calculation results in the performance curve of the fan.

The performance improvement that is predicted by the numerical simulation is further compared to the result that is obtained from the wind-tunnel testing, as shown in Table 12. Both the difference in the pressure drop and that in the air-flow rate are within 5%, which indicates a small difference between the simulation and the experiment results.

Table 12. Comparison between results from the simulation and the experiment.

	Results from the Simulation	Results from the Experiment
Rotation speed at the maximum flow rate (RPM)		2000
Maximum static pressure Ps (mm-Aq)	1.92	1.75
Maximum flow rate Q (CFM)	40.4	38.3

5. Results and Discussion

In this study, the gray design GRA was used to determine the important design parameters for improving the design performance of the fan with the best solution for symmetry or asymmetry. Based on the results obtained by GRA, the priority of design parameters for improving performance was determined, and GRA stated that the second design can provide better performance than the other two designs. The effects of these design parameters are further studied through numerical simulations and experiments. The simulation results also showed that the static pressure of the No. 2 design was 1.92 mm-Aq, and its flow rate was 40.4 CFM. Compared to the other two designs, it was obvious that performance-wise, No. 2 was the best of the three fan designs. By taking the weighted averages of the correlation degrees for the design parameters, the resulting maximum static pressures were No. 1: 0.3182, No. 2: 0.3491, and No. 3: 0.3327. Moreover, the maximum flow rates were No. 1: 0.3218, No. 2: 0.3486, and No. 3: 0.3296. It is clear that the maximum flow rate of No. 2 was the largest among these three designs. The most important design parameters can be determined by GRA at an earlier stage of fan design.

The results of the case study indicated that among fan parameters, the one with the greatest influence was the leading-edge radius. The outside diameter is another design parameter that shows a higher correlation degree. However, it is already known that an oversized fan could cause a stall, which leads to the phenomenon of rapidly deteriorating fan performance. Moreover, in space-constrained applications, the outside diameter is typically not selected as one of the design variables, because increasing the outside diameter leads to a bigger fan, which might fail to fit into the available space.

Author Contributions: The author contributed to the paper. H.-H.L. collected and organized the data and acts as the corresponding author, J.-H.C. and C.-H.C. proposed the methods. All authors have read and agreed to the published version of the manuscript.

Funding: This work was supported by the Ministry of Science and Technology of the Republic of China under grant MOST-108-2221-E-468-003.

Conflicts of Interest: The author declares no conflict of interest.

References

1. Bredell, J.R.; KrÃÃger, D.G.; Thiart, G.D. Numerical investigation of fan performance in a forced draft air-cooled steam condenser. *Appl. Therm. Eng.* **2006**, *26*, 846–852. [CrossRef]
2. Hurault, J.; Kouidri, S.; Bakir, F.; Rey, R. Experimental and numerical study of the sweep effect on three-dimensional flow downstream of axial flow fans. *Flow Meas. Instrum.* **2010**, *21*, 155–165. [CrossRef]
3. Rhee, D.H.; Cho, H.H. Effect of vane/blade relative position on heat transfer characteristics in a stationary turbine blade: Part 2. Blade surface. *Int. J. Therm. Sci.* **2008**, *47*, 1544–1554. [CrossRef]
4. Lai, H.H.; Chen, C.H.; Chen, Y.C.; Yeh, J.W.; Lai, C.F. Product design evaluation model of child car seat using gray relational analysis. *Adv. Eng. Inform.* **2009**, *23*, 165–173. [CrossRef]
5. Wei, G.W. Grey relational analysis method for 2-tuple linguistic multiple attribute group decision making with incomplete weight information. *Expert Syst. Appl.* **2011**, *38*, 4824–4828. [CrossRef]
6. Qiu, B.; Wang, F.; Li, Y.; Zuo, W. Research on Method of Simulation Model Validation Based on Improved Grey Relational Analysis. *Phys. Procedia* **2012**, *25*, 1118–1125. [CrossRef]
7. Li, Q.X. Grey dynamic input–output analysis. *J. Math. Anal. Appl.* **2009**, *359*, 514–526. [CrossRef]

8. Trivedi, H.V.; Singh, J.K. Application of Grey System Theory in the Development of a Runoff Prediction Model. *Biosyst. Eng.* **2005**, *92*, 521–526. [CrossRef]
9. Li, L.; Huang, G.; Chen, J. Aerodynamic characteristics of a tip-jet fan with a large blade pitch angle. *Aerosp. Sci. Technol.* **2019**, *91*, 49–58. [CrossRef]
10. Wang, Z.; Wu, Y.; Lu, S.; Meng, X.; Zhang, J. A study on model experiment and aerodynamic match of Wind Energy Fan. *Sustain. Cities Soc.* **2019**, *49*, 101618. [CrossRef]
11. Wu, Y.; Pan, D.; Peng, Z.; Hua, O. Blade force model for calculating the axial noise of fans with unevenly spaced blades. *Appl. Acoust.* **2019**, *146*, 429–436. [CrossRef]
12. Villiers, D.J.; Mathews, M.J.; Maré, P.; Kleingeld, M.; Arndt, D. Evaluating the impact of auxiliary fan practices on localised subsurface ventilation. *Int. J. Min. Sci. Technol.* **2019**, *29*, 933–941. [CrossRef]
13. Tournier, J.M.; El-Genk, M.S. Axial flow, multi-stage turbine and compressor models. *Energy Convers. Manag.* **2009**, *51*, 16–29. [CrossRef]
14. Qian, X.; Deba, D. Design of heterogeneous turbine blade. *Comput. Aided Des.* **2003**, *35*, 319–329. [CrossRef]
15. Lin, H.H. Application of Fuzzy Decision Model to the Design of a Pillbox for Medical Treatment of Chronic Diseases. *Appl. Sci.* **2019**, *9*, 4909. [CrossRef]
16. Lin, S.C.; Huang, C.L. An integrated experimental and numerical study of forward–curved centrifugal fan. *Exp. Therm. Fluid Sci.* **2002**, *26*, 421–434. [CrossRef]
17. Li, Y.L.; Liu, J.; Ou, H.; Du, Z.H. Internal flow mechanism and experimental research of low pressure axial fan with forward-skewed blades. *J. Hydrodyn. Ser. B* **2008**, *20*, 299–305. [CrossRef]
18. Niu, M.; Zang, S. Experimental and numerical investigations of tip injection on tip clearance flow in an axial turbine cascade. *Exp. Therm. Fluid Sci.* **2011**, *35*, 1214–1222. [CrossRef]
19. Yao, C. Application of Gray Relational Analysis Method in Comprehensive Evaluation on the Customer Satisfaction of Automobile 4S Enterprises. *Phys. Procedia* **2012**, *33*, 1184–1189.
20. FLUENT User's Guide. Available online: https://www.ansys.com/products/fluids/ansys-fluent (accessed on 1 January 2019).
21. Lin, H.H.; Huang, Y.Y. Application of ergonomics to the design of suction fans. In Proceedings of the 1st IEEE International Conference on Knowledge Innovation and Invention, Jeju Island, South Korea, 23–27 July 2018; pp. 203–206.
22. Hsiao, S.W.; Lin, H.H.; Lo, C.H.; Ko, Y.C. Automobile shape formation and simulation by a computer-aided systematic method. *Concurr. Eng. Res. Appl.* **2016**, *24*, 290–301. [CrossRef]
23. Lin, H.H. Improvement of Human Thermal Comfort by Optimizing the Airflow Induced by a Ceiling Fan. *Sustainability* **2019**, *11*, 3370. [CrossRef]
24. Lin, H.H.; Hsiao, S.W. A Study of the Evaluation of Products by Industrial Design Students. *Eurasia J. Math. Sci. Technol. Educ.* **2018**, *14*, 239–254. [CrossRef]
25. Hsiao, S.W.; Lin, H.H.; Lo, C.H. A study of thermal comfort enhancement by the optimization of airflow induced by a ceiling fan. *J. Interdiscip. Math.* **2016**, *19*, 859–891. [CrossRef]
26. Lin, H.H.; Cheng, J.H. Application of the Symmetric Model to the Design Optimization of Fan Outlet Grills. *Symmetry* **2019**, *11*, 959. [CrossRef]

Article

Applying Educational Data Mining to Explore Students' Learning Patterns in the Flipped Learning Approach for Coding Education

Hui-Chun Hung [1], I-Fan Liu [2], Che-Tien Liang [3] and Yu-Sheng Su [4,*]

[1] Graduate Institute of Network Learning Technology, National Central University, Taoyuan City 320, Taiwan; hch@cl.ncu.edu.tw
[2] Center for General Education, Taipei Medical University, Taipei City 110, Taiwan; ifanliu@tmu.edu.tw
[3] Graduate Institute of Data Science, Taipei Medical University, Taipei City 110, Taiwan; m946105010@tmu.edu.tw
[4] Department of Computer Science and Engineering, National Taiwan Ocean University, Keelung City 202, Taiwan
*   Correspondence: ntoucsiesu@mail.ntou.edu.tw

Received: 31 December 2019; Accepted: 28 January 2020; Published: 2 February 2020

Abstract: From traditional face-to-face courses, asynchronous distance learning, synchronous live learning, to even blended learning approaches, the learning approach can be more learner-centralized, enabling students to learn anytime and anywhere. In this study, we applied educational data mining to explore the learning behaviors in data generated by students in a blended learning course. The experimental data were collected from two classes of Python programming related courses for first-year students in a university in northern Taiwan. During the semester, high-risk learners could be predicted accurately by data generated from the blended educational environment. The f1-score of the random forest model was 0.83, which was higher than the f1-score of logistic regression and decision tree. The model built in this study could be extrapolated to other courses to predict students' learning performance, where the F1-score was 0.77. Furthermore, we used machine learning and symmetry-based learning algorithms to explore learning behaviors. By using the hierarchical clustering heat map, this study could define the students' learning patterns including the positive interactive group, stable learning group, positive teaching material group, and negative learning group. These groups also corresponded with the student conscious questionnaire. With the results of this research, teachers can use the mid-term forecasting system to find high-risk groups during the semester and remedy their learning behaviors in the future.

Keywords: blended learning; learning behaviors; learning performance; machine learning; online programming course

---

1. Introduction

With the development of the Internet, emerging forms of distance education can eliminate the geographical and temporal separation between two learners, and the knowledge can be transmitted to all corners of the world through the teaching environment of the online platform. Furthermore, another significant advantage of the Internet is that the teaching will be transformed from teacher-centered to learner-centered [1]. Distance learning enables learners to more flexibly manage their time and progress, and choose the time and place to learn. Therefore, it also improves the shortcomings of the traditional educational environment such as a lack of flexibility, limited delivery distance, and inability to repeat learning [2].

However, traditional asynchronized distance teaching has its disadvantages. For example, the learner's problem can be answered by message or by mail, however, it takes a longer time when compared to the face-to-face environment where learners can ask questions and obtain answers immediately, as the teacher cannot grasp it instantly. With the advancement of technology, the synchronous distance learning (live) environment has become another choice for the teaching environment. The advantages of the live broadcast environment include providing innovative learning models, motivating learners, providing formal multi-learning materials, and making the learner reflect [3]. In the live broadcast environment, learners can ask questions to the instructor promptly under the live broadcast, and the instructor can respond promptly. Compared with asynchronous learning, students can ask questions more freely. Discussing the teaching content of the teacher with other students does not affect the learning quality of other students, and the students can respond to the teaching content of the teacher so that the instructor can immediately see whether the teaching content is correctly transmitted. Immediate response is unattainable in traditional face-to-face and non-synchronous distance learning [4].

Additionally, a chat box in the learning management system can become a conduit for communication between learners. The live environment breaks this gap, and learners can instantly exchange ideas and explain questions with others in the chat box. This is followed by the possibility of learners chatting with each other [5]. The advantages and disadvantages of synchronous and asynchronous learning are different, so the blended learning environment has become one of the choices of today's learning approaches. The blended learning environment proposed by this study integrates traditional face-to-face courses, asynchronous, and synchronized online learning. Therefore, it can provide students with the most flexible learning environment. In the literature review of blended learning, there have been few studies on the learning approach integrated with Facebook live. Therefore, this study hopes to explore the learner's learning experience and learning achievement with the use of educational data mining through traditional distance education, face-to-face teaching, and learning via Facebook live.

This study aims to use machine learning and symmetry-based learning algorithms to explore the relationship between the data generated by the learning process in a blended learning environment and learning achievement. The research questions in this study are as follows:

1. In the blended learning environment, can we use the data generated in the learning process to forecast learners' performances?
2. Can we apply the generated model to predict the data of the other class?
3. Can we find a specific learning model from the learner's learning behavior? How can the learning group be defined and which variables should they be based on?

To solve these research questions, this paper explored which variables were related to the learner's learning performance in a mixed-education environment of mixed face-to-face courses. This study collected the learning records generated by students in a blended learning environment such as the degree of synchronous and asynchronous participation, the submission of assignments, and the discussion in the online forum. Through educational data exploration, predictions can be made in the interim period so that the instructors are in the second half of the semester. Personalized guidance could then be given to high-risk students. Therefore, this study is able to predict learners' learning performance and provide personalized guidance or reminders for high-risk learners to enhance the efficiency and effectiveness of future teaching.

2. Literature Review

2.1. Blending Learning Environment

In the last few decades, information and communication technology has revolutionized the processes of learning. The blending learning environment can be defined as the combination of traditional face-to-face courses and online learning environments that can complement each other's shortcomings [2]. The implementation is complicated and challenging as the proportion of face-to-face and online learning will lead to an unlimited number of combinations [6]. The definition and classification of the blended learning environment mainly include two aspects: one is the transmission method (offline and online), and the other is the learning method (educator-oriented and learner center) [7]. Therefore, from these two aspects, there are four possible blended learning environments: (1) Mostly face-to-face courses and significant online interactions; (2) mostly online courses and offline group discussions; (3) mostly face-to-face courses and online resources provided; and (4) mostly online courses and optional face-to-face discussions. The blended learning environment of this study was similar to the fourth type. Most of the online courses and online assignments, discussions, quizzes, and live online courses were synchronized, while the remaining few face-to-face courses were mainly for the start of course introduction, environment building, follow-up question discussions, and examinations. Furthermore, Francis, and Raftery [8] also proposed three digital learning models using hybrid education: (1) Basic course management and helping learners; (2) Hybrid learning brings significant improvements in the teaching and learning process; and (3) The first two modes achieve personalized guidance through multiple online courses and modules. Based on the above three stages, it is recommended that the academic management staff of the university should have sufficient awareness of the strategies, structure, and support of each stage in order to improve the higher education and learning environment [6]. The course in this study was the blended learning environment of online courses, live online interactions, and face-to-face courses. To bring about significant improvement in learning to learners, this research established a mid-term prediction model, so that teachers could find high-risk groups of learners in the mid-term, and provide personalized guidance and reminders to them, in order to achieve the process of improving the learning effectiveness.

Moreover, thematic research on the blended learning environment has pointed out that 41% of the research in the past decade has raised questions about education design including education models, strategies, best practices, learning environments, and curriculum [9,10]. Sikder, Herold, Meinel, and Lorenzen-Zabel [10] combined theoretical knowledge and practice-oriented education to present e-Learning platforms, which included lectures, tests, and practical exercises aside from short teasers and technical tutorials as the major learning modules components. Keržič, Tomaževič, Aristovnik, and Umek [11] explored the critical factors of blended learning for higher education students and indicated that e-learning was positively perceived when the teacher was engaged in an e-course and students' attitude to the subject had a direct impact.

2.2. Educational Data Mining

In the field of educational data mining, predicting the performance of learners is one of the most practical applications. According to the definition of the Educational Data Mining Community website [12], educational data mining is "a rapidly emerging discipline that focuses on developing methods that can explore specific information in the educational environment and uses its methods to gain a deeper understanding for students' learning performance and set the goals for them." In addition, many leading experts in educational data exploration divide educational data exploration into the following sections: statistics and visualization, prediction (classification, regression, and density estimation), clustering, correlation analysis, outlier detection, and semantic analysis [5,12–14]. The goal is to understand learners' learning behaviors and predict their knowledge absorption [14]. However, predicting learner performance is not easy, and a large number of factors or personal characteristics may affect learner performance. Factor characteristics include the learner's background, past learning performance, and interactions between learners and educators [15]. When predicting learner performance, the method used will vary depending on the predictive variables [16]. The application of educational data mining in student learning performance is improving the learning process and guiding learners to learn, providing feedback suggestions based on learner learning behavior, evaluating learning materials and course equipment, early detection of abnormal learning behaviors and problems, and overall a deeper understanding of the learning environment [14,17].

In recent years, the application of predicting the performance of learners is mostly in higher education [18,19]. The main reasons are the popularity of learning management systems (LMS) such as Moodle, Claroline, and Blackboard. The reason why such a learning management system can be quickly popularized is mainly because it can effectively, flexibly, and merely manage the experience of online courses. In addition, the learning management system can accumulate a large amount of information including the number of times the learner visits the webpage, the time, the time and number of viewing the resources, the status and performance of the assignment, and even the interaction record with others in the chat room and discussion area. Therefore, this kind of information is crucial for analyzing the learner's behavior and predicting the learner's performance. This allows the teacher to find any inappropriate parts in the course, or the deficiency to improve, so that future teaching will be better suited [17,20].

In supervised learning, random forest (RF) is one of the statistical learning theories, and the approach is applied to make predictions with multiple decision trees and uses voting to obtain the final prediction results. To effectively train the random forest model, the number of trees in the random forest needs to be reduced [21]. Scholars currently compare various decision trees and random forest algorithms for performance predictions. Random forests have been proven to demonstrate the best possible performances when all of the features are included in the model [22].

In unsupervised learning, the clustering approach is a basic exploratory tool in data mining. The clustering approach attempts to classify data when the actual group membership classification is not known. There are also cases in which the clustering approach is applied to educational data mining such as while using the clustering and sequential approach to simulate learner behavior patterns in games [23].

With the learning management system, the data collection of learners becomes more and more convenient, but the information is more complicated. Therefore, it is difficult to analyze the current learning behaviors using traditional research methods. This study applies classification, grouping, data visualization, and other educational data exploration methods to analyze the learning behavior of complex learners, and hopes to explore the variables that affect the learners' learning performance and thus help the overall learning efficiency.

2.3. Visual Analysis

Data visualization is an emerging field that aims to address a growing database of scale and complexity. The visualization of data developed from the fields of statistics, probability, and data presentation is to understand the large datasets that exist in the database. Moreover, data visualization techniques are mathematical tools that aggregate large datasets into a single representation or numerical value. Such models include time-series graphs, heat maps, etc. [24]. In the era when computers were still not widespread, there was already data visualization [25], like the weather map of Francis-Galton in the 1980s. There are many complicated technologies used in data nowadays. Data visualization is mainly used in business and science. Unlike data mining, data visualization usually deals with raw materials such as numbers or letters [26], which makes the process of visualizing data consume a lot of computing energy and time. Large database management systems often encounter such problems.

The use of data visualization is part of the critical trend of educational data mining [27]. Researchers have pointed out the representative power of data visualization. Data visualization is not a neutral presentation, but magnifies the meaning or persuasiveness of data so that it can be used to generate discourse and opinions. Additionally, it can persuade others to make the same belief in others' opinions [27]. Therefore, Beer [28] indicates that researchers need to examine the process of data visualization in detail so that everyone can take these visual effects seriously.

3. Method

3.1. Participants

In this study, we present an exploratory study to conduct practical teaching experimental research in universities. This study plans to conduct empirical evidence-based research on a compulsory course. The study was conducted in the Python programming course at a university in northern Taiwan (from now on referred to as Class A). This course is a general education course with a total of 38 students that come from many various departments. To protect the personal information of students, the data was coded. This course utilizes a the blended learning education approach that includes a weekly online lecture available on the learning management system. Students are required to watch instructional videos. In addition, there are face-to-face courses, which allow students to interact with their peers in person, the online quiz, test, and the Facebook live class twice in the semester. To extrapolate our research model, this study also collected the data from another Python course in the same university (hereinafter referred to as Class B). The total number of students in this course was 34. Its weekly learning approach was the same as Class A.

3.2. Data Collection

The collection of course materials analyzed in this study included (1) the students' asynchronous online learning behavior; (2) the students' synchronous online learning behavior; and (3) the students' self-evaluation.

(1)   The source of the students' asynchronous online learning behavior was obtained from the learning management system log file.

The information included the student's necessary information (student number, department, name), the results of 11 regular homework assignments, the order of payment, the time of payment, whether to submit late, the results of the final report, three times the average test scores, and the usual class interaction scores. In this study, the data were collected from the learning management system including the total score of the semester, the grades of the students, the 11 assignment scores, the number of unsubmitted assignments, and the number of delayed submitted assignments. There were 38 rows of data, each with 80 fields. The teaching log file showed what teaching content each student clicked at any time, so the variables were courseID, userID, click content ID, logTime, exitTime, and account. There were a total of 4575 items (clicks).

(2)   The source of the students' synchronous learning behavior was from the Facebook live platform.

Using the Facebook Graph API (Application Programming Interface) to obtain the platform information as Creat_time (message generation time), live_broadcast_timestamp (message generated at the time of the broadcast), message (message content), and NAME (message publisher). There were two live sessions with 275 responses for the first session and 125 responses for the second session. Therefore, there were a total of 400 articles and four fields each after the merger.

(3)   The source of the students' course evaluation was from the learner questionnaire.

This questionnaire was submitted to the students after the semester finished. There were four major themes, namely, personal background, teaching platform and curriculum design planning, actual platform usage, and open questions.

3.3. Research Tools

The tools used in this research environment included the learning management system (My2TMU), Facebook live platform, Facebook Graph API, as follows:

(1)   Learning Management System

The Learning Management System collects the learner's course of study in this course for one semester including basic materials, weekly work assignments, quiz results, and click on the content on the platform.

(2)   Facebook Live Platform, Facebook Graph API

Using the Facebook live platform provides an environment for learners to do distance learning, as learners can enjoy the advantages of distance learning and the advantages of instant interaction, so that learners can use the communication device to learn and interact with the teacher at any place, and also interact with their peers.

The Facebook Graph API is the primary method for applications to read and write Facebook social relationships by using Python to connect to the Facebook Graph API to obtain the information generated by learners on Facebook live and organize them appropriately. The version used in this study was v2.12. This study used Facebook graph API to obtain live platform information including creat_time (message generated time), live_broadcast_timestamp (message generated time), message (message content), and NAME (message publisher).

(3)   Final learner questionnaire

At the end of the semester, the student was asked to fill out a questionnaire, which was an online test with a total of 23 questions.

3.4. Data Analysis

After the environment was built, Python was used to analyze the data in this study. The main version of Python used in this study was 3.6, with the following kits: Pandas, the Numpy suite for data collation, Matplotlib, Seaborn, the SciPy kit for data visualization, and the Scikit-learn kit for data analysis. This study used Python to obtain the learner's answer data during Facebook live teaching and to analyze the data including machine learning methods such as classification and clustering. The original data were preprocessed and standardized as Z-scores , which had a mean of zero and a standard deviation of 1.

In supervised learning, we explored the learner's learning outcomes at the end of the term, and the target variable was whether the final grade was passed or not. Three algorithms were used, namely logistic regression, decision trees, and random forests. This study first applied these three models to make mid-term predictions and find the best one. After that, to evaluate the model, this study applied the best model of the three models to another class.

In unsupervised learning, we used the clustering approach to find learners with different learning behaviors. This study applied Euclidean distance (European distance) and the hierarchical grouping method to generate the tree diagram. Euclidean distance is the most common distance measure and is suitable to measure the distance of individuals in space. The hierarchical grouping method is a hierarchical structure, which repeatedly splits the data or aggregation, and finally becomes a tree structure.

4. Results

In this study, the curriculum with flipped learning for the programming course was proposed. Moreover, an 18-week Python programming online course was designed in the general education curriculum.

4.1. Mid-Term Forecast

This study applied the following machine learning classification methods: (1) logistic regression, (2) decision tree, and (3) random forest to make mid-term predictions and applied the best model to another class. Among the three models that evaluate the y variable scores, the f1-score of the random forest model was 0.83, which was higher than the f1-score of the logistic-regression and decision trees, as shown in Table 1. Therefore, the best model in this study was the random forest model.

Table 1. Model evaluation and comparison between the three models.

Data Label	Logistic Regression			Decision Tree			Random Forest		
	Precision	Recall	F1-Score	Precision	Recall	F1-Score	Precision	Recall	F1-Score
0/Fail	0.33	0.50	0.40	0.33	0.50	0.40	0.53	0.50	0.50
1/Pass	0.89	0.80	0.84	0.89	0.80	0.84	0.90	0.90	0.90
Avg/Total	0.80	0.75	0.77	0.80	0.75	0.77	0.83	0.83	0.83

To evaluate the predicting model trained by Class A, we further applied the model to Class B. The time of the B class data was also processed the same as Class A. This study further used this model to predict Class B. The results indicate that the F1-score was 0.77, as shown in Table 2. Therefore, this model also had a successful interim prediction for Class B.

Table 2. Model evaluation for Class B.

Data Label	Precision	Recall	F1-Score
0/Fail	0.33	0.50	0.40
1/Pass	0.89	0.80	0.84
Avg/Total	0.80	0.75	0.77

4.2. Learning Behavior Grouping

To further explore the different learning behaviors, this study applied hierarchical clustering to measure the distance of individuals in space. Moreover, to have a closer understanding of the tree structure, the tree diagram was added to the group heat map to generate a hierarchical clustering heat map, as shown in Figure 1. The closer the distance, the smaller the difference between the individuals.

The color depth of the heat map represents the original value. Since the data were standardized, the maximum value was about 4.5, and the minimum value was about −1.5. From the hierarchical structure on the left, we can see the learner's grouping and understand which learners' learning behaviors were similar. Moreover, the relationship between the variables can be seen from the hierarchical structure above; so we can understand the degree of association between the variables. The correlation between the variables is one of the links that the instructor wants to explore. Therefore, the variable hierarchy above the hierarchical group heat map is represented separately from the disguised name below, as shown in Figure 2.

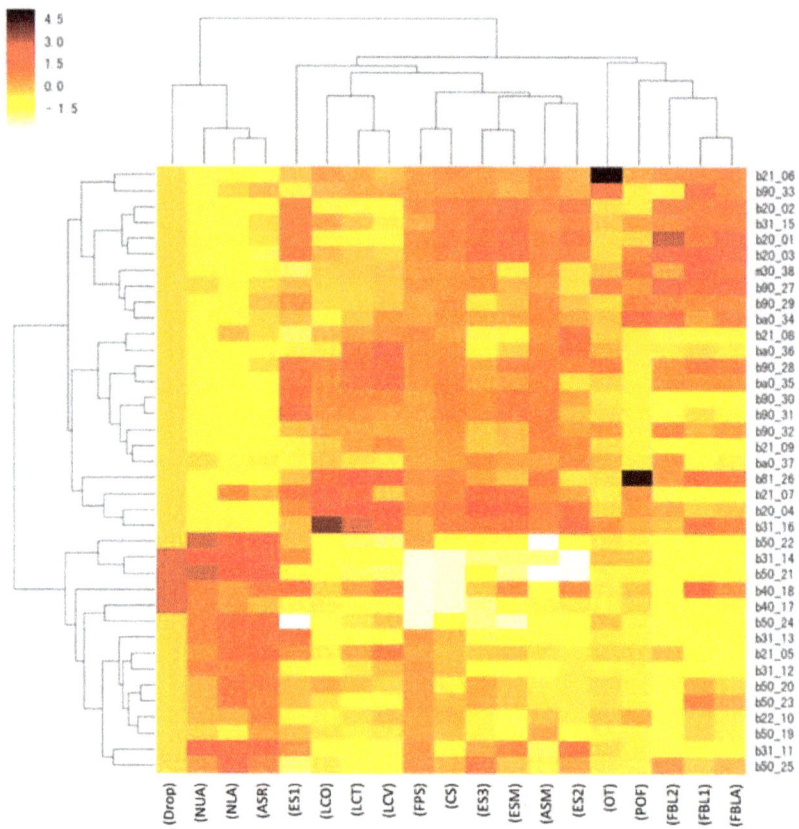

Figure 1. Hierarchical group heat map.

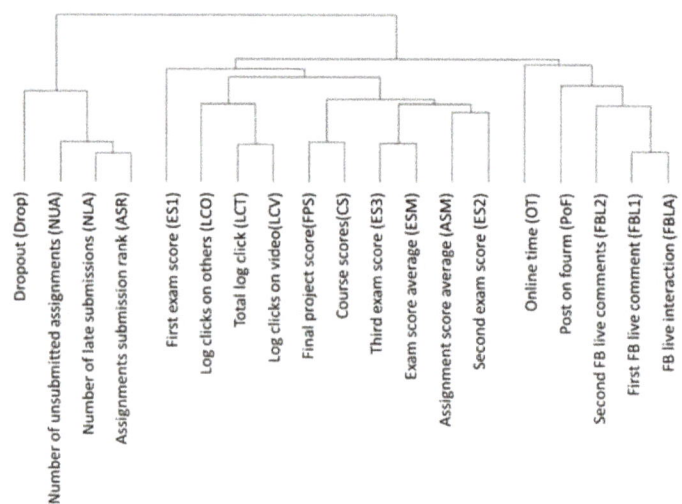

Figure 2. Hierarchical group heat map variable group.

In Figure 2, there are four variables on the left: Dropout (Drop), the number of unsubmitted assignments (NUA), the number of late submissions (NLA), and the assignment submission rank (ASR). In this part, the higher these features are, the more negative the students are. The features in the middle section of Figure 2 are the behaviors on the learning management server including the first exam score (ES1), the log clicks on others (LCO), the total log click (LCT), the log clicks on video (LCV), final project score (FPS), course scores (CS), the third exam score (ES3), the exam score average (ESM), the assignment score average (ASM), and the second exam score (ES2). Most of the features in the right section of Figure 2 are the interaction of learners including online time (OT), post on forum (PoF), second FB live comments (FBL2), first FB live comment (FBL1), and FB live interaction (FBLA). From Figure 3, the hierarchical group heat map can further explore the similarity of learning behavior among learners. From the left part, the first order is divided into green and red groups.

The learning behavior of these two groups of students is very different. The color of CS and ASM in group R is lighter than that in group G. The learning participation of group G such as log clicks, reading time, and live response was much better than group R. Then, the red group and the green group were divided into the next level, as shown in Figure 3. It can be seen that the green group could be further divided into three groups: upper (G1), middle (G2), and lower (G3), while the red group could be divided into the upper group (R1) and the lower group (R2).

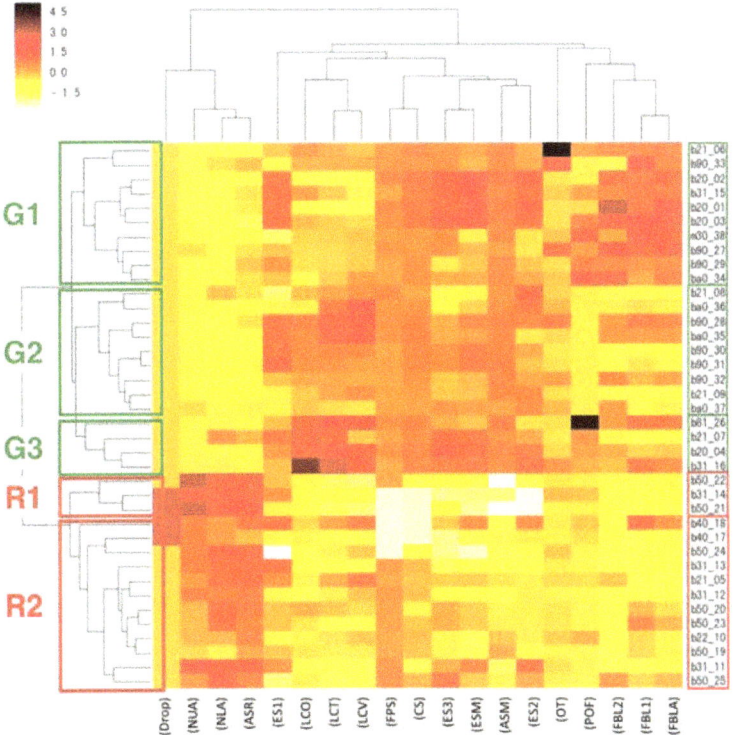

Figure 3. Fine grouping of the hierarchical group heat map.

The upper group (R1) and the lower group (R2) of the red groups showed that the learning performance of the two groups was quite shallow in the part of the log click and the part of the live response. The difference is that the three students of R1 were divided into a group. The number of non-delivery assignments, late submissions, and total payment priorities was particularly deep, as shown in Figure 3, and the learning performance was the lowest. From the CoI theory proposed by

Garrison, Anderson, and Archer [29], it can be seen that the R1 and R2 students had a low degree of social presence, teaching presence, and cognitive presence. For learners in these groups, their interactive performance, teaching content clicks, small test scores, and assignments were quite weak. Therefore, we can define red groups as learning negative groups based on their learning behavior.

The green group can be divided into three groups: upper (G1), middle (G2), and lower (G3). G1 was relatively light in the part of the log clicks, but the color of the live answer was quite dark. From the CoI theory proposed by Garrison, Anderson, and Archer [29], it can be seen that the G1 group of learners had a higher degree of social and emotional connection with others. Therefore, we can define this group as synchronized interactive active groups, according to their learning behavior.

4.3. Discussion on Conscious Learning Attitude

In the questionnaire, the question "I think my learning model for this semester" had four options, as shown in Table 3, namely "Active learning", "Regular learning", "On-demand learning", and "Negative learning". Active learners are not only able to learn the learning content uploaded by the teacher, but also actively learn additional knowledge and can actively ask questions or actively help students to answer questions. Regular learners learn the learning materials weekly and submit their assignments on time; on-demand learners are unable to complete the learning content on time every week, and negative learning refers to students who are too busy to undertaken this course. We then cross-analyzed the four types of student self-identified learning patterns and hierarchical group heat maps, as shown in Figure 4.

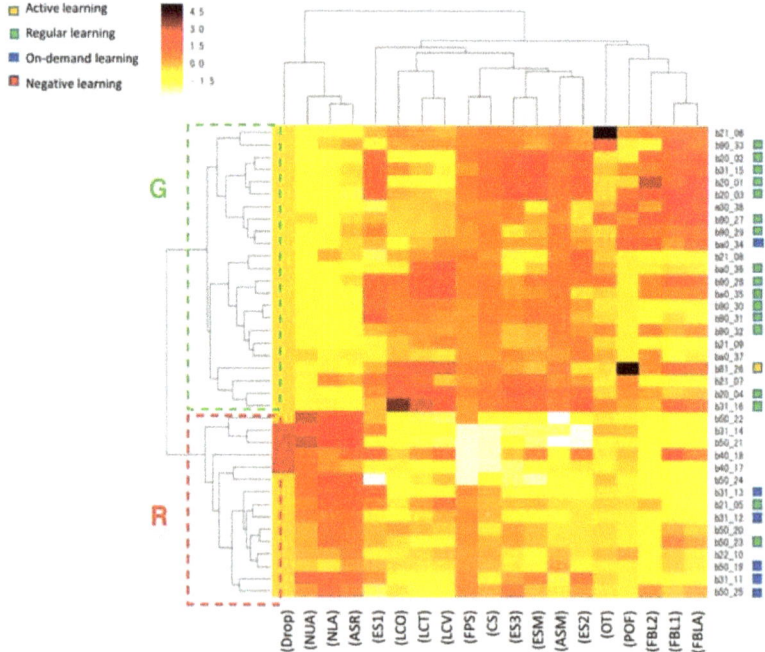

Figure 4. Self-conscious learning mode and hierarchy cross-analysis of the group heat map.

Table 3. Learning mode questionnaire.

Question	Active Learning	Regular Learning	On-Demand Learning	Negative Learning
I think my learning mode for this course	1 (4%)	19 (73%)	6 (23%)	0 (0%)

As shown in Figure 4, the four conscious learning modes are represented by yellow, green, blue, and red, respectively. The invalid questionnaires are indicated by blanks. We received 17 questionnaires for the 23 students in the green group (G) where one was an active learner, up to 15 students were regular quantitative learners, and one was a demand-based learner. In the red group (R), there were 15 students from whom we obtained seven valid questionnaires, among which two students were regular quantitative learners, and five students were demand-based learners. It can be seen from the learner's conscious learning model that there were roughly the same conclusions as this study. The green group (G) had better scores among the 17 valid questionnaires, up to 16 were regular quantitative or active learners. Therefore, this study believes that learners who have a stable learning pace can have a better learning performance in a mixed education environment. In contrast, the red group (R) with poor grades consisted of only two of the seven valid questionnaires. The other five were demand-based learners, that is, learners who do not have a stable learning pace, and cannot complete the learning content provided by the teacher on time every week.

5. Conclusions and Discussion

In this paper, we explored which variables were related to the learner's learning performance in a mixed-education environment of mixed face-to-face courses and live-action real-time teaching in a traditional online learning environment through educational data exploration, and which could make predictions in the interim period so instructors could be present in the second half of the semester for personalized guidance to be given to high-risk students. The pace of learning, synchronization, and non-synchronization activity had a significant impact on learning performance, and then methods of logistic regression, decision tree, and other methods were used to train the mid-term prediction model. The class also had a high accuracy rate. Finally, the content of the self-conscious questionnaire was discussed. The learner's thoughts also had complementary effects on the overall research. The conclusions can be divided into the following three major points:

5.1. Explore the Impact of Overall Learning Behavior on Learning Performance in a Mixed Learning Environment

(1) Synchronous/non-synchronized participants have better learning performance.

The study found that active students in an online education environment will have a better learning performance. Since this mixed education environment included traditional online courses and live teaching, the activity level included the clicks of online teaching content and the live broadcast environment. Under the interaction with others, high-score learners had at least one of these two variables relatively prominent, and even some high-score learners were prominent at the same time.

(2) Learners with stable pace have better learning performance.

The learning pace also plays a very important relationship for end-of-term learning performance. In the online teaching environment, the learning step changes discussed in this study included the frequency of clicks on the teaching content, and whether the click-through teaching content was stable from the beginning of the semester to the end of the semester. Moreover, from the submission of the assignment including the time of late submission, it could be found that the high-score learners were relatively unsuccessful in their clicks of teaching content, were stable learners, and often handed in their assignments on time.

5.2. Use Machine Learning to Establish an Interim Warning System to Predict High-risk Group Learners

(1) Using a mixed education environment to generate data can accurately predict high-risk group learners during the period

After understanding the impact of learning behavior on learning performance, this study hopes to develop a mid-term warning model so that teachers can know which learners are high-risk groups during the study period. Compared to logistic regression and decision trees, the random forest model had a 0.83 f1-score and 0.83 accuracy in predicting learner pass or fail variables, which shows that the model in the study was more accurate than the other two models. This study contributes empirical evidence to support the study results of Osmanbegović, Suljic, and Agić [22], who used classification algorithms to determine dominant factors for the students' performance prediction and found that random forest had better accuracy.

(2) The model trained in this study can be extrapolated and applied to other courses to predict learner performance.

In order to understand whether it is possible to extrapolate the model, this study built the model for other courses, so that the same pedagogic could also be applied in the other class (B class). When we extrapolated the model to the data collected from Class B to make predictions, there was a 0.90 f1-score and 0.91 model accuracy. In the mid-term, students can explore their learning behaviors for the predicted high-risk groups, improve their learning pace and lack of activity, and thus improve their learning performance.

5.3. Model Analysis Definitions for Learner Learning Behavior

(1) Using the clustering method can explore the fixed learning mode for the learner's learning behavior.

Through the hierarchical group heat map, it is possible to understand the learner's learning behavior at a glance and see that different groups of learners have different learning modes, some being relatively prominent in the teaching content click, and some active in the live teaching environment. This result supports the research findings of Hou [23], which addressed the analysis of learners' potential clusters and the behavioral patterns of each cluster. Moreover, this study addressed a more in-depth analysis integrating clustering and the heat map.

(2) Using hierarchical clustering and heat map can further define learning mode grouping from multi-dimensional user data variables.

According to the hierarchical group heat map, according to the synchronous teaching participation degree, the teaching content click degree and the homework paying situation, it can be seen that the learner is roughly divided into a green group with better scores that could be further subdivided, and a red group with poor scores. The three subdivisions in the green group were the "Interactive Active Group", "Stable Learning Group", and "Teaching Content Active Group", while the red group was the "Learning Negative Group". In the future, the instructor may be advised to provide different reminders and guidance for different groups of learners such as being able to set additional reminders for groups that have been promising their assignments, or for those who have been less engaged in face-to-face classes. The additional interaction is believed to have a significant improvement in learning performance and learning.

5.4. Limitations of This Study

While this study adds new insights into the application of educational data mining to explore students' learning patterns in the flipped learning approach, some limitations may be the subject of future research. The questionnaires relied on self-reporting, which may not have been answered accurately, so the sample could be biased [30]. Another limitation is that the use of a digital solution for collecting data might have led to a selection bias for students. Finally, this study was carried out in specific higher education environments that use a particular LMS. The behavior data of students' learning activities were acquired and limited from the LMS student log files and Facebook Graph API. Thus, further generalizations to other blended learning environments must be made with care. Therefore, based on the results we obtained in this study, in the future, educators will be able to use the mid-term forecasting system to find high-risk groups during the semester.

With the advancement of technology and the development of social media, the possibility of learning environments has become more diverse, which also brings different benefits to learners and educators as learners can access the content at their own pace. Future research can refer to the course design of this study and make more use of the interactive advantages of synchronous/asynchronous online teaching. In addition, qualitative data can be collected and analyzed. For example, future work will explore whether the interactive data of the learner is related to learning. Moreover, a larger sample data and the addition of new data exploration clusters and classification algorithms might be conducted to provide additional evidence.

Author Contributions: Conceptualization, methodology and writing-original manuscript, H.-C.H., C.-T.L. and Y.-S.S. Review and editing, I.-F.L. Furthermore, Y.-S.S. acted as a corresponding author. All authors have read and agreed to the published version of the manuscript.

Funding: This work was supported by the Ministry of Science and Technology of Taiwan under contract numbers MOST 106-2511-S-038-009-, MOST 108-2511-H-019-002, MOST 108-2511-H-019-003, and MOST 108-2511-H-008-017-. The authors would like to thank all the people who took part in this study.

Conflicts of Interest: The authors no conflicts of interest.

References

1. Ni, A.Y. Comparing the effectiveness of classroom and online learning: Teaching research methods. *J. Public Aff. Educ.* **2013**, *19*, 199–215. [CrossRef]
2. Graham, C.R.; Woodfield, W.; Harrison, J.B. A framework for institutional adoption and implementation of blended learning in higher education. *Internet High. Educ.* **2013**, *18*, 4–14. [CrossRef]
3. Wang, Q.Y. A generic model for guiding the integration of ICT into teaching and learning. *Innov. Educ. Teach. Int.* **2008**, *45*, 411–419. [CrossRef]
4. He, W. Examining students' online interaction in a live video streaming environment using data mining and text mining. *Comput. Hum. Behav.* **2013**, *29*, 90–102. [CrossRef]
5. Romero, C.; López, M.-I.; Luna, J.-M.; Ventura, S. Predicting students' final performance from participation in on-line discussion forums. *Comput. Educ.* **2013**, *68*, 458–472. [CrossRef]
6. Garrison, D.R.; Kanuka, H. Blended learning: Uncovering its transformative potential in higher education. *Internet High. Educ.* **2004**, *7*, 95–105. [CrossRef]
7. Park, Y.; Yu, J.H.; Jo, I.H. Clustering blended learning courses by online behavior data case study in a Korean higher education institute. *Internet High. Educ* **2016**, *29*, 1–11. [CrossRef]
8. Francis, R.; Raftery, J. Blended learning landscapes. *Brookes Ejournal Learn. Teach.* **2005**, *1*, 1–5.
9. Halverson, L.R.; Graham, C.R.; Spring, K.J.; Drysdale, J.S.; Henrie, C.R. A thematic analysis of the most highly cited scholarship in the first decade of blended learning research. *Internet High. Educ.* **2014**, *20*, 20–34. [CrossRef]
10. Sikder, S.; Herold, H.; Meinel, G.; Lorenzen-Zabel, A. Blessings of open data and technology: E-learning examples on land use monitoring and e-mobility. In Proceeding of the STS Conference, Graz, Austria, 6–7 May 2019.
11. Keržič, D.; Tomaževič, N.; Aristovnik, A.; Umek, L. Exploring critical factors of the perceived usefulness of blended learning for higher education students. *PLoS ONE* **2019**, *14*, e0223767. [CrossRef] [PubMed]
12. Baker, R.; Yacef, K. The State of educational data mining in 2009: A review and future visions. *J. Educ. Data Min.* **2009**, *1*, 3–17. [CrossRef]
13. Baker, R. Data Mining for Education. In *International Encyclopedia of Education*, 3rd ed.; Peterson, P., Baker, E., Mcgaw, B., Eds.; Elsevier: Oxford, UK, 2012; pp. 112–118. [CrossRef]

14. Romero, C.; Ventura, S. Educational data mining: A review of the state of the art. *IEEE Trans. Syst. ManCybern. Part C Appl. Rev.* **2010**, *40*, 601–618. [CrossRef]
15. Araque, F.; Roldan, C.; Salguero, A. Factors influencing university drop out rates. *Comput. Educ.* **2019**, *53*, 563–574. [CrossRef]
16. Hämäläinen, W.; Vinni, M. *Classifiers for Educational Data Mining*; Chapman & Hall/CRC: London, UK, 2011. [CrossRef]
17. Romero, C.; Ventura, S.; Garcia, E. Data mining in course management systems: Moodle case study and tutorial. *Comput. Educ.* **2008**, *51*, 368–384. [CrossRef]
18. Romero, C.; Ventura, S. Data mining in education. *Wiley Interdiscip. Rev. Data Min. Knowl. Discov.* **2013**, *3*, 12–27. [CrossRef]
19. Romero, C.; Espejo, P.; Romero, R.; Ventura, S. Web usage mining for predicting final marks of students that use Moodle courses. *Comput. Appl. Eng.* **2013**, *21*, 135–146. [CrossRef]
20. Romero, C.; Ventura, S.; Espejo, P.; Hervás, C. Data mining algorithms to classify students. In Proceedings of the Educational Data Mining, Montréal, QC, Canada, 20–21 June 2008; pp. 20–21.
21. Kulkarni, V.Y.; Sinha, P.K. Pruning of random forest classifiers: A survey and future directions. In Proceedings of the 2012 International Conference on Data Science & Engineering (ICDSE 2012), Cochin, Kerala, India, 18–20 July 2012; pp. 64–68.
22. Osmanbegović, E.; Suljic, M.; Agić, H. Determing dominant factors for students performance prediction by using data mining classification algorithms. *Tranzicija* **2015**, *16*, 147–158.
23. Hou, H.-T. Computers in human behavior integrating cluster and sequential analysis to explore learners' flow and behavioral patterns in a simulation game with situated-learning context for science courses: A video-based process exploration. *Comput. Hum. Behav.* **2015**, *48*, 424–435. [CrossRef]
24. Friendly, M. A Brief History of Data Visualization. In *Handbook of Data Visualization*; Springer Handbooks Comp. Statistics; Springer: Berlin/Heidelberg, Germany, 2008. [CrossRef]
25. Tufte, E.R. The visual display of quantitative information. *Am. J. Phys.* **1986**, *53*, 1117. [CrossRef]
26. Kochevar, P. Database Management for Data Visualization. In *Database Issues for Data Visualization*; Lee, J.P., Grinstein, G.G., Eds.; Springer Lecture Notes in Computer Science: Berlin/Heidelberg, Germany, 1994; Volume 871. [CrossRef]
27. Gitelman, L.; Jackson, V. Introduction. In *Raw data. Is an Oxymoron*; Gitelman, L., Ed.; MIT Press: Cambridge, MA, USA, 2013.
28. Beer, D. *Popular Culture and New Media: The Politics of Circulation*; Palgrave: London, UK, 2013.
29. Garrison, D.; Anderson, R.T.; Archer, W. Critical inquiry in a text-based environment: Computer conferencing in higher education. *Internet High. Educ.* **2000**, *2*, 87–105. [CrossRef]
30. Shaw, R.S. A study of the relationships among learning styles, participation types, and performance in programming language learning supported by online forums. *Comput. Educ.* **2012**, *58*, 111–120. [CrossRef]

Article

Problems of Creation and Usage of 3D Model of Structures and Theirs Possible Solution

Dalibor Bartonek * and **Michal Buday**

Institute of Geodesy, Faculty of Civil Engineering, Brno University of Technology, Veveri 95, 602 00 Brno, Czech Republic; buday.m@fce.vutbr.cz
* Correspondence: bartonek.d@fce.vutbr.cz; Tel.: +420-605-912-767

Received: 29 November 2019; Accepted: 13 January 2020; Published: 20 January 2020

Abstract: This article describes problems that occur when creating three-dimensional (3D) building models. The first problem is geometric accuracy; the next is the quality of visualization of the resulting model. The main cause of this situation is that current Computer-Aided Design (CAD) software does not have sufficient means to precision mapping the measured data of a given object in field. Therefore the process of 3D model creation is mainly a relatively high proportion of manual work when connecting individual points, approximating curves and surfaces, or laying textures on surfaces. In some cases, it is necessary to generalize the model in the CAD system, which degrades the accuracy and quality of field data. The article analyzes these problems and then recommends several variants for their solution. There are described two basic methods: using topological codes in the list of coordinates points and creating new special CAD features while using Python scripts. These problems are demonstrated on examples of 3D models in practice. These are mainly historical buildings in different locations and different designs (brick or wooden structures). These are four sacral buildings in the Czech Republic (CR): the church of saints Johns of Brno-Bystrc, the Church of St. Paraskiva in Blansko, further the Strejc's Church in Židlochovice, and Church of St. Peter in Alcantara in Karviná city. All of the buildings were geodetically surveyed by terrestrial method while using total station. The 3D model was created in both cases in the program AUTOCAD v. 18 and MicroStation.

Keywords: structure surveying; 3D model; reverse engineering; CAD

---

1. Introduction

A geodetic survey of buildings and other construction is one of the common activities in the field of surveying. Satellite methods, photogrammetry, or currently popular laser scanning are also used in addition to classical terrestrial measurements. Very often, three-dimensional (3D) models of buildings and structures are created from the measured data for the purpose of reconstruction, evaluation of the current state or just visualization.

A number of technologies have been developed for this activity. [1,2] using the latest knowledge of related fields. There are still problems that have not yet been satisfactorily resolved, despite the facts in the flow chart (Figure 1). The article tries to identify, analyze, present, and try to propose an acceptable solution. These problems can be divided into two basic categories:

1. display geometry (position and height accuracy) of the geographic object (GO) in the terrain into appropriate software, and
2. visualization of the created 3D model in suitable software.

Problem ad (1) is related to obtaining the data by measurement. Today's geodesy data collection technologies enable the user to focus the selected object with the required accuracy. These technologies integrate many features and provide high user comfort. The next step in the processing chain (Figure 1) poses a problem. It is a creation of a 3D model from a list of coordinates of detailed points, which is based on connecting points with general lines being selected from the element libraries in the software. In many cases, it is necessary to generalize the model to make interconnection possible. In fact, this means degrading the geodesist's work in the field while locating the object.

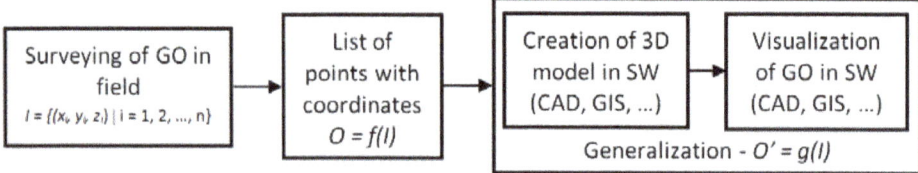

Figure 1. Flow chart of three-dimensional (3D) model creation.

The problem ad (2) mainly depends on the supply of surface materials in selected software that represent the surface of a focused GO in the field. Basically, it is about surface textures or the implementation of photographs on the partial surfaces of an object. The essence of the problem, in this case, is that the real surfaces of the object are in general not planar, which, in current applications, is solved by decomposing the general surface into a system of planar surfaces. As a result, the resulting image is rendered when the object is visualized.

This paper presents the problems encountered in the creation of 3D models that the authors have encountered in solving projects in [3–6]. These are mainly historical church buildings, in the case of [3] wooden buildings. The text also contains a solution proposal in the context of current possibilities of available applications.

The paper is organized, as follows. In Section 2, we mention some related work for 3D models. In Section 3, we describe problems regarding 3D model creation and propose the method of a possible solution. Section 4 demonstrates the experiments. Section 5 contains discussion of the given topic. Section 6 addresses the conclusions.

2. Related Works

A number of papers have been published on this topic, of which only those are directly related to our article are described, as follows. Previous publications can be divided into the following categories:

1. Building information modelling (BIM).
2. Reverse engineering.
3. New technologies for data processing from TLS (Terrestrial Laser Scanning).
4. New methods for 3D models creation.
5. Visualization of 3D objects.

2.1. BIM

The topic of the paper [1] is the integration of BIM with the issue of reverse engineering. It is about improving the sub-phases of the information flow throughout the entire project cycle. The aim is to reduce the errors and increase efficiency by supporting technologies, such as prefabrication, virtual reality, 3D printing, etc. The proposed methodology also includes tools for managing and organizing the entire workflow. The results can be used in projects for the renewal of towns and municipalities. The article [7] deals with Historic Building Information Modeling (HBIM). It is special library of historical architectural elements, from which it is possible to reconstruct entire historical buildings and complexes. The library was created in ArchiCAD GDL (Geometric Description Language). These are parametric elements whose specific geometry is defined by the user. The resulting model is composed of laser scanning surveying or photogrammetric data and is completed with elements from library. The resulting model serves for conservation purposes. The publication [8] is a continuation of the research that was presented in the previous paper [7]. The method for HBIM creation was supplemented by algorithms of data segmentation from point clouds that were obtained by ground laser scanning of buildings. This is a difficult task to be solved by algorithm, therefore a heuristic method was used.

2.2. Reverse Engineering

Article [2] deals with reverse engineering technology. The topic is terrestrial 3D laser scanning. There are latest technologies of point cloud processing by powerful technical and software tools described. The work [9] proposes a special procedure for laser scanning of buildings. The method consists in optimizing the arrangement of devices in the space by means of a telecommunication device located on the roofs of buildings. It is a virtual simulation of antenna sites, which generates a 3D scenario of the process. The method was verified at a project at the Technical University of Madrid. [10] is another work that belongs to the field of reverse engineering. It proposes an algorithm for comparing the actual state of the pipe design with the state in the Computer-Aided Design (CAD) system. 3D model is obtained from laser scanning. The method is used for reconstruction of buildings or verification of quality in construction. In the work [11] is described a new method of reverse engineering combined with knowledge engineering of construction. It deals with a definition of inverse CAD process while using topology and tree structure of design process. A specific geometry of the model is then created from this general concept.

2.3. Terrestrial Laser Scanning

The study [12] presents a new methodology for creating a 3D model of wooden structures. This is a quick procedure based on generative algorithms. Terrain data are obtained while using Terrestrial Laser Scanning. The method was verified in the framework of a research project of wooden roof structures in Bologna. A new algorithm for transforming point clouds into a 3D model while using parametric tools was developed. The model is created, in general, and other building elements can be modeled by changing of input parameters. The work [13] deals with the creation of a 3D model of the church that was obtained by Terrestrial Laser Scanning. There are described methods of point cloud analysis and digitization in CAD system.

2.4. New Methods for 3D Models Creation

In the paper [14] is proposed a method for reconstruction of geometry of 3D object and its components. The object is surveyed by geodetic terrestrial methods and obtained list of points is processed by object methods. The paper describes strategies for recognizing elements in objects and develops new algorithms that improve existing methods. The article [15] describes the method of reconstruction of reinforced concrete arch bridge. The technology is based on the smallest element method. The aim is to identify the structure of the building and create a 3D model of its actual execution. The work also presents an analysis of the accuracy of the geometric method that was used in object surveying. The content of the article [16] is a new technology for the surveying of historical buildings in hard to reach places. It is a photogrammetric method that uses fish-eye lenses. The advantage of the method is the speed of data acquisition and optimization of the data volume. It is an alternative method to Unmanned Aerial Vehicles (UAV). The paper [17] describes the monitoring of buildings by using of the methods of engineering geodesy. Buildings are geodetically surveyed and 2D or 3D models can be obtained from the data. The models are also used for the stage of structure protection and safety of buildings. The method of laser scanning of buildings is described in detail. It is also possible to create drawings, which include views and sections, from the data. Close-range photogrammetry can also be used to create orthoimage and linear drawing. This method is particularly suitable for surveying historical buildings that do not have building documentation.

Ref. [18] describes a special approach to 3D modeling. It is a new method of hybrid 3D reconstruction of objects, which is a combination of building elements, and computer graphics methods.

This integrated method takes advantage of the geometry, topology, and visualization of building objects in the process of 3D model creation.

2.5. Visualization of 3D Objects

The paper [19] deals with the analysis of 3D objects visualization. A new method that is based on the analysis of topology and time series of the object is proposed. The method does not need its own 3D model; the information is directly obtained from the data. Visualization is used to obtain new functional relationships within an object. The method is verified on a case study. Article [20] deals with the reconstruction of architectural elements of historical buildings from a cloud of points that were obtained by laser scanning. It is a high resolution 3D model in Triangular Irregular Network (TIN) format. The benefit is the high speed of algorithm and realistic visualization of the object. The method is used in architecture for the reconstruction of historical buildings. The paper [21] describes a method for approximating the surface of 3D models in CAD software. The method allows for creating a surface from suitably randomly selected points that were obtained from reverse engineering or from the design process. Geodesy algorithms motivated the technology, e.g., the creation of a digital terrain model. The paper [22] presents a design of an algorithm based on Gaussian map. It is a procedure that is suitable for visualization of ancient architecture. Technology has been proven during archaeological research of ancient cities. Experiments show that the method is accurate enough, with minimum noise, and no need for user intervention.

3. Materials and Methods

3.1. Problems Description

The process of creating a 3D model of a geographic object (GO) is simplified according to the diagram presented in Figure 1. The GO can represent any real object of interest (e.g., building or other construction). A general and not yet fully solved problem is to create a 3D model in suitable software (SW) with given accuracy.

Let $I = \{(x_i, y_i, z_i) | i = 1, 2, \ldots, n\}$ is a set of vectors (input) that represent the points of the GO in the real world, $O = \{(x_o, y_o, z_o) | o = 1, 2, \ldots, n\}$ is a set of vectors (output), describing the points in a digital database of a 3D model in suitable software. Subsequently, mapping function f from real space to the digital database of 3D model

$$O = f(I) \tag{1}$$

must fulfill the following conditions:

1. The mapping f must be homomorphism for capturing the topology of GO.
2. f must satisfy the level of geometric accuracy that was reached with surveying the GO.

After generalization in software (Figure 1) the mapping function can be described with following equation:

$$O' = g(I) \tag{2}$$

where $O' = U(O)$ and U represents neighborhood of vectors that display the input vectors of GO from the real space to the digital database. The neighborhood size of U then represents the true accuracy of the 3D model in used SW. This accuracy is less, and then the a priori accuracy of the geodetic survey of GO in field. The mapping function g then describes the functional repository of the used SW (CAD, GIS), while $g \subset f$ is valid.

A closed set of functions g though to describe the topology of GO, but it is necessary to generalize the 3D model in many cases. That means $O' \subset O \subseteq I$. In fact, we lose not only the accuracy of the position of points, but also some details of GO, which is a big disadvantage for further use of 3D models in engineering practice.

Another problem when creating a 3D model is the visualization of the results. This mainly involves rendering surfaces that represent the surface of an object. The problem can be divided into two categories:

1. Texture mapping onto surface areas,
2. The quality of the appearance of texture.

3.2. Posibility of Solutions

Replacing complex elements with simple entities where there is no further need for generalization can solve the problem of the geometric accuracy of a 3D model in suitable software. E.g. composite arcs or larger degree curves can be replaced with polyline elements. For the model [6], the authors used the method of topological coding [23]. The main principle consists in adding special codes into list of coordinates of points, according to field sketch:

topological code: **L/x,y or S/x**

where,

> **L** refers to line
> **S** refers to surface (polygon),
> **x** is unique identifier of the line (integer), and
> **y** is order number of the point in the line segment.

Format of the list of coordinates is following:

Point ID, Y, X, Z [, code$_1$, code$_2$, ... , code$_n$]

Every attribute is separated by comma, while topological codes in square brackets are optional. The application in the form of the script in Python language was created for points input into graphic editors. This script offers automatic creation of topologically correct drawing in CAD or GIS-based software. The script also checks duplicities of entities and provides full topology of the drawing. Figure 2 presents the chart diagram.

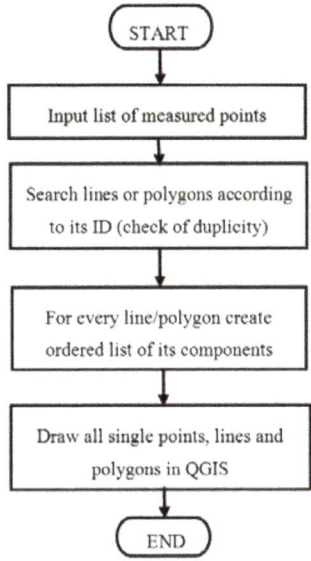

Figure 2. Flow chart of Python script for topological drawing in Quantum GIS.

The problem of visualization of the produced 3D model can be solved in two ways:

1. Using available visualization tools in commonly available CAD products (e.g., AUTOCAD, MicroStation, etc.). In this case, there are two options:

    (a)   used build-in modules with customary texture models in given software, and
    (b)   create your own set of textures and import them into the software (if the program product has the appropriate features for this purpose).

2. Placing the textures on 3D surfaces of given GO while using the special applications—see [24,25].

   In models [3–5], variant 1 (a) was used to represent the surface, in model [6] variant 1 (b).

3.3. Transformation of Point Cloud into 3D Model

Creating a 3D model by terrestrial laser scanning technology has several phases:

1. Data acquisition. Scanning an object in the field results in a raw point cloud, which, in many cases, is made up of several partial clouds.
2. Transformation of individual measured point clouds from the coordinate system of the scanner into a project system to create a homogeneous spatial model. Subsequently, transformation from the project system to a global coordinate system—national grids and an altimetry system. In the Czech Republic, it is S-JTSK (Datum of Uniform Trigonometric Cadastral Network) and Bpv (Baltic Vertical Datum After Adjustment).
3. Subsequent levelling of measured data using the ICP (iterative closest point) technology correlation of mutual position of individual clouds.
4. Create a 3D/BIM (Building Information Modelling) model. The core of this phase is to detect objects (edges, faces) in the point cloud and link these elements to the resulting model. Generally, it is necessary that modelling complied with agreed standards. The phase of focusing and plotting of the 3D/BIM model of the current state should be part of the so-called BIM Execution Plan if the whole project will be processed by the BIM methodology. LOD (Level of Detail), i.e., the level of detail and detail that the 3D/BIM model will carry from the phase of surveying the current state.

4. Results

The problems that are described in the previous chapter, the authors encountered in the practical implementation of 3D models of existing GO. These are four sacral buildings in the Czech Republic (CR): the church of saints Johns of Brno-Bystrc (Figures 3 and 4), the Church of St. Paraskiva in Blansko (Figures 4 and 5), further the Strejc's Church in Židlochovice and Church of St. Peter in Alcantara in Karviná city. All of the buildings were geodetically surveyed by the terrestrial method while using total station. The 3D model was created in both cases in the program AUTOCAD v. 18 and MicroStation.

Figure 3. Photo of Church of St. Jonh the Baptist and John the Apostle in Brno-Bystrc district (CR)-zoom.

Figure 4. Three-dimensional (3D) model of Church of St. Jonh the Baptist and John the Apostle in Brno-Bystrc district (CR).

Figure 5. Photo of wooden Church of St. Paraskiva in Blansko (CR)–zoom.

Geometric inaccuracy became evident when transforming the measured points from the terrain into suitable software (AUTOCAD) in several details. Of all cases, we will show as a demonstration example in Figures 3 and 4 during the 3D model creation of church of St. Jonh the Baptist and John the Apostle in Brno-Bystrc district (CR). Figure 3 shows a detail of the rotunda with the upper and lower arches indicated. Both arcs actually have different radii of curvature. When creating a wireframe, it was not possible to connect these arcs with vertical edges in AUTOCAD. The connection of the same edge to the lower arc was disconnected and vice versa when joining an edge to the upper arc. It was necessary to generalize the model in order to connect both arcs with vertical edges. The result after generalization is evident in Figure 4. The disadvantage is that similar inaccuracies limit the further practical use of the 3D model, e.g., making sections for the purpose of object reconstruction etc.

Problems with visualization are demonstrated on the 3D model of the Church of St. Paraskiva in Blansko (CR). In general, it involves laying textures or patterns with the real appearance of the building material used (e.g., roof tiles or shingles) on 3D model surfaces. In existing programs, this is only possible when the surfaces are planar. However, for real objects, the surfaces of buildings rarely meet that requirement. Most of these are general areas in space—see Figure 5. In this case, it is very difficult to apply realistic textures to these surfaces. Usually, it is necessary to decompose the surface into a numerous of planar patterns and then cover them with textures or photographs. Usually, triangles or planar quadrilaterals are used. However, this process is very time-consuming and laborious and in many cases the result does not correspond to the exerted effort. Figures 5 and 6 show the difference.

Figure 6. Visualization of 3D model of wooden Church of St. Paraskiva in Blansko (CR).

A similar situation can be seen in the models Strejc Church in Židlochovice—Figures 7 and 8—and the Church of St. Peter of Alcantara in Karviná (CZE)—Figures 9 and 10. The visualization of the Strejc Church (Figure 8) was done while using standard tools in the AUTOCAD program. The difference between the actual situation (Figure 7) and the display in AutoCAD (Figure 8) of the two images is obvious. For another 3D model of the Church of St. Peter of Alcantara in Karviná—Figures 9 and 10—visualization was performed while using own set of textures. The result is evident from the comparison of Figures 9 and 10. The visualization quality of the 3D model is many times higher than that of the Strejc's Church (Figures 7 and 8). However, the bases of both visualizations are the same—the textures in both models were placed on generalized (planar) surfaces. If we left the model without generalization, one of the methods described in [24] or [25] would have to be used for visualization.

Figure 7. Real view of Strejc's Church in Židlochovice (CR).

Figure 8. Visualization of 3D model of Strejc's Church in Židlochovice (CR).

Figure 9. Photo of Church of St. Peter of Alcantara in Karviná (CR).

Figure 10. Visualization of 3D model Church of St. Peter of Alcantara in Karviná (CR).

The described problems occur in all of the 3D models of real objects that the authors have encountered. These examples are only selected typical demonstration examples.

5. Discussion

Practically, the above-mentioned problems manifested in two aspects:

1. geometric accuracy of the model, and
2. quality of model visualization.

The solution of the above problems is dealt with in several works, of which the most important ones are mentioned.

The Geometric accuracy of the model is further explained in publications [11,18,23]. In paper [11] the geometry of the 3D model is complemented by a knowledge database that was obtained from a real object. It includes, for example, a tree that captures the GO topology. The article [18] is dealing with the problem of the precision issue with hybrid modeling. The elements of the building are extracted and then formally saved to the library as an object for further use. Another solution is presented in [23]. It is a complement of the list of coordinates with topological codes directly when surveying the object in the field. From the list of point coordinates, the exact drawing in the CAD program is then automatically displayed while using a Python script.

Visualization quality is discussed in [16,21,24,25]. The article [16] presents a visualization of historical objects that were captured by photogrammetrically special fish-eye camera. The object is measured by laser scanning. An interesting approach to surface modeling is described in [21]. The object's surface is approximated by curves that are defined by randomly selected points from the point cloud from laser scanning. Curves are created by special VB.NET applications in the AUTOCAD program. The work [24] proposes a method of classification of the surface of a 3D object that is based on the skeleton metric of this object. The result of the classification is a set of classes of segments that can be used for the whole surface. The most appropriate display method is then selected for each class. The publication [25] presents a design of the CatSurf system for displaying 3D objects in CAD (Computer-Aided Design). It is the surface texture information system, which is a part of the integrated CAD surface texture platform. The disadvantage of the latter two applications is that they are highly specialized systems that are difficult for ordinary users available.

6. Conclusions and Future Work

The main problems for creating 3D models of existing GOs were identified and described. The experimental results show that the currently used CAD programs are relatively outdated in its repertoire when compared to the quality and possibilities of geodetic surveying of real objects. Current methods of data acquisition in the field use modern technologies that allow for surveying the object with high accuracy. In addition, data collection devices have a number of built-in features, which allow for people with basic training to use them.

The authors proposed a solution concerning the first part of the problem mentioned in Section 4, namely the geometric accuracy of the 3D model. A Python application was created to produce a wireframe 3D model from a list of coordinate points with topological codes in suitable software. This procedure will significantly speed up the whole process-see Figure 1 and make the work easier for users. The script can be added as a plug-in to CAD software.

Further research in this area will be focused on solving the problem of quality visualization of the 3D model of GO.

Author Contributions: D.B. has provided support materials, elaborated literature review and create system model, M.B. conducted an overall editorial of the whole article and a professional translation. All authors have read and agreed to the published version of the manuscript.

Funding: This research was funded by the Grant No. FAST-J-19-5994 of the Brno University of Technology, Czech Republic.

Acknowledgments: We greatly appreciate the careful reviews and thoughtful suggestions by reviewers.

Conflicts of Interest: The authors declare no conflict of interest.

References

1. Ding, Z.; Liu, S.; Liao, L.; Zhang, L. A digital construction framework integrating building information modeling and reverse engineering technologies for renovation projects. *Autom. Constr.* **2019**, *102*, 45–58. [CrossRef]
2. Mesaros, P.; Kozlovska, M.; Hruby, K. Potential of IT Based Reverse Engineering in Civil Engineering and architecture. In Proceedings of the 3rd International Multidisciplinary Scientific Conference on Social Sciences and Arts SGEM 2016, Vienna, Austria, 6–9 April 2016; pp. 765–772.
3. Michalková, J. 3D Model Dřevěného Kostela sv. Paraskivy v Blansku (3D Model of Wooden Church of st. Paraskiva in Blansko). Master's Thesis, Brno University of Technology, Brno, Czech Republic, 2018; p. 49.

4.   Vilčeková, L. 3D Model Kostela sv. Janů v Brně—Bystrci (3D Model of st John's Church in Brno-Bystrc). Master's Thesis, Brno University of Technology, Brno, Czech Republic, 2018; p. 46.

5.   Kaličiaková, J. 3D Model Objektu Strejcův sbor v Židlochovicích. (3D Model of Strejc's Church in Židlochovice). Master's Thesis, Brno University of Technology, Brno, Czech Republic, 2018; p. 45.

6.   Giemza, L. 3D model kostela sv. Petra z Alkantary v Karviné (3D model of church of st. Peter of Alkantara in Karviná). Master's Thesis, Brno University of Technology, Brno, Czech Republic, 2019; p. 45.

7.   Prati, D.; Zuppella, G.; Mochi, G.; Guardigli, L.; Gulli, R. Wooden trusses reconstruction and analysis through parametric 3d modeling. In Proceedings of the 8th International Workshop on 3D Virtual Reconstruction and Visualization of Complex Architectures (3D-ARCH), Bergamo, Italy, 6–8 February 2019; Volume 42-2, pp. 623–629.

8.   Nieves-Chinchilla, J.; Martinez, R.; Farjas, M.; Tubio-Pardavila, R.; Cruz, D.; Gallego, M. Reverse engineering techniques to optimize facility location of satellite ground stations on building roofs. *Autom. Constr.* **2018**, *90*, 156–165. [CrossRef]

9.   Meidow, J.; Uslaender, T.; Schulz, K. Obtaining as-built models of manufacturing plants from point clouds. *Automatisierungstechnik* **2018**, *66*, 397–405. [CrossRef]

10.   Lee, M.; Lee, S.; Kwon, S.; Chin, S. A Study on Scan Data Matching for Reverse Engineering of Pipes in Plant Construction. *KSCE J. Civ. Eng.* **2017**, *21*, 2027–2036. [CrossRef]

11.   Perkins, M.; Daniels, K. Visualizing Dynamic Gene Interactions to Reverse Engineer Gene Regulatory Networks using Topological Data Analysis. In Proceedings of the 21st International Conference on Information Visualisation (IV), London, UK, 11–14 July 2017; pp. 384–389.

12.   Pawlowicz, J.A. Importance of Laser Scanning Resolution in the Process of Recreating the Architectural Details of Historical Buildings. In Proceedings of the World Multidisciplinary Civil Engineering-Architecture-Urban Planning Symposium (WMCAUS), Prague, Czech Republic, 12–16 June 2017.

13.   Pawlowicz, J.A.; Szafranko, E. Application of reverse engineering in modelling of rural buildings of religious worship. In Proceedings of the 15th International Scientific Conference on Engineering for Rural Development, Jelgava, Latvia, 25–27 May 2016; pp. 762–766.

14.   Artese, S.; Lerma, J.L.; Zagari, G.; Zinno, R. The survey, the representation and the structural modeling of a dated bridge. In Proceedings of the 8th International Congress on Archaeology, Computer Graphics, Cultural Heritage and Innovation (ARQUEOLOGICA), Valencia, Spain, 5–7 September 2016; pp. 162–168.

15.   Covas, J.; Ferreira, V.; Mateus, L. 3D Reconstruction with Fisheye Images Strategies to Survey Complex Heritage Buildings. In Proceedings of the Digital Heritage International Congress, Granada, Spain, 28 September–2 October 2015; pp. 123–126.

16.   Cazacu, R.; Grama, L. Constructing 3D Surfaces from Random Sets of Points by Means of Level Curves. In Proceedings of the 17th International Conference on Innovative Manufacturing Engineering, Iasi, Romania, 23–24 May 2013; Volume 371, pp. 483–487.

17.   Dore, C.; Murphy, M. Semi-automatic modelling of building facades with shape grammars using historic building information modelling. In Proceedings of the Conference on 3D Virtual Reconstruction and Visualization of Complex Architectures (3D-ARCH), Trento, Italy, 25–26 February 2013; Volume 40-5-W1, pp. 57–64.

18.   Musat, C.C.; Herban, I.S. Study on Reverse Engineering of Historical Architecture in Timisoara BASED on 3D Laser Point Technologies. *J. Environ. Prot. Ecol.* **2012**, *13*, 1107–1116.

19.   Murphy, M.; McGovern, E.; Pavia, S. Historic building information modelling—Adding intelligence to laser and image based surveys. In Proceedings of the 4th ISPRS International Workshop 3D-ARCH—3D Virtual Reconstruction and Visualization of Complex Architectures, Trento, Italy, 2–4 March 2011; Volume 38-5, pp. 1–7.

20.   Zhao, J.; Wu, J.; Wang, Y. Ancient Architecture Point Cloud Data Segmentation Based on Gauss Map. In Proceedings of the International Conference on Ecological Protection of Lakes-Wetlands-Watershed and Application of 3S Technology (EPLWW3S 2011), Nanchang, China, 25–26 June 2011; Volume 3, pp. 402–405.

21.   Durupt, A.; Remy, S.; Ducellier, G.; Guyot, E. A new reverse engineering process, the combination between the knowledge extraction and the geometrical recognition techniques. In Proceedings of the International Conference on Computers and Industrial Engineering (CIE39), Troyes, France, 6–9 July 2009; Volume 1–3, pp. 1367–1372.

22. De Luca, L.; Veron, P.; Florenzano, M. Reverse engineering of architectural buildings based on a hybrid modeling approach. *Comput. Graph. UK* **2006**, *30*, 160–176. [CrossRef]
23. Bartoněk, D.; Bureš, J.; Ježek, J.; Vacková, E. Automatic creation of field survey sketch by using of topological codes. In Proceedings of the 19th International Multidisciplinary Scientific GeoConference SGEM 2019, Albena, Bulgaria, 28 June–9 July 2019; Volume 19, pp. 777–784, ISBN 978-619-7408-79-9. [CrossRef]
24. Beynea, T.B.; Morina, G.; Leonardb, K.; Hahmann, S.; Carliera, A. A salience measure for 3D shape decomposition and sub-parts classification. *Graph. Models* **2018**, *99*, 22–30. [CrossRef]
25. Qi, Q.; Scott, P.J.; Jiang, X.; Lu, W. Design and implementation of an integrated surface texture information system for design, manufacture and measurement. *Comput. Aided Des.* **2014**, *50*, 41–53. [CrossRef]

Article

A Balance Interface Design and Instant Image-based Traffic Assistant Agent Based on GPS and Linked Open Data Technology

Fu-Hsien Chen [1] and Sheng-Yuan Yang [2,*]

[1]   Department of Electrical Engineering, St. John's University, New Taipei City 25135, Taiwan;
      fu@mail.sju.edu.tw
[2]   Department of Information and Communication, St. John's University, New Taipei City 25135, Taiwan
*   Correspondence: ysy@mail.sju.edu.tw; Tel.: +886-2-2801-3131 (ext. 6396)

Received: 16 October 2019; Accepted: 15 December 2019; Published: 18 December 2019

Abstract: Taiwan is a highly informational country, and a robust traffic network is not only critical to the national economy, but is also an important infrastructure for economic development. This paper aims to integrate government open data and global positioning system (GPS) technology to build an instant image-based traffic assistant agent with user-friendly interfaces, thus providing more convenient real-time traffic information for users and relevant government units. The proposed system is expected to overcome the difficulty of accurately distinguishing traffic information and to solve the problem of some road sections not providing instant information. Taking the New Taipei City Government traffic open data as an example, the proposed system can display information pages at an optimal size on smartphones and other computer devices, and integrate database analysis to instantly view traffic information. Users can enter the system without downloading the application and can access the cross-platform services using device browsers. The proposed system also provides a user reporting mechanism, which informs vehicle drivers on congested road sections about road conditions. Comparison and analysis of the system with similar applications shows that although they have similar functions, the proposed system offers more practicability, better information accessibility, excellent user experience, and approximately the optimal balance (a kind of symmetry) of the important items of the interface design.

Keywords: linked open data; GPS; traffic assistant agents; balance interface design

---

1. Introduction

Within the context of the artificial intelligence, big data, and cloud computing (ABC) era, 2019 is the first year to have fifth generation (5G) mobile Internet service, meaning big data based on the cloud computing environment must rely on artificial intelligence technology for processing. Without solutions for the volume, velocity, variety, and veracity (4V) data characteristics, the problem of garbage in, garbage out (GIGO), which is often mentioned in computer science, would continue to occur. Therefore, open data can be a valuable tool to solve this problem. Taiwan is a densely-populated country with heavy traffic, especially in metropolitan areas; thus, an open traffic database provided by government units is very important. This study aims to solve the problem of "true" and "false" in big data floods, which are commonly found in cloud information systems, by using the linked open data (LOD) technology of government units.

Smart mobile devices have become prevalent in recent years. According to the eMarketer survey data of 2016 [1], Taiwan has the highest penetration rate of smartphones, as compared with Singapore and South Korea. According to the 2018 Household Digital Opportunity Survey Report of the National Development Council [2], the mobile Internet usage rate among Taiwanese grew from 41.9% in 2009 to

98.2% in 2018, and mobile Internet usage has long been the main trend of Taiwanese Internet usage. The survey also pointed out that 84.9% of Taiwanese people over 12 years old use mobile Internet, while more than 60% of Taiwanese people under sixty-five years old use mobile Internet; therefore, the use of information systems or applications on various mobile devices by Taiwanese people via the Internet is undoubtedly a fairly mature digital skill. Thus, appropriate, convenient, and correct mobile or cloud information systems or applications are in great demand.

Taiwan is a highly informational country, and transportation is one of the most important issues related to people on a daily basis. The transportation network is critical to the national economy, and is the benchmark index of public construction. Building a sound, smooth traffic network is an important infrastructure for the country's economic development, while traffic has a negative association with a large influx of vehicles, road construction, and traffic accidents [3–6], which are the reasons for road congestion. However, Taiwan is a country with dense population and heavy traffic, which is one of the main causes of road congestion. Therefore, there is a growing demand for information systems that can facilitate real-time and dynamic queries, and present current traffic information, especially for mobile application users.

With the prevalence of smartphones, people have easy access to communication applications, such as LINE, WeChat, Telegram, and Skype. As of March 9, 2018, the monthly active users of LINE reached nineteen million in Taiwan, with an opening rate of about twenty times a day [7], which also represents the dependence of users on smartphones. In addition, the Taiwanese have a large number of private cars [8], and self-help travel is a common mode of transport [9]. Therefore, most of them rely on car navigation systems, resulting in a higher probability of road congestion [10], such as the problem mentioned by the Waze system—because there is no car owner feedback mechanism, this leads to a dilemma in avoiding the "A" road section, crowding drivers into the "B" road section, just like the problem caused by using only Google Maps [11]. For this reason, if users can use mobile devices to access the real-time traffic information on communication applications, they can be updated on the latest road conditions, and thus avoid traffic congestions. This study aims to develop a traffic assistant agent with a simplified reporting procedure on an intuition-based graphical interface for smartphone use.

Therefore, this study used LOD and global positioning system (GPS) technology to develop an intelligent traffic assistant agent with user-friendly interfaces and a user reporting mechanism to instantly check traffic flow. This study further explored the feasibility of developing a traffic assistant agent that does not require users to download the application, provides cross-platform services, saves system development costs, and has practicability, accessibility of information, excellent user experience, and approximately optimal balance of the important interface design.

2. Literature Review and Development Technologies

This study first reviewed the various traffic assistant information systems available in Taiwan, including Road Condition Autotoll [12], Real-Time Traffic Image—RoadCam [13], Police Broadcast Real-time Traffic [14], and New Taipei City Advanced Traveler Information System [15], as shown in Table 1. In summary, although those systems can use the responsive web design (RWD) technology to present the information page with the most suitable size and can automatically retrieve the users' position, they still need improvement on providing the correct corresponding information, and most of them lack the user reporting function.

Table 1. Four traffic assistant information systems explored in this study.

No	System Name	System Picture	Advantages	Disadvantages
1	Road Condition Autotoll (capture time: December 19, 2018)		Provide comprehensive applications for national expressways, hot spot areas, roads to and from airports, important urban roads, and scenic spots.	GPS function is still not perfect; often unable to read location information. Lack of user reporting function in this study.
2	Real-Time Traffic Image (capture time: October 19, 2018)		Provide real-time image application program for road conditions in the counties and cities of Taiwan, which is practical.	Lack of user reporting function in this study.
3	Police Broadcast Real-Time Road Conditions (capture time: October 18, 2018)		The website display includes date, time, road section description, category, etc. At this stage, it is still necessary to report road condition information by telephone or retrieve relevant government data to present relevant traffic information.	Lack of user reporting function in this study.
4	New Taipei City Advanced Traveler Information System (capture time: December 24, 2018)		The website uses responsive web design technology and can view relevant information pages in an optimal size on an intelligent handheld device.	The user position cannot be automatically obtained, and only the longitude and latitude positions originally set by the map are presented. Furthermore, entering the website still needs to be checked manually. Lack of user reporting function in this study.

Open data, especially government open data, is a huge resource that has not been fully developed. Since 2011, Taiwan's government has also started to actively promote open government and open data platforms to provide public access. The JavaScript Object Notation (JSON) format is most commonly used on the government open data platform, which can store and transmit data in plain text. For more complex types of data, transmission can be achieved through objects or arrays [16]. LOD is an applied open database that organizes the data set according to the principle of linked data, and there are many research studies related to open data or LOD in the literature. For example, Wang [17] combined common classification methods, K nearest neighbor (KNN), support vector machines (SVM), and decision trees to explore and link open data in order to achieve automatic classification of news articles. Yan [18] took the government open data platform as the database, and explored the technologies related to building the resource description framework (RDF) based on a prototype platform. Yu [19] studied Taipei City's open data platform's traffic parking lot, smiling bicycle (a public bicycle rental system in Taipei), and mass rapid transit (MRT, a rapid transit system in Taipei) data to explore relevant technologies for LOD applications. Albert et al. [20] presented the research project census data open linked (CEDAR) data set, which is the historical census of the Netherlands in LOD format, to explore a more accessible, better connected, and history-related integration technology. Alobaidi et al. [21] used LOD to enrich the query content to improve search effectiveness and ranking. Selvam et al. [22] proposed a systematic approach using ontology and LOD with significance to semantic links in the social event detection (SED) task. Pourhomayoun et al. [23] proposed an effective end-to-end system for traffic vision, detection, and counting on real-time traffic open data. Most of the above literature used LOD support to add and evaluate subsequent information services. This study used the Azure cloud platform support to extract appropriate corresponding LOD information (as detailed in Sections 3.2 and 3.3) after referencing GPS position capture and conversion comparison, which supports the operation of the overall information system and enhances the correctness, authenticity, and integrity of location-based service (LBS) information service provision.

GPS is a satellite network that can measure satellite distance to accurately capture the position on earth by continuously transmitting coded information to satellites, thus providing corresponding information services. The GPS information is free and open to users around the world. There are many research studies on the GPS architecture and information systems in literature. For example, Juan [24] quoted GPS signals from mobile devices to explore relevant technologies to analyze the space–time data of office workers' lives. Liou [25] integrated the GPS and Beidou satellite navigation system, put forward an adaptive multitime algorithm, and explored relevant technologies to achieve single-frequency, real-time, precise single-point positioning. Tu [26] connected the electronic collars of pets to smartphones as an invisible dog leash, and used the BLE (Bluetooth Low Energy) received signal strength indicator with GPS to explore the relevant technologies of real-time pet tracking. Fridman et al. [27] used styles, applications, web browsing, and GPS positioning to realize the related technologies of active authentication on mobile devices. Based on the survey of GPS-assisted travel in Beijing suburbs, Ta et al. [28] studied the related technologies of personal commuting efficiency according to the difference in commuting distance and route selection efficiency between the morning and evening peak traffic hours. Aliprantis et al. [29] described a concept with image identification and matching from the Europeana platform, which can link the LOD cloud from cultural institutes around Europe and mobile augmented reality applications for cultural heritage without accurate geo-based locations. Khaghani et al. [30] proposed a platform for dynamic performance assessment of roadway networks, leveraging coarse GPS data from probe vehicles, such as taxis, to quantify the resilience of road networks using a multidimensional approach. As discussed above, it is a development trend of contemporary information systems to quote GPS to explore corresponding information services (as detailed in Section 3.2). This study used GPS location acquisition and conversion to explore how to quote appropriate LOD and related technologies to provide corresponding value-added LBS information services.

In the age of the Internet, mobile devices have been widely used to deliver personalized services. For example, LBS can be integrated with various technologies, such as GPS positioning and mobile data, to provide comprehensive applications for services related to spatial location. In fact, LBS was first provided as a rescue service by an American operator in 1996, then called E911 [31]. Syu [32] used the message queuing telemetry transport protocol (MQTT) to carry out data transmission and realize mobile phone application push and broadcast function. They applied relevant technologies to calculate the value of taxi drivers' LBS in the Taoyuan area of Taiwan. Chou [33] combined the Internet of Things (IoT) and LBS, using Arduino and mobile devices to identify the common causes of accidents in China, including not paying attention while driving, not keeping a safe distance, careless driving, drunk driving, etc. They also implemented various technologies to integrate sensors with government public data to provide suitable services. Sun et al. [34] used the location label to mark the sensitive and normal locations of mobile users, and designed an algorithm based on the location label to explore relevant technologies to protect users' location privacy. Wang et al. [35] discussed the issues of location awareness and privacy protection, and explored technologies related to location-based services according to users' requirements, as based on different locations. Ukrit et al. [36] proposed an LBS architecture PROFILER (a framework for constructing location centric profiles), which was centered on the discrete locations visited by users, and explored the related application technologies. Lin et al. [37] described the methods for efficiently finding the links across maps, converting the data into RDF, and querying the resulting knowledge graphs to solve the problem of how to convert vector data extracted from multiple historical maps into linked data. Sansonetti et al. [38] proposed a research study on integrating the recommendation process of nearby points of artistic and cultural interest (POIs) with related multimedia content, exploiting the potential offered by LOD by following semantic links in the LOD graph with the specific POI to provide personalized suggestions. As discussed above, location-based services, mobile positioning services, and location services can obtain the location information (geographic coordinates) of mobile users through the mobile operator's radio communication network (e.g., GSM—Global System for Mobile Communications, network, CDMA—Code Division Multiple Access, network) or external location method (e.g., GPS). With the support of the back-end geographic information system (GIS) platform, it provides corresponding value-added information services. This study utilized the Azure cloud platform to explore how to intercept appropriate LOD information after referencing local GPS location acquisition and conversion to support the overall system operation, thus effectively adding value to the quality of cloud information consultation and the sharing function of this system.

Microsoft Azure [39,40] is Microsoft's public cloud service platform. At present, Microsoft Azure can support up to 30 kinds of service contents, including computing, storage, analysis, network, management, and identification services. In addition, Microsoft Azure has data centers around the world, which have tens of thousands of servers to provide users with applications and research. If one server suddenly fails, another server can take its place to operate in real time to ensure the sustainability of website services. Many studies have used Azure as the research tool. Liou [41] proposed a set of air quality index (AQI) deterioration real-time early warning systems, as based on the Azure cloud computing platform, including a prediction model, evaluation model, and other system operation modules. Ho [42] used the decision tree model on the Microsoft Azure platform to analyze data, and produced a customer recommendation list with high purchase probability. Lin [43] developed an integrated entrance guard system using a face recognition system and Raspberry Pi 3 in combination with Microsoft Azure cloud services. Richard [44] introduced all information services provided on Microsoft Azure. Diaz and Freato [45] mentioned that Microsoft Azure has supported data administrators and developers to provide a rich platform for big data workloads, such as linked services with Azure data and Azure storage. Färber [46] presented the Microsoft Academic Knowledge Graph (MAKG), a large RDF data set with over eight billion triples with information about scientific publications and related entities, used to solve problems in the LOD cloud. Based on the above systems, and with the support of the Azure cloud platform, this study explored and developed relevant technologies for cloud information consultation and sharing in order to automatically balance system operation efficiency and corresponding system stability.

The Google Maps API (Application Programming Interface) is a copyrighted and chargeable development kit developed by the Google company. Users can use the application functions provided by Google Maps (e.g., JavaScript API) by typing the kit and serial key into the webpage program. The users receive the data through the mobile client or the webpage, which presents the map and positioning information. The use of Google Maps API as a map in the literature is quite extensive. Chang [47] combined Google Maps API with a generic algorithm to generate an approximate optimal solution, explored the mathematical vehicle route model, and provided an effective route guidance plan, thereby greatly shortening drive time. Wang [48] analyzed the data of the old 119 system and the new 119 system, explored the related issues and research, and developed technologies for shortening emergency rescue response time after the emergency rescue service data was imported into the geographic information system. Rahmi et al. [49] consolidated MySQL as the main data storage space, and used Firebase to store additional data, where Firebase's real-time database processes chat data to provide corresponding notifications, along with Google Maps API to support GIS to explore and construct relevant technologies to meet the needs of mobile Android and web applications between doctors and patients. Xia et al. [50] introduced population grid data into a new gravity radiation model, used Google Maps API to obtain grid-level travel costs, and explored relevant technologies for population flow estimation. Tan et al. [51] adopted SPARQL (SPARQL Protocol and RDF Query Language) to query useful data from the DBpedia LOD database to acquire related data nodes and used the page rank algorithm to calculate the importance of each data node in order to build a concept map for awareness training in cybersecurity. Nyo and Hein [52] presented a technique for guiding and controlling autonomous vehicles by using the Google Maps API with GPS for localization of the vehicle on the Google Maps application via WiFi module. As mentioned above, the system also used Google Geolocation API to display the converted address information on the system page according to the location reported by the users' device and WiFi signal. Furthermore, Google Traffic [53] was used to indicate color stratification as the congestion degree of road sections, thus presenting the authenticity of the proposed system by verifying road condition information in multiple ways.

In summary, this study used a service-oriented information service system, which integrates LOD and GPS to achieve a mobile information service system, with the aim of solving all of the aforementioned problems. The proposed system serves no borders and uses a cross-platform approach, meaning users do not have to repeat learning for different programs. With the integration of Google Maps, the proposed system allows users to obtain the current real-time traffic information according to their current location. In the case of traffic emergencies, they can also inform rear road users to pay special attention, suggesting diversions through the system's registration form. The real-time information can be sent to the police units so the appropriate manpower can be deployed or other measures could be taken to reduce road congestion. In addition, the proposed system can facilitate user reporting of traffic conditions and can quickly check the nearest intersection monitor and sensors, so as to present the information on the most suitable page. Moreover, it can access cloud resources more quickly, and use and combine the open data provided by the government units to obtain the users' first-hand information regarding traffic conditions. In summary, the proposed system has the advantages of real-time information, simple and convenient usage, cross-platform use-related information services, and optimal balance of the important items of the interface design, without the need to update applications,

3. Proposed System Architecture

This study used Visual Studio and ASP.NET to develop the web applications. The Microsoft SQL server was used as the system data storage space, and Microsoft Azure was especially used to set up the cloud servers. Government open data was screened and analyzed according to specific conditions by using the circulation method of R language, and then traffic monitoring data content was built in the corresponding cloud database. The R & D of hardware equipment included an ASUS MD710 computer mainframe and ACER E5-572G notebook computer, which were used for information system development and testing. Finally, the receiver was tested on an ASUS ZenFone 3 intelligent handheld device.

3.1. Overall System Architecture

The intelligent traffic assistant agent is a comprehensive application with a location-based service as the main body, which includes a front-end agent web application (web app), a back-end database (SQL Server), and a corresponding cloud database (Azure SQL Database). The system architecture is shown in detail in Figure 1. The web application can present all of the operating functions of the proposed system: real-time information, endpoint information, and historical analysis. When a user opens the system through an intelligent networking handheld device, the system reads the longitude and latitude of the user's location through the GPS function of the device and presents it on the map, which can display the current road condition data in color; for example, green means smooth and red means congestion. Drivers are reminded to obtain the latest road condition information through the graphical interface in advance. The user reporting subsystem provides up-to-date road condition information, including member verification, addition of reports, and report records.

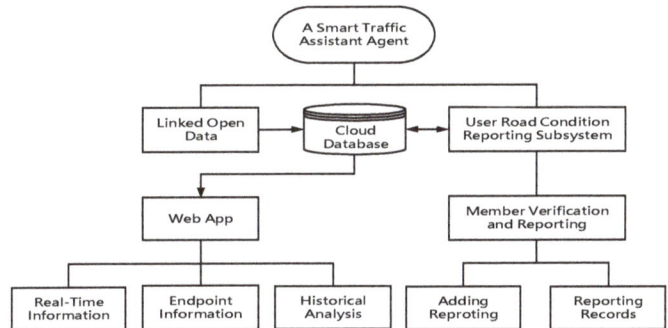

Figure 1. Overall system architecture.

3.2. LOD

The dynamic vehicle information of the proposed system is retrieved via the API of the New Taipei City Government open data platform through R language. The API updates data once every five minutes, randomly returns a record with eight fields of real traffic data once at a single sensing point, and transmits data in the JSON format, as shown in Figure 2. Each field has the following corresponding meanings.

1. vdid: endpoint number;
2. datacollecttime: collection time;
3. status: endpoint status;
4. vsrid: road number;
5. vsrdir: road direction;
6. speed: detect the current speed;
7. carid: detect the types;
8. volume: detect the number of vehicles.

```
[{"vdid":"65000V049350","datacollecttime":"2018/08/24
13:04:00","status":"0","vsrid":"1","vsrdir":"0","speed":"61","laneoccupy":"4","carid":"L","volume":"0"},
{"vdid":"65000V049350","datacollecttime":"2018/08/24
13:04:00","status":"0","vsrid":"1","vsrdir":"0","speed":"61","laneoccupy":"4","carid":"S","volume":"12"},
{"vdid":"65000V049350","datacollecttime":"2018/08/24
13:04:00","status":"0","vsrid":"0","vsrdir":"0","speed":"63","laneoccupy":"5","carid":"L","volume":"2"},
{"vdid":"65000V049350","datacollecttime":"2018/08/24
13:04:00","status":"0","vsrid":"0","vsrdir":"0","speed":"63","laneoccupy":"5","carid":"S","volume":"7"},
{"vdid":"65000V049350","datacollecttime":"2018/08/24
13:04:00","status":"0","vsrid":"3","vsrdir":"1","speed":"57","laneoccupy":"8","carid":"L","volume":"0"},
{"vdid":"65000V049350","datacollecttime":"2018/08/24
13:04:00","status":"0","vsrid":"3","vsrdir":"1","speed":"57","laneoccupy":"8","carid":"S","volume":"17"},
{"vdid":"65000V049350","datacollecttime":"2018/08/24
13:04:00","status":"0","vsrid":"2","vsrdir":"1","speed":"42","laneoccupy":"15","carid":"L","volume":"1"},
{"vdid":"65000V049350","datacollecttime":"2018/08/24
13:04:00","status":"0","vsrid":"2","vsrdir":"1","speed":"42","laneoccupy":"15","carid":"S","volume":"16"}
]
```


Figure 2. Dynamic vehicle data (Origin URL: https://data.ntpc.gov.tw/od/detail?oid=875D5555-A881-4561-8E9D-93B96C959384).

This study only stored the relatively important dynamic vehicle data. Figure 3 shows the traffic data extraction concept, as developed for the proposed LOD subsystem. It is mainly based on road direction (vsrdir) as the water shed, access to a single kind of vehicle (carid), and a single road (vsrid). The corresponding cloud database stores the LOD, based on the principle of low host load and saving database space. Figure 4 shows the open data captured by the subsystem at random time intervals. Based on the error message, meaning that the sensors cannot frequently read the traffic flow that occurs on the open data platform, the subsystem filters and analyzes the data through the program loop of R language, which captures and links the vdid, longitude, and latitude of the vehicle static data of the traffic sensing endpoint. Figure 5 illustrates the linked data content. The LOD are captured by the vehicle static information of the open information platform through R language and stored in the cloud database. Figure 6 shows the fragment program. The API updates data once a day. The web application automatically retrieves the database every five minutes and presents it on the webpage, thus greatly improving the real-time accuracy of the proposed system.

```
data.inital <- fromJSON("http://data.ntpc.gov.tw/od/data/api/92DC5275-ABE0-4AF0-ADEF-1A7A5487617491?$format=json")
data.inital.tb3 <- data.inital[which(data.inital$vsrdir == 0 &
                                     data.inital$vsrid == 0 &
                                     data.inital$carid == 'S'),]
data.inital.tb4 <- data.inital[which(data.inital$vsrdir == 1 &
                                    (data.inital$vsrid == 1 |
                                     data.inital$vsrid == 2 |
                                     data.inital$vsrid == 4) &
                                     data.inital$carid == 'S'),]

sqlConnString <- "driver={SQL Server};server=.;database=OpenDataDB;trusted_connection=true"
conn <- odbcDriverConnect(sqlConnString)
sqlSave(conn, data.inital.tb3, tablename = "Tb3_Final", append = TRUE, rownames = FALSE)
sqlSave(conn, data.inital.tb4, tablename = "Tb4_Final", append = TRUE, rownames = FALSE)
close(conn)
```

Figure 3. Vehicle dynamic data capture program segment.

vdid	datacollecttime	status	vsrid	vsrdir	speed	laneoccupy	carid	volume
65000V026160	2018/8/14 15:59	0	0	0	63	3	S	7
65000V017200	2018/8/14 15:59	0	0	0	48	2	S	7
65000V078810	2018/8/14 15:59	0	0	0	61	2	S	4
65000V008130	2018/8/14 15:59	0	0	0	55	1	S	1
65000V190740	2018/8/14 15:59	0	0	0	57	1	S	7
65000VT16170	2018/8/14 15:59	0	0	0	35	1	S	5
65000VT16510	2018/8/14 15:59	0	0	0	80	1	S	6
65000VT01420	2018/8/14 15:59	0	0	0	0	0	S	0

Figure 4. Example of reading real open data at a single time.

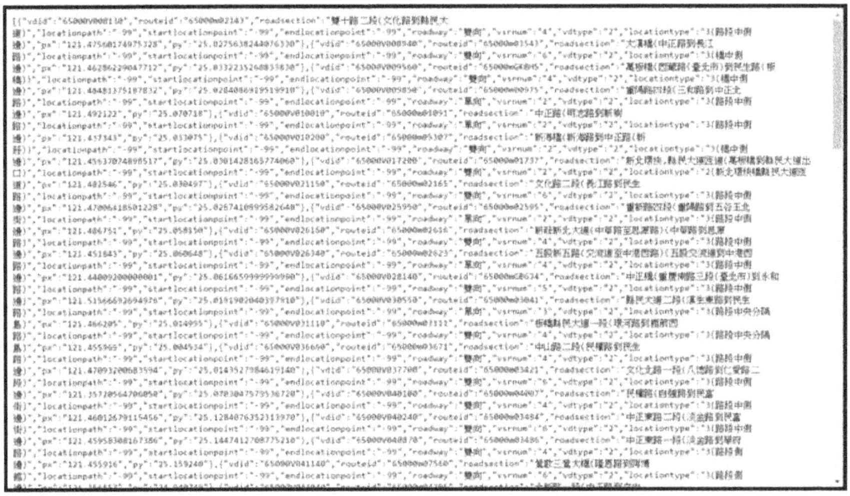

[{"vdid":"65000V00011U","routeid":"65000m02143","roadsection":"雙十路二段(文化路到龍民大
道)","locationpath":"-99","startlocationpoint":"-99","endlocationpoint":"-99","roadway":"雙向","vsrnum":"4","vdtype":"2","locationtype":"3(路段中間
道)","px":"121.4756017497532B","py":"25.0275638244076330"},{"vdid":"65000V00034U","routeid":"65000m01543","roadsection":"大業路(中正路到興三
路)","locationpath":"-99","startlocationpoint":"-99","endlocationpoint":"-99","roadway":"雙向","vsrnum":"6","vdtype":"2","locationtype":"3(樓中側
道)","px":"121.4628622904712","py":"25.0332235260033630"},{"vdid":"65000V00956U","routeid":"65000mG0045","roadsection":"萬壽橋(西濱路(臺北市)到民生路1樓
域)","locationpath":"-99","startlocationpoint":"-99","endlocationpoint":"-99","roadway":"雙向","vsrnum":"4","vdtype":"2","locationtype":"3(樓中側
道)","px":"121.4848137510783U","py":"25.0240096925199510"},{"vdid":"65000V00950U","routeid":"65000m00935","roadsection":"重陽橋往N(三和路到中正北
道)","locationpath":"-99","startlocationpoint":"-99","endlocationpoint":"-99","roadway":"單向","vsrnum":"2","vdtype":"2","locationtype":"3(路段中間
道)","px":"121.492123","py":"25.070718"},{"vdid":"65000V01001U","routeid":"65000m01093","roadsection":"中正路(明光路到新
路)","locationpath":"-99","startlocationpoint":"-99","endlocationpoint":"-99","roadway":"單向","vsrnum":"4","vdtype":"2","locationtype":"3(路段中間
路)}","px":"121.457343","py":"25.053075"},{"vdid":"65000V01020U","routeid":"65000m05307","roadsection":"新海路(新海路到中正路(新
道)","locationpath":"-99","startlocationpoint":"-99","endlocationpoint":"-99","roadway":"雙向","vsrnum":"2","vdtype":"2","locationtype":"3(樓中側
U)","px":"121.45633074889517","py":"25.0301428165774060"},{"vdid":"65000V01720U","routeid":"65000m01737","roadsection":"新北環快.鶯民大道到道士樓到鶯民大橋出
入口)","px":"121.482546","py":"25.030497"},{"vdid":"65000V02115U","routeid":"65000m02165","roadsection":"文化路二段(長江路到林
路)","locationpath":"-99","startlocationpoint":"-99","endlocationpoint":"-99","roadway":"雙向","vsrnum":"6","vdtype":"2","locationtype":"3(路段中間
道)","px":"121.484751","py":"25.050150"},{"vdid":"65000V02610U","routeid":"65000m02616","roadsection":"新莊新北大道(中華路至重慶路到民眾
道)","locationpath":"-99","startlocationpoint":"-99","endlocationpoint":"-99","roadway":"雙向","vsrnum":"2","vdtype":"2","locationtype":"3(路段中間
道)","px":"121.47006418501228","py":"25.0257410099582640"},{"vdid":"65000V02595U","routeid":"65000m02595","roadsection":"重新路四段(重陽橋到五谷王北
街)","locationpath":"-99","startlocationpoint":"-99","endlocationpoint":"-99","roadway":"雙向","vsrnum":"2","vdtype":"2","locationtype":"3(路段中間
路)","px":"121.453843","py":"25.060648"},{"vdid":"65000V02640U","routeid":"65000m02623","roadsection":"五股新五路(交流道至中圈西廊到五股交流道到中圈
路)","locationpath":"-99","startlocationpoint":"-99","endlocationpoint":"-99","roadway":"雙向","vsrnum":"4","vdtype":"2","locationtype":"3(路段中間
路)","px":"121.44009200000001","py":"25.0616659999999990"},{"vdid":"65000mG0634","roadsection":"中正線(蘆慶州路三段(臺北市)到永和
路)","locationpath":"-99","startlocationpoint":"-99","endlocationpoint":"-99","roadway":"單向","vsrnum":"5","vdtype":"2","locationtype":"3(路段中間
道)","px":"121.5156659269497U","py":"25.0191900040379107"},{"vdid":"65000m050","routeid":"65000m0041","roadsection":"鶯民大道二段(淡東路到民生
路)","px":"121.466205","py":"25.014995"},{"vdid":"65000V01110U","routeid":"65000m01113","roadsection":"榆幡鶯民大道一段(後湖路到民生
路)","px":"121.455965","py":"25.004534"},{"vdid":"65000V03660U","routeid":"65000m03671","roadsection":"中山路二段(民權路到民生
路)","locationpath":"-99","startlocationpoint":"-99","endlocationpoint":"-99","roadway":"單向","vsrnum":"2","vdtype":"2","locationtype":"3(路段中間
路)","px":"121.470932000003594","py":"25.014352785461914U"},{"vdid":"65000m03421","roadsection":"文化北路一段(八德路到仁愛路)到復興路二
路)","locationpath":"-99","startlocationpoint":"-99","endlocationpoint":"-99","roadway":"單向","vsrnum":"4","vdtype":"2","locationtype":"3(路段中間
路)","px":"121.35720564700058","py":"25.0703047579536720"},{"vdid":"65000V04010U","routeid":"65000m04007","roadsection":"民權路(自強路到復興路
路)","locationpath":"-99","startlocationpoint":"-99","endlocationpoint":"-99","roadway":"單向","vsrnum":"4","vdtype":"2","locationtype":"3(路段中間
路)","px":"121.46012679115456","py":"25.1284076352113970"},{"vdid":"65000V04240U","routeid":"65000m03484","roadsection":"中正東路二段(自強路到民
路)","locationpath":"-99","startlocationpoint":"-99","endlocationpoint":"-99","roadway":"單向","vsrnum":"6","vdtype":"2","locationtype":"3(路段中間
道)","px":"121.45958308167386","py":"25.1447412708775210"},{"vdid":"65000V04007U","routeid":"65000m03486","roadsection":"中正東路一段(自強路到雙鳳
路)","locationpath":"-99","startlocationpoint":"-99","endlocationpoint":"-99","roadway":"雙向","vsrnum":"4","vdtype":"2","locationtype":"3(路段中側
道)","px":"121.455916","py":"25.159240"},{"vdid":"65000V01140U","routeid":"65000m07560","roadsection":"集慶三重大橋(環西路到河民
路)","locationpath":"-99","startlocationpoint":"-99","endlocationpoint":"-99","roadway":"單向","vsrnum":"2","vdtype":"3","locationtype":"3(路段中側
道)"...

Figure 5. Linked static vehicle data. "雙向" is two-way; "單向" is one-way; The rest of the Chinese characters are the addresses of the corresponding road sections.

```
5    data.location <- fromJSON("http://data.ntpc.gov.tw/od/data/api/D4ACABED-5960-4A2B-9AF1-C62D7AF17622?format=json")
6    data.loca.tb1 <- data.location[c("vdid","px","py","roadsection")]
7    write.csv(data.loca.tb1, file = "Tb1_Sensor.csv", row.names = FALSE)
8    conn.azure <- odbcDriverConnect("Driver={ODBC Driver 13 for SQL Server};
9                                     Server=tcp:ywsite.database.windows.net,1433;Database=OPDataDB;
10                                    Uid=m622@ywsite;Pwd=Isleisle509c;Encrypt=yes;
11                                    TrustServerCertificate=no;Connection Timeout=30;")
12   #conn <- odbcDriverConnect("driver={SQL Server};server=.;database=OpenDataDB;trusted_connection=true")
13   #sqlSave(conn,data.loca.tb1,tablename = "Tb1_Sensor10",rownames = FALSE,append = TRUE)
14   sqlUpdate(conn.azure,data.loca.tb1,tablename="Tb1_Sensor",index = "vdid",fast = TRUE)
15   close(conn.azure)
```

Figure 6. Fragment program for storing linked open data (LOD) in the cloud database.

3.3. Establishment of Cloud Server and Corresponding Databases

The proposed system uses Microsoft Azure cloud services to set up a web server, as installation in the cloud can reduce the consumption of physical equipment. The system service can be diversified and open, and the website server is not restricted by physical attacks, such as a power failure in the computer room, network interruption, etc. In the case of power failure and network interruption, it can also quickly switch to hosts in other regions through cloud services to continuously maintain the adequacy rate of this cloud system, thus improving the practicability and sustainability of the cloud system. The cloud database stores the data of the New Taipei City Government, including open information, report status information, reporter information, and related information used by the corresponding subsystem, such as web applications and the user road condition reporting subsystem. When the users open the system, the system displays their longitude and latitude, as obtained through the Google Geolocation API on the system webpage, and converts it to the corresponding physical address. When users want to report new road condition information, such as information regarding congestion status, they can log the current situation information through the user road condition reporting subsystem, and other drivers and police units can know the situation and take follow-up actions. The positioning function used in this subsystem adopts HTML 5 to obtain the position longitude and latitude. Table 2 presents the description of the main parameters of GetPosition. However, as a cloud database has the advantage of flexibility, it can select the most suitable service content according to the current requirements, thereby reducing costs. The disadvantage lies in the safety problem; however, as traffic information is not personal data and is part of public information, the information service can be taken as the first priority.

Table 2. Description of main parameters and functions of GetPosition.

Parameter	Function Description
Coords.Latitude	Latitude
Coords.Longitude	Longitude
Accuracy	Accuracy (error range between detected position and actual position)
Maximum Age	Time to reacquire location information

3.4. User Reporting Mechanism on Road Conditions

At present, when traffic-related information systems in Taiwan are confronted with traffic emergencies, most people would call the police broadcast hotline, while people on the national highway would call the 1968 traffic voice service line for traffic inquiries, traffic notifications, road rescues, and other services. In addition, some road users often use the Google Maps application to check traffic information. However, neither the system nor the mobile phone application has a real-time information user reporting function for traffic emergencies. The establishment of the road condition user reporting mechanism requires users to enter the member mode in order to check the identity of the reporter and ensure the accuracy of the reported data. This can improve the system's ability to present the comprehensive real-time traffic data to other road users, and further improve the real-time information deficiency faced by government units in making decisions when considering road condition assessment. Figure 7a shows the operating flow of the road condition user reporting mechanism of the proposed system. First, member registration is required to ensure that the users are road users. As mentioned above, when users encounter road congestion (e.g., accidents, heavy traffic, processions, control, etc.) while driving on a road section and their path is blocked, the users can open the road condition reporting function of the system to add report information (e.g., congestion type and current situation). After completing the corresponding report form, the report form can be sent out and stored in the cloud database of the system. The processing flow of the added report information is shown in Figure 7b. Subsequent responsible units, such as police units and their related responsible units, receive the report information and then dispatch manpower to the location for road condition elimination in a more real-time and accurate manner, thereby solving the problems efficiently.

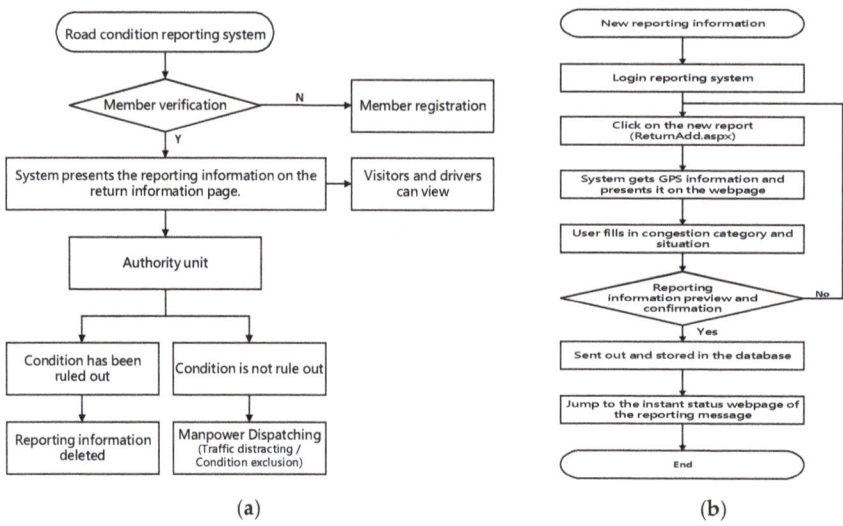

(a) (b)

Figure 7. Road condition user reporting mechanism and processing flow chart: (**a**) operation process of road condition user reporting mechanism; (**b**) processing flow chart of newly added information.

4. System Presentation and Efficiency Analysis

4.1. System Presentation

As mentioned above, the proposed system uses intelligent handheld devices and computer devices to present an intelligent traffic assistant agent using GPS and LOD technology with the optimal page size. The real-time display images include real-time information, a real-time image list, endpoint analysis information, and real-time reporting, which are listed and explained as follows. The web application of the agent presents all the operational functions of the proposed system. When the users turn on the system, the system takes the longitude and latitude of the users' location and presents it on a map through the GPS function of the intelligent handheld device held by the users, and then displays the "real-time status" data, such as the green color indicating smooth conditions and red indicating congested road conditions. Figure 8 shows the system execution screens, as opened by a computer and a handheld device, respectively.

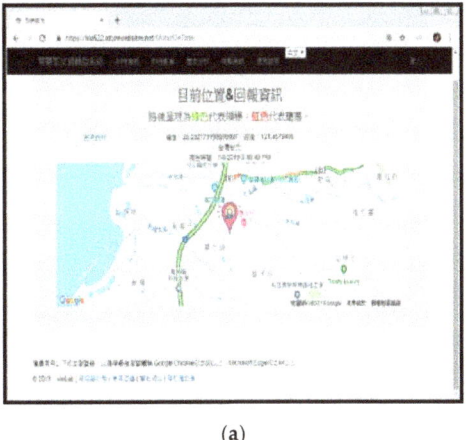 

(a)   (b)

Figure 8. Real-time traffic information displays for a (**a**) computer and (**b**) mobile phone. In this step we take the longitude and latitude of the user's location and present it on a map through the GPS function of the intelligent handheld device held by the user, and then, display the "real-time status" data, such as the green color indicating the smooth conditions and red indicating congested road conditions.

Regarding the traffic "real-time image" function, the proposed system combines many supported Google Chrome browsers and Apple Safari browsers and matches RWD web page technology for seamless connection to achieve a cross-platform and diversified traffic assistant agency system. When the users open this system, the system uses the GPS function of the intelligent handheld device held by the user, which reads the users' location and displays the nearby monitor screen, and then verifies the road section. Figure 9 shows the execution screens of a computer and a handheld device, respectively; users can click the "view" hyperlink to display the real-time image of the endpoint; users can also click endpoint number 3 to show the real-time image of Zhongzheng Road and Zhongshan Road. The computer and handheld device displays are shown in Figure 10 (capture time: 7 April 2019).

(a)   (b)

Figure 9. Real-time traffic image (Take the Tamsui District, New Taipei City, Taiwan as an example) displays for a (**a**) computer and (**b**) mobile phone. In this step we use the GPS function of the intelligent handheld device held by the user, which will read the user's location and display the nearby monitor screen, and then, look at the road section.

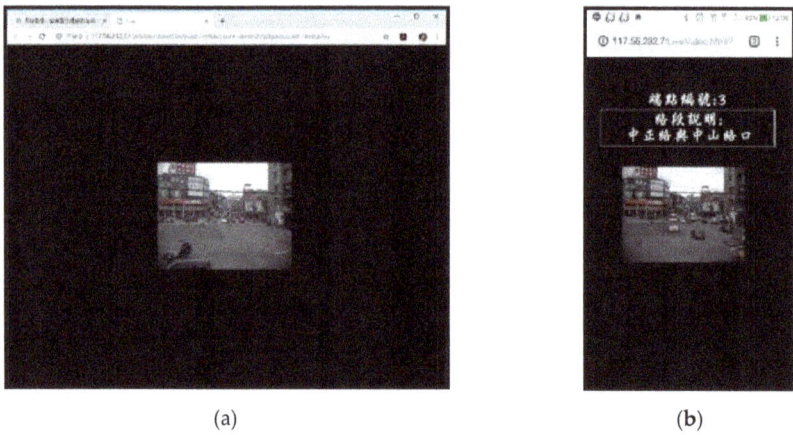

(a)   (b)

Figure 10. Real-time traffic image displays for endpoint number 3 for a (**a**) computer and (**b**) mobile phone. In this step we Click the "View" hyperlink to display the real-time image of the endpoint; for example, click endpoint number 3 to show the real-time image of Zhongzheng Road and Zhongshan Road.

As mentioned above, the "endpoint information" of this system is extracted from the dynamic vehicle information of the New Taipei City open data platform as the real-time data of the sensing endpoint. Due to the poor contact of the reported data of some sensing endpoints, the inability to sense data, and other reasons, the data is incomplete. In this study, JSON is retrieved through R language and the data content required for screening is automatically retrieved and stored in the cloud database, which is then presented through the traffic assistance agent web application. Users can use the intelligent handheld devices to view the most real-time endpoint road condition information (including speed and vehicle detection). Figure 11 shows the execution screens of a computer and handheld device, respectively.

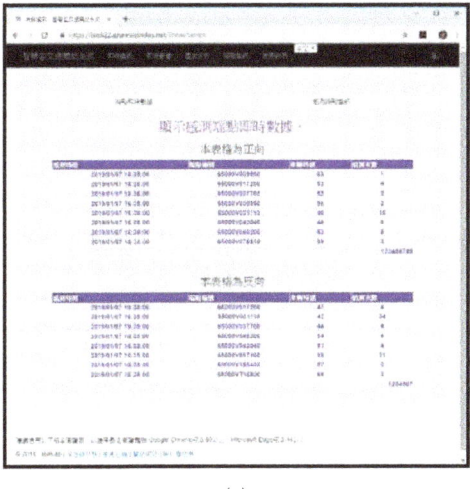 

(a)   (b)

Figure 11. Real-time data displays of endpoint information for a (**a**) computer and (**b**) mobile phone. The "endpoint information" of this system is extracted from the dynamic vehicle information of the New Taipei City Open Data Platform as the real-time data of the sensing endpoint, including forward and reverse road information.

Based on the above, as sensing endpoints often fail to sense data and report error messages, the proposed system uses the longitude and latitude positions of the endpoints of the vehicles not in motion and the vdid of the sensing endpoints on the New Taipei City open data platform. The obtained data are filtered through R language, and the data of each sensing endpoint is updated in the system's cloud database. If there are additional sensors, the system stores them in the cloud database. The users can view the "sensing endpoint information" of each sensed endpoint through the system, so that government information can be openly and transparently displayed. At that time, the users or government work units can use a smart handheld device to view and attend to the endpoint to carry out relevant work services, such as inspection and repair. Figure 12 shows the execution screen of the system opened by a computer and a handheld device, respectively. Users click the "navigation" hyperlink to open the icon of the Google Maps display endpoint. Taking endpoint number 65000V008130 as an example, the displays of the computer and handheld devices are shown in Figure 13.

Regarding the "road condition reporting function", after logging into the system, when users arrive at an accident emergency site, their intelligent handheld device can read the longitude and latitude of their location through the built-in GPS function and present it in real time on the reporting page. If the users click on the relevant information (e.g., category, event, date, and time) of the emergency site, the system will immediately store it in the cloud database, and road users behind them can see the information. The police and public works units can arrive at the site promptly to eliminate the reported traffic condition, thus shortening the congestion time of the traffic section. Figure 14 shows the execution screens opened for a computer and a mobile phone, respectively, to complete the "add road condition reporting information".

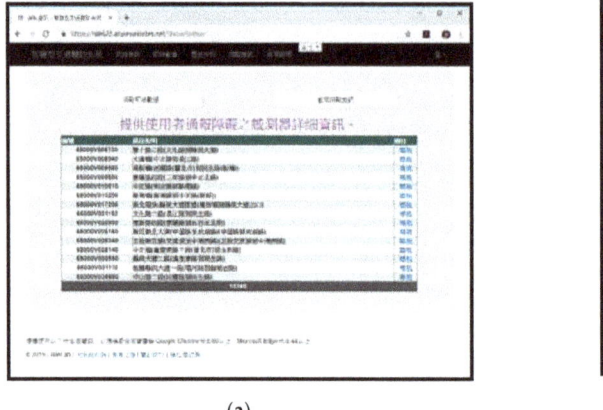

(a)	(b)

Figure 12. Sensing information displays of endpoints for a (**a**) computer and (**b**) mobile phone. The user will view the "sensing endpoint information" of each sensed end through the system, at that time, the user or government work unit will hold an intelligent handheld device to view and attend the endpoint to carry out relevant work services.

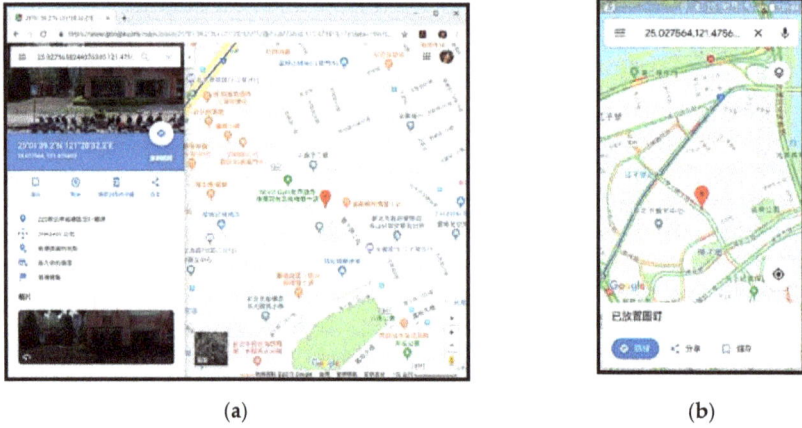

(a)	(b)

Figure 13. Navigating to endpoint number 65000V008130. Displays for a (**a**) computer-side and (**b**) mobile phone. The user will click the "navigation" hyperlink to open the icon of the Google Map display endpoint. Taking endpoint number 65000V008130 as an example.

After being verified as a member and completing the "add road condition reporting information", the "history analysis" page of the system obtains that reporting information through the cloud database, as shown in Figure 15; that is, clicking on the navigation function of the page will open the icon of the Google Maps display endpoint.

The proposed system establishes a back-end management system for user reporting information, which is mainly responsible for management, such as report information resolution, member management, real-time image settings, etc. The identities of users are visitors, registered members, and administrators. The hierarchical information of relevant identities is detailed in Table 3. Visitors only have the function of viewing information, and cannot use any of the reporting functions. Registered members can enter the member area to add new reporting information. Administrators can enter the management area, browse and modify all member data, and delete members. Figure 16 shows the page of the back-end system management.

Figure 14. Road condition user reporting function displays for a (**a**) computer and (**b**) mobile phone. After logging into the system, when a user arrives at an accident emergency site, their intelligent handheld device can read the longitude and latitude of their location through the built-in GPS function, and present it in real time on the reporting page to complete add road condition reporting information.

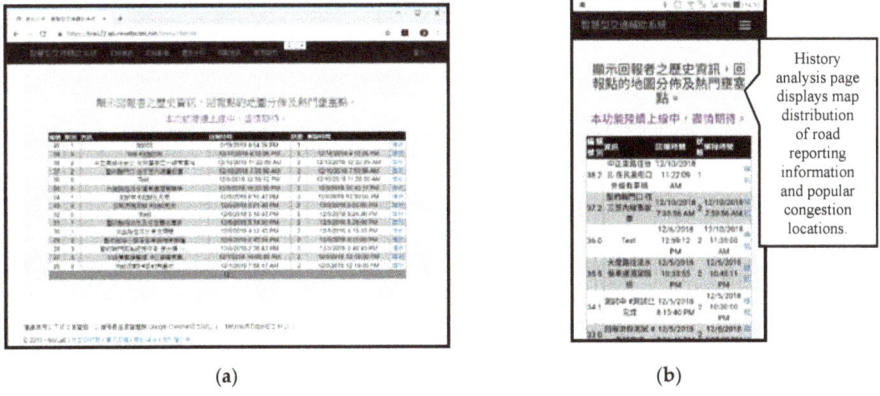

Figure 15. History analysis page displays map distribution of road reporting information and popular congestion locations for a (**a**) computer and (**b**) mobile phone. History analysis page of the system will obtain the return information through the cloud database; that is, clicking on the navigation function of the page will open the icon of the Google Maps display endpoint and go there.

Table 3. Hierarchical authority list for system members.

Authority Items	Visitor	Registered Member	Administrator
View real-time information	✓	✓	✓
View live images	✓	✓	✓
View reporting information	✓	✓	✓
Add reporting information	✗	✓	✓
Modify reporting information	✗	✗	✓
Real-time image setting	✗	✗	✓
Membership management	✗	✗	✓

Legend: "✓" means to have this function; while "✗" means none.

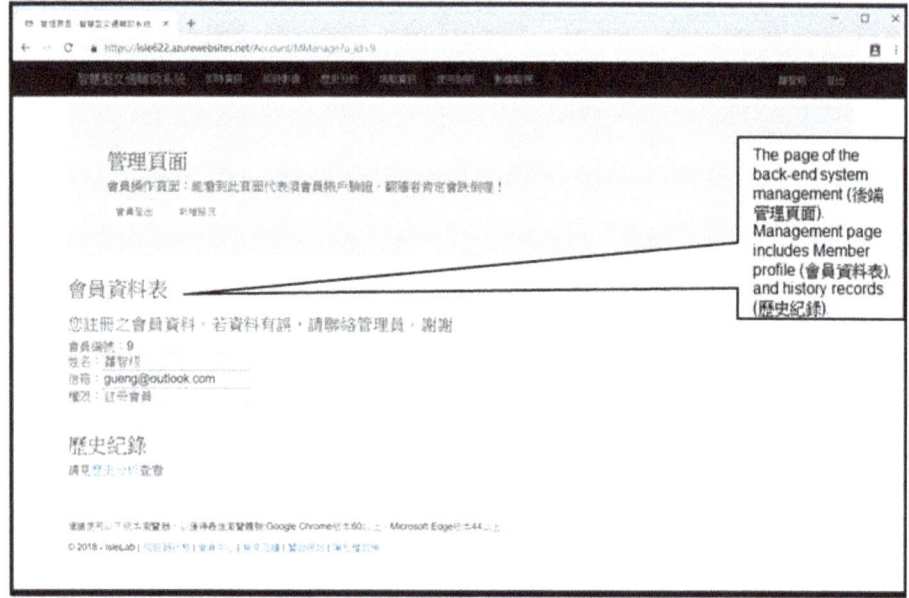

Figure 16. Back-end system management page.

The reason for the hierarchical management of the system's modification of report information is that different authorities and responsible units (e.g., police units, public works units, etc.) have slightly different authority. For example, the real-time image setting is within the jurisdiction of police units. Table 4 and Figure 17 present the hierarchical authority table of power and the responsible units, the page of power, and responsible unit management area in the system. In addition, the public works unit, police unit, and back-end management can add and modify the information related to the road condition report. Police units and back-end management can add and modify settings related to real-time imaging devices. Finally, back-end management has all the rights of the system, and can delete additional settings, such as members.

Table 4. Hierarchical authority list of power and responsible units.

Administrative Rights Items	Public Works Unit	Police Unit	Back-end Management
Add reporting information	✓	✓	✓
Modify reporting information	✓	✓	✓
Real-time image setting	✗	✓	✓
Membership management	✗	✗	✓

Legend: "✓" means to have this function; while "✗" means none.

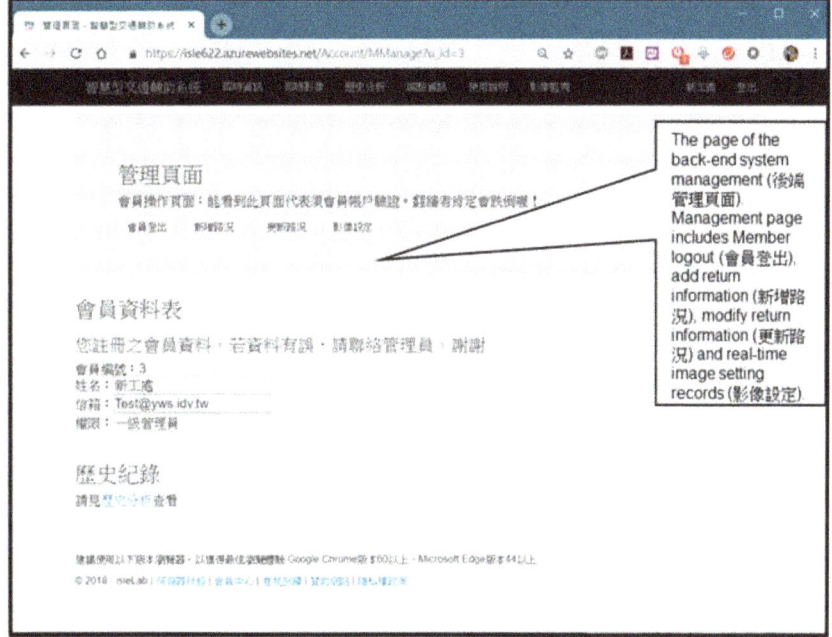

Figure 17. Authority unit management page.

4.2. Performance Analysis

The following comparisons and experiments involved users from related departments (Department of Information and Communication Engineering), unrelated departments (Department of Business Management), and users from the middle–high age group (over 45 years old). There were ten persons in each group, and the three groups were used as a reference frame for analysis. All comparisons and evaluation results were based on opinions, for which there was 75% or greater agreement in the three groups, showing that the majority of the subjects were in agreement.

The main comparison objects of this system are four traffic information systems related applications, namely Police Broadcast Real-Time Traffic, Road Condition Autotoll, Real-time Traffic Image—RoadCam, and New Taipei City Advanced Traveler Information System. The following experiment presents the analysis of data accessibility and system performance. This is the first condition for a user-friendly interface of information systems—rapid response information accessibility. The content of the first experiment is to explore the data accessibility analysis of the system, as well as its related systems. This study compared the following items: open real-time information, real-time images, road condition reporting, real-time weather, and other information for analysis and comparison. Table 5 presents the comparison between the proposed system and its related applications for the corresponding information, as obtained by the number of clicks with the same target. The "X" in the table indicates that the system does not have this function, while the numbers in the table means the number of clicks for the corresponding data. This study performed the click count experiment in real time, and recorded the data for analysis and comparison with other applications. This study developed a total of eight comparison items for the average number of clicks to perform experiments, and developed Equation (1), where S is the comparison system, K is the comparison system number, n is the number of system comparison items, Click is the number of clicks of the comparison item, x denotes that the comparison item of the system has no such function and is not included in the calculation, and N_x is the number of such functions available. In the end, the data acquisition clicks of the proposed system averaged 1.6 times, while both the RoadCam and New Taipei City Advanced Traveler Information System averaged 1.8 times, and the remaining Police Broadcast Real-Time Traffic and Autotoll both averaged more than 2 times. It is obvious that the proposed system exhibits excellent performance in the accessibility of the information acquisition interface, which is the first condition for building an effective information system with user-friendly interfaces.

$$\text{Average}_S = \mathop{S}_{1}^{K} \left(\frac{\sum\limits_{i \notin x}^{n} \text{Click}_i}{n - N_x} \right) \tag{1}$$

The second experiment is the content of system operating environment and user experience analysis. This is also one of the important topics for constructing contemporary information systems with user-friendly interfaces. In order to explore the function test of this system, three categories are planned: positioning, multiplatform, and user experience, including eight items for function comparison (see Table 6 for details). The comparison objects of this experiment are the four systems mentioned above. Through experimental tests, the proposed system can accurately obtain the current location, both indoors and outdoors, with a handheld device with the GPS function, and the other two can also obtain the current location. The system can be used normally in multiplatform browsers. At present, there are only IOS and Android platforms in Autotoll, and there is no computer version. Finally, the user experience part of the comparison system is discussed. The proposed system uses the Microsoft Azure web server and database applications. The traffic can be automatically allocated to optimize the operation of the system, and the multiplatform RWD responsive webpage technology [54] is adopted. Thus, users do not have to repeatedly learn the webpage version or the mobile phone version of the proposed system.

The third experiment is the content and function balance of the important items of the interface design. The design preference to importance ratio (DIR) (Equation (2)), as proposed by Ha [55], is taken as the design principle of the human–machine interface (HMI) of this system. The HMI is ideal when combined with the balancing index (BI) (see Equation (3)) to define the user interface. In short, if BI is zero, all the HMI elements in the interface design are balanced and perfect. Its physical meaning is that the HMI design satisfies the principle of the design preference for importance, and the interface operation of the system is more consistent with user demand:

$$\text{DIR}_{ijk} = \frac{\frac{DP_{ij}}{\sum_{i=1}^{n} DP_{ij}}}{\frac{I_{ik}}{\sum_{i=1}^{n} I_{ik}}} \tag{2}$$

$$\text{BI}_{jk} = \frac{\left| \sum_{i=1}^{n} \log_{10} \text{DIR}_{ijk} \right|}{n} \tag{3}$$

where DIR_{ijk} is the DIR of design attribute j and importance attribute k of HMI interface element i. DP_{ij} is the design preference of design attribute j of HMI interface element i. I_{ik} denotes design importance k of the importance attribute of HMI interface element i. BI_{jk} is the balance index of design attribute j and importance attribute k, while n is the total number of HMI interface elements.

Table 5. Analysis results of data acquisition accessibility.

Click	The Proposed System	Police Broadcast Real-Time Traffic	Road Conditions Autotoll	Real-Time Traffic Image: RoadCam	New Taipei City Advanced Traveler Information System
Open real-time information	1	2	3	2	1
Open real-time image	1	2	3	2	2
Open traffic report	4	6	X	X	X
Road conditions reporting content	1	2	X	X	X
Real-time weather information	X	1	X	2	2
Personalized subscription information	X	X	1	X	X
Disaster prevention monitoring information	X	X	X	2	3
Other information links	1	1	1	1	1
Average information acquisition	**1.6**	**2.3**	**2**	**1.8**	**1.8**

Table 6. Analysis results of system performance.

Comparison Items	Comparison Items	The proposed System	Police Broadcast Real-Time Traffic	Road Conditions Autotoll	Real-Time Traffic Image: RoadCam	New Taipei City Advanced Traveler Information System
Positioning	Continuous positioning	✓	✓	✓	✓	✓
	Outdoor positioning	✓	✓	✓	✓	✓
Cross platform	Android	✓	✓	✓	✓	✓
	iOS	✓	✗	✓	✓	✓
	Chrome/Safari	✓	✓	✗	✗	✗
User experience	Easy to update	✓	✓	✓	✓	✓
	Easy to operate	✓	✓	✗	✗	✓
	Graphical interface	✓	✗	✓	✓	✓

Legend: "✓" means to have this function; while "✗" means none.

A five-scale evaluation scheme is used to evaluate each design preference (DP_{ij}). The design preferences include very good, good, moderate, weak, and very weak, and their corresponding values are five, four, three, two, and one, respectively. Table 7 displays the evaluation of the HMI elements in our design attributes, as well as the corresponding informational importance of the proposed agents, as evaluated using the analytic hierarchy process [55]. Table 8 illustrates the evaluation results of DIR with average BI = 0.012534 (BI should approach zero for the best balance) of the proposed agents. "Needs to improve a little bit" means that the proposed system interface design must be slightly adjusted, but will not affect the operations of the proposed system. The verification results show that the human-machine interfaces of our proposed agents can meet important design preferences and provide approximately optimal balance.

Table 7. Evaluation of human–machine interface (HMI) elements and their corresponding importance.

HMI Elements	Description	Design Preference	Informational Importance	Remarks
RTILL	Real-time information label	5	0.204082	Label in Text
ARTILT	A real-time image list	3	0.265306	List in Text
EAILL	Endpoint analysis information label	3	0.265306	Label in Text
RTTLL	Real-time reporting label	3	0.265306	Label in Text

Table 8. Evaluation results of design preference to importance ratio (DIR) where balancing index (BI) = 0.012534.

HMI Elements	DIR	Description	BI
RTILL	1.228055	Needs to improve a little bit	
ARTILT	0.939958	Needs to improve a little bit	0.012534
EAILL	0.944658	Needs to improve a little bit	
RTTLL	0.939958	Needs to improve a little bit	

Finally, this study did not analyze the correctness and satisfaction of the information provided by the proposed system because it adopted the LOD of government units, which provide accurate and reliable information. Moreover, this study developed an instant image-based information system that is user-friendly and offers nearly perfect design interfaces to present the LOD of government units. The answer to the above question is, hence, self-evident.

5. Conclusions and Discussions

Based on GPS and LOD technology, this study developed a multiplatform, cloud-based, and instant image-based traffic assistance agent with user-friendly interfaces, which provides various functions, including real-time information, real-time images, endpoint analysis information, and real-time user reporting. It also has quite good user experience in the system interface design, and the system files and required databases are cloud-based. On a smart handheld device with Internet connection, users can access the proposed system and the condition reporting system. The proposed system does not have any demand on the device capacity, as all systems are developed on a cloud-based smart traffic assistance system. The comprehensive system operation and presentation to performance comparison confirmed that the proposed system has excellent accessibility to the information acquisition interface and approximately optimal balance of the important items of the interface design. Moreover, due to the application of the Microsoft Azure web server and database, traffic can be automatically allocated to optimize system operation. In addition, this study adopted RWD responsive webpage technology, which makes it easier for users to use various devices without having to learn the relevant interfaces of the proposed system.

Although the proposed system has been successfully combined with Microsoft Azure cloud services to achieve cloud-based, instant image-based, and traffic LOD of government units, there are still many unfinished, missing, and yet to be improved areas, which are listed and described as follows:

1. At present, the time for reporting road conditions is modified manually. In the future, a new system can be added to automatically judge the subsequent elimination time or cooperate with relevant government units, so that such units can carry out linked modification actions.
2. Future studies can strengthen the introduction of the data analysis function, and present periodic analysis charts to facilitate exploration, planning, and overall review of traffic flows before being updated by cities in the future;
3. Future studies can add an automatic reporting function for endpoint maintenance, where users can report sensor endpoint failures to relevant units;
4. Multiple accounts (e.g., LINE login, Google login, etc.) can be authorized to save users' login time before reporting road conditions.
5. As mentioned above, Taiwan is narrow and crowded with people and cars. Depending on the navigation system, it is prone to the dilemma of avoiding road section "A" and entering the congested road section "B". Therefore, the introduction of the proposed system is unique; however, Taiwan (including its capital city, Taipei) is an international tourist destination, and the interface version of future applications should adopt a multilingual model. In addition, future studies could explore how combining the proposed system with international traffic databases, such as a NoSQL database approach for processing traffic-related big data [56] and a real approach on open data and databases in analysis of traffic accidents [57]. Finally, further research can target the setting up of traffic information-related urban development strategies, data privacy rights, and urban data plans (e.g., four data-driven algorithms extracting useful information from high resolution traffic data [58], providing another level of automation in processing mechanisms and deserve more attention.

Author Contributions: The research article was completed by two authors, F.-H.C. and S.-Y.Y. F.-H.C. and S.-Y.Y. jointly designed the overall architecture and related algorithms, and also conceived and designed the experiments, however S.-Y.Y. coordinated the overall plan and direction of the experiments and related skills. F.-H.C. and S.-Y.Y. contributed analysis tools and also analyzed the data. F.-H.C. performed the experiments, and S.-Y.Y. wrote this paper and the related reply. All authors have read and agreed to the published version of the manuscript.

Funding: This research is partly sponsored under grants 106-2221-E-129-008 and 107-2632-E-129-001 by the Ministry of Science and Technology, Taiwan.

Acknowledgments: The authors would like to thank Yu-Wei Wu for his assistance in earlier system implementation and preliminary experiments. The authors feel deeply indebted to the Department of Electrical Engineering and Department of Information and Communication, St. John's University, Taiwan, for all aspects of assistance provided. All authors have read and agreed to the published version of the manuscript.

Conflicts of Interest: The authors declare no conflict of interest.

References

1. Sodano, D. Mobile Taiwan: A Look at a Highly Mobile Market. 2016. Available online: https://www.emarketer.com/Article/Mobile-Taiwan-Look-Highly-Mobile-Market/1014877 (accessed on 11 July 2019).
2. National Development Council, Taiwan. Investigation Report on Individual and Household Digital Opportunity in 2018. Available online: https://ws.ndc.gov.tw/Download.ashx?u=LzAwMS9hZG1pbmlzdHJhdG9yLzEwL2NrZmlsZS81NzQyMWI3YS0xMjhmLTRmMzAtYTY0ZC04MzdiYTkzNmRkZDUucGRm&n=MTA35bm05YCL5Lq65a625oi25pW45L2N5qmf5pyD6Kq%2f5p%2bl5aCx5ZGKLnBkZg%3d%3d&icon=.pdf (accessed on 26 March 2019).
3. He, W.Z. Taichung City Government Transcends Data Framework and Discovers Hidden Emotions to Make Traffic Management No Longer a Problem. 2017. Available online: https://www.ithome.com.tw/news/117280 (accessed on 11 July 2019).

4.   Ministry of the Interior Police Department, Taiwan. Traffic Accident Main Data Inquiry. 2018. Available online: http://stat.motc.gov.tw/mocdb/stmain.jsp?sys=100&funid=b3303 (accessed on 8 April 2018).

5.   Works Bureau of New Construction Office of the Taipei Municipal Government, Taiwan. Taipei City Road Management Center Promotes Road Construction Management Innovation Measures have Achieved Remarkable Results. 2018. Available online: https://nco.gov.taipei/News_Content.aspx?n=BB8CF6F431E10 630&s=6F5661BD113EBD54 (accessed on 8 April 2018).

6.   New Taipei City Tourism Bureau, Taiwan. *Annual Report of Tourism Bureau in 2017*; New Taipei City Government: Taiwan, China, 2017. Available online: http://cdn.tour-ntpc.com/site/a3be84a9-7283-40a3-b4c7 -758bf39c7828/Content/Upload/ContentPageFile/d4aebfd7-72ea-40d8-b87e-40c7e277e524.pdf (accessed on 26 March 2019).

7.   Su, W.B. The Number of Monthly Active Users in Line has Grown to 19 Million, and the Service Data has been Revealed! 2018. Available online: https://www.ithome.com.tw/news/121717 (accessed on 10 March 2018).

8.   Taiwan Ministry of Communications, China. Self-Use Passenger Car Usage Survey. 2019. Available online: http://ipgod.nchc.org.tw/dataset/315000000h-000004 (accessed on 24 November 2019).

9.   Taiwan Tourism Bureau, Ministry of Communications, China. Survey of Taiwanese Tourists in 2018. Available online: https://admin.taiwan.net.tw/FileUploadCategoryListC003340.aspx?CategoryID=7b8dffa9-3b9c-4 b18-bf05-0ab402789d59&appname=FileUploadCategoryListC003340 (accessed on 24 November 2019).

10. National Police Agency, Taiwan Ministry of Interior, China. Interior Information Open Platform: Historical Traffic Accident Information. 2019. Available online: https://data.moi.gov.tw/MoiOD/Data/DataDetail.aspx ?oid=67781E29-8AAD-46A9-A2C8-C3F339592C27 (accessed on 24 November 2019).

11. Flynn, D.B.; Gilmore, M.M.; Sudderth, E.A. *Estimating Traffic Crash Counts Using Crowdsourced Data: Pilot Analysis of 2017 Waze Data and Police Accident Reports in Maryland*; Technical Report: DOT-VNTSC-BTS-19-01; Bureau of Transportation Statistics: Washington, DC, USA, 2018.

12. Chunghwa Telecom, Taiwan. Road Condition Quick Easy Pass Plus. 2018. Available online: https://play.google.com/store/apps/details?id=com.chttl.android.traffic.plus (accessed on 19 December 2018).

13. Antyi, Taiwan. Real-Time Image of Traffic. 2018. Available online: https://play.google.com/store/apps/detai ls?id=com.sy.twcctv (accessed on 19 October 2018).

14. Lin, Z.Y. Police New. 2018. Available online: https://play.google.com/store/apps/details?id=tw.myself.oceanl in.newsbroadcast (accessed on 19 December 2018).

15. New Taipei City Government, Taiwan. New Taipei City Instant Traffic Information Network. 2018. Available online: http://atis.ntpc.gov.tw/ (accessed on 9 November 2018).

16. Cai, Y.H. Development of an Omni-Restaurant Information Consultation & Sharing Agent System with GPS & Linked Open Data Technologies. Master's Thesis, Department of Information and Communication Engineering, St. John's University, New Taipei City, Taiwan, 2015.

17. Wang, Y.C. Automatic News Classification Using Machine Learning and Linked Open Data. Master's Thesis, Department of Information Management, National University of Kaohsiung, Kaohsiung, Taiwan, 2018.

18. Yan, Y.H. Building a Prototyping Platform of Open Government Linked Data. Master's Thesis, Institute of Information Management, Chung Yuan Christian University, Taoyuan, Taiwan, 2018.

19. Yu, C.Y. Research on Linked Open Data Applications Using Dynamic Context Data Analysis. Master's Thesis, Department of Computer Science and Information Engineering, National Taiwan Normal University, Taipei, Taiwan, 2016.

20. Albert, M.P.; Ashkan, A.; Christophe, G.; Stefan, S. CEDAR: The Dutch historical censuses as Linked Open Data. *Semant. Web* **2017**, *8*, 297–310.

21. Alobaidi, M.; Mahmood, K.; Sabra, S. Semantic Enrichment for Local Search Engine using Linked Open Data. In Proceedings of the 25th International Conference Companion on World Wide Web, Montreal, QC, Canada, 11–15 April 2016; pp. 631–634.

22. Selvam, S.; Balakrishnan, R.; Ramakrishnan, B. Social Event Detection-A Systematic Approach using Ontology and Linked Open Data with Significance to Semantic Links. *Int. Arab J. Inf. Technol.* **2018**, *15*, 729–738.

23. Pourhomayoun, M.; Wang, H.; Vahedi, M.; Mazari, M.; Smith, J.; Owens, H.; Chernicoff, W. Real-Time Big Data Analytics for Traffic Monitoring and Management for Pedestrian and Cyclist Safety. In Proceedings of the Fifth International Conference on Big Data, Small Data, Linked Data and Open Data, Valencia, Spain, 24–28 March 2019; p. 33.

24. Juan, Y.C. Analysis of Spatio-temporal Data of Office Workers Based on GPS Signals Collected from Mobile Device. Master's Thesis, Department of Computer Science and Information Engineering, Feng Chia University, Taichung, Taiwan, 2018.

25. Liou, G.T. A Multi-Epoch Algorithm for GPS/BDS Single Frequency Precise Point Positioning. Master's Thesis, Institute of Applied Mechanics, National Taiwan University, Taipei, Taiwan, 2016.

26. Tu, T.Y. Method of Monitoring and Locating Pets by GPS and BLE. Master's Thesis, Institute of Electrical Engineering and Computer Science Master's In-service Program, National Taipei University of Technology, Taipei, Taiwan, 2016.

27. Fridman, L.; Weber, S.; Greenstadt, R.; Kam, M. Active Authentication on Mobile Devices via Stylometry, Application Usage, Web Browsing, and GPS Location. *IEEE Syst. J.* **2017**, *11*, 513–521. [CrossRef]

28. Ta, N.; Zhao, Y.; Chai, Y. Built environment, peak hours and route choice efficiency: An investigation of commuting efficiency using GPS data. *J. Transp. Geogr.* **2016**, *57*, 161–170. [CrossRef]

29. Aliprantis, J.; Kalatha, E.; Konstantakis, M.; Michalakis, K.; Caridakis, G. Linked Open Data as Universal Markers for Mobile Augmented Reality Applications in Cultural Heritage. *Digit. Cult. Herit. Lect. Notes Comput. Sci.* **2018**, *10605*, 79–90.

30. Khaghani, F.; Rahimi-Golkhandan, A.; Jazizadeh, F.; Garvin, M.J. Urban Transportation System Resilience and Diversity Coupling Using Large-scale Taxicab GPS Data. In Proceedings of the 6th ACM International Conference on Systems for Energy-Efficient Buildings, Cities, and Transportation, New York, NY, USA, 13–14 November 2019; pp. 165–168.

31. Bellavista, P.; Kupper, A.; Helal, S. Location-based services: Back to the Future. *IEEE Pervasive Comput.* **2008**, *7*, 85–89. [CrossRef]

32. Syu, Y.C. The Framework of Location Based Service Using IoT Subscription Agreement—Taxi Service in Taoyuan Area as a Case. Master's Thesis, Department of Computer Science and Information Engineering, Yuan Ze University, Taoyuan, Taiwan, 2018.

33. Chou, C.C. The Study of Driving Safety Enhancement based on IoT and LBS. Master's Thesis, Department of Information Communication, Chinese Culture University, Taipei, Taiwan, 2017.

34. Sun, G.; Liao, D.; Li, H.; Yu, H.; Chang, V. L2P2: A location-label based approach for privacy preserving in LBS. *Future Gener. Comput. Syst.* **2017**, *74*, 375–384. [CrossRef]

35. Wang, Y.; Xu, D.; Li, F. Providing Location-Aware Location Privacy Protection for Mobile Location-Based Services. *Tsinghua Sci. Technol.* **2016**, *21*, 243–259. [CrossRef]

36. Ukrit, M.F.; Venkatesh, B.; Suman, S. Location Based Services with Location Centric Profiles. *Int. J. Electr. Comput. Eng.* **2016**, *6*, 3001–3005.

37. Lin, C.; Su, H.; Knoblock, C.A.; Chiang, Y.Y.; Duan, W.; Leyk, S.; Uhl, J.H. Building Linked Data from Historical Maps. In Proceedings of the Second Workshop on Enabling Open Semantic Science Co-Located with 17th International Semantic Web Conference, Monterey, CA, USA, 8–12 October 2018; pp. 59–67.

38. Sansonetti, G.; Gasparetti, F.; Micarelli, A. Cross-Domain Recommendation for Enhancing Cultural Heritage Experience. In Proceedings of the 27th Conference on User Modeling, Adaptation and Personalization, New York, NY, USA, 9–12 June 2019; pp. 413–415.

39. Microsoft. The Developer's Guide to Azure. 2018. Available online: https://go.microsoft.com/fwlink/?LinkI d=862819&clcid=0x409 (accessed on 20 December 2018).

40. Wikipedia. Microsoft Azure. 2018. Available online: https://en.wikipedia.org/wiki/Microsoft_Azure (accessed on 20 December 2018).

41. Liou, W.X. An Azure ACES System for Air Quality Index Deteriorating Real-Time Early Warning. Master's Thesis, Department of Information Management, National Yunlin University of Science and Technology, Yunlin, Taiwan, 2018.

42. Ho, Y.C. A Case Study on Machine Learning for Customer Relationship Management in Service Industry. Master's Thesis, Department of Management Information Systems, National Chengchi University, Taipei, Taiwan, 2018.

43. Lin, Y.Z. Implementation of an Access Control System by Using Raspberry PI and Cloud Face Recognition Services. Master's Thesis, Department of M-Commerce and Multimedia Applications, Asia University, Taichung, Taiwan, 2017.

44. Richard, H. Check Out This Interactive Map of Azure Platform Services. 2016. Available online: https://www.itprotoday.com/microsoft-azure/check-out-interactive-map-azure-platform-services (accessed on 24 November 2018).
45. Diaz, F.; Freato, R. Azure Data Lake Store and Azure Data Lake Analytics. In *Cloud Data Design, Orchestration, and Management Using Microsoft Azure*; Apress: Berkeley, CA, USA, 2018; pp. 327–392.
46. Färber, M. The Microsoft Academic Knowledge Graph: A Linked Data Source with 8 Billion Triples of Scholarly Data. In Proceedings of the 18th International Semantic Web Conference, Auckland, New Zealand, 26–30 October 2019; pp. 113–129.
47. Chang, W.C. Integration of Google Map API and Evolutionary Algorithm for Route Planning in Road Flatness Inspection. Master's Thesis, Department of Civil Engineering, National Chung Hsing University, Taichung, Taiwan, 2016.
48. Wang, J.W. Using Google Maps in Ambulance Dispatch—A Case Study of the 119 Emergency System of the New Taipei City Fire Department. Master's Thesis, Department of Computer Science, National Taipei University of Education, Taipei, Taiwan, 2018.
49. Rahmi, A.; Piarsa, I.N.; Buana, P.W. FinDoctor-Interactive Android Clinic Geographical Information System Using Firebase and Google Maps API. *Int. J. New Technol. Res.* **2017**, *3*, 8–12.
50. Xia, N.; Cheng, L.; Chen, S.; Wei, X.; Zong, W.; Li, M. Accessibility based on Gravity-Radiation model and Google Maps API: A case study in Australia. *J. Transp. Geogr.* **2018**, *72*, 178–190. [CrossRef]
51. Tan, Z.; Hasegawa, S.; Beuran, R. Concept Map Building from Linked Open Data for Cybersecurity Awareness Training. In Proceedings of the Japanese Society for Artificial Intelligence Special Interest Group on Advanced Learning Science and Technology Workshop, Ishikawa, Japan, 14 July 2018; pp. 1–6.
52. Nyo, M.T.H.; Hein, W.Z. Design and Construction of Navigation Based Auto Self-Driving Vehicle using Google Map API with GPS. *Int. J. Trend Sci. Res. Dev.* **2019**, *3*, 65–68.
53. Wikipedia. Google Traffic. 2018. Available online: https://en.wikipedia.org/wiki/Google_Traffic (accessed on 21 December 2018).
54. Ryan, B. Responsive Web Design vs. Mobile App Development. 2013. Available online: https://www.techrepublic.com/blog/web-designer/responsive-web-design-vs-mobile-app-development/ (accessed on 15 December 2018).
55. Ha, J.S. A Human-machine Interface Evaluation Method Based on Balancing Principles. *Procedia Eng.* **2014**, *69*, 13–19. [CrossRef]
56. Kaur, M.; Singh, S.; Aggarwal, N. Study of NoSQL Databases for Storing and Analyzing Traffic Data. In Proceedings of the 2nd International Conference on Communication, Computing and Networking, NITTTR, Chandigarh, India, 29–30 March 2018; pp. 577–587.
57. Gladović, P.V.; Deretić, N.N. Open data and databases in analysis of traffic accidents in Belgrade. *Tehnika* **2018**, *73*, 247–253. [CrossRef]
58. Amini, Z. Data-Driven Approaches for Robust Signal Plans in Urban Transportation Networks. Ph.D. Thesis, Engineering-Civil and Environmental Engineering, Graduate Division, University of California, Berkeley, CA, USA, 2018.

Article

Investigation of High-Efficiency Iterative ILU Preconditioner Algorithm for Partial-Differential Equation Systems

Yan-Hong Fan [1], Ling-Hui Wang [1,*], You Jia [1], Xing-Guo Li [1], Xue-Xia Yang [1] and
Chih-Cheng Chen [2,*]

1   Department of Mechanics, School of Applied Science, Taiyuan University of Science and Technology,
    Taiyuan 030024, China; yhfan0509@tyust.edu.cn (Y.-H.F.); YouJia@tyust.edu.cn (Y.J.);
    XingGuoLi@tyust.edu.cn (X.-G.L.); XueXiaYang@tyust.edu.cn (X.-X.Y.)
2   Information and Engineering College, Jimei University, Xiamen 361021, Fujian, China
*   Correspondence: LingHuiWang@tyust.edu.cn (L.-H.W.); 201761000018@jmu.edu.cn (C.-C.C.)

Received: 14 October 2019; Accepted: 21 November 2019; Published: 28 November 2019

Abstract: In this paper, we investigate an iterative incomplete lower and upper (ILU) factorization preconditioner for partial-differential equation systems. We discretize the partial-differential equations into linear equation systems. An iterative scheme of linear systems is used. The ILU preconditioners of linear systems are performed on the different computation nodes of multi-central processing unit (CPU) cores. Firstly, the preconditioner of general tridiagonal matrix equations is tested on supercomputers. Then, the effects of partial-differential equation systems on the speedup of parallel multiprocessors are examined. The numerical results estimate that the parallel efficiency is higher than in other algorithms.

Keywords: iterative ILU; preconditioner; partial-differential equations; parallel computation

---

1. Introduction

In applied sciences, such as computational electromagnetics, the solving of partial-differential equation systems is usually touched upon. Many variables need to be sought for solving engineering problems. These often need to be transformed into a solution of partial differential equations. When solving partial differential equations, the equations need to be discretized. When discretizing partial differential equations, symmetric systems of equations are usually gotten. Hence, it is necessary to use the idea of symmetry to solve partial differential equations. Several studies on multi-computers have appeared. For instance, Eric Polizzi and Ahmed H. Sameh [1] contributed a spike algorithm as a parallel solution to hybrid banded equations. The algorithm firstly decomposes banded equations into block-tridiagonal form and then makes full use of the divide and conquer technique. However, by increasing the bandwidth, the parallel computation becomes much more complex, leading to a decrease in the parallel efficiency. Obviously, the highly efficient parallelism of banded systems is of great importance. Methods for block-tridiagonal linear equations contain iterative algorithms such as the multi-splitting algorithm [2,3]. The multi-splitting algorithm (MPA) [2] can be used to solve large band linear systems of equations; however, it sometimes has lower parallel efficiency. In [4], a method for working out block-tridiagonal equations is provided by the authors. Any incomplete type preconditioner will be appropriate for the algorithm. Based on the Galerkin principle, the parallelism solution for large-scale banded equations is investigated in [5]. In [6], a parallel direct algorithm is used on multi-computers. In [7], a parallel direct method for large banded equations is presented. A preconditioner of large-scale banded equations is discovered in [8–14]. The block successive over-relaxation method (BSOR) [10] can be adopted to solve large-scale systems of equations, but different parallel efficiencies will be presented because of the different optimal relaxation factors. These algorithms use parallelism to solve banded equations but they cannot contain solving partial differential equations. From better provision of a computing environment, a highly efficient preconditioner can be carried out on multi-computers [15–23]. Simultaneously, Krylov subspace solvers [24–30] and preconditioners [31–38] for large-scale banded equations are commonly used, including the generalized minimal residual (GMRES) [39]. The pseudo-elimination method with parameter k (PEk) [40] can be applied on multi-processors; however, the setting of parameter k will influent the speedup and parallel efficiency. These are mostly preconditioners for sparse linear systems or partial differential equation problems in Graphics Processing Unit (GPU) computation. However, these methods consume great computational effort. The development of a new algorithm which needs less calculation among every iteration and has more speedup and higher parallel efficiency is required. This paper is based on the symmetry subject of solving partial differential equation systems. The systems of equations are usually symmetric. In the process of solving them, the systems of equations need to be divided into blocks. The block equations may be symmetric or asymmetric, so this paper considers the general form of block equations. Of course, for symmetric block equations, the incomplete lower and upper factorization preconditioner (ILUP) algorithm is suitable. This paper is concerned with partial-differential equation systems of the form $Ax = b$. The associated iterative form $Mx^{(k+1)} = Nx^{(k)} + b$ is used. The linear tridiagonal special form is tested on multi-processors. Then, the iterative ILUP for partial differential equation systems is used to examine multi central processing unit (CPU) cores.

The outline is as mentioned hereunder. Section 2 describes a decomposition strategy of a parallel algorithm. Section 3 documents the analysis of convergence. Section 4 introduces the parallel implementation of this algorithm. The analysis of results computations with numerical examples including a large-scale system of equations and partial-differential equations are presented in Section 5. Finally, we conclude the paper in Section 6.

2. Decomposition Strategy

Consider large-scale band equations

$$Ax = b \tag{1}$$

that is

$$
\begin{pmatrix}
A_1 & B_1 & & & & \\
C_2 & A_2 & B_2 & & & \\
& \ddots & \ddots & \ddots & & \\
& & C_{n-1} & A_{n-1} & B_{n-1} \\
& & & C_n & A_n
\end{pmatrix}
\begin{pmatrix}
x_1 \\ x_2 \\ \vdots \\ x_{n-1} \\ x_n
\end{pmatrix}
=
\begin{pmatrix}
b_1 \\ b_2 \\ \vdots \\ b_{n-1} \\ b_n
\end{pmatrix}
$$

where A_i, B_i, and C_i are $d_i \times d_i$, $d_i \times d_{i+1}$, and $d_i \times d_{i-1}$, and x_i, b_i are the d_i- vectors of the unknowns and the right–hand side,

The coefficient matrix A can be approximately decomposed as

$$A \approx GH \tag{2}$$

Generally, supposing $n = pm(m \geq 2, m \in \mathbf{Z})$, where p represents the processors, let

$$M = GH$$

where

$$G = \begin{pmatrix} I_1 & & & & & & & & \\ L_2 & I_2 & & & & & & & \\ & \ddots & \ddots & & & & & & \\ & & L_m & I_m & S_m & & & & \\ & & & & I_{m+1} & & & & \\ & & & & L_{m+2} & I_{m+2} & & & \\ & & & & & \ddots & \ddots & & \\ & & & & & & L_{2m} & I_{2m} & S_{2m} \\ & & & & & & & I_{2m+1} & \\ & & & & & & & L_{2m+2} & I_{2m+2} \\ & & & & & & & & \ddots & \ddots \\ & & & & & & & & & L_n & I_n \end{pmatrix}$$

$$H = \begin{pmatrix} U_1 & S_1 & & & & & & & \\ & \ddots & \ddots & & & & & & \\ & & U_{m-1} & S_{m-1} & & & & & \\ & & & U_m & & & & & \\ & & & L_{m+1} & U_{m+1} & S_{m+1} & & & \\ & & & & & \ddots & \ddots & & \\ & & & & & & U_{2m-1} & S_{2m-1} & \\ & & & & & & & U_{2m} & \\ & & & & & & & L_{2m+1} & U_{2m+1} & S_{2m+1} \\ & & & & & & & & \ddots & \ddots \\ & & & & & & & & & U_{n-1} & S_{n-1} \\ & & & & & & & & & & U_n \end{pmatrix}$$

(3)

in which

$$S_i = B_i, \ i = m(q-1)+1, \cdots, m(q-1)+m-1, \ q = 1, \cdots, p$$
$$S_i = B_i A_{i+1}^{-1}, \ i = mq, \ q = 1, 2 \cdots, p-1$$
$$L_i = C_i, \ i = m(q-1)+1, q = 2, 3, \cdots, p$$
$$L_i = C_i U_{i-1}^{-1}, \ i = m(q-1)+2, \ \cdots, m(q-1)+m, \ q = 1, \cdots, p$$
$$U_i = A_i, \ i = m(q-1)+1, \ q = 1, \cdots, p$$
$$U_i = A_i - L_i S_{i-1}, \ i = mq+2, \ \cdots, mq+m-1, \ q = 0, \cdots, p-1; i = mq+m, q = p-1$$
$$U_i = A_i - L_i S_{i-1} - S_i L_{i+1}, \ i = m(q-1)+m, \ q = 1, \cdots, p-1$$
and I_i is a $d_i \times d_i$ unit matrix, $i = 1, \cdots, n$.

Then

$$N = M - A$$

that is

$$N = \begin{pmatrix} (O) & & & & & & & \\ & & & O & S_m S_{m+1} & & & \\ & & O & & & & & \\ & L_{m+2} L_{m+1} & & & & & & \\ & & & (O) & & & & \\ & & & & O & S_{2m} S_{2m+1} & & \\ & & & & & \ddots & & \\ & & & & & & O & \\ & & & & & L_{(p-1)m+2} L_{(p-1)m+1} & & \\ & & & & & & & (O) \end{pmatrix}$$

where (O) is the $\sum\limits_{i=1}^{m} d_i \times \sum\limits_{i=1}^{m} d_i$ zero matrix. Therefore, the new iterative scheme for the large-scale band system of equations is

$$GHx^{(k+1)} = Nx^{(k)} + b \tag{4}$$

where the iterative matrix is

$$T = H^{-1}G^{-1}N$$

Obviously, GH is nonsingular, which is the necessary condition that the algorithm holds. In terms of the structure of G and H, the parallelism of the iterative algorithm is preferable.

The strategy is an ILUP algorithm. Compared with published algorithms [2,10,40], the ILUP algorithm requires less multiplication and adds calculation among every iteration, meaning this algorithm has more speedup and higher parallel efficiency. It is appropriate for solving the large-scale system of equations and partial-differential equations for multi-core processors.

3. Analysis of Convergence

3.1. Preliminary

Here, some notations are introduced. Two definitions and one lemma are mentioned.

Definition 1. ([39]) A real $n \times n$ matrix $A = (a_{i,j})$ with $a_{i,j} \leq 0$ for all $i \neq j$ is an M-matrix if A is nonsingular and $A^{-1} \geq O$.

Definition 2. ([39]) The matrix $A, M, N, A = M - N$ is a regular splitting of A if M is nonsingular, $M^{-1} \geq O$, $N \geq O$.

Lemma 1. ([39]) Presume $A = M - N$ is a regular splitting of A. Then, A is nonsingular and $A^{-1} \geq O$, if and only if $\rho(M^{-1}N) < 1$.

3.2. Proposition and Theorem

Note that the inverse matrix of the following matrix is gained by the algorithm of the Gaussian elimination. Firstly, from the definitions and lemma, a proposition is obtained as follows.

Proposition 1. If A is an M-matrix, in this way, the matrices U_i $(i = 1, 2, 3, \cdots, n)$ defined by Expression (3) satisfy $U_i^{-1} \geq O$.

Proof. From Expression (3), in terms of the contracture of A, G, H and $M = GH, N = M - A$, we have

$U_i = A_i, i = m(q-1)+1, q = 1, \cdots, p$

$U_i = A_i - L_i S_{i-1} = A_i - C_i U_{i-1}^{-1} B_{i-1}, i = m(q-1)+2, \cdots, m(q-1)+m-1, q = 1, \cdots, p-1;$

$i = n - m + 1, \cdots, n, q = p;$

$U_i = A_i - L_i S_{i-1} - S_i L_{i+1}, i = m(q-1)+m, q = 1, \cdots, p-1.$

As A is an M-matrix, then $U_i^{-1} \geq O$ for $i = m(q-1)+1, q = 1, \cdots, p$

Let $W_i = \begin{pmatrix} A_{(i-1)m+1} & B_{(i-1)m+1} & & & \\ C_{(i-1)m+2} & A_{(i-1)m+2} & \ddots & & \\ & & \ddots & \ddots & B_{im-1} \\ & & & C_{im} & A_{im} \end{pmatrix}$, then $W_i^{-1} \geq O$.

Since the block on the m-th row and m-th column of W_i^{-1} is U_i^{-1} for $i = m(q-1)+2, \cdots, m(q-1)+m-1$ and $q = 1, \cdots, p-1;$

Hence, $U_i^{-1} \geq O$ for $i = m(q-1)+2, \cdots, m(q-1)+m-1$ and $q = 1, \cdots, p-1;$

Furthermore,

$V_i = \begin{pmatrix} A_{(i-1)m+1} & B_{(i-1)m+1} & & \\ C_{(i-1)m+2} & A_{(i-1)m+2} & \ddots & \\ & \ddots & \ddots & B_{(i-1)m+m} \\ & & C_{(i-1)m+m+1} & A_{(i-1)m+m+1} \end{pmatrix},$

Similarly, the block on the m-th row and m-th column of V_i^{-1} is U_i^{-1} for $i = m(q-1)+m, q = 1, \cdots, p-1$ by inducing. Therefore, $U_i^{-1} \geq O$ for $i = m(q-1)+m, q = 1, \cdots, p-1$. Then, we have $U_i^{-1} \geq O (i = 1, \cdots, n)$.

Secondly, taking advantage of the above lemma and proposition, a theorem is given. ☐

Theorem 1. *If A is an M-matrix, then the approximate factorization of matrix A can be represented by Expression (2), and the iterative scheme Algorithm (4) converges to $X^* = A^{-1}b$.*

Proof. From the above proposition, the approximate factorization of matrix A can be represented by Expression (2).

Firstly, prove $N \geq O$.

As A is an M-matrix, then $A_{im+1}^{-1} \geq O$, $B_{im+1} \leq O$, $B_{im} \leq O$, $C_{im+1} \leq O$, $C_{im+2} \leq O$, for $i = 1, \cdots, p-1$. Hence, $B_{im} A_{im+1}^{-1} B_{im+1} \geq O$, $C_{im+2} A_{im+1}^{-1} C_{im+1} \geq O$, for $i = 1, \cdots, p-1$. Therefore, $N \geq O$.

Secondly, prove $M^{-1} \geq O$.

Since $M^{-1} = \widetilde{U}^{-1} \widetilde{L}^{-1}$, provided $\widetilde{L}^{-1} = \begin{pmatrix} \widehat{L}_1 & -\widehat{S}_1 & & & \\ & \widehat{L}_2 & -\widehat{S}_2 & & \\ & & \ddots & \ddots & \\ & & & \widehat{L}_{p-1} & -\widehat{S}_{p-1} \\ & & & & \widehat{L}_p \end{pmatrix},$

where

$\widehat{L}_i = \begin{pmatrix} I_{(i-1)m+1} & & & \\ -L_{(i-1)m+2} & I_{(i-1)m+2} & & \\ & \ddots & \ddots & \\ & & -L_{im} & I_{im} \end{pmatrix}, i = 1, \cdots, p, -\widehat{S}_i = \begin{pmatrix} O \\ \vdots \\ O \\ -S_{im} \end{pmatrix}, i = 1, \cdots, p-1,$

and

$$\widetilde{u}^{-1} = \begin{pmatrix} \widehat{u}_1 & & & & \\ \widehat{C}_2 & \widehat{u}_2 & & & \\ & \ddots & \ddots & & \\ & & \widehat{C}_{p-1} & \widehat{u}_{p-1} & \\ & & & \widehat{C}_p & \widehat{u}_p \end{pmatrix},$$

where

$$\widehat{u}_i = \begin{pmatrix} u^{-1}_{(i-1)m+1} & -u^{-1}_{(i-1)m+1}B_{(i-1)m+1}u^{-1}_{(i-1)m+2} & \cdots & (-1)^{m-1}\prod_{j=1}^{m-1}u^{-1}_j B_j u^{-1}_{im} \\ & \ddots & \ddots & \vdots \\ & & u^{-1}_{(i-1)m+m-1} & -u^{-1}_{(i-1)m+m-1}B_{(i-1)m+m-1}u^{-1}_{im} \\ & & & u^{-1}_{im} \end{pmatrix},$$

$$\widehat{C}_i = \begin{pmatrix} -C_{(i-1)m+1}u^{-1}_{(i-1)m} \\ \vdots \\ O \\ O \end{pmatrix}. \quad i = 2, \cdots, p.$$

According to the proposition, $u^{-1}_i \geq O(i = 1, \cdots, n)$. Since

$$L_{(i-1)m+j} = C_{(i-1)m+j}u^{-1}_{(i-1)m+j-1}, j = 2, \cdots, m, i = 1, \cdots, p$$

we have $-L_{(i-1)m+j} \geq O, i = 1, \cdots, p, j = 2, \cdots, m$. Therefore, $\widetilde{L}^{-1} \geq O, \widetilde{u}^{-1} \geq O M^{-1} \geq O$.

Finally, based on $M^{-1} \geq O, N \geq O$ and Lemma 1, we conclude that $\rho(M^{-1}N) < 1$. That is, this algorithm converges. □

This section shows that the condition in the theorem is a sufficient condition for convergence of the algorithm. If A is not an M-matrix, Algorithm (4) is sometimes convergent, as is shown in the following section (Example 1).

4. Parallel Implementations

4.1. Storage Method

For the i-th processor $P_i(i = 1, \cdots, p)$, allocate $A_{(i-1)m+j}$, $B_{(i-1)m+j}$, $C_{(i-1)m+j}$ $(i \neq p, j = 1, \cdots, m, m+1; i = p, j = 1, \cdots, m)$, $b_{(i-1)m+j}$ $(j = 1, \cdots, m)$, the initial vector $x^{(0)}_{(i-1)m+j}$, and the convergence tolerance ε.

4.2. Circulating

(1) $Gy = b + Nx^{(k)}$ is solved to obtain y.

P_i $(i = 1, \cdots, p-1)$ acquires $x^{(k)}_{(i+1)m+2}$ from P_{i+1} and then computes to obtain $y_{(i-1)m+q}, q = 1, \cdots, m-1, i = 1, \cdots, p$ and y_n. P_i $(i = 1, \cdots, p-1)$ gains $y_{(i+1)m+1}$ from P_{i+1} and then obtains $y_{im}, i = 1, \cdots, p-1$.

(2) $Hx^{(k+1)} = y$ is solved to obtain $x^{(k+1)}$.

P_i $(i = 1, \cdots, p)$ computes to obtain $x^{(k+1)}_{(i-1)m+q}$ $(q = 2, \cdots, m, i = 1, \cdots, p)$ and $x^{(k+1)}_1$. The -ith processor P_i $(i = 2, \cdots, p)$ gains $x^{(k+1)}_{im}$ from P_{i-1} and then computes to obtain $x^{(k+1)}_{(i-1)m+1}, i = 2, \cdots, p$.

(3) On $P_i(i = 1, \cdots, p)$, judge $\|x^{(k+1)}_{(i-1)m+j} - x^{(k)}_{(i-1)m+j}\| \leq \varepsilon$. Following this, stop if correct, or otherwise, go back to step (1).

5. Results Analysis of Numerical Examples

For testing the new algorithm, some results on the Inspur TS10000 cluster have been given by the new algorithm and order 2 multi-splitting algorithm [2], which is a well-known parallel iterative algorithm. The PEk method [40] is used on the inner iteration of the order 2 multi-splitting algorithm. Suppose $d_i = d_{i-1} = d_{i+1} = t$, $x_i^{(0)} = (0, \cdots, 0)^T_{t \times 1}$, $\|x^{(k+1)} - x^{(k)}\|_\infty < \bar{\varepsilon}$, $\bar{\varepsilon} = 10^{-10}$.

In the tables, P is the number of processors, l is the inner iteration time, k is the parameter of the PEk method, T is the run time (in seconds), I is the iterative time, S is the speedup and E is the parallel efficiency (E = S/P). In the following figures, ILUP, BSOR, PEk, and MPA, respectively, denote the iterative incomplete lower and upper factorization preconditioner, the block successive over-relaxation method, the PEk method, and the multi-splitting algorithm.

5.1. Results Analysis of the Large-Scale System of Equations

Example 1. *A in Expression (1) represents*

$$A_i = \begin{bmatrix} 12 & -2 & & & \\ -3 & 12 & -2 & & \\ & \ddots & \ddots & \ddots & \\ & & -3 & 12 & -2 \\ & & & -3 & 12 \end{bmatrix}_{t \times t}, B_i = \begin{bmatrix} 2.2 & -1.3 & & & \\ -3 & 2.2 & -1.3 & & \\ & \ddots & \ddots & \ddots & \\ & & -3 & 2.2 & -1.3 \\ & & & -3 & 2.2 \end{bmatrix}_{t \times t},$$

$$C_i = \begin{bmatrix} 2 & 2 & & & \\ -1 & 2 & 2 & & \\ & \ddots & \ddots & \ddots & \\ & & -1 & 2 & 2 \\ & & & -1 & 2 \end{bmatrix}_{t \times t}, b_i = \begin{bmatrix} (i-1)k+1 \\ (i-1)k+2 \\ \vdots \\ ik-1 \\ ik \end{bmatrix}_{t \times 1 \ldots}, and (i = 1, 2, \cdots, n),$$

where $B_n = C_1 = O$, $n = 300$, and $t = 300$. The numerical results are shown in Tables 1–5, and in Figures 1 and 2.

The first example is not a numerical simulation regarding any partial differential equations (PDE); we use this example in order to test the correctness of the iterative incomplete lower and upper factorization preconditioner algorithm. The first example can build a good foundation for the second example regarding PDE. The solutions to the large-scale system of equations for Example 1 by the ILUP are shown in Table 1 and the details of these are as follows: This problem requires solving with more than eight processors and the number of iterations is 238. When increasing the number of processors, time and parallel efficiency all decrease. The number of processors for solving Example 1 transforms from 4 to 64 and the parallel efficiency changes from 91.14% to 73.80%. All of the parallel efficiency values are higher than those in published works, including Cui et al.'s [10], Zhang et al.'s [40], and Yun et al.'s [2] methods, with the values being above 73%. No matter how many processors are used to calculate the problem, the error tolerance of this example is the same: 6.897×10^{-11}.

Table 1. The iterative incomplete lower and upper factorization preconditioner (ILUP) for Example 1.

P	1	4	8	16	32	64
T	119.1036	32.6697	17.5870	9.6202	4.8371	2.5217
I	233	237	238	238	238	238
S		3.6457	6.7723	12.3806	24.6231	47.2324
E		0.9114	0.8465	0.7738	0.7695	0.7380
Δ	6.897×10^{-11}	6.897×10^{-11}	6.897×10^{-11}	6.897×10^{-11}	6.897×10^{-11}	6.897×10^{-11}

The results of Example 1 when using the BSOR method [10] are listed in Table 2. When more than four processors are used to resolve the problem of Example 1, the number of iterations is 216. When increasing the number of processors, the time and parallel efficiency decrease. The cost of the time of every iteration and communication is more than that found when using the ILUP algorithm for the large-scale system of equations. Hence, the speedup, which is less than that found when using the ILUP algorithm, decreases. Thus the parallel efficiency is not better than that found when using the ILUP algorithm for the large-scale system of equations. When the number of processors for solving Example 1 is four, the parallel efficiency is 59.56%; however, the parallel efficiency is 91.14% for four processors when using the ILUP algorithm. When increasing the number of processors, the parallel efficiency decreases to 44.81%, which is lower than that found when using the ILUP algorithm.

Table 2. The key to the block successive over relaxation method (BSOR) method for Example 1 ($\omega = 2.0$).

P	1	4	8	16	32	64
T	112.0383	47.0284	25.0183	14.0130	7.5833	3.9065
I	211	216	216	216	216	216
S		2.3824	4.4783	7.9953	14.7743	28.6800
E		0.5956	0.5598	0.4997	0.4617	0.4481

The results of Example 1 when using the PEk method published by Zhang et al. [40] are described as Table 3. When more than four processors are used to resolve the problem of Example 1, the number of iterations is 227. When increasing the number of processors, the time and parallel efficiency decrease. The cost of the time of every iteration and communication is more than that when using the ILUP algorithm for the large-scale system of equations. Hence, the speedup, which is less than that found when using the ILUP algorithm, decreases. Therefore, the parallel efficiency is poorer than that found when using the ILUP algorithm for the large-scale system of equations. When the number of processors used when solving Example 1 is four, the parallel efficiency is 64.08%; however, the parallel efficiency is 91.14% for four processors when using the ILUP algorithm. When increasing the number of processors, the parallel efficiency decreases to 44.79%, corresponding to the parallel efficiency when using the BSOR method, which is lower than that found when using the ILUP algorithm, 73.80%.

Table 3. Answers for the pseudo-elimination method with parameter k (PEk) for Example 1 (k = 1.6).

P	1	4	8	16	32	64
T	114.3098	44.5992	24.7489	14.2286	7.6159	3.9878
I	224	227	227	227	227	227
S		2.5630	4.6188	8.0338	15.0094	28.6649
E		0.6408	0.5773	0.5021	0.4690	0.4479

The results of Example 1 when using the multi-splitting algorithm (MPA) published by Yun et al. [2] are introduced in Table 4. As seen in Table 4, when more than four processors are used to solve the problem of Example 1, the number of iterations is 174. When increasing the number of processors, the time and parallel efficiency decrease. The cost of the time of every iteration and communication is more than that when using the ILUP algorithm for the large-scale system of equations. Hence, the speedup, which is less than that found when using the ILUP algorithm, decreases. Thus, the parallel efficiency is poorer than when using the ILUP algorithm for the large-scale system of equations. When the number of processors for solving Example 1 is four, the parallel efficiency is 55.64%, 33.50% less than that that found when using the ILUP algorithm. When increasing the number of processors, the parallel efficiency decreases to 40.82%, about 4% less than the parallel efficiency obtained with the BSOR method, which is 23% lower than that that found when using the ILUP algorithm.

Table 4. The solutions to the multi-splitting algorithm (MPA) used for Example 1.

P	1	4	8	16	32	64
T	103.597	46.547	24.716	13.717	7.472	3.9564
I	172	174	174	174	174	174
S		2.2256	4.1915	7.5525	13.8647	26.1254
E		0.5564	0.5239	0.4720	0.4333	0.4082

This section compares the speedup and parallel efficiency performance of the ILUP algorithm with methods in other recently published works, including Cui et al.'s [10], Zhang et al.'s [40], and Yun et al.'s [2] methods. Table 5 introduces a summary and comparison of the speedup and parallel efficiency with the different methods used for Example 1 on 64 CPU cores, which is better than other works [2,10,40]. As seen in Table 5, the speedup obtained with our method for Example 1 on 64 CPU cores is 47.2324, and the parallel efficiency is 73.80%. The parallel efficiency obtained with the ILUP algorithm is about 29% higher than that obtained using the BSOR method. The parallel efficiency is 29.01% more than that obtained using the PEk method. The parallel efficiency obtained with the BSOR method corresponds to the parallel efficiency obtained with the PEk method. The parallel efficiency is 23% higher than that obtained using the MPA algorithm.

Table 5. Comparison speedup and parallel efficiency with the different methods used for Example 1 on 64 central processing unit (CPU) cores.

Compared List	ILUP Algorithm	Block Successive over Relaxation Method [10]	Pseudo-Elimination Method with Parameter k [40]	Multi-Splitting Algorithm [2]
Speedup	47.2324	28.6800	28.6649	26.1254
Parallel Efficiency	0.7380	0.4481	0.4479	0.4082

Figure 1 illustrates the speedup performances obtained with the ILUP algorithm and the other three methods for Example 1 at different CPU cores. As seen from Figure 1, when increasing the number of processors, the speedup obtained using all the methods increases. No matter how great the number of processors, the speedup obtained using the ILUP algorithm is significantly higher than that obtained using the other three methods, especially when the number of processors is more. Regardless of the number of processors, the speedup values obtained using the BSOR method, the PEk method, and the MPA algorithm are close, particularly those obtained with the BSOR method and the PEk method.

Figure 2 shows the parallel efficiency performance of the ILUP algorithm and the other three methods for Example 1 at different CPU cores. As seen from Figure 2, when increasing the number of processors, the parallel efficiency obtained using all the methods decreases. Regardless of the number of processors, the parallel efficiency obtained using the ILUP algorithm is much higher than that found using the other three methods, maintaining a value of more than 70%. No matter the number of processors, the parallel efficiency values obtained using the PEk method, the BSOR method, and the MPA algorithm are lower and nearer, especially those found using the BSOR method and the PEk method. In particular, when the number of processors is 64, the parallel efficiency obtained using the ILUP algorithm rises above 73%; however, the parallel efficiencies obtained using the BSOR method, the PEk method, and the MPA algorithm are only about 40%. The ILUP algorithm has the clear superiority of producing exceedingly higher parallel efficiency values.

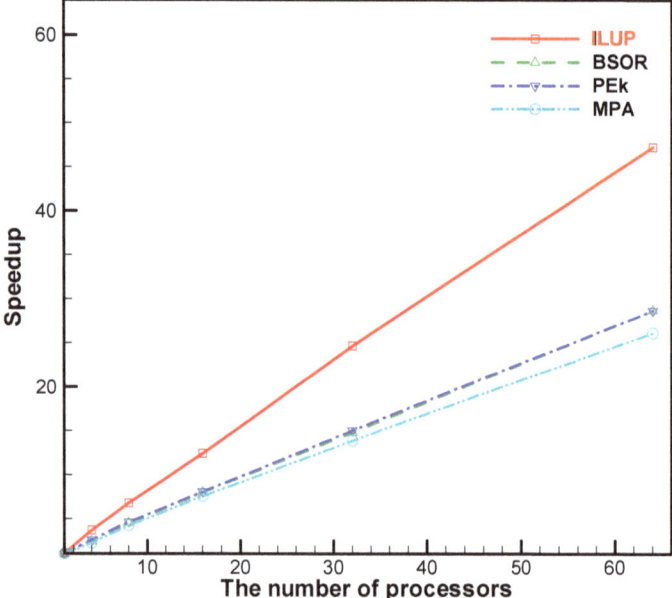

Figure 1. The speedup values for Example 1.

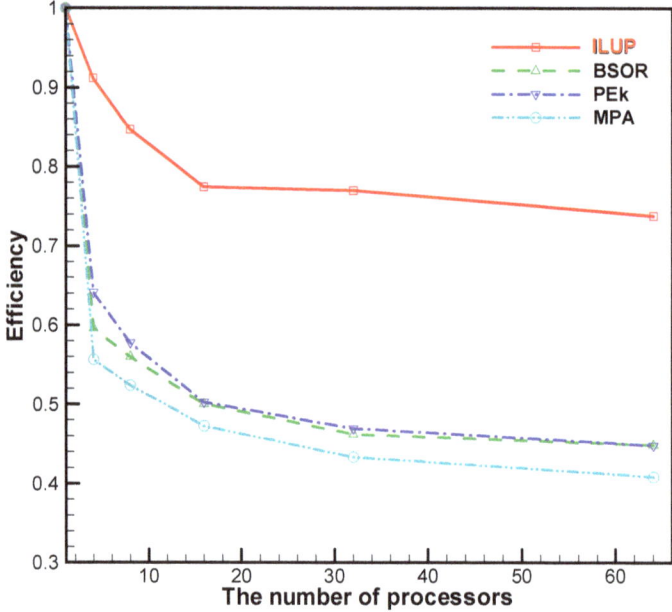

Figure 2. The parallel efficiency values for Example 1.

5.2. Results Analysis of the Partial-Differential Equations

Example 2. *Given the equations*

$$C_x \frac{\partial^2 u}{\partial x^2} + C_y \frac{\partial^2 u}{\partial y^2} + (C_1 \sin 2\pi x + C_2)\frac{\partial u}{\partial x} + (D_1 \sin 2\pi y + D_2)\frac{\partial u}{\partial y} + Eu = 0$$
$$0 \le x \le 1,\ 0 \le y \le 1$$

and

$$u|_{x=0} = u|_{x=1} = 10 + \cos \pi y$$
$$u|_{y=0} = u|_{y=1} = 10 + \cos \pi x,$$

$C_x, C_y, C_1, C_2, D_1, D_2$ *and E are invariants. Let* $C_x = C_y = E = 1, C_1 = C_2 = D_1 = D_2 = 0$ *and* $h = 1/101$. *The results are given in Tables 6–10 and in Figures 3 and 4.*

The finite difference method is used to discretize Example 2 in the tests. We adopt second-order central difference schemes to discretize Example 2 and then converse the format for numerical simulation; lastly, we test the iterative incomplete lower and upper factorization preconditioner algorithm on different processors. The results to the partial-differential equations for Example 2 obtained using the ILUP are listed in Table 6. The details are thus: This problem was solved with more than four CPU cores and the number of iterations was 560. When increasing the number of processors, the time and the parallel efficiency can be seen to all decrease. When the number of processors used for solving Example 2 changes from 4 to 64 the parallel efficiency changes from 89.48% to 71.64%. All of the parallel efficiency values are higher than in the published works [2,10,40], being above 71%. Regardless of how many processors are used to compute Example 2, the error allowance of this problem can be seen to be equally 3.158×10^{-11}.

Table 6. The iterative incomplete lower and upper factorization preconditioner for Example 2.

P	1	4	8	16	32	64
T	121.7960	34.0280	19.6270	10.2140	5.1830	2.6565
I	578	560	560	560	560	560
S		3.5793	6.2055	11.9244	23.4991	45.8483
E		0.8948	0.7757	0.7453	0.7343	0.7164
Δ	3.158×10^{-10}	3.158×10^{-10}	3.158×10^{-10}	3.158×10^{-10}	3.158×10^{-10}	3.158×10^{-10}

The results for Example 2 obtained with the BSOR method [10] are listed in Table 7. When more than four processors are used to resolve the problem of Example 2, the number of iterations is 793. When increasing the number of processors, the time and parallel efficiency decrease. The cost of the time of every iteration and communication is more than that obtained using the ILUP algorithm for the large-scale system of equations. Hence, the speedup, which is less than that found when using the ILUP algorithm, decreases. Thus, the parallel efficiency is not as good as that found using the ILUP algorithm for the partial-differential equations. When the number of processors used for solving Example 2 is four, the parallel efficiency is 86.24%, 3.24% lower than that found when using the ILUP algorithm for the partial-differential equations. With increasing the number of processors, the parallel efficiency decreases to 52.42%, which is less than that obtained using the ILUP algorithm, 71.64%.

Table 7. The key to the BSOR method for Example 2 ($\omega = 2.0$).

P	1	4	8	16	32	64
T	144.8230	41.9830	26.6220	14.1590	7.6370	4.3165
I	779	793	793	793	793	793
S		3.4496	5.4400	10.2283	18.9633	33.5510
E		0.8624	0.6800	0.6393	0.5926	0.5242

The results obtained for Example 2 using the PEk method [40] are given in Table 8. When more than four processors are used to resolve the problem of Example 2, the number of iterations is 798. When increasing the number of processors, the time and parallel efficiency decrease. The cost of the time of every iteration and communication is more than that obtained using the ILUP algorithm for the large-scale system of equations. Hence, the speedup, which is less than that obtained when using the ILUP algorithm, decreases. Thus, the parallel efficiency is poorer than that found when using the ILUP algorithm for the partial-differential equations. When the number of processors used for solving Example 2 is four, the parallel efficiency is 80.59%, which is 8.89% lower than that found when using the ILUP algorithm. When increasing the number of processors, the parallel efficiency decreases to 48.40%, which is 23.24% lower than that obtained with the ILUP algorithm.

Table 8. Answers to the PEk method for Example 2 (k = 2.7).

P	1	4	8	16	32	64
T	157.7210	48.9280	29.4860	16.0790	9.3640	5.0917
I	786	798	798	798	798	798
S		3.2235	5.3490	9.8091	16.8433	30.9764
E		0.8059	0.6686	0.6131	0.5264	0.4840

The results for Example 2 obtained with the multi-splitting algorithm [2] are introduced in Table 9. As seen in Table 9, when more than four processors are used to solve the problem of Example 2, the number of iterations is 838. When increasing the number of processors, the time and parallel efficiency decrease. The cost of the time of every iteration and communication is more than that found when using the ILUP algorithm for the partial-differential equations. Hence, the speedup, which is less than that found using the ILUP algorithm, decreases. Thus, the parallel efficiency is poorer than that obtained using the ILUP algorithm for the large-scale system of equations. When the number of processors used for solving Example 2 is four, the parallel efficiency is 78.25%, 11.23% less than that obtained using the ILUP algorithm. When increasing the number of processors, the parallel efficiency decreases to 46.34%, about 6% less than the parallel efficiency obtained with with the BSOR method, corresponding to the parallel efficiency obtained with the PEk technique, which is 25.3% lower than that found using the ILUP algorithm.

Table 9. The solutions to the multi-splitting algorithm for Example 2.

P	1	4	8	16	32	64
T	180.6459	57.7139	32.2524	17.7462	10.9967	6.0917
I	824	838	838	838	838	838
S		3.1300	5.6010	10.1794	16.4273	29.6547
E		0.7825	0.7001	0.6362	0.5134	0.4634

This section compares the speedup and parallel efficiency performance of the ILUP algorithm with methods in other recently published works, including Cui et al.'s [10], Zhang et al.'s [40], and Yun et al.'s [2] methods. Table 10 provides a summary and comparisons of speedup and parallel efficiency obtained using the different methods for Example 2 on 64 CPU cores, which is better than other published works. As seen in Table 10, the speedup in our method for Example 2 on 64 CPU cores is 45.8483 and the parallel efficiency is 71.64%. The parallel efficiency obtained using the ILUP algorithm is 19.22% higher than found using the BSOR method. The parallel efficiency is 23.24% more than that found using the PEk method. The parallel efficiency is 25.3% higher than that obtained using the MPA algorithm.

Table 10. Comparison of speedup and parallel efficiency values obtained using the different methods for Example 2 on 64 CPU cores.

Compared List	ILUP Algorithm	Block Successive over Relaxation Method [10]	Pseudo-Elimination Method with Parameter k [40]	Multi-Splitting Algorithm [2]
Speedup	45.8483	33.5510	30.9764	29.6547
Parallel Efficiency	0.7164	0.5242	0.4840	0.4634

Figure 3 compares the speedup performance of ILUP algorithm and the other three methods for Example 2 at different CPU cores. As seen from Figure 3, when increasing the number of processors, the speedup values of all the methods increase. Regardless of the number of processors, the speedup obtained using the ILUP algorithm is much higher than that found using the other three methods, in particular when the number of processors is greater. No matter the number of processors, the speedup values found using the BSOR method, the PEk method, and the MPA algorithm are close, especially for those found using the PEk technique and the MPA algorithm. For example, when the number of processors is 64, the speedup found using the ILUP algorithm rises above 45; however, the speedup values obtained using the BSOR method, the PEk method, and the MPA algorithm are only about 30. Obviously, the ILUP algorithm has the advantage of producing higher speedup values.

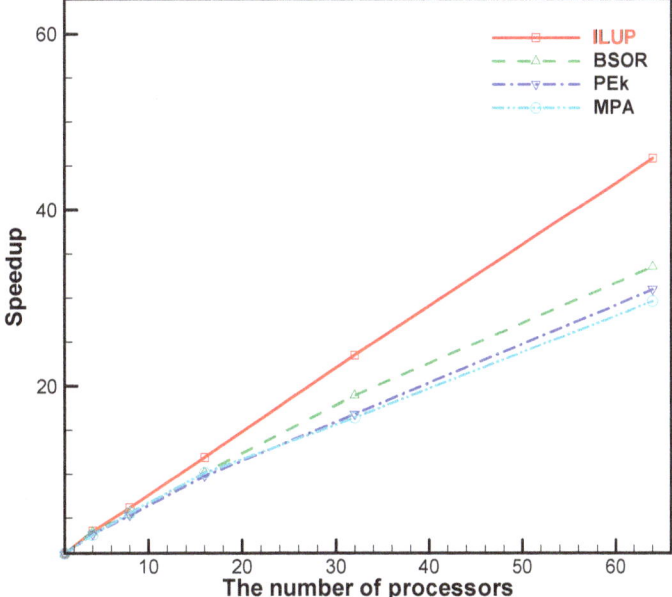

Figure 3. The speedup values for Example 2.

Figure 4 shows the parallel efficiency performance of the ILUP algorithm and the other three methods for Example 2 at different CPU cores. As seen from Figure 4, when increasing the number of processors, the parallel efficiency of all the methods decreases. Regardless of the number of processors, the parallel efficiency obtained using the ILUP algorithm is much higher than that found using the other three methods, maintaining a value of more than 70%. When increasing the number of processors, the parallel efficiency values obtained using the BSOR method, the PEk method, and the MPA algorithm are lower and sustain a descent, especially for those found using the MPA algorithm. In particular, when the number of processors is 64, the parallel efficiency obtained using the ILUP algorithm rises above 71%; however, the parallel efficiency values found using the BSOR method, the PEk method, and the MPA algorithm are only about 50%. The ILUP algorithm is clearly beneficial in its production of exceedingly high parallel efficiency values.

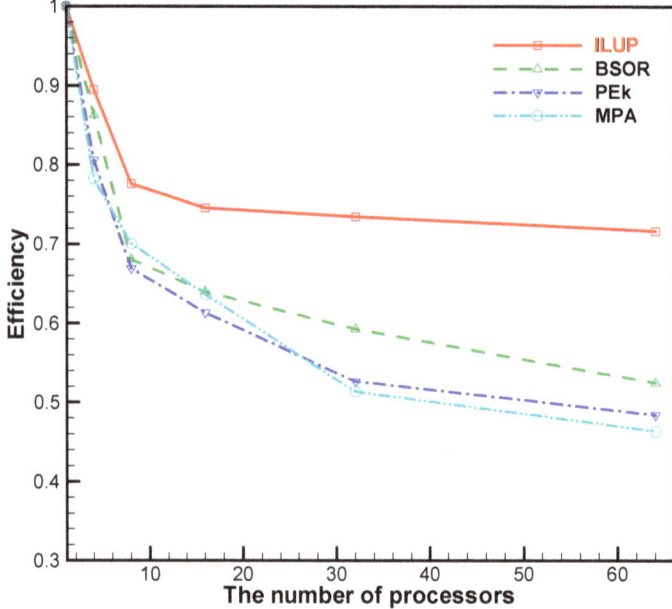

Figure 4. The parallel efficiency values for Example 2.

6. Conclusions

In this work, an iterative incomplete LU factorization preconditioner for partial-differential equation systems has been presented. The partial-differential equations were discretized into linear equations with the form Ax = b. An iterative scheme of linear systems was used. The iterative ILU preconditioners of linear systems and partial-differential equations systems were performed on different computation nodes of multi-CPU cores. From the above numerical results in the tables and figures, we can obtain the following conclusions:

1. The ILUP algorithm for the large-scale system of equations and partial-differential equation systems was performed on different multi-CPU cores. The numerical results show that the solutions are consistent with the theory.
2. From Example 1, when *A* is neither positive nor an M-matrix, the ILUP algorithm still converges.
3. At any multi-CPU cores, the speedup of the ILUP algorithm for the system of equations is far higher than that found using the BSOR method [10], the PEk method [40], and the MPA algorithm [2]. Evidently, the ILUP algorithm has the advantage of producing higher speedup values.
4. No matter the number of processors, the parallel efficiency of the ILUP algorithm is preferable. The parallel efficiency of the ILUP algorithm is higher than that of the other three algorithms. For example, the parallel efficiency of the ILUP algorithm achieves a value of above 73.8% (as seen in Table 5), which is higher than that for any other algorithm, including the BSOR method [10], the PEk method [40], and the MPA algorithm [2]. Obviously, the ILUP algorithm has the superiority of producing exceedingly high parallel efficiency values.

Author Contributions: Conceptualization, Y.-H.F. and L.-H.W.; methodology, L.-H.W.; software, Y.J.; validation, Y.J., X.-G.L., and X.-X.Y.; formal analysis, Y.-H.F.; investigation, Y.-H.F.; resources, L.-H.W.; data curation, Y.J.; writing—original draft preparation, X.-G.L. and C.-C.C.; writing—review and editing, Y.-H.F.; visualization, Y.-H.F.; supervision, Y.-H.F.; project administration, L.-H.W. and C.-C.C.; funding acquisition, X.-X.Y.

Funding: This research was funded by the Natural Science Foundation of Shanxi Province, China (201801D221118), the National Natural Science Foundation of China (grant nos. 11802194 and 11602157) and the Taiyuan University of Science and Technology Scientific Research Initial Funding (TYUST SRIF. 20152027, 20162037).

Conflicts of Interest: The authors declare no conflict of interest. We confirm that the manuscript has been read and approved by all named authors and that there are no other persons who satisfied the criteria for authorship but are not listed. We further confirm that the order of authors listed in the manuscript has been approved by all of us. We confirm that we have given due consideration to the protection of intellectual property associated with this work and that there are no impediments to publication, including the timing of publication, with respect to intellectual property. In so doing we confirm that we have followed the regulations of our institutions concerning intellectual property.

References

1. Polizzi, E.; Sameh, A.H. A parallel hybrid banded system solver: The spike algorithm. *Parallel Comput.* **2006**, *32*, 177–194. [CrossRef]
2. Yun, J.H. Convergence of nonstationary multi-splitting methods using ILU factorizations. *J. Comput. Appl. Math.* **2005**, *180*, 245–263. [CrossRef]
3. Zhang, B.L.; Gu, T.X. *Numerical Parallel Computational Theory and Method*; Defense Industry Press: Beijing, China, 1999; pp. 161–168.
4. Wu, J.; Song, J.; Zhang, W.; Li, X. Parallel incomplete factorization pre-conditioning of block-tridiagonal linear systems with 2-D domain decomposition. *Chin. J. Comput. Phys.* **2009**, *26*, 191–199.
5. Duan, Z.; Yang, Y.; Lv, Q.; Ma, X. Parallel strategy for solving block-tridiagonal linear systems. *Comput. Eng. Appl.* **2011**, *47*, 46–49.
6. Fan, Y.; Lv, Q. The parallel iterative algorithms for solving the block-tridiagonal linear systems. *J. Basic Sci. Text. Coll.* **2010**, *23*, 174–179.
7. El-Sayed, S.M. A direct method for solving circulant tridiagonal block systems of linear equations. *Appl. Math. Comput.* **2005**, *165*, 23–30. [CrossRef]
8. Bai, Z. A class of parallel decomposition-type relaxation methods for sparse systems of linear equations. *Linear Algebra Appl.* **1998**, *282*, 1–24.
9. Cui, X.; Lv, Q. A parallel algorithm for block-tridiagonal linear systems. *Appl. Math. Comput.* **2006**, *173*, 1107–1114. [CrossRef]
10. Cui, X.; Lv, Q. A parallel algorithm for band linear systems. *Appl. Math. Comput.* **2006**, *181*, 40–47.
11. Akimova, E.; Belousov, D. Parallel algorithms for solving linear systems with block-tridiagonal matrices on multi-core CPU with GPU. *J. Comput. Sci.* **2012**, *3*, 445–449. [CrossRef]
12. Terekhov, A.V. A fast parallel algorithm for solving block-tridiagonal systems of linear equations including the domain decomposition method. *Parallel Comput.* **2013**, *39*, 245–258. [CrossRef]
13. Ma, X.; Liu, S.; Xiao, M.; Xie, G. Parallel algorithm with parameters based on alternating direction for solving banded linear systems. *Math. Probl. Eng.* **2014**, *2014*, 752651. [CrossRef]
14. Terekhov, A. A highly scalable parallel algorithm for solving Toeplitz tridiagonal systems of linear equations. *J. Parallel Distrib. Comput.* **2016**, *87*, 102–108. [CrossRef]
15. Gu, B.; Sun, X.; Sheng, V.S. Structural minimax probability machine. *IEEE Trans. Neural Netw. Learn. Syst.* **2017**, *28*, 1646–1656. [CrossRef]
16. Zhou, Z.; Yang, C.N.; Chen, B.; Sun, X.; Liu, Q.; Wu, Q.M.J. Effective and efficient image copy detection with resistance to arbitrary rotation. *IEICE Trans. Inf. Syst.* **2016**, *E99*, 1531–1540. [CrossRef]
17. Deng, W.; Zhao, H.; Liu, J.; Yan, X.; Li, Y.; Yin, L.; Ding, C. An improved CACO algorithm based on adaptive method and multi-variant strategies. *Soft Comput.* **2015**, *19*, 701–713. [CrossRef]
18. Deng, W.; Zhao, H.; Zou, L.; Li, G.; Yang, X.; Wu, D. A novel collaborative optimization algorithm in solving complex optimization problems. *Soft Comput.* **2016**, *21*, 4387–4398. [CrossRef]
19. Tian, Q.; Chen, S. Cross-heterogeneous-database age estimation through correlation representation learning. *Neurocomputing* **2017**, *238*, 286–295. [CrossRef]

20. Xue, Y.; Jiang, J.; Zhao, B.; Ma, T. A self-adaptive artificial bee colony algorithm based on global best for global optimization. *Soft Comput.* **2018**, *22*, 2935–2952. [CrossRef]

21. Yuan, C.; Xia, Z.; Sun, X. Coverless image steganography based on SIFT and BOF. *J. Internet Technol.* **2017**, *18*, 435–442.

22. Qu, Z.; Keeney, J.; Robitzsch, S.; Zaman, F.; Wang, X. Multilevel pattern mining architecture for automatic network monitoring in heterogeneous wireless communication networks. *China Commun.* **2016**, *13*, 108–116.

23. Chen, Z.; Ewing, R.E.; Lazarov, R.D.; Maliassov, S.; Kuznetsov, Y.A. Multilevel preconditioners for mixed methods for second order elliptic problems. *Numer. Linear Algebra Appl.* **1996**, *3*, 427–453. [CrossRef]

24. Liu, H.; Yu, S.; Chen, Z.; Hsieh, B.; Shao, L. Sparse matrix-vector multiplication on NVIDIA GPU. *Int. J. Numer. Anal. Model. Ser. B* **2012**, *2*, 185–191.

25. Liu, H.; Chen, Z.; Yu, S.; Hsieh, B.; Shao, L. Development of a restricted additive Schwarz preconditioner for sparse linear systems on NVIDIA GPU. *Int. J. Numer. Anal. Model. Ser. B* **2014**, *5*, 13–20.

26. Chen, Z.; Liu, H.; Yang, B. Parallel triangular solvers on GPU. In Proceedings of the International Workshop on Data-Intensive Scientific Discovery (DISD), Shanghai, China, 2–4 June 2013.

27. Yang, B.; Liu, H.; Chen, Z. GPU-accelerated preconditioned GMRES solver. In Proceedings of the 2nd IEEE International Conference on High Performance and Smart Computing (IEEE HPSC), Columbia University, New York, NY, USA, 8–10 April 2016.

28. Liu, H.; Yang, B.; Chen, Z. Accelerating the GMRES solver with block ILU (K) preconditioner on GPUs in reservoir simulation. *J. Geol. Geosci.* **2015**, *4*, 1–7.

29. Barrett, R.; Berry, M.; Chan, T.F.; Demmel, J.; Donato, J.; Dongarra, J.; Eijkhout, V.; Pozo, R.; Romine, C.; Vander, V.H. *Templates for the Solution of Linear Systems: Building Blocks for Iterative Methods*, 2nd ed.; SIAM: Philadelphia, PA, USA, 1994.

30. Saad, Y. *Iterative Methods for Sparse Linear Systems*, 2nd ed.; SIAM: Philadelphia, PA, USA, 2003.

31. Cai, X.C.; Sarkis, M.A. restricted additive Schwarz preconditioner for general sparse linear systems. *Math. Sci. Fac. Publ.* **1999**, *21*, 792–797. [CrossRef]

32. Ascher, U.M.; Greif, C. Computational methods for multiphase flows in porous media. *Math. Comput.* **2006**, *76*, 2253–2255.

33. Hu, X.; Liu, W.; Qin, G.; Xu, J.; Yan, Y.; Zhang, C. Development of a fast auxiliary subspace pre-conditioner for numerical reservoir simulators. In Proceedings of the Society of Petroleum Engineers SPE Reservoir Characterisation and Simulation Conference and Exhibition, Abu Dhabi, UAE, 9–11 October 2011.

34. Cao, H.; Tchelepi, H.A.; Wallis, J.R.; Yardumian, H.E. Parallel scalable unstructured CPR-type linear solver for reservoir simulation. In Proceedings of the SPE Annual Technical Conference and Exhibition, Dallas, TX, USA, 9–12 October 2005.

35. NVIDIA Corporation. CUSP: Generic Parallel Algorithms for Sparse Matrix and Graph. Available online: http://code.google.com/p/cusp-library (accessed on 25 December 2008).

36. Chen, Z.; Zhang, Y. Development, analysis and numerical tests of a compositional reservoir simulator. *Int. J. Numer. Anal. Mode.* **2009**, *5*, 86–100.

37. Klie, H.; Sudan, H.; Li, R.; Saad, Y. Exploiting capabilities of many core platforms in reservoir simulation. In Proceedings of the SPE RSS Reservoir Simulation Symposium, Bali, Indonesia, 21–23 February 2011.

38. Chen, Y.; Tian, X.; Liu, H.; Chen, Z.; Yang, B.; Liao, W.; Zhang, P.; He, R.; Yang, M. Parallel ILU preconditioners in GPU computation. *Soft Comput.* **2018**, *22*, 8187–8205. [CrossRef]

39. Varga, R.S. *Matrix Iterative Analysis*; Science Press: Beijing, China, 2006; pp. 91–96.

40. Zhang, K.; Zhong, X. *Numerical Algebra*; Science Press: Beijing, China, 2006; pp. 70–77.

Article

Parameter Optimization for Computer Numerical Controlled Machining Using Fuzzy and Game Theory

Kai-Chi Chuang [1], Tian-Syung Lan [1,2], Lie-Ping Zhang [1,*], Yee-Ming Chen [3] and Xuan-Jun Dai [1]

[1] College of Mechanical and Control Engineering, Guilin University of Technology, Guilin 541004, Guangxi, China; s1038901@gmail.com (K.-C.C.); tslan888@yahoo.com.tw (T.-S.L.); daixuanjun@163.com (X.-J.D.)
[2] Department of Information Management, Yu Da University, Miaoli County 36143, Taiwan
[3] Department of Industrial Engineering and Management, Yuan Ze University, Taoyuan City 32003, Taiwan; chenyeeming@gmail.com
* Correspondence: zlp_gx_gl@163.com

Received: 20 October 2019; Accepted: 21 November 2019; Published: 25 November 2019

Abstract: Under the strict restrictions of international environmental regulations, how to reduce environmental hazards at the production stage has become an important issue in the practice of automated production. The precision computerized numerical-controlled (CNC) cutting process was chosen as an example of this, while tool wear and cutting noise were chosen as the research objectives of CNC cutting quality. The effects of quality optimizing were verified using the depth of cut, cutting speed, feed rate, and tool nose runoff as control parameters and actual cutting on a CNC lathe was performed. Further, the relationships between Fuzzy theory and control parameters as well as quality objectives were used to define semantic rules to perform fuzzy quantification. The quantified output value was introduced into game theory to carry out the multi-quality bargaining game. Through the statistics of strategic probability, the strategy with the highest total probability was selected to obtain the optimum plan of multi-quality and multi-strategy. Under the multi-quality optimum parameter combination, the tool wear and cutting noise, compared to the parameter combination recommended by the cutting manual, was reduced by 23% and 1%, respectively. This research can indeed ameliorate the multi-quality cutting problem. The results of the research provided the technicians with a set of all-purpose economic prospective parameter analysis methods in the manufacturing process to enhance the international competitiveness of the automated CNC industry.

Keywords: CNC machining; semantic rules; fuzzy quantification; fuzzy inference; Game theory

1. Introduction

Under strict international environmental regulations, although there are various cutting conditions related to environmental protection quality, tool wear and cutting noise are always considered preferentially as because of their green environmental protection quality in the practice of machining of cutting. There are often sophisticated nonlinear relationships in the problem of parameter optimization in multi-quality precision CNC production. The industry often selects appropriate machining parameters that rely on the program of the numerically controlled machine tool or the technicians' experience, but the results are not necessarily optimal and are not guaranteed to be optimal under multi-quality (more than two target qualities). Most of the cutting parameter optimization literature obviously does not meet the needs of the industry as it either considers only a single quality (only one target quality) or has overly costly research.

According to the research on cutting parameters, using Analytic Hierarchy Process (AHP) to combine the innovative thinking model of Teoriya Resheniya Izobretatelskikh Zadatch (TRIZ) and the concept of green production reduces the impact on the environment [1]. The optimal turning parameters obtained by using fuzzy semantic quantification can indeed be used as a method of analyzing parameters for practical cutting operations under environmental and cost considerations [2]. Considering the problem of cutting noise, Lan, Chuang, and Chen analyzed the combination of the optimal factor level with the Taguchi method [3]. However, the research only explored the noise target and thus was research of a single quality. Zhang et al. analyzed the influence of cutting parameters on noise with the variance. The cutting cost model was proposed after the analysis results showed that the cutting depth was the main factor affecting the cutting noise. However, the results of the study also applied only to a single quality material [4]. Hossein and Kops's research showed that the cutting temperature increases under larger cutting depth and higher cutting speed, which in turn shortens the tool life. However, the research took time to carry out the cutting work and belonged solely to the research of a single quality and not a multi-quality research [5]. The research of Schultheiss et al. which pointed out that reducing the tool wear can shorten the time of the production cycle and reduce energy consumption, was also a single quality research and not a multi-quality research [6]. Weng obtained the optimal cutting parameters by using fuzzy quantification. The parameters can reach 10% of the level prior the whole experiment under the Technique for Order Preference by Similarity to Ideal Solution (TOPSIS) arrangement. This proved that the experiment is unnecessary in reaching this level and can result in cost and labor savings. While this research obtained optimized combinations of multi-quality parameters, it did not take into account the conflict of quality objectives [7]. Li et al. established a multi-quality optimization model scheme by applying game theory to machining cutting parameters. The research showed that game theory was suitable for multi-quality optimization design, but failed to obtain the best combination [8]. Zhou et al. reduced and optimized the carbon footprint of the cutting process through game theory, but multi-quality was not taken into account [9]. Tian et al. considered the tool wear conditions and optimized the cutting parameters through game theory. The research discussed the conditions of tool wear, which was also a single-quality optimization research [10]. The above-mentioned local or abroad researches on cutting parameters are either discussions of only a single quality or optimization plans with specific conditions. Not only is there no further explanation and analysis of the conflict between the production qualities, but there is a need to be achieved through the actual operation of the cutting equipment, which is a waste of material resources, time, and labor and has an influence on the surrounding environment. Different control parameters are required when the processing conditions (materials, equipment, and tools) are different, which troubles the CNC industry. Therefore, developing a set of general production optimal mechanisms with green innovation by analyzing the inference method of green product design without equipment operation will be positive for the competitiveness and development of the precision CNC turning industry.

Based on the shortcomings of the above-mentioned researches, this research integrate fuzzy theory and game theory. Through the method of semantic quantification, a set all-purpose prediction models provides the fuzzy value of each goal for the selection of cutting parameters without actual cutting by the machine. The research also resolves the conflict problem between production qualities and control parameters by using game theory. The best quality strategy was obtained through statistic to help improve the understanding of engineering science by technicians as a consideration in the design or manufacture of future products. Through the result of this research, a set of optimal, all-purpose economic prospective parameter analysis methods could be provided to the technicians to enhance the overall competitiveness of the automated CNC cutting industry.

2. Research Background

2.1. Tool Wear

From Taylor's tool life formula, the wear of high-speed steel tools refers to the use time in the upper limit of the low wear rate area, which was used to record the characteristics of High Speed Steel

(HSS) tools and to obtain Formula (1) after rearrangement [11]. The relationship between feed rate and cutting speed must be properly matched during cutting. For instance, friction phenomenon instead of cutting might occur with overly slow speed. However, unexpected high-speed might break the cutting edge or roughen the transient surface.

$$TV^{1/n} f^{1/m} d^{1/l} = C'$$

(1)

T: tool function.
V: cutting speed.
f: feed rate.
D: diameter of milling cutter.
n, m: constant of tool material properties (acquired by experiment or experience).
l: cutting length.
C': cutting speed of tool life in 1 minute (supplied by tool manufacturer).

(1) According to the formula, a larger depth of cut and higher cutting speed lead to less tool wear.
(2) From expert experience, higher cutting speed and feed rate lead to less tool wear.
(3) According to the formula, lower cutting speed and feed rate lead to less tool wear.

2.2. Cutting Noise

All the noise values produced by the measurement experiment, including the noise values produced by motor idling and cutting experiments, were substituted in the formula, since their differences are lower than three Decibels, as shown in Formula (2) [12].

$$LPC = 10 \log \left[10^{\frac{LPB}{10}} - 10^{\frac{LPA}{10}} \right]$$

(2)

LPB: the measured value of motor running with no cutting.
LPA: the measured value of motor running with cutting.

(1) Smaller depth of cut, less noise.
(2) Slower cutting speed, less noise.
(3) Improving the pressure of the tool, less noise.

2.3. Fuzzy Theory

In 1965, Professor Zadeh of the University of California, Berkeley proposed fuzzy theory, which is a kind of fuzzy concept quantification based on fuzzy sets. It is mainly focused on making a correct judgment without going through complicated calculation processes of the fuzzy message of the human brain or incomplete information [13]. The language 'IF ... THEN ... ' is used in fuzzy theory to represent the fuzzy relationship. A language represents a qualitative conditional sentence and an uncertain rule, which is quantified by fuzzy mathematical tools. Fuzzy logic control was used to convert the input language to a fuzzy set. The fuzzy logic control architecture included the fuzzification interface, interface engine, defuzzification interface, and the fuzzy rule-based system, as shown in Figure 1 [14].

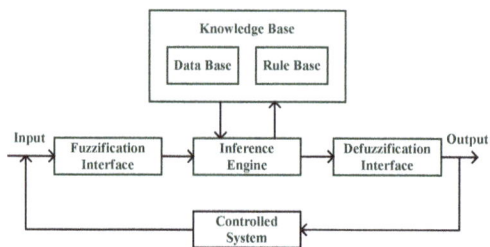

Figure 1. Architecture of fuzzy logic control.

2.4. Game Theory

Game theory was proposed in 1928 [15,16] before being promoted by the economist John Nash. Von Neumann and Oskar Morgenstern co-authored *The Theory of Games and Economic Behavior* in 1944, which analyzed game theory and economic behavior in detail and explained zero-sum games, the league, and the cooperative game, which further laid the theoretical foundation of game theory. Shapley contributed significantly to the development of core theory in game theory through the development of a prisoner's dilemma game [17]. Nash proved the Nash equilibrium existence theorem in 1953, which set a milestone for the current non-cooperative game theory.

Game theory is a state of confrontation for two or more contestants in a rational situation, with the pursuit of their own interests as the greatest goal. The conflict and cooperative relationship between rational contestants, using mathematical model simulation, has been widely used in various types of study.

2.4.1. Elements of a Game

The setting of contestants is rational in a game; however, the result could be quite the contrary or could have a Pareto principle, which is just in line with the current economic development trend since the smart contestant sits first in the best strategy of others to greater their own payoff function. Therefore, the result of the game is not necessarily rational or efficient but is closer to the economic situation. The main elements of the game are:

1. Player: The actor who makes decisions with the greatest goal of pursuing his own interest.
2. Nature: If not a contestant, the action taken is determined by a well-known probability.
3. Action set: A collection of all possible actions taken by a contestant.
4. Payoff Function: The remuneration that a contestant receives when the results of a game are shown, which is generally affected by all participants.

2.4.2. Information Structure

The information structure was divided into four types by Rasmusen, which are perfect information, complete information, certain information, and symmetric information [18]. Within a game with perfect information, each information set is a single node, which means that the players are clear about at which decision point the decision is made. If not, the game is called an imperfect information game. A game in which the players sit aware of the following three situations is called a complete information game. If not, the game is called an incomplete information game.

1. The identities of the players.
2. The moves could be taken by all players.
3. The utility function of all players.

Within a game with certain information, players will not act naturally after acting. If not, the game is called an uncertain information game. In a game with symmetric information, the information a player gets at the move node or at the end is the same with other players. If not, the game is called an asymmetric information game. Based on the player's simultaneous move (static game) or sequential move (dynamic game), and prior information (strategy and playoff) a player has or does not have, the game is divided into four types, as shown in Table 1.

Table 1. Four main types of game.

	Perfect Information	**Imperfect Information**
Static	Nash Equilibrium	Bayesian Nash Equilibrium
Dynamic	Sub-dame Perfect Nash Equilibrium	Perfect Bayesian Nash Equilibrium

2.4.3. Bargaining Games

The largest difference between bargaining game theory and decision theory is that the problems faced by a group of decision makers in a given situation can solve many economic problems. Therefore, game theory, which is widely used by economic, political, and financial experts, not only has the rigor of a mathematical model, but also simplifies the complex interaction phenomena in a real environment, and provides the strategic behavior analysis method for decision makers. In 1950, Nash assumed that a group of axioms would only get a solution to a set of bargaining models based on a non-cooperative game, which was divided into four parts [19,20].

1. Pareto efficiency

The outcome of the contestants' bargaining is beneficial to both parties; in other words, there is no other bargaining outcome that can increase the interests of all participants at the same time.

2. Independence of the irrelevant alternatives

Add things that do not matter in the game, and the outcome of the bargain is not affected.

3. Symmetry

If there is symmetry in the contestants' negotiation questions, the two contestants will receive an equal result.

4. Invariance under strategically equivalent representations

The utility function after the monotonic transformation still indicates that the participants have the same preference, and the monotonic transformed utility function does not affect the bargaining result.

Nash's suggestion, as shown in Formula (3), proved that the bargaining solution exists and is unique if these four axioms are satisfied.

$$\max_{s1s2}(S_1 - d_1)(S_2 - d_2) \tag{3}$$

d_1, d_2: the payoff that both players can get when there is no agreement of the bargain.
S_1, S_2: the payoff that both players can get when there is a agreement of the bargain.

Only if the result of the bargain is better than the one before the bargain can the players be motivated to bargain, so that $(d_1, d_2) \le (s_1, s_2)$.

Bargaining game theory has been used universally in economics, international relationships, calculator science, military strategy, and other disciplines. Some general topics that had used bargaining game theory are about the efficiency and rationality of solving supplier selection problems [21], the demand-response resource allocation between distribution networks [22], the reduction of environmental risks to enterprises in production processes [23], the solutions to the upload transmission power optimization problems in the multilateral bargaining model [24], and more.

3. Research Design

Precision CNC cutting was taken as an example in this research, while tool wear and cutting noise were selected as the green production quality of CNC cutting. The depth of cut, cutting speed, feed rate, and tool nose runoff were taken as control parameters. Fuzzy theory was used to define the semantic rules of the relationship of control parameters and production quality to carry out the fuzzy quantification. The quantified output values were introduced into game theory to resolve the conflict among the two production qualities and four control parameters. The strategy probability statistics of the game result and the strategy option with the highest sum of probability as the best strategy of that production quality were taken.

3.1. Fuzzy Rules Establishment

In the selection of the fuzzy membership function, different membership functions, based on each rule, were compared by entering the three factors: Cutting speed, cutting depth, and feed rate. The minimum membership function was calculated by the intersection, and the maximum value of the union was selected as the output part of the set to calculate the value of the center of gravity of the largest area in order to obtain the fuzzy value. The triangular membership function was used as the fuzzy pattern and the defuzzification was calculated by the center of gravity. Tool wear and cutting noise were chosen as the production qualities in this research. According to the literature, relevant cutting experience level range, and the suggestion of cutting parameters from tool manuals, was determined as low, medium, or high. The cutting characteristics of the target were obtained by using semantic quantification and were divided into five levels: Greatest, large, moderate, small, and minimal.

3.1.1. Tool Wear

Tool wear is a vital factor affecting cutting quality in precision machining. Changing the cutting tool before the end of the tool's life may result in higher production cost, lower production efficiency, and many disposals of tool inserts, which cause environmental pollution. Therefore, this study established fuzzy rules using cutting speed, cutting depth, and feed rate to minimize the tool wear, as shown in Table 2.

Table 2. Tool wear fuzzy rule table.

Rule	Parameter	Cutting Speed	Cutting Depth	Feed Rate	Tool Wear Rate
1		low	low	low	high
2		low	low	moderate	maximum
3		low	low	high	high
4		low	moderate	low	moderate
5		low	moderate	moderate	high
6		low	moderate	high	high
7		low	high	low	minimum
8		low	high	moderate	minimum
9		low	high	high	low
10		moderate	low	low	maximum
11		moderate	low	moderate	maximum
12		moderate	low	high	maximum
13		moderate	moderate	low	moderate

Table 2. *Cont.*

Rule	Parameter	Cutting Speed	Cutting Depth	Feed Rate	Tool Wear Rate
14		moderate	moderate	moderate	moderate
15		moderate	moderate	high	high
16		moderate	high	low	low
17		moderate	high	moderate	minimum
18		moderate	high	high	low
19		high	low	low	high
20		high	low	moderate	maximum
21		high	low	high	maximum
22		high	moderate	low	moderate
23		high	moderate	moderate	low
24		high	moderate	high	low
25		high	high	low	low
26		high	high	moderate	minimum
27		high	high	high	low

3.1.2. Cutting Noise

The noise during the cutting process is mainly caused by the vibration phenomenon, which not only interferes with the entire cutting process, but also seriously influences the quality of the work piece. The noise might even influence the mood of the technicians during work, which has a certain negative impact on production quality. In order to reduce the vibration frequency, it is necessary to reduce the cutting speed, depth of cutting, and feed rate of the tool, which in turn reduces productivity. Therefore, the fuzzy rules were established with cutting speed, depth of cutting, and feed rate as the factors based on the semantic considerations, as shown in Table 3.

Table 3. Cutting noise fuzzy rule table.

Rule	Parameter	Cutting Speed	Cutting Depth	Feed Rate	Cutting Noise
1		low	low	low	minimum
2		low	low	moderate	minimum
3		low	low	high	low
4		low	moderate	low	minimum
5		low	moderate	moderate	low
6		low	moderate	high	moderate
7		low	high	low	low
8		low	high	moderate	low
9		low	high	high	low
10		moderate	low	low	moderate
11		moderate	low	moderate	moderate
12		moderate	low	high	moderate
13		moderate	moderate	low	low
14		moderate	moderate	moderate	high
15		moderate	moderate	high	high

Table 3. *Cont.*

Rule	Parameter	Cutting Speed	Cutting Depth	Feed Rate	Cutting Noise
16		moderate	high	low	moderate
17		moderate	high	moderate	moderate
18		moderate	high	high	moderate
19		high	low	low	maximum
20		high	low	moderate	maximum
21		high	low	high	maximum
22		high	moderate	low	maximum
23		high	moderate	moderate	maximum
24		high	moderate	high	maximum
25		high	high	low	maximum
26		high	high	moderate	maximum
27		high	high	high	maximum

3.2. Variability of the Input and Output Domains

The operation had three inputs and one output. The input target was the control factor, and the output target was the default result. The input domain of the variables was in the interval [0,5] and was divided into five equal parts. The output domain of the variables was in the interval [0,40] and was divided into 40 equal parts.

1. Input target (1): The degree of membership of cutting speed as the control factor (Figure 2).

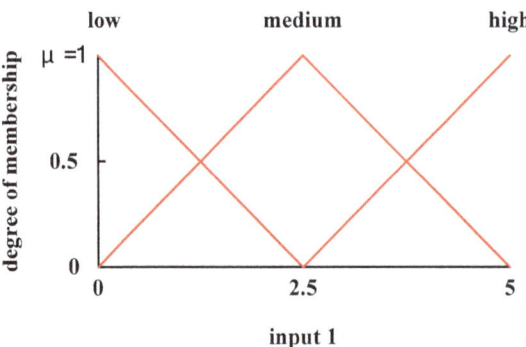

Figure 2. Degree of membership of the cutting speed.

Fuzzy terms: The degree of membership presented in Figure 2 is listed in Table 4.

Table 4. Input membership values of cutting speed.

Fuzzy Term	0	1.25	2.5	3.75	5
Low	1	0.5	0	0	0
Medium	0	0.5	1	0.5	0
High	0	0	0	0.5	1

2. Input target (2): The degree of membership of cutting depth as the control factor (Figure 3).

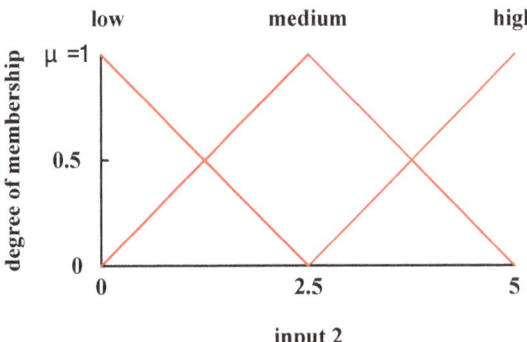

Figure 3. Degree of membership of the cutting depth.

Fuzzy terms: The degree of membership presented in Figure 3 is listed in Table 5.

Table 5. Input membership values of cutting depth.

Fuzzy Term	0	1.25	2.5	3.75	5
Low	1	0.5	0	0	0
Medium	0	0.5	1	0.5	0
High	0	0	0	0.5	1

3. Input target (3): The degree of membership of feed rate as the control factor (Figure 4).

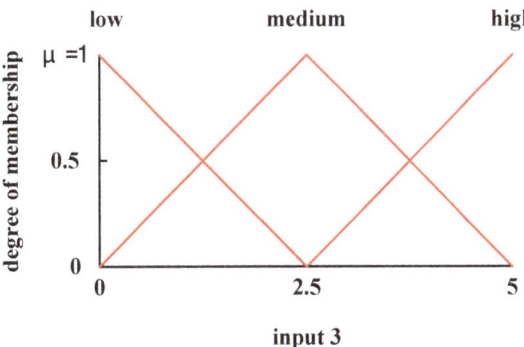

Figure 4. Degree of membership of the feed rate.

Fuzzy terms: The degree of membership presented in Figure 4 is listed in Table 6.

Table 6. Input membership values of feed rate.

Fuzzy Term	0	1.25	2.5	3.75	5
Low	1	0.5	0	0	0
Medium	0	0.5	1	0.5	0
High	0	0	0	0.5	1

4. Output target: Membership functions of the output variable (Figure 5).

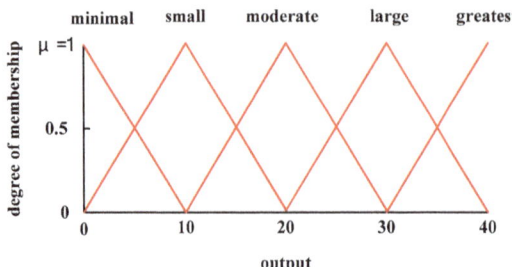

Figure 5. Degree of membership of output variables.

Fuzzy terms: The degree of membership presented in Figure 5 is detailed in Table 7.

Table 7. Output membership values.

No.	Minimal	Small	Moderate	Large	Greatest
0	1	0	0	0	0
1	0.84	0	0	0	0
2	0.68	0	0	0	0
3	0.52	0	0	0	0
4	0.36	0	0	0	0
5	0.2	0.04	0	0	0
6	0.04	0.2	0	0	0
7	0	0.36	0	0	0
8	0	0.52	0	0	0
9	0	0.84	0	0	0
10	0	1	0	0	0
11	0	0.68	0	0	0
12	0	0.52	0	0	0
13	0	0.36	0	0	0
14	0	0.2	0.04	0	0
15	0	0.04	0.2	0	0
16	0	0	0.36	0	0
17	0	0	0.52	0	0
18	0	0	0.68	0	0
19	0	0	0.84	0	0
20	0	0	1	0	0
21	0	0	0.84	0	0
22	0	0	0.68	0	0
23	0	0	0.52	0	0
24	0	0	0.36	0	0
25	0	0	0.2	0	0
26	0	0	0.04	0.04	0
27	0	0	0	0.2	0
28	0	0	0	0.36	0
29	0	0	0	0.52	0
30	0	0	0	0.68	0
31	0	0	0	0.84	0
32	0	0	0	1	0
33	0	0	0	0.84	0
34	0	0	0	0.68	0.04
35	0	0	0	0.52	0.2
36	0	0	0	0.36	0.36
37	0	0	0	0.2	0.52
38	0	0	0	0.04	0.68
39	0	0	0	0	0.84
40	0	0	0	0	1

3.3. Combination of Rules and Fuzzy Operation

According to the level range (low, medium, and high), the corresponding membership functions were the highest point of each fuzzy area, and the membership functions of input targets were determined by way of the intersection. The fuzzy operations of each target-preset result are shown below.

1. When the fuzzy region denotes "minimal" tool wear:

$$Average\ value = \frac{1 \times 0.84 + 2 \times 0.68 + 3 \times 0.52 + 4 \times 0.36 + 5 \times 0.2 + 6 \times 0.04}{1 + 0.84 + 0.68 + 0.52 + 0.36 + 0.2 + 0.04} = 1.1769$$

2. When the fuzzy region denotes "small" tool wear:

$$Average\ value$$
$$= \frac{5 \times 0.04 + 6 \times 0.2 + 7 \times 0.36 + 8 \times 0.52 + 9 \times 0.84 + 10 \times 1 + 11 \times 0.68 + 12 \times 0.52 + 13 \times 0.36 + 14 \times 0.2 + 15 \times 0.04}{0.04 + 0.2 + 0.36 + 0.52 + 0.84 + 1 + 0.68 + 0.52 + 0.36 + 0.2 + 0.04}$$
$$= 9.966$$

3. When the fuzzy region denotes "moderate" tool wear:

$$Average\ value$$
$$= \frac{14 \times 0.04 + 15 \times 0.2 + 16 \times 0.36 + 17 \times 0.52 + 18 \times 0.68 + 19 \times 0.84 + 20 \times 1 + 21 \times 0.84 + 22 \times 0.68 + 23 \times 0.52 + 24 \times 0.36 + 25 \times 0.2 + 26 \times 0.04}{0.04 + 0.2 + 0.36 + 0.52 + 0.68 + 0.84 + 1 + 0.84 + 0.68 + 0.52 + 0.36 + 0.2 + 0.04}$$
$$= 20$$

4. When the fuzzy region denotes "large" tool wear:

$$Average\ value$$
$$= \frac{26 \times 0.04 + 27 \times 0.2 + 28 \times 0.36 + 29 \times 0.52 + 30 \times 0.68 + 31 \times 0.84 + 32 \times 1 + 33 \times 0.84 + 34 \times 0.68 + 35 \times 0.52 + 36 \times 0.36 + 37 \times 0.2 + 38 \times 0.04}{0.04 + 0.2 + 0.36 + 0.52 + 0.68 + 0.84 + 1 + 0.84 + 0.68 + 0.52 + 0.36 + 0.2 + 0.04}$$
$$= 32$$

5. When the fuzzy region denotes "greatest" tool wear

$$Average\ value$$
$$= \frac{34 \times 0.04 + 35 \times 0.2 + 36 \times 0.36 + 37 \times 0.52 + 38 \times 0.68 + 39 \times 0.84 + 40 \times 1}{0.04 + 0.2 + 0.36 + 0.52 + 0.68 + 0.84 + 1}$$
$$= 38.23$$

3.4. Optimal Strategies of Games

A bargaining game for the two production qualities that are often considered in precision machining of cutting, tool wear, and cutting noise was conducted, and an innovative optimal mechanism was development afterward. The conflict among two production qualities and four control parameters was resolved through the perfect Bayesian equilibrium of game theory with one production quality as one individual player. The main strategy was chosen according to different production qualities. The probability value of strategies generated by the game was calculated to select the one with the highest sum of probability as the optimal strategy of each production quality. The optimal strategy chosen was also used to obtain the optimal plan of multi-quality and multi-strategy.

3.4.1. Establishment of the Game Model

1. Player (target)

Tool wear and cutting noise were set as players. The experimental data of the qualities are shown in Tables 8 and 9. Player A is referred to as the tool wear (the smaller, the better) and player B as the cutting noise (the smaller, the better) in the following.

Table 8. Test data of tool wear.

Cutting Speed (m/min)	Depth of Cut (mm)	Feed Rate (mm/rev)	Tool Nose Runoff (mm)	Tool Wear (μm^{-2})
2	2	2	2	4.38
3	2	2	1	4.13
3	2	2	2	3.87
1	2	2	1	4.21
2	3	2	2	2.97
1	2	2	3	4.13
2	2	1	1	4.38
2	2	2	1	4.04
2	1	3	3	4.13
1	2	1	3	4.55
1	3	3	2	3.38

Table 9. Test data of cutting noise.

Cutting Speed (m/min)	Depth of Cut (mm)	Feed Rate (mm/rev)	Tool Nose Runoff (mm)	Cutting Noise (dB)
2	2	2	2	82.83
1	2	2	2	81.73
3	2	2	2	85.97
2	1	2	2	82.61
2	3	2	2	82.91
2	2	1	2	82.55
2	2	3	2	82.93
2	2	2	1	82.79
2	2	2	3	82.81
1	1	1	1	81.5
1	3	3	2	81.94

2. Strategic planning (control parameter)

(1) Cutting speed
(2) Depth of cut
(3) Feed rate
(4) Tool nose runoff

3.4.2. Target of Bargaining Games

The overall optimal improvement strategy was prioritized to obtain important control parameters considered preferentially by each production quality and was used to improve the turning process to obtain the best multi-quality and multi-strategy optimization. In order to take both the production qualities into account to develop a multi-quality and multi-strategy optimization, the numbers of the appearance of each strategy of the production qualities were counted. Four main strategies were selected, and the output values of their corresponding semantic rules after quantification were imported into game theory. The initial payoff matrix (Z1) was constructed under consideration of the strategies of the two production qualities, as shown in Table 10.

Table 10. Initial payoff matrix Z1.

		B			
		B-1	B-2	B-3	B-4
A	A-1				
	A-2				
	A-3				
	A-4				

3.4.3. Mixed Strategies Game

In the initial payoff matrix initially established, the payoff value of all strategic combinations were filled in the corresponding spaces, resulting in the two-player multi-strategy game payoff matrix Z2, as shown in Table 11. Matrix Z2 was analyzed to establish whether the dominant strategy (one player's strategies are always better than the other player's strategies) existed. If positive, the matrix must first be simplified, as shown in Table 12. Finally, the probability values generated by all the games were statistically analyzed with their strategy probability, and the strategy with the highest probability sum was chosen to be the optimal strategy of that production quality. The optimal strategy of each production quality and its adoption probability are listed in Table 13 to obtain the optimal multi-quality and multi-strategy strategies.

Table 11. Payoff matrix Z2.

		B			
		B-1	B-2	B-3	B-4
A	A-1	(P_{a1}, P_{b1})	(P_{a2}, P_{b2})	(P_{a3}, P_{b3})	(P_{a4}, P_{b4})
	A-2	(P_{a5}, P_{b5})	(P_{a6}, P_{b6})	(P_{a7}, P_{b7})	(P_{a8}, P_{b8})
	A-3	(P_{a9}, P_{b9})	(P_{a10}, P_{b10})	(P_{a11}, P_{b11})	(P_{a12}, P_{b12})
	A-4	(P_{a13}, P_{b13})	(P_{a14}, P_{b14})	(P_{a15}, P_{b15})	(P_{a16}, P_{b16})

P_a: the payoff value of quality A under different situation; P_b: the payoff value of quality B under different situation.

Table 12. Simplified payoff matrix Z3.

		B	
		B-1	B-2
A	A-1	(P_{a1}, P_{b1})	(P_{a2}, P_{b2})
	A-2	(P_{a3}, P_{b3})	(P_{a4}, P_{b4})

Table 13. Optimal multi-quality and multi-strategy strategies.

Player	Optimal Strategy	Adoption Probability (%)
Tool wear (S)		
Cutting noise (Z)		

4. Experimental Verification

4.1. Experimental Condition

As a precision turning experiment, medium-carbon steel S45C with Ø45 mm × 250 mm, 100 mm clamping length, and a disposable tool were, respectively, used as the research targets and the cutting tool. The cutting blade was model NX2525, manufactured by Mitsubishi, and the tool holder was model WTJNR2020K16, manufactured by Toshiba. With the control parameter range recommended by the blade manufacturer, the cutting speed was between 150–300 m per minute, the cutting depth was 1–4.5 mm, and the feed rate was 0.17–0.45 mm per revolution, the experiment setting was listed in below.

1. Cutting depth: 0.5 mm, 1 mm, and 1.5 mm.
2. Cutting speed: The highest CNC lathe rotational speed of the tool was 3000 rpm, the diameter of the medium-carbon steel S45C used in the turning experiment was Ø45 mm, and its highest cutting speed was 339.292 m per minute. The cutting speed was set as 250 m per minute, 200 m per minute, and 150 m per minute, according to the recommendations given by the disposable blade.
3. Feed rate: The feed rate of the precision turning experiment was 0.02 mm per revolution, 0.06 mm per revolution, and 0.1 mm per revolution.

The cutting parameters, according to the above, are shown in Table 14.

Table 14. Cutting parameters.

Control Parameter	Level 1	Level 2	Level 3
A: Depth of cut (mm)	0.5	1	1.5
B: Cutting speed (m/min)	150	200	250
C: Feed rate (mm/rev)	0.02	0.06	0.1
D: Tool nose runoff (mm)	−0.1	±0.03	0.1

4.2. Result of Single Target Production Quality Verification

The median of the experimental results was used for comparative analysis. The median of the tool wear was 4.38 μm^{-2}, as shown in Table 15. The median of the cutting noise was 82.83 dB, as shown in Table 16. According to the median and the comparative analysis of the two production qualities, the data obtained in this research was better than the median, which showed that the innovative strategies of both production qualities were optimized, as shown in Table 17.

Table 15. Median values of tool wear.

Cutting Speed (m/min)	Depth of Cut (mm)	Feed Rate (mm/rev)	Tool Nose Runoff (mm)	Tool Wear (μm^{-2})
200	1	0.06	±0.03	4.38

Table 16. Median values of cutting noise.

Cutting Speed (m/min)	Depth of Cut (mm)	Feed Rate (mm/rev)	Tool Nose Runoff (mm)	Cutting Noise (dB)
200	1	0.06	±0.03	82.83

Table 17. Data of single quality optimization.

Tool Wear	Cutting Speed (m/min)	Depth of Cut (mm)	Feed Rate (mm/rev)	Tool Nose Runoff (mm)	(μm^{-2})
Cutting speed	250	1	0.06	±0.03	3.87
Depth of cut	200	1.5	0.06	±0.03	2.97
Feed rate	200	1	0.02	±0.03	4.55
Median	200	1	0.06	±0.03	4.38
Cutting Noise	Cutting Speed (m/min)	Depth of Cut (mm)	Feed Rate (mm/rev)	Tool Nose Runoff (mm)	(dB)
Cutting speed	150	1	0.06	±0.03	81.73
Depth of cut	200	0.5	0.06	±0.03	82.61
Feed rate	200	1	0.02	±0.03	82.55
Median	200	1	0.06	±0.03	82.83

4.3. Multi-Quality Optimal Strategy

4.3.1. Establish Initial Payoff Matrix Z2

Four preferred groups of the strategy were chosen through the experimental combination and fuzzy quantified. The output values were input into matrix Z2, as shown in Table 18. The parameters of the matrix were defined as follows.

Table 18. Multi-quality payoff matrix Z2.

		B			
		B-1	**B-2**	**B-3**	**B-4**
A	**A-1**	(9.966,9.966)	(9.966,9.966)	(9.966,20)	(9.966,20)
	A-2	(20,9.966)	(20,9.966)	(20,20)	(20,20)
	A-3	(9.966,9.966)	(9.966,9.966)	(9.966,20)	(9.966,20)
	A-4	(1.769,9.966)	(1.769,9.966)	(1.769,20)	(1.769,20)

Player (target)

A: Tool wear

B: Cutting noise

Strategy planning

A-1: Cutting speed is "low", cutting depth is "high", and feed rate is "high". (Rule9)
A-2: Cutting speed is "medium", cutting depth is "medium", and feed rate is "low". (Rule13)
A-3: Cutting speed is "medium", cutting depth is "high", and feed rate is "low". (Rule16)
A-4: Cutting speed is "medium", cutting depth is "high", and feed rate is "high". (Rule18)
B-1: Cutting speed is "low", cutting depth is "high", and feed rate is "high". (Rule9)
B-2: Cutting speed is "medium", cutting depth is "medium", and feed rate is "low". (Rule13)
B-3: Cutting speed is "medium", cutting depth is "high", and feed rate is "low". (Rule16)
B-4: Cutting speed is "medium", cutting depth is "high", and feed rate is "high". (Rule18)

4.3.2. Mixed Strategy as the Problem Solver

Since the initial payoff matrix Z2 cannot obtain the equilibrium solution or the approximate equilibrium solution, the cycle repeated continuously in some strategy combinations and a mixed strategy was needed for problem solving. As shown in the simplified payoff matrix Z3 (Table 19), two strategies remained, respectively, in both production quality A and B. However, the values of the strategies were output after fuzzy quantification, the differences of the values couldn't be distinguished clearly. To solve the problem, the strategy values were restored to the corresponding experimental values, as shown in Table 20. The optimal strategy combination was A1 and B1, as shown in Table 20. The optimal strategy of the two production qualities and its adoption probability are shown in Table 21.

Table 19. Simplified multi-quality payoff matrix Z3.

		B	
		B-1	**B-2**
A	**A-1**	(9.966,9.966)	(9.966,9.966)
	A-3	(9.966,9.966)	(9.966,9.966)

Table 20. Restored data of simplified payoff matrix Z3.

		B	
		B-1	**B-2**
A	**A-1**	(3.38,81.94)	(3.38,82.55)
	A-3	(4.38,81.94)	(4.38,82.55)

Table 21. Multi-quality optimization.

Player	Optimal Strategy	Adoption Probability (%)
Tool wear (S)	Increasing the cutting depth	100
Cutting noise (Z)	Reducing the cutting speed	100

4.3.3. Analysis of the Results of Multi-Quality Optimization

The conflict between production qualities and control parameters was aimed to be solved through the game matrix with the green production issue, which was internationally concerned and was selected as the research target. Multi-quality optimization was obtained through game theory. The optimal strategies of tool wear and cutting noise were, respectively, increasing the cutting depth and decreasing the cutting speed. The optimization obtained was further compared to the median commonly used in the industry, as shown in Table 22. The results of the comparison show that the improvement of the multi-quality cutting problem can indeed be achieved even without the operation of the equipment, and further develop a set of universal green innovative production optimization mechanism, which can provide technical personnel with a set of all-purpose economic prospective parameter analysis methods to stimulate alternative, innovative considerations of the industry.

Table 22. Comparison of multi-quality optimization and median data.

	Cutting Speed (m/min)	Depth of Cut (mm)	Feed Rate (mm/rev)	Tool Nose Runoff (mm)	Comparison	
Multi-Quality Optimization	150	1.5	0.1	±0.03	Tool wear	3.38 (μm^{-2})
					Cutting noise	81.94 (dB)
Median	200	1	0.06	±0.03	Tool wear	4.38 (μm^{-2})
					Cutting noise	82.83 (dB)

5. Conclusions

Nowadays, the industrial production design is getting more and more complicated, and with the increasingly demanding machining requirements, the setting of cutting parameters must be extremely strict to prevent changes to some parameters that could influence other production qualities. The most difficult breakthrough of CNC turning was the difficulty in setting the turning parameter. Due to the considerations of cost and time, the quality characteristics were judged by expert experience with a trial and error method, which might cause the doubts of improper use of quality measurement indicators.

Coupled with the environmental awareness and international regulation in recent years, reducing environmental harm in the product design stage avoids being labeled as a high pollution industry and prevents being forced to move or even close down factories. It is necessary for the automated CNC turning industry to use an easy-to-use quality-improving analysis program. In view of the inability of the operators to optimize the turning quality, fuzzy theory was used in the research to define the semantic rule of the relationship between control parameters and production qualities for fuzzy quantification. The output value after quantification was input into game theory to resolve the conflict between control parameters and production qualities for carrying out the game of multi-quality. With the statistic of the strategy probability, the strategy with the highest sum of probability was selected to obtain the multi-quality and multi-strategy optimization.

The results show that, within the parameter combination of multi-quality optimization, compared with the parameter combination recommended in the cutting manual, the tool wear reduced by 23% and the cutting noise reduced by 1%. The cutting problem of multi-quality is indeed improved by the research. In order to enhance the international competitiveness of the automated CNC cutting industry, the method used in the research can further be promoted and applied to the process or other industries.

Symmetry 2019, 11, 1450

Author Contributions: Conceptualization, T.-S.L.; Formal Analysis, K.-C.C., T.-S.L., L.-P.Z. and X.-J.D.; Investigation, L.-P.Z., Y.-M.C. and X.-J.D.; Methodology, K.-C.C. and T.-S.L.; Project administration, K.-C.C.; Software, K.-C.C., L.-P.Z. and Y.-M.C.; Validation, K.-C.C. and X.-J.D.; Visualization, Y.-M.C.; Writing-Original Draft, T.-S.L., L.-P.Z. and X.-J.D; Writing-Review & Editing, K.-C.C. and Y.-M.C.

Funding: This research received no external funding.

Acknowledgments: The authors would like to thank for the comments from many reviewers to improve this work.

Conflicts of Interest: The authors declare no conflicts of interest.

References

1. Lan, T.S.; Chuang, K.C.; Chen, Y.M. Automated Green Innovation for CNC Machining Design. *J. Adv. Mech. Eng.* **2018**, *10*, 1–11.
2. Lan, T.S.; Chuang, K.C.; Chen, Y.M. Optimization of Machining Parameters Using Fuzzy Taguchi Method for Reducing Tool Wear. *J. Appl. Sci.* **2018**, *8*, 1011. [CrossRef]
3. Lan, T.S.; Chuang, K.C.; Chen, Y.M. Optimal Production Parameters under Considerations of Noise Using Fuzzy Taguchi Method. In Proceedings of the 4th IEEE International Conference on Applied System Innovation, Chiba, Japan, 13–17 April 2018; pp. 354–357.
4. Zhang, L.; Zhang, B.; Hong, B.; Huang, H.H. Optimization of Cutting Parameters for Minimizing Environmental Impact: Considering Energy Efficiency, Noise Emission and Economic Dimension. *J. Precis. Eng. Manuf.* **2018**, *19*, 613–624. [CrossRef]
5. Hossein, K.A.E.; Kops, N. Investigation on the use of cutting temperature and tool wear in the turning of mild steel bars. *J. Mech. Eng. Sci.* **2017**, *11*, 3038–3045. [CrossRef]
6. Schultheiss, F.; Zhou, J.; Gröntoft, E.; Ståhl, J.E. Sustainable machining through increasing the cutting tool utilization. *J. Clean. Prod.* **2013**, *59*, 298–307. [CrossRef]
7. Weng, Y.Z. Multi-Objective Optimization of CNC Turning Parameters Using Fuzzy Analysis. Master's Thesis, Tatung University, Taipei City, Taiwan, 2007.
8. Li, Y.; Wang, H.P.; Wang, F.Y. Multi-objective Design of Process Parameter Based On Game Theory. *J. Appl. Mech. Mater.* **2011**, *121–126*, 964–967. [CrossRef]
9. Zhou, G.; Lu, Q.; Xiao, Z.; Zhou, C.; Tian, C. Cutting parameter optimization for machining operations considering carbon emissions. *J. Clean. Prod.* **2019**, *208*, 937–950. [CrossRef]
10. Tian, C.; Zhou, G.; Zhang, J.; Zhang, C. Optimization of cutting parameters considering tool wear conditions in low-carbon manufacturing environment. *J. Clean. Prod.* **2019**, *226*, 706–719. [CrossRef]
11. Hung, L.T. *Cutting Tool Science (Revised Edition)*; Chuan-Hwa Book Co., Ltd.: Taipei City, Taiwan, 2016.
12. Chiu, M.C.; Lan, T.S. *Noise control Theory and engineering design*; Wu-Nan Book Inc.: Taipei City, Taiwan, 2014.
13. Zadeh, L.A. Fuzzy sets. *Inf. Control.* **1965**, *8*, 338–353. [CrossRef]
14. Chiu, T.Y. Micro-environment Control by Using Fuzzy Theory. Master's Thesis, National Chung Hsing University, Taichung City, Taiwan, 2015.
15. Chen, S.Y. A Comparative Study of Allocation of Projects Benefit with Cooperative Game Theory. Master's Thesis, National UNITED University, Miaoli County, Taiwan, 2015.
16. Lan, C.H. Construction of Deduction Fuzzy Optimization System for Multiple-quality Production Using TRIZ and Game Theory. Master's Thesis, Yu Da University, Miaoli County, Taiwan, 2008.
17. Shapley, L.S. *A Value for n-Person Games. Contributions to the Theory of Games Volume II*; Kuhn, H.W., Tucker, A.W., Eds.; Princeton University Press: Princeton, NJ, USA, 1953.
18. Gao, Y.; Li, Z.; Wang, F.; Wang, F.; Tan, R.R.; Bi, J.; Jia, X. A game theory approach for corporate environmental risk mitigation. *J. Resour. Conserv. Recycl.* **2018**, *130*, 240–247. [CrossRef]
19. Nash, J. The Bargaining Problem. *J. Econom.* **1950**, *18*, 155–162. [CrossRef]
20. Rubinstein, A. Perfect Equilibrium in a Bargaining Model. *J. Econom.* **1982**, *50*, 97–109. [CrossRef]
21. Liu, T.; Deng, Y.; Chan, F. Evidential Supplier Selection Based on DEMATEL and Game Theory. *J. Fuzzy Syst.* **2018**, *20*, 1321–1333. [CrossRef]
22. Marzband, M.; Javadi, M.; Pourmousavi, S.A.; Lightbody, G. An advanced retail electricity market for active distribution systems and home microgrid interoperability based on game theory. *J. Electr. Power Syst. Res.* **2018**, *157*, 187–199. [CrossRef]

23. Rasmusen, E. *Games and Information: An Introduction to Game Theory*; Blackwell Publishers: Hoboken, NJ, USA, 1989.

24. Tsiropoulou, E.E.; Kapoukakis, A.; Papavassiliou, S.J. Energy-efficient subcarrier allocation in SC-FDMA wireless networks based on multilateral model of bargaining. In Proceedings of the 2013 IFIP Networking Conference, Brooklyn, NY, USA, 22–24 May 2013.

Article

An Efficient Data Transmission with GSM-MPAPM Modulation for an Indoor VLC System

Jing-Jing Bao [1,*], Chun-Liang Hsu [2] and Jih-Fu Tu [3]

[1] School of Electronics and Electrical Engineering, Dong Guan Polytechnic, No. 3, University Road, Songshan Lake District, Dongguan 523808, Guangdong, China

[2] Department of Electrical Engineering, St. John's University, New Taipei City 25135, Taiwan; liang112953@gmail.com

[3] Department of Industrial Engineering and Management, St. John's University, New Taipei City 25135, Taiwan; tu@mail.sju.edu.tw

*   Correspondence: baojj@dgpt.edu.cn; Tel.: +86-135-807-597-29 or +86-076-923-306-282

Received: 31 July 2019; Accepted: 16 September 2019; Published: 2 October 2019

Abstract: As an emerging wireless communication technique, visible light communication is experiencing a boom in the global communication field, and the dream of accessing to the Internet with light is fast becoming a reality. The objective of this study was to put forward an efficient and theoretical scheme that is based on generalized spatial modulation to reduce the bit error ratio in indoor short-distance visible light communication. The scheme was implemented while using two steps in parallel: (1) The multi-pulse amplitude and the position modulation signal were generated by combining multi-pulse amplitude modulation with multi-pulse position modulation using transmitted information, and (2) certain light-emitting diodes were activated by employing the idea of generalized spatial modulation to convey the generated multi-pulse amplitude and position modulation optical signals. Furthermore, pulse width modulation was introduced to achieve dimming control in order to improve anti-interference ability to the ambient light of the system. The two steps above involved the information theory of communication. An embedded hardware system, which was based on the C8051F330 microcomputer and included a transmitter and a receiver, was designed to verify the performance of this new scheme. Subsequently, the verifiability experiment was carried out. The results of this experiment demonstrated that the proposed theoretical scheme of transmission was feasible and could lower the bit error ratio (BER) in indoor short-distance visible light communication while guaranteeing indoor light quality.

Keywords: generalized spatial modulation; indoor visible light communication; multi-pulse amplitude and position modulation; dimming control

---

1. Introduction

The white light-emitting diode (LED), which is known as the future star of green lighting, is replacing traditional lighting lamps and is widely used in indoor lighting applications. A promising characteristic of LEDs is their ability to rapidly flash on or off, which makes it possible to convey information; this is called visible light communication (VLC) and it is receiving global attention [1,2]. As VLC has advantages that traditional radio frequency (RF) communication lacks, it is considered to be a mutual enhancement to RF, especially for indoor applications. However, VLC also has flaws in its current development, such as the limitation of emission power using existing LEDs, which increases the bit error ratio (BER) and lowers the transmission rate of the system. Furthermore, using visible light to convey information is very susceptible to ambient light. Some solutions were proposed to address this, including the adoption of key technologies or algorithms in existing RF communication into the VLC system, so long as the structure or circuit involving the transmitter and receiver was slightly modified or upgraded.

One solution was to introduce effective modulations. In recent years, many modulations were put forward to achieve excellent communication and illumination performances regarding indoor VLC. In [3], multi-pulse position modulation (M-PPM), with the joint purpose of dimming and communication, was proposed to increase the spectrum efficiency and dimming control for VLC. In [4], an adaptive M-ary pulse amplitude modulation (M-PAM) for an indoor VLC system coordinated multiple LED lamps to provide users with the highest rate of data transmission. In [5], an optical spatial multiple pulse position modulation, which combined a high spectral efficiency space shift setting with high energy efficiency multiple pulse position modulation, was proposed to provide a balance between complexity, achievable spectral efficiency, and energy efficiency in an indoor VLC system. In [6], a VLC system that is based on the offset pulse position modulation (Offset-PPM) was demonstrated using a single commercial highpower white LED (30 W) and a new coding scheme; the results of this experiment showed that this modulation attained good performance regarding the BER within a certain distance. In [7], a multi-LED phaseshifted on-off keying (OOK) modulation was designed to overcome the two key challenges of the limited modulation bandwidth and the non-linearity nature of LEDs in VLC. In [8], an asymmetric frequency shift keying (FSK) modulation technique was proposed to mitigate flickering and dimming in square and rectified waves in a VLC system. In [9], an optical software-defined radio VLC system was introduced, with variable pulse position modulation being considered at the transmitter. This system attained a better data transmission rate through the simulation. In addition to the single carrier modulation above, many multi-carrier modulations were proposed. In [10], a DC-biased optical orthogonal frequency division multiplexing (DCO-OFDM) was used for optical wireless communication, which reduced the complexity of the conventional OFDM and improved the BER. However, the DCO-OFDM signal needed to be clipped at zero power and at peak power of the LED, which distorted the signal. In [11], an asymmetrically clipped optical OFDM (ACO-OFDM) was used to reduce the peak to average power ratio (PAPR) when only the odd subcarriers were modulated, resulting in low bandwidth utilization efficiency. The OFDM, which is a multi-carrier modulation, was very promising regarding its resistance to inter-symbol interference and its high spectral efficiency. However, due to the nonlinear response of the LED, when the OFDM was applied to the VLC, it caused a higher BER and PAPR than the single carrier modulation. Single carrier modulation might be a better alternative due to the requirements for the BER and PAPR in an indoor VLC system. Generally speaking, indoor LED lamps consist of multiple wicks, so the spatial dimension offered by these wicks to convey information can be utilized. Over the past 10 years, generalized spatial modulation (GSM) has received worldwide attention due to its better performance in the reduction of the BER than traditional spatial modulation [12,13].

The main objective of this paper was to put forward an efficient and theoretical scheme that is based on generalized spatial modulation with multi-pulse amplitude and position modulation (GSM-MPAPM) for an indoor VLC system to achieve a lower BER and pulse width modulation (PWM) dimming. This scheme was implemented on an embedded hardware system, which was based on the C8051F330 microcomputer.

The main contributions of this paper are as follows:

(1)   An efficient GSM-MPAPM modulation with PWM dimming for an indoor VLC system.
(2)   An embedded hardware system, including a transmitter and a receiver, which were both based on the C8051F330 microcomputer. The key innovation of this paper was to employ the idea of GSM to convey MPAPM optical signals. Moreover, an embedded hardware system was designed to achieve the theoretical scheme.

Firstly, theoretical and mathematical models for the VLC system were set up, which were mainly related to the information theory of communication. Secondly, the experiment was carried out with the help of computers and the embedded hardware that was designed for this purpose. This paper is organized, as follows. Section 2 presents related knowledge and the theory regarding MPAPM and GSM. Section 3 describes a typical VLC system and introduces the proposed modulation. Section 4 details the scheme implementation. Analysis of the experimental results is found in Section 5. Finally, the conclusions of this paper are summarized in Section 6.

2. Related Knowledge and Theory

2.1. MPAPM

The MPAPM is a compound modulation and it combines two basic modulations that are commonly employed in a VLC system, namely, MPPM and MPAM. For MPPM, the L continuous signals of the binary are modulated in a time period consisting of Q time-slots and optical pulses that appear in P time slots. Moreover, the P optical pulses are arranged according to a certain regularity, so it is generally recorded as (Q,P)MPPM. The relationship of Q, P, and L are expressed in Equation (1), where C_Q^P means the gain of MPPM [14].

$$C_Q^P = Q!/P!(Q-P)! \geq 2^L \tag{1}$$

According to the definition of MPPM, it can be derived that the information that every symbol of MPPM can transmit is $log_2 c_Q^P$ bits. Hence, the capability of the data transmission increases as the number of time-slots increases, which means that MPPM has a significant advantage regarding bandwidth utilization, as demonstrated in multiple analyses. MPPM is better than PPM in these two aspects.

For MPAM, all of the information is encoded into the amplitude of the signal; the amplitude of the transmitted signal takes M different values, which implies that each pulse conveys $log_2 M$ bits per symbol time [15]. MPAM is superior to the other modulations when it comes to the data transmission rate and implementation complexity.

In this study, we designed a modulation method where the data transmission rate was the product of MPPM and MPAM, namely MPAPM. For analysis and comparison, it was expressed as (Q,P,M)MPAPM.

2.2. Generalized Spatial Modulation

As a special form of spatial modulation, GSM is receiving increased attention. Similarly to the multiple input and multiple output (MIMO) method, GSM activates multiple transmitting LEDs at the same time to simultaneously transmit the same information. The BER and the transmission rate both have thresholds because the number of LEDs cannot infinitely increase, which ultimately affects the performance of the system. However, unlike MIMO, GSM conveys transmitted information while using the activated combination of the transmitting LEDs, which are selected from a regular table. As a result, the number of transmitting LEDs required to achieve a certain spectral efficiency or BER is reduced as compared to MIMO. Transmitting the same information from more than one LED at a time not only retains the core advantage of MIMO, but also offers an increase in spatial diversity. Moreover, GSM completely avoids inter-carrier interference (ICI) at the receiver, which improves the reliability of the communication system [16,17]. Figure 1 depicts the GSM system model.

GSM employs multiple transmitting LEDs, which are activated to send the same information. Hence, a cluster of LEDs needs to be defined as the spatial constellation points. The combination number of LEDs is $N = C_{N_t}^{N_s}$, theoretically, where N_t is the number of transmitting LEDs and N_s is the number of LEDs activated at one time. However, the combination number of LEDs that can be considered for transmission must be a power of two, which removes doubles. Finally, the combination number is reduced to $N_x = 2^n$, where $n = \lfloor log_2 C_{N_t}^{N_s} \rfloor$ and $\lfloor \bullet \rfloor$ is the floor operation.

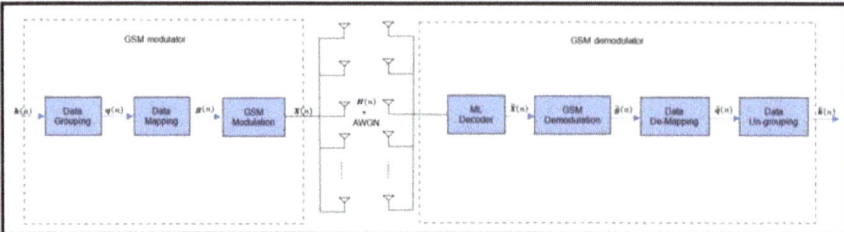

Figure 1. The generalized spatial modulation (GSM) system model.

3. The VLC System Based on GSM-MPAPM

3.1. GSM-MPAPM

Figure 2 depicts the indoor VLC system model based on GSM-MPAPM; an example of the data mapping and transmission for every time slot is also shown. For simplicity, eight information bits are transmitted for every time slot. Hence, in every time slot, eight bits are first selected from the input data. Subsequently, they are divided into two groups, i.e., the group consisting of the combination of transmitting LEDs and the group consisting of the information bits. Finally, the two groups are separately mapped according to mapping tables. For simplicity, $N_t = 5$ and $N_s = 2$ are assumed. The mapping procedure modulates the first five bits while using MPAPM, and the remaining three bits are mapped according to the activated combination of LEDs.

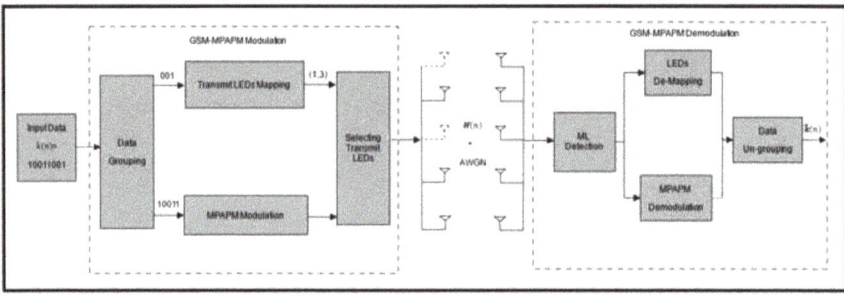

Figure 2. The indoor visible light communication (VLC) system model based on generalized spatial modulation with multi-pulse amplitude and position modulation (GSM-MPAPM).

The resultant combinations of the transmitting LEDs are listed in Table 1 and the MPAPM mapping rules are listed in Table 2. Table 1 shows that eight combinations of transmitting LEDs were obtained with the last three bits at one time. In Table 2, every time period is divided into 16 time slots, with the first four information bits are conveyed in one of the 16 time slots. The remaining bit is used to indicate the level of transmission power. Figure 3 shows the function of each bit in a time slot. For instance, if the eight information bits transmitted at one time are [10011001], then five information bits are transmitted from LED 1 and LED 3 in time slot 10 with a transmission power of U_2. In each time slot, the information bits are transmitted from the exiting five transmitting LEDs, while only two LEDs are activated at any given time. If MIMO is used instead of this modulation, the number of transmitting LEDs is increased to eight to maintain the same BER and spectral efficiency.

Table 1. Light-emitting diode (LED) combinations.

Grouped Bits	LED Combinations
000	(1,2)
001	(1,3)
010	(1,4)
011	(1,5)
100	(2,3)
101	(2,4)
110	(2,5)
111	(3,4)

Table 2. The MPAPM mapping rules.

Information Bits		Mapping Rules	Information Bits		Mapping Rules
	0000	Transmit in time slot 1		1000	Transmit in time slot 9
	0001	Transmit in time slot 2		1001	Transmit in time slot 10
	0010	Transmit in time slot 3		1010	Transmit in time slot 11
Front_4bits	0011	Transmit in time slot 4	Front_4bits	1011	Transmit in time slot 12
	0100	Transmit in time slot 5		1100	Transmit in time slot 13
	0101	Transmit in time slot 6		1101	Transmit in time slot 14
	0110	Transmit in time slot 7		1110	Transmit in time slot 15
	0111	Transmit in time slot 8		1111	Transmit in time slot 16
Middle_1bit	0	The amplitude is U_1			
	1	The amplitude is U_2			

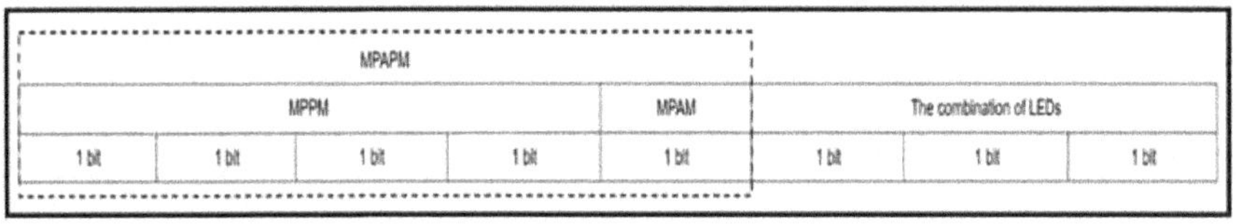

Figure 3. The function of each bit in a time slot.

3.2. Information Bits Transmitted per Symbol

In Section 2, MPAPM modulation is expressed as (Q,P,M)MPAPM, where Q is the total number of time slots in a time period, P is the time slot that is used to convey the information bits, and M is the level number of amplitude. Therefore, in the proposed GSM-MPAPM, the information bits transmitted per symbol are represented by $log_2\left(C_Q^P \times M \times C_{N_t}^{N_s}\right)$. GSM-MPAPM conveys more information bits per symbol and has a higher transmission rate when compared to the other modulations. Table 3 lists the information bits transmitted per symbol according to the different modulations.

Table 3. Comparison of information bits per symbol.

Modulation Schemes	Bits Transmitted per Symbol (bits)
MPAM	$log_2 M$
MPPM	$log_2 C_Q^P$
VPAPM	$log_2(Q \times M)$
VMPAPM	$log_2\left(C_Q^P \times M\right)$
GSM-MPAPM	$log_2\left(C_Q^P \times M \times C_{N_t}^{N_s}\right)$

3.3. Dimming Strategy

In the proposed GSM-MPAPM, the dimming mechanism is implemented while using the principle that is presented in Table 4. The PWM dimming strategy is employed to change the intensity of LED illumination according to the duty cycle ratio of the pulse signals transmitted in time slot i. This is expressed as the differentiated dimming. In Table 4, p_i (i = 1, 2, 3, 4) is the dimming coefficient, which has four different values throughout the 16 time slots. When p_1 = 0.2, the information bits are conveyed in time slot 1, time slot 5, time slot 9, and time slot 13. When p_2 = 0.4, the information bits are conveyed in time slot 2, time slot 6, time slot 10, and time slot 14. When p_3 = 0.6, the information bits are conveyed in time slot 3, time slot 7, time slot 11, and time slot 15. When p_4 = 0.8, the information bits are conveyed in time slot 4, time slot 8, time slot 12, and time slot 16. The dimming is implemented by choosing different coefficients. T represents a time period, including 16 time slots, while U represents the dimming voltage that is responsible for transmitting the information bits. In the actual system, if the dimming resolution of the PWM is k, the value increases four-fold in the innovative differentiated dimming. Undoubtedly, a more precise dimming is implemented with this strategy.

Table 4. The dimming strategy.

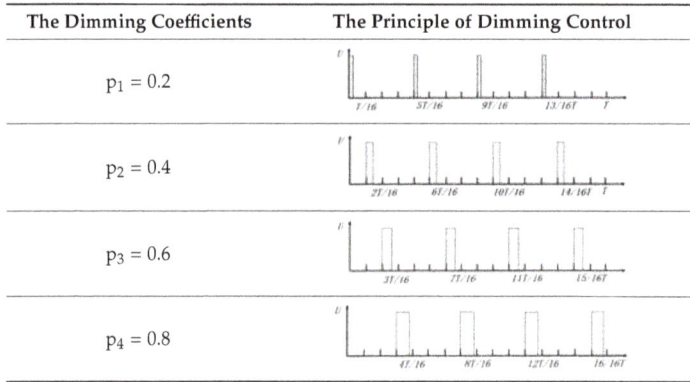

The Dimming Coefficients	The Principle of Dimming Control
p_1 = 0.2	
p_2 = 0.4	
p_3 = 0.6	
p_4 = 0.8	

4. System Design and Implementation

An embedded hardware system was designed in order to test the performance of the proposed GSM-MPAPM in indoor VLC system. The overall system architecture includes two parts, the transmitter and the receiver. The optical signals of the information bits are conveyed in the free space. Figure 4 shows the overall VLC system architecture in detail.

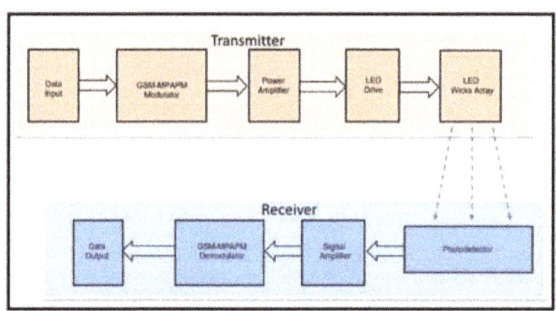

Figure 4. The overall VLC system architecture.

4.1. Design of the Transmitter

The main modules of the transmitter circuit are the GSM-MPAPM modulator, the power amplifier, the LED driver, and the array of LED wicks. All of the modules work normally under the control of the C8051F330 microcomputer, which was designed by Silicon Labs. Although it is an 8-bit microcomputer, the C8051F330 has a high-speed core of CIP-51, which is fully compatible with the 8051. This microcomputer was our first choice for communication design.

The input data processed by the transmitter are firstly generated by the Tera_Term, which is usually used to produce serial data in a communication system. Secondly, the serial data is separated into two data stream groups, namely, the LED combinations and the information bits. The bits are processed according to mapping tables. This procedure is performed while using the GSM-MPAPM modulator, of which the C8051F330 plays a key role.

A power amplifier is set as the primary amplification before the driver circuit to improve the drive capability of the transmitter to the wicks since the power of the white LED wicks used in the system is 2 W. The corresponding circuit is shown in Figure 5. In the power amplifier, the LM386 is chosen as an amplifier due to its low power consumption and weak harmonic distortion. By adding an external capacitor between pin 1 and pin 8, the voltage gain can be adjusted to any value up to 200. The power consumption of the LM386 at a static state is only 24 mW at the supply voltage of 5 V.

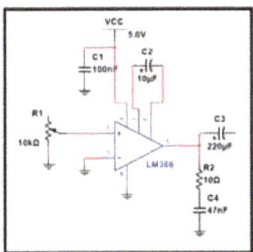

Figure 5. The power amplifier circuit.

The triode S9013 is employed to form a common-shot amplification and drive circuit in the single LED driver circuit, which is shown as Figure 6. The output signals from the power amplifier are inputted into the circuit through the SEND pin. When there is high input, the S9013 is turned on and the corresponding LED is activated to light up. When there is low input, there is no current flowing through the S9013 and the LED is off.

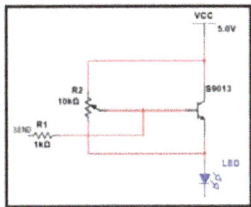

Figure 6. The single LED driver circuit.4.2. Design of the Receiver.

The receiver circuit mainly consists of three parts, i.e., the photodetector, the amplifier, and the GSM-MPAPM demodulator. Figure 7 shows that the resistance value of the photodiode LED1 decreases if it detects optical signals, which thereby forces the positive input voltage of the dual voltage comparator LM393 to decrease; the inverted input voltage is determined by the variable resistance R_2. Therefore, when the photosensitive detector detects obvious optical signals, the voltage of the positive input is less than that of the inverted input, the output of LM393 is low, and LED2 is on. If the photosensitive detector does not detect obvious optical signals, then the voltage of pin 3 is more than that of pin 2, the output is high, and LED2 is off. The sensitivity of the whole photodetector system can be changed by adjusting the value of R_2, according to indoor light interference.

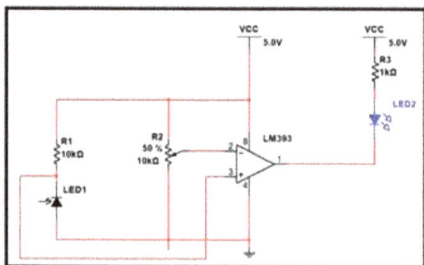

Figure 7. The photodetector.

The amplitude of the photodetector's output signals cannot be satisfied with the demodulation threshold; therefore, the signals must be amplified. In Figure 8, The high frequency triode 2SC1815 is introduced to form a secondary amplifier circuit, which is not only cost-effective, but it also obtains better gain than the operational amplifier.

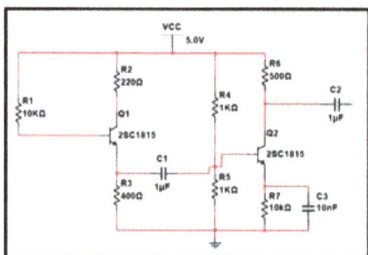

Figure 8. The signal amplifier.

4.2. Frame Format

Figure 9 shows the frame format of the information bits, which consists of four parts: The synchronization header of the frame, the training level, the information bits, and the terminator. The synchronization header of the frame uses a high level of six time slots and a low level of one time slot. To eliminate the effects of ambient light, the training levels are added before the information bits. Finally, the signal $0 \times 0D$ is used to represent the end of a frame.

7 time slots	3 time slots	128 time slots	0x0D
The Frame Synchronization Header	The Training Level	The Information Bits	The Terminator

Figure 9. The frame format of the information bits.

The character sent by the serial debugging assistant at the transmitter is displayed in an ASCII format. The eight information bits are conveyed with (16,1,2)GSM-MPAPM. Figure 10 shows the processes for generating and receiving information bits at the transmitter and receiver.

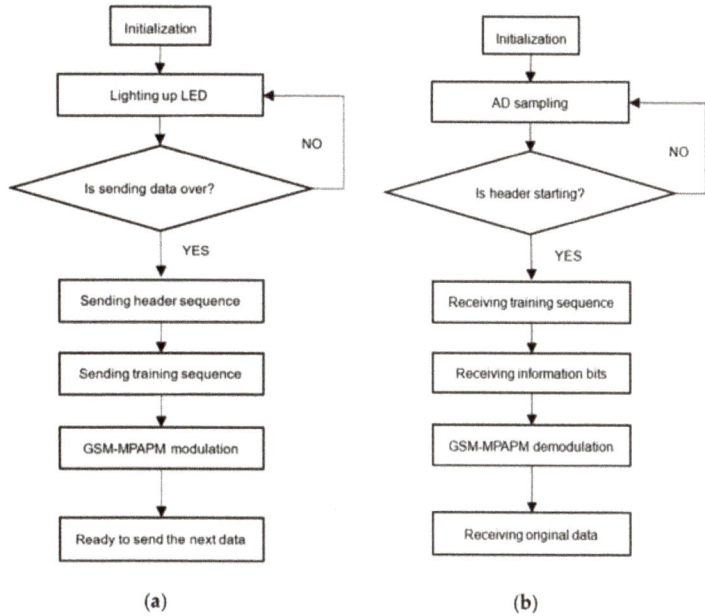

Figure 10. The process for the generation and reception of information bits (**a**) at the transmitter; and, (**b**) at the receiver.

5. Results and Analysis

Measurements were made in a laboratory that was 4 m long, 3 m wide, and 4 m high, in the presence of other light sources in an indoor environment. It was supposed that the light emitting from the LEDs was on line of sight. At the transmitter, there were five LEDs, each with 2 W of power. The maximum output current of the pins on the C8051f330 was 25 mA, and the drive capability of the designed LED drive circuit was limited; the luminous power in the measurement was only 1.5 W. Furthermore, the C8051f330 clock influenced signal generation at the transmitter. Thus, 25 kb/s was chosen as the transmission rate of the information bits in the experiment.

This experiment firstly analyzed two activated LEDs which transmitted eight information bits at a time in the system. The different peak values of the voltage were obtained by changing the distance between the photodetector and the LEDs. The photodetector faced LED1, but all LEDs were at the same level and each was a different distance from the photodetector. Therefore, the different values were obtained according to LED combinations. In Figures 11–14 seven high-level synchronization headers were followed by three training levels. The measured peak value of the voltage decreased from 4.15 V, 2.30 V, 1.40 V, to 0.8 V when the distance between the photodetector and the LEDs increased from 10 cm, 20 cm, 30 cm, to 40 cm. It could be seen clearly that in Figure 14, when the distance between the photodetector and the LEDs was 40 cm, the measured peak value of the voltage was almost close to that of the noise. The received signals were drowned in the noise at this moment. The BER performance was also analyzed in this experiment. Figure 15 shows that the BER remained below the value of 1.5×10^{-4} when the distance was within 40 cm. However, when the distance exceeded 40 cm, the value rapidly increased. The results of the peak value of the voltage and the BER performance of the received signals both demonstrated that the distance between the photodetector and the LEDs was a very crucial parameter for the proposed scheme in the indoor short-distance VLC. This was mainly due to the use of a single photodetector for the intensity modulation/direct detection demodulation at the receiver. The illumination intensity decreased as the distance increased. Once the distance exceeded a threshold that guaranteed normal communication, multiple levels of interference appeared in the received signals. From the five figures above, it was drawn that the threshold of the distance was 40 cm. Ultimately, the aliasing signals prevented the original signals from being modulated. Furthermore, the luminous power of 1.5 W in this experiment might have been a factor that influenced the BER performance.

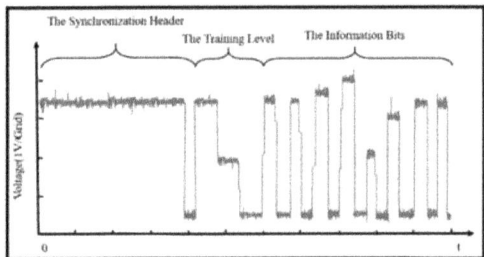

Figure 11. The amplified waveform of the received signals (10 cm).

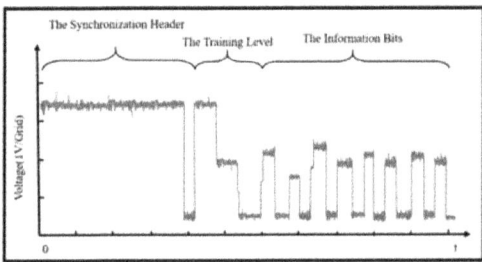

Figure 12. The amplified waveform of the received signals (20 cm).

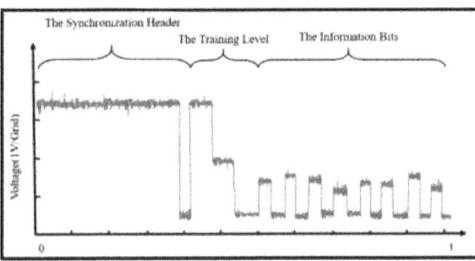

Figure 13. The amplified waveform of the received signals (30 cm).

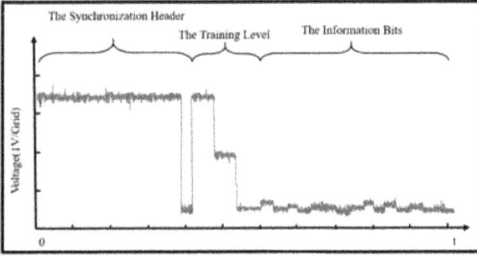

Figure 14. The amplified waveform of the received signals (40 cm).

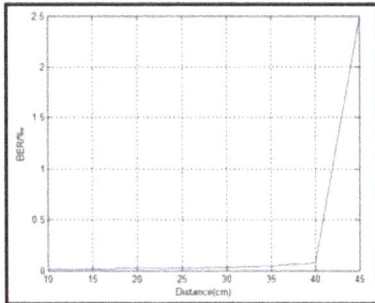

Figure 15. The relationship between the BER (‰) and the distance.

When the string of the information bits with the value of 1010010111000011001111001100110 was transmitted by Tera_Term at the transmitter, the receiver correctly received the string. In this test, the BER performances of GSM-MPAPM, V-MPAPM, MPAPM, and ACO-OFDM were compared. Among them, the latter three were mentioned in [5,9,11], respectively. Figure 16 shows that, within the range of the observation distance that was set between 15 cm and 23 cm, as the distance increased, the proposed (16,1,2)GSM-MPAPM, V-MPAPM, and MPAPM all had a stable performance in the BER, while the BER of ACO-OFDM gradually raised. This was because when the OFDM was applied to the VLC, due to the nonlinear response of the LED, it caused a higher BER than the single carrier modulation, such as GSM-MPAPM, V-MPAPM, and MPAPM, which were compared in Figure 16. At the same time, the (16,1,2)GSM-MPAPM had a better BER performance as compared to V-MPAPM and MPAPM obviously. That is because, as the distance increased, the channel fading was getting more and more serious, the (16,1,2)MPAPM provided an increase in spatial diversity via the two activated LEDs, which causes an improvement in the BER.

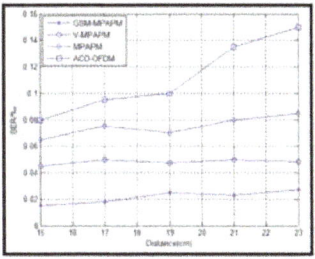

Figure 16. Comparisons of the bit error ratios (BERs) of different modulations.

Three dimming coefficients were chosen and an illuminance curve was drawn through measurements in Figure 17 to verify the relationship between the illuminance and the distance of the (16,1,2)MPAPM. It can be seen from the curve that the illuminance decreased as the distance increased over the three different dimming coefficients. This caused a decrease in the peak value of the voltage and an increase in the BER. For p_1, the system had a stable dimming effect when it was kept within the measurement distance, but the illuminances of all of the distances were slightly lower than the other two values. For p_2 and p_3, there were sudden changes at some values, i.e., 14 cm, 20 cm. Therefore, p_1 was the best choice when a stable dimming was considered.

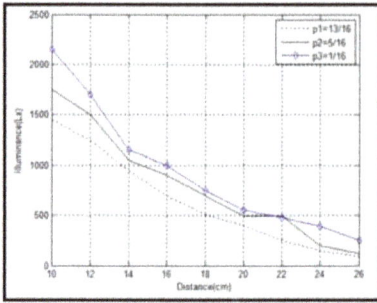

Figure 17. The relationship between the illuminance and the distance under three dimming coefficients.

6. Conclusions

In this paper, an efficient data transmission scheme was successfully implemented in a designed indoor VLC system. The experiments demonstrated that the proposed GSM-MPAPM modulation worked normally so long as the distance between a transmitter and a receiver was within 40 cm, and the BER performance was inversely proportional to the distance. Furthermore, GSM-MPAPM had a better BER performance than VMPAPM, MPAPM, and ACO-OFDM when the same distance was used. This was because activated LEDs, of which there are usually multiple, provide an increase in spatial diversity to resist communication fading. Additionally, the differentiated dimming strategy caused a stable dimming effect on the variety of light in the environment.

Author Contributions: This research article contains three authors, including J.-J.B., who graduated from Communication and Information System, was responsible for conceiving and designing the modulation algorithm, designing the embedded hardware system together with C.-L.H., doing the experiments, and writing the paper and the replies. C.-L.H., was charged of analyzing the data and designing the embedded hardware system together with J.-J.B., J.-F.T., who was charged for proposing many helpful suggestions on the conception, format and writing of this paper. With the author's help, this study can be completed as schedule.

Funding: This research was funded by [Social Science and Technology Development of Dongguan Science and Technology Bureau] grant number [2019507156746], [School-Enterprise Cooperation Horizontal Research of Dongguan Polytechnic] grant number [2018H64], [Technical R&D and service team of Dongguan Polytechnic] grant number [CXTD201802].

Conflicts of Interest: The authors declare no conflict of interest.

References

1. Elgala, H.; Mesleh, R.; Haas, H. Indoor Optical Wireless Communication: Potential and State-of-Art. *IEEE Commun. Mag.* **2011**, *49*, 56–62. [CrossRef]
2. Grubor, J.; Randel, S.; Langer, K.D. Broadband Information Broadcasting Using LED-Based Interior Lighting. *J. Light. Technol.* **2008**, *26*, 3883–3892. [CrossRef]
3. Lou, S.; Gong, C.; Wu, N. Joint Dimming and Communication Design for Visible Light Communication. *IEEE Commun. Lett.* **2017**, *21*, 1043–1046. [CrossRef]
4. Jie, L.; Maite, B. Adaptive M-PAM for Multiuser MISO Indoor VLC Systems. In Proceedings of the 2016 IEEE Global Communications Conference, Washington, DC, USA, 4–8 December 2016; pp. 1–6.
5. Thai-Chien, B.; Mauro, B.; Suwit, K. Theoretical Analysis of Optical Spatial Multiple Pulse Position Modulation. In Proceedings of the 2018 Global Communications Conference, Abu Dhabi, UAE, 9–13 December 2018; pp. 112–120.
6. Mostafa, H.; Martin, J.; Peter, J. Visible Light Communication based on Offset Pulse Position Modulation Using High Power. In Proceedings of the 32nd General Assembly and Scientific Symposium of the International Union of Radio Science, Montreal, QC, Canada, 19–26 August 2017; pp. 223–228.
7. Zhang, D.-F.; Zhu, Y.-J.; Zhang, Y.-Y. Multi-LED Phase-Shifted OOK Modulation Based Visible Light Communication Systems. *IEEE Photonics Technol. Lett.* **2013**, *25*, 2251–2254. [CrossRef]

8. Yamga, G.M.; Ndjionge, A.R.; Ouahada, K. Low Complexity Clipped Frequency Shift Keying for Visible Light Communications. In Proceedings of the IEEE 7th International Conference on Adaptive Science & Technology, Accra, Ghana, 22–24 August 2018; pp. 62–70.
9. Chumchewkul, D. Performance Evaluation of VPPM Visible Light Communications based on Simulation with Experiment's Parameters. In Proceedings of the 15th International Conference on Electrical Engineering, Electronics, Computer, Telecommunications and Information Technology, Chiang Rai, Thailand, 18–21 July 2018; pp. 26–35.
10. Zhang, M.; Zhang, Z. An optimum DC-Biasing for DCO-OFDM system. *IEEE Commun. Lett.* **2014**, *18*, 1351–1354. [CrossRef]
11. Armstrong, J.; Lowery, A.J. Power efficient optical OFDM. *Electron. Lett.* **2006**, *42*, 370–372. [CrossRef]
12. Su, S.; Chung, W.; Wu, C. Exploiting Entire GSSK Antenna Combinations in MIMO Systems. *IEEE Commun. Lett.* **2015**, *19*, 719–722. [CrossRef]
13. Olanrewaju, H.; Thompson, J.; Popoola, W. Generalized Spatial Pulse Position Modulation for Optical Wireless Communications. In Proceedings of the 2016 IEEE 84th Vehicular Technology Conference, Montreal, QC, Canada, 18–21 September 2016; pp. 245–249.
14. Qin, Y.L. Performance Analysis of MPPM Based Wireless Optical Communication System. Master's Thesis, Xidian University of Electronic Technology, Xi'an, China, 2018.
15. Effros, M.; Goldsmith, A.; Liang, Y. Generalizing Capacity: New Definitions and Capacity Theorems for Composite Channels. *IEEE Trans. Inf. Theory* **2010**, *56*, 3069–3087. [CrossRef]
16. Younis, A.; Serafimovski, N.; Mesleh, R. Generalised Spatial Modulation. In Proceedings of the 44th Asilomar Conference on Signals, Systems and Computers, Pacific Grove, CA, USA, 7–11 November 2010; pp. 1498–1502.
17. Younis, A.; Basnayaka, D.A.; Hass, H. Performance Analysis for Generalised Spatial Modulation. *Eur. Wirel.* **2014**, *45*, 1–6.

Article

Fault Diagnosis System for Induction Motors by CNN Using Empirical Wavelet Transform

Yu-Min Hsueh, Veeresh Ramesh Ittangihal, Wei-Bin Wu, Hong-Chan Chang and Cheng-Chien Kuo *

Department of Electrical Engineering, National Taiwan University of Science and Technology, Taipei 10607, Taiwan; D10307009@mail.ntust.edu.tw (Y.-M.H.); valiant.veeru@gmail.com (V.R.I.); andytouchwork@gmail.com (W.-B.W); hcchang@mail.ntust.edu.tw (H.-C.C.)
* Correspondence: cckuo@mail.ntust.edu.tw

Received: 5 August 2019; Accepted: 27 September 2019; Published: 29 September 2019

Abstract: Detecting the faults related to the operating condition of induction motors is a very important task for avoiding system failure. In this paper, a novel methodology is demonstrated to detect the working condition of a three-phase induction motor and classify it as a faulty or healthy motor. The electrical current signal data is collected for five different types of fault and one normal operating condition of the induction motors. The first part of the methodology illustrates a pattern recognition technique based on the empirical wavelet transform, to transform the raw current signal into two dimensional (2-D) grayscale images comprising the information related to the faults. Second, a deep CNN (Convolutional Neural Network) model is proposed to automatically extract robust features from the grayscale images to diagnose the faults in the induction motors. The experimental results show that the proposed methodology achieves a competitive accuracy in the fault diagnosis of the induction motors and that it outperformed the traditional statistical and other deep learning methods.

Keywords: empirical mode decomposition; pattern recognition; wavelet; empirical wavelet transform; convolutional neural network; induction motor; fourier transform; fault diagnosis

1. Introduction

Because of the simple design, low cost, low maintenance, and easy operation, induction motors are one of the most commonly used rotating machines in the industry. In spite of the fact that these machines are more reliable and robust in nature, failure of induction motors is expected, due to the various stresses they encounter during their operating conditions. The most responsible factors behind such failure conditions could be either from mechanical or electrical forces. Different types of machinery faults, like broken bars, bearing faults, an unbalanced rotor, and stator faults and winding faults, have been discussed in the literature [1,2]. Many studies have been conducted on fault diagnosis in recent years. Early detections of the problems are vital to save time and costs, so as to take remedial measures to avoid an entire system failure [3]. The fault diagnosis methods can be classified widely into signal-based, model-based, active/hybrid and knowledge-based methods [4,5]. The knowledge-based methods, also called data-driven methods, require a huge amount of historical data to find the signal patterns for the fault diagnosis of the system.

The predictive maintenance and the data-driven methods are commonly used to analyze signals such as the current, temperature, electrical tension and vibrations, which are captured by the use of sensors [6,7]. The signal-based features are extracted for the fault diagnosis. However, the extracted features need to undergo the feature selection techniques to avoid repeated information and also to significantly reduce the feature dimensions, which can improve the performance by retaining important features [8]. Finally, the selected features are used for the fault diagnosis via various methods based on traditional statistical and machine learning models. [9–11]. The traditional methods have achieved significant results. However, the feature selection for the methods depends heavily on the knowledge and expertise of signal processing methods with respect to the diagnosis. Furthermore, the traditional machine learning methods are not capable enough to produce distinguishable features of original data and constantly require a process of feature extraction from the signal [10,12–15].

Apart from the significant development in machine learning, deep learning has emerged as an effective study that can overcome the above-mentioned drawbacks in fault diagnosis. It can avoid manual feature extraction and automatically learn the abstract features from the raw data [12]. Various deep learning methods have been studied and applied to fault diagnosis, such as the stacked sparse auto-encoder [13], sparse auto-encoder [15], deep belief network (DBN) [16], denoising auto-encoder [17], and sparse filtering [18]. Deep learning has achieved significant results in comparison to traditional machine learning methods. The convolutional neural network (CNN), known as one of the proven deep learning models, has delivered promising results in learning useful features [14,19]. However, there are still many studies to be done on the application of deep learning on fault diagnosis.

In the real world, most of the pattern recognition tasks deal with time-series data. Weather and forecasting, video processing, biomedical signal processing, stock and currency exchange rate data processing have been studied with time-series data [20–22]. Similarly, electrical industry devices such as induction motors also often deal with time-series data like the current, voltage, temperature and vibration signals. Since machinery data signals belong to time-domain signals, a one-dimensional (1-D) CNN is studied to diagnose motor faults [23]. However, in a few cases, the machinery data can be viewed in two dimensions (2-D), such as the time-frequency domain, to avoid redundant data, representing the data as 2-D images using the empirical wavelet transform [24].

As part of this study, the current signals from 3 phase induction motors are considered for a fault diagnosis. All three currents are the same with a difference in the phase, and each phase current was used as one sample for representation in empirical wavelet transform. The main contributions of this work are summarized as follows. First, we propose a well-defined data preprocessing method, in which 2-D features are extracted and represented as an image from the current signal. Second, an effective CNN model is proposed to extract and learn the features automatically from the images. Finally, the proposed CNN-based method achieves promising results compared to other deep learning and traditional methods.

The remainder of this paper is structured as follows: Section 2 reviews the related works. Section 3 introduces the proposed framework, in which the data preprocessing and proposed CNN model are discussed. Section 4 presents the experimental results. Finally, the conclusion and future research works are presented in Section 5.

2. Related Works

In recent years, many signal processing techniques have been studied in the frequency domain, time domain, and time-frequency domain to extract the full features and detect the machine operating condition using classification methods. Time-frequency domain methods are preferred, among others, to analyze and extract the features from the non-stationary signals. Wang et al. [25] applied wavelet scalogram images as an input to CNN to learn the features and detect the faults. Lee et al. [26] analyzed a corrupted raw signal and the effect of the noise on training the CNN model. Ge et al. [27] studied and theoretically analyzed the empirical mode decomposition (EMD) method. Lei et al. [28] used the EMD method to extract features from vibration signals and discussed a kurtosis-based method for fault diagnosis. Pandya et al. [11] constructed an efficient KNN classifier using an asymmetric proximity function for fault diagnosis. Yang et al. [10] proposed an SVM-based method to diagnose the fault patterns of roller bearings. Ngaopitakkul et al. [9] proposed a decision algorithm based on ANN for a fault diagnosis using discrete wavelet transform (DWT) and backpropagation neural networks. The high-frequency component of the current signals is decomposed by using a mother wavelet called Daubechies (db4). The DWT extracts the high-frequency component from the fault current signals and the coefficients of the first scale from the DWT are used to detect the fault. Ma et al. [29] proposed a method to extract the features of bearing faults based on the complete ensemble EMD (CEEMD) by enhancing the mode characteristic and via the introduction of adaptive noise to diagnose the bearing faults of rotatory machines. Ge et al. [30] proposed a fault diagnosis method based on an empirical wavelet transform sub modal hypothesis test and ambiguity correlation classification to diagnose the rolling bearing faults using vibration signals. However, the authors concentrated only on rolling bearing faults. Deng et al. [31] studied a fault diagnosis method to extract a new feature by combining Hilbert transform coefficients, the correlation coefficients and the ensemble empirical mode decomposition (EEMD). The vibration signal is decomposed into a list of multiple intrinsic mode functions (IMFs) with distinct frequencies using the EEMD. Agarawal et al. [32] presented a comparative study of ANN and SVM using continuous wavelet transforms and energy entropy methods to diagnose and classify the rolling element bearing faults. Mother base wavelet is selected from four real-valued base wavelets based on the entropy criterions and the energy. The statistical features are extracted from the wavelet coefficients of real signals. The extracted statistical features are provided to ANN and SVM as input for the classification of the bearing faults. These comparative results show SVM giving a better performance than ANN. Jayaswal et al. [33] provided a brief review of recent studies on ANN, fuzzy logic and wavelet transform, used to diagnose rotating machinery faults using raw vibration signals. However, special attention is only given to rolling element bearing faults. Bin et al. [34] studied a method using wavelet coefficients and empirical mode decomposition to extract features and classify faults using a multi-layer perceptron network. However, the ANN study found two main concerns: (1) A large dependency on a prior knowledge of signal processing methods and an expertise in the diagnostic process; and (2) the ANNs studied for the fault diagnosis of induction motors might be limited in their learning capacity from learning complex and nonlinear relationships because of the large information on motor currents. Thus, it is essential to study the deep architecture network for fault diagnosis.

Deep learning is more advanced when compared to traditional machine learning methodology. Due to its potential ways of featuring representation, it has been extensively used in machine health monitoring systems [35]. Jia et al. [36] proposed a neural network-based method to diagnose faults using an auto-encoder. Cho et al. [37] used recurrent neural networks and dynamic Bayesian modeling for fault detection in induction motors. However, with RNN, the information flows via the hidden states and is much slower than with CNN. Deep learning models like deep auto-encoders (DAE), deep belief networks (DBN) and CNNs have been studied for fault diagnosis [13,14,16]. Ince et al. [20] used a one-dimensional (1-D) CNN for a real-time motor fault diagnosis. Xu et al. [38] proposed a study based on the Gabor wavelet and the neural network to detect the image intelligence. The authors employed the Gabor wavelet transform to extract the features of information from images. Abdeljaber et al. [39] proposed a 1-D CNN for real-time structural damage detection. Furthermore, there are various ways to represent machinery data in the 2-D format. Chong [40] proposed an effective way to extract the features by converting 1-D vibration signals into 2-D grayscale images. Gaowei et al. [41] proposed a method based on deep CNN and random forest ensemble learning with a remarkable performance; however, they only focused the bearing fault diagnosis. Lu et al. [42] used a probabilistic neural network as an image classifier by converting signals to images using a bispectrum. Kang et al. [43] used 2-D greyscale images created using Shannon wavelets for an induction motor fault diagnosis. However, an expert's knowledge is necessary for these conversion methods. Although methods such as neural networks, using raw data signals, are considered in many studies in order to diagnose and classify faults, data preprocessing is a highly important action in deep learning. Processing huge quantities of data and examining several qualities of parameters leads to a lot of troubles in data preprocessing. Data with distinct characteristics need distinct methods to extract their characteristics. Many studies use frequency, time-frequency, and histograms to convert signals into images for classification. Similarly, in the proposed study, a two-dimensional matrix generated from wavelet coefficient values is represented as an image. The benefit of presenting an image instead of the raw one-dimensional current signal is that the image can provide spatial and temporal dependencies. Moreover, CNN has been a popular deep learning algorithm for working with image datasets, and traditionally it is two dimensional. The benefit of using CNN over a neural network is its ability to develop an internal representation of a two-dimensional image or a matrix of values. It helps the model to learn the position and scale of different structures in the image data or in the two-dimensional matrix data. It also helps to reduce the number of parameters involved by learning high-level features and via the reusability of weights. In this study, an efficient 1-D signal to 2-D greyscale image representation is proposed by using an empirical wavelet transform. This method is free of any predefined parameters and eliminates the expert's interference.

3. Proposed Methodology

This section describes the proposed EWT-CNN-based fault diagnosis methodology. As part of data preprocessing, the raw current signal is converted into images using EWT modes. Then, a deep CNN model is presented to extract and learn the features for the fault diagnosis.

3.1. Pattern Recognition Technique

As the most common data-driven methods are unable to deal with direct original signals for the fault diagnosis, preprocessing the raw signal is necessary. In recent years, an empirical mode decomposition (EMD) algorithm proposed by Huang et al. [44], and has gained a great interest in signal analysis due to its ability to separate stationary and non-stationary components from a signal. However, although its adaptability seems appreciable, the lack of a mathematical theory is the main issue with this approach. To deal with this problem, an Ensemble EMD (EEMD) is proposed to compute several EMD decompositions of the original signal, averaging the decompositions to get a final EEMD. This method seems appreciable, but it increases the computational cost [45].

Currently, wavelet analysis is classified as one of the most used tools to analyze signals. An extensive literature about wavelet theory [46–48] can be referenced for further details. In the temporal domain, with a scaling factor $s > 0$ and a translation factor $u \in \mathbb{R}$, the wavelet dictionary $\{\psi_{u,s}\}$ is defined as:

$$\psi_{u,s}(t) = \frac{1}{\sqrt{s}}\psi\left(\frac{t-u}{s}\right) \tag{1}$$

The scaling factor s is used to stretch or compress the wavelet function in order to change the oscillating frequency, and the translation facto u is used to change the position of the time window. The wavelet functions define the focal features and time-frequency properties, which can effectively capture the non-stationary characteristics of the signal. There are many wavelets functions that are studied, such as Morlet, Meyer, Symlet, Gabor, Coiflet, and Haar [49–52]. All these methods use either a prescribed scale subdivision or use the output of the classic wavelet output smartly. However, they failed to build a full adaptive wavelet transform. Thus, the proposed method uses a new approach called empirical wavelet transform (EWT) to build a family of wavelets adapted to the processed signal [24,30]. The empirical wavelet transform is defined in a step-by-step manner rather than in a single mathematical formulation as is the case of the classic wavelet transform. The main idea behind the EWT is to extract the different modes of a signal based on Fourier supports detected from the spectrum information of the processed signal.

The following steps summarize the empirical wavelet transform proposed in [24]:

Step 1: Find the Fourier transform of the processed input signal.
Step 2: Segment the Fourier spectrum by detecting the local maxima in the spectrum.
Step 3: Sort the local maxima in decreasing order
Step 4: Define the boundaries of every segment as the center between two successive maxima.
Step 5: Follow the construction idea of Meyer's wavelet to obtain a tight frameset
Step 6: Obtain the corresponding signal filters (modes as defined in [24]).

The proposed empirical wavelets correspond to the dilated version of a single mother wavelet in the temporal domain. However, the corresponding dilatation factors do not follow a prescribed scheme but are detected empirically. For further details on the EWT, we refer the reader to the literature [24]. A three-phase current signal from the induction motor is collected. Ten cycles (one full cycle having 167 data points) for each phase current signal, i.e., 1670 continuous points, are sampled.

Then, the 1670 points are converted into a 1670 × N time-frequency spectrum, which consists of the coefficient matrices via the empirical wavelet transform. N stands for the number of modes, and the sufficient raw signal characteristics can be obtained by choosing the appropriate value. Finally, the grayscale image is represented from the time-frequency spectrum.

The raw current signals collected from the different induction motors working at different faulty/healthy condition and operating on the same load condition are shown in Figure 1. However, they are non-distinguishable, and it is almost impossible to diagnose the fault condition of the motors by using the raw current signals. Figure 2 shows the same set of raw signals that are processed by EWT, and they look absolutely distinguishable from each other. Hence, it is indeed necessary to preprocess the raw current signals by EWT in order to find the distinguishable patterns.

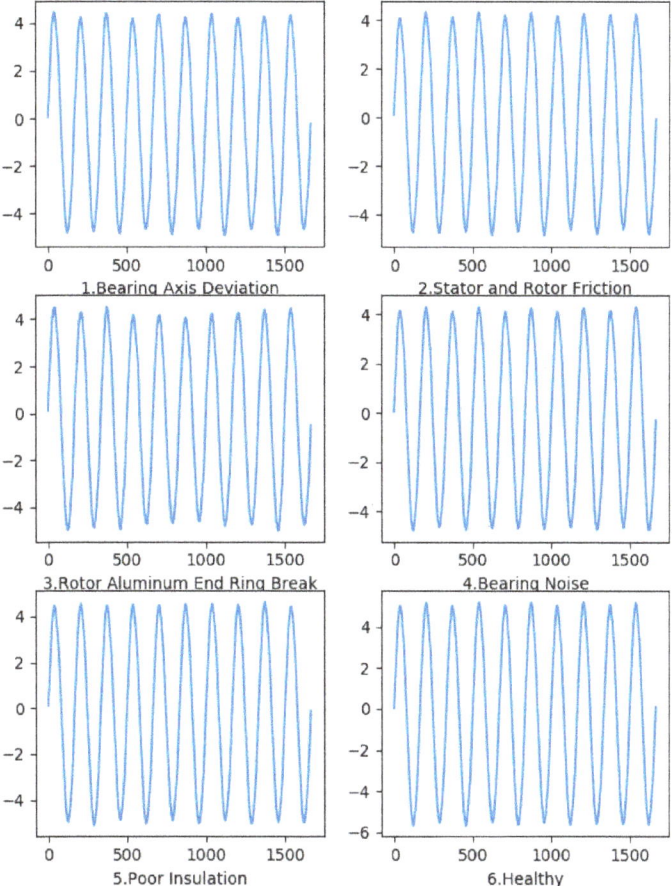

Figure 1. The induction motor current signals: 10 cycles of current signals for each fault and healthy conditions motors (non-distinguishable patterns).

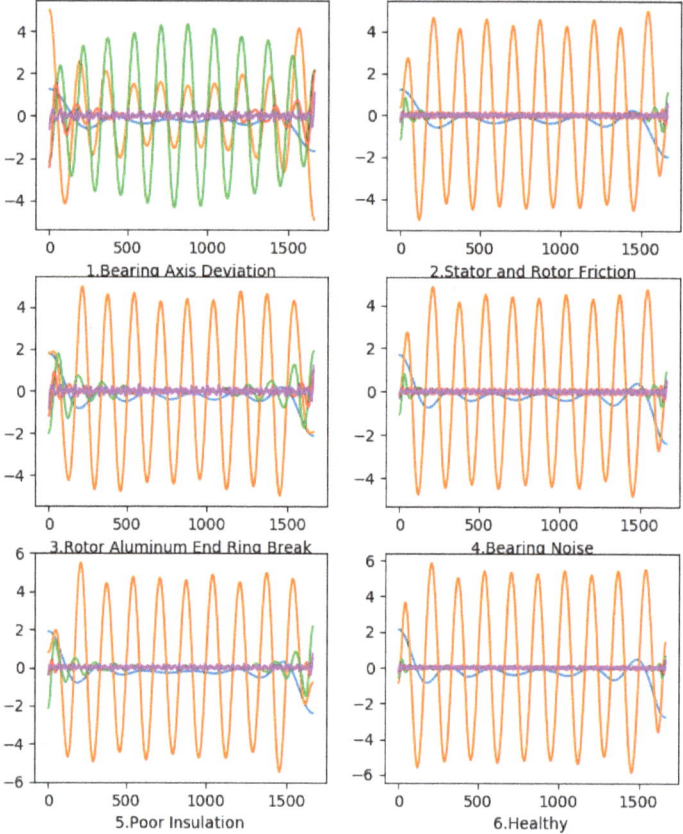

Figure 2. The EWT modes of the current signals: The EWT modes plot for the respective fault and healthy conditions (distinguishable patterns).

The CNN model training will be difficult with the 1670 × N image, as the latter results in computational complexity. A simple image resizing method based on scikit-image processing [53] is used to decrease the image size. Figure 3 illustrates the entire workflow of the proposed method. Figure 4 shows the distinguishable grayscale resized (32 × 32) images for each fault type and the healthy type of motors data.

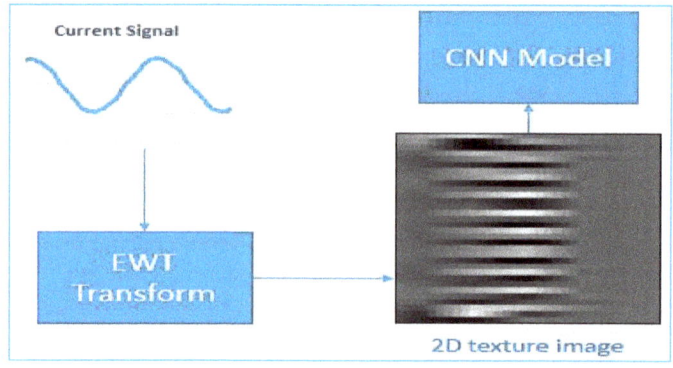

Figure 3. The architecture of the proposed methodology.

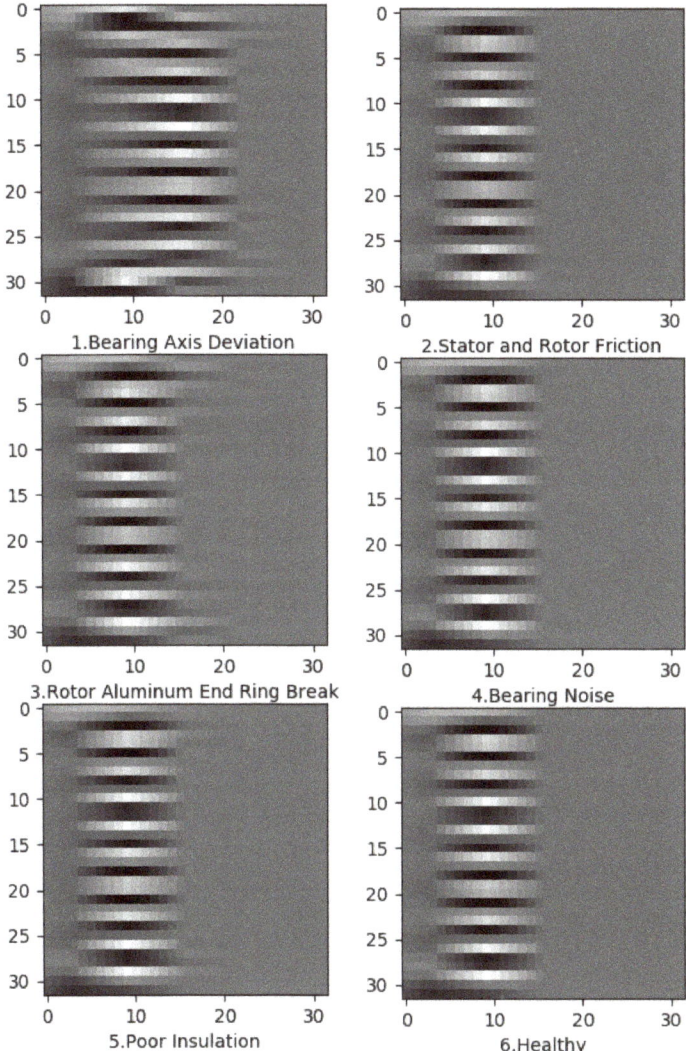

Figure 4. Grayscale images (32 × 32) for the EWT modes of each fault and the healthy conditions of the motor.

3.2. Proposed Deep Convolutional Neural Network

After converting the raw current signals into grayscale images, a deep CNN model is designed and pre-trained for feature learning. The proposed deep CNN has a three-stage structure. Each stage represents a feature learning stage with a different feature-level, which includes convolution, activation, and pooling layers.

Figure 5 illustrates the architecture of the proposed CNN model, which consists of three convolutional layers with filters 32–3 × 3, 64–3 × 3 and 128–3 × 3, respectively. In addition to that, there are three max-pooling layers of size 2 × 2. The most commonly-used activation functions are the hyperbolic tangent, softmax, ReLU, and sigmoid function [54]. Among them, ReLU has proven to be more effective than the others. However, during the training, ReLU units can die, and this could occur when a large gradient flows through a ReLU neuron. This causes the weights to update, so that the neuron will never activate again on any data point. A leaky ReLU is an attempt to solve this problem [55,56]; thus, the leaky ReLU (Rectified Linear Units) is applied as an activation function to introduce non-linearity into each stage, allowing the CNN to learn complex models. Pooling is used to reduce the resolution of the input image via the process of subsampling, and Max Pooling is used in the proposed model.

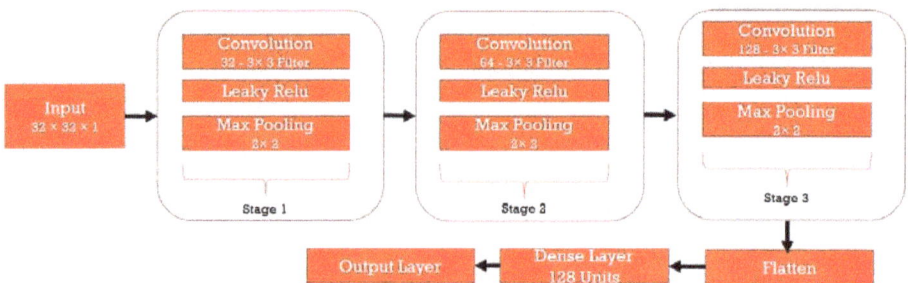

Figure 5. The proposed 3-stage Convolutional Neural Network Architecture diagram. 32 × 32 size grayscale images are fed into the CNN model. The architecture consists of three convolutions layers followed by pooling layers and two fully connected layers.

Training the CNN model involves learning all the weights and biases, and it is important to optimize these parameters for an efficient feature learning. Apart from the training parameters, the CNN also needs to optimize the hyperparameters, such as the learning rate and dropout. The dropout is an important property of CNN, which can greatly help in preventing the overfitting by generalizing the model [57]. A dropout of size 0.4 is used for a better regularization in the proposed CNN. The adapted moment estimation (ADAM), which is a backpropagation algorithm, is used to optimize the learning rate and other hyperparameters. The ADAM adapts the learning rate scale through different layers and avoids the manual assignment for choosing the best learning rate [58]. At the end of the three stages, the feature maps are flattened and classified via a fully connected layer for 6 types of classifications.

4. Experimental Results and Discussion

To assess the performance of the proposed methodology, the raw current signal data from an experimental setup involving a total of six induction motors with the same specifications are used. This includes one healthy and five fault types of raw current data signals, which are collected from the experimental setup. The six types of current signals are studied and analyzed for the healthy condition of the motor, as well as for the following five faulty conditions of the motor [59]. The data preprocessing and the CNN model are written in Python 3.6 with TensorFlow and run on the Windows 64 bit operating system.

4.1. Faults in Induction Motors

The motors undergo various types of failure modes, mostly due to electrical and mechanical forces. These failure modes eventually break the entire system from its normal working condition. This section deals mainly with the five types of faults, namely: bearing axis deviation, stator and rotor friction, rotor aluminum end ring break, bearing noise and poor insulation.

1. Bearing Axis Deviation: The structure of the bearing is precise. If it is disturbed by some external forces, the structure of the bearing may be affected. After connecting the motor to the load, an earthquake, collision, and the assembly process may introduce an offset of midpoints on both ends of the connection, which causes heating problems and unwanted noise. A normal motor with a full load is used, and, for this experiment, the coupling is shifted 0.5 mm upward to imitate the deviation condition. The experimental motor model is shown in Figure 6d.

2. Stator and Rotor Friction and Poor Insulation: Because of friction, overheating, insulation aging, dampness and corona, the stator or rotor coil is short-circuited, and hence it will break down if not diagnosed. The insulation of the adjacent turns in the stator coil will be damaged, causing a short circuit, as shown in the Figure 6a. When the motor is started, the short-circuit current value will be high due to the difference in excessive voltage caused by different wound turns in the stator, and the motor will be burnt. The experimental motor model is shown in Figure 6a.

3. Rotor Aluminum End Ring Break: The outer ring damage is one of the most common faults. If the starting frequency is very high and/or the motor is overloaded, the rotor bar will break due to the excessive current. For this experiment, a hole with a diameter of 7 mm and a depth of 30 mm is made on the rotor bar to simulate the fault condition. The experimental motor model is shown in Figure 6b.

4. Bearing Noise: Damage to the bearing's outer race is considered one of the constant faults observed in bearings. The structure of the bearing is always kept precise. However, if the structure is disturbed by an external force or some other structures of bearing, this causes messy and numerous harmonics in the measured spectrum. A hole with a diameter and depth of 1 mm is made in the outer race to simulate the fault condition for this experiment. The experimental motor model is shown in Figure 6c.

The proposed method uses the motor raw current signal values to analyze and find patterns for the fault diagnosis of the above-listed motor faults.

Figure 6. The experimental motor model: (**a**) stator and rotor friction and poor insulation; (**b**) rotor aluminum end ring break; (**c**) bearing noise; and (**d**) bearing axis deviation.

4.2. Dataset

The collected dataset from the experiment consists of 900 samples [60]. 50 samples from the healthy condition motor and 50 samples from each kind of faulty condition motor on a 100% load (full load) are collected and analyzed. As three-phase induction motors are used in this study, there are three current signals, with differences in the phase, and each phase current is considered when preparing the dataset. Hence, a total of 150 raw current data samples are prepared for the healthy motor and for each of the five faulty motors, as described in Table 1.

Table 1. The dataset samples used for the evaluation.

Bearing Axis Deviation	Stator and Rotor Friction	Rotor Aluminum End Ring Break	Bearing Noise	Poor Insulation	Healthy	Total
150	150	150	150	150	150	900

The data set is divided into three parts, as described in Table 2. 70% of the dataset (630 samples for training) and 15% of the dataset (135 samples for validation) are used simultaneously to train the CNN model. The remaining 15% (135 image samples) are used to test the trained CNN model. Cross-validation techniques are often used for simple models having few trainable parameters like linear regression, logistic regression, small neural networks and support vector machines. A CNN model having many parameters will lead to too many possible changes in the architecture. However, in this study, the proposed CNN model is trained and evaluated using a k-fold cross-validation with the data split ratio shown in Table 2.

Table 2. The dataset samples used for the evaluation.

Data Split Ratio		
Training	70%	630
Validation	15%	135
Test	15%	135

4.3. CNN Performance Evaluation Results

The proposed CNN model is trained over 150 epochs to learn the robust features for each type of faulty condition motor and one normal operating condition motor. A k-fold cross-validation technique with five folds is applied manually to evaluate the model training and testing. The CNN model is trained to extract and learn the features from 630 samples of the training dataset, simultaneously validated against 135 samples of the validation dataset during each iteration for the five folds of the dataset split. The trained CNN model is evaluated against 135 samples of the test-dataset. The model is cross-validated over five folds with the dataset split ratio being described in Table 2, after which the averages of all the accuracies and losses during each fold are collected in order to observe the accuracies and losses during the training, as shown in Figure 7.

The proposed CNN model is trained and tested with batch sizes of 16, 32 and 64, and we found the best results to be with a size of 32. The CNN model is trained over 50 to 200 epochs to learn the robust features and analyze the classification performance, in order to choose the number of epochs. The average accuracies and losses (training and validation) are collected at each iteration while training the CNN model with a k-fold cross-validation technique and are then plotted, as shown in Figure 7. The CNN model hit the training accuracy by almost 100% with a validation accuracy of around 91%. Over the 150 epochs, the proposed CNN model was able to learn the robust and generalized features of the EWT grayscale images, in order to diagnose the motor faults and classify them into faulty or healthy categories.

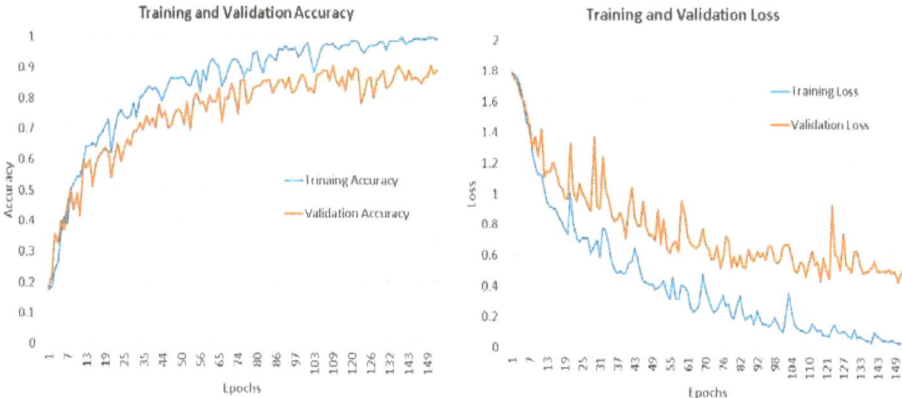

Figure 7. The accuracy and loss curves over 150 epochs of the CNN training.

To evaluate the performance of the trained CNN model, 135 samples of the test dataset are used. The performance result of the trained CNN model looks interesting, with an average accuracy of 97% on the test dataset, as described in the classification report (Table 3). From the classification report, it can be clearly seen that the proposed CNN model is capable of extracting and learning the features from the test dataset and of classifying the features for the respective faulty and healthy conditions. The proposed model is able to classify the healthy condition, bearing axis deviation fault, rotor aluminum end ring break fault and bearing noise fault more effectively than the other faults. However, the model needs to be tuned in the case of the motor with poor insulation faulty condition. Poor insulation can also be observed due to stator and rotor friction and bearing axis deviation. Hence, there are some misclassifications with other types. Figure 8 illustrates the confusion matrix, which explains the classification results on the test dataset (135 samples) using the well-trained CNN model. Almost all the test samples are correctly classified, with a few misclassifications involving the poor insulation condition and other faulty conditions.

Confusion Matrix							
class	Fault 0	Fault 1	Fault 2	Fault 3	Fault 4	Fault 5	
Fault 0	23	0	0	0	0	0	23
Fault 1	0	22	0	0	0	0	22
Fault 2	0	0	21	0	0	1	22
Fault 3	0	0	0	23	0	1	24
Fault 4	0	0	1	0	21	0	22
Fault 5	0	0	0	0	1	21	22
	23	22	22	23	22	23	135
Actual							

(Predicted — row axis label)

Figure 8. The confusion matrix for the test dataset. (Fault 0: Healthy, Fault 1: Bearing Axis Deviation, Fault 2: Stator and Rotor Friction, Fault 3: Rotor Aluminum End Ring Break, Fault 4: Bearing Noise, Fault 5: Poor Insulation).

In order to assess the performance metrics of the proposed deep CNN model, a few of the other statistical and deep learning models are chosen to compare them with the proposed deep CNN model. The experiment of comparing this model with the traditional methodologies is conducted with the same dataset that was considered to evaluate the proposed model. The collected dataset is used to evaluate the traditional and other deep learning models listed in Table 4. The proposed methodology is compared with the deep belief network (DBN) [16], SVM [61], sparse filter [18], ANN [16] and adaptive deep convolutional neural network (ADCNN) [62]. Similar to the proposed CNN model, a k-fold cross-validation technique with five folds is used to train these methods. The test data (135) samples are used to evaluate these models. The prediction accuracy for the test dataset is collected for each of these methods and presented in Table 4.

Table 3. The classification report of the test dataset.

Classification Report				
CLASS	**Precision**	**Recall**	**F1-Score**	**Support**
Healthy	1.00	1.00	1.00	23
Bearing Axis Deviation	0.97	0.96	0.94	22
Stator and Rotor Friction	0.95	0.97	0.96	22
Rotor Aluminum End Ring Break	1.00	1.00	1.00	23
Bearing Noise	0.93	1.00	0.96	22
Poor Insulation	0.96	0.89	0.95	23
Accuracy			0.97	135
Macro avg	0.97	0.97	0.97	135
Weighted avg	0.97	0.97	0.97	135

Table 4. The comparison results.

Methods	Accuracy (%)
DBN	92.2
SVM	89.8
Sparse filter	96.4
ANN	81.8
ADCNN	96.2
Proposed CNN	**97.37**

The comparison results explain that the proposed deep CNN model attains a prominent result when compared to the other methods. The prediction accuracy is 97.37%, which is better than all the other methods; this shows the significant performance of the proposed deep CNN model.

5. Conclusions and Future Work

An effective methodology was presented to diagnose the faults in a three-phase induction motor based on EWT and deep CNN. The main contributions of this study are that we propose a method to convert time-series data, such as current signals, into grayscale images, using EWT and applying the proposed deep CNN model to classify the EWT grayscale images for a fault diagnosis. The proposed methodology was tested for five fault types of the induction motor, including bearing axis deviation, stator and rotor friction, rotor aluminum end ring break, bearing noise and poor insulation, and it achieved a significant accuracy of 97.37%. The proposed methodology performed better than the other traditional and deep learning methods. We demonstrated that the proposed methodology, which took into account a single variable as the input feature, yielded promising results when compared to rule-based diagnosis systems that take into account multiple features for a fault diagnosis.

The limitations of the proposed methodology are as follows. First, the dataset that was considered for the experiment was comparatively small, and a huge number of data samples need to be collected for different load conditions, such as no load, half load or full load. Second, data from motors with different specifications need to be collected in order to learn more generalized features. Third, the most common faults in induction motors need to be detected in order to avoid misclassifications. Based on the limitations described above, our future work is focused on collecting more data samples from induction motors having different specifications and working at different loads, as well as investing in collecting information on the most common fault types in induction motors in order to avoid misclassification. Furthermore, CNN-based transfer learning can be studied to reduce training costs.

Author Contributions: Y.-M.H. has generated the data and analyzed the faults in induction motors. W.-B.W. validated the data for each kind of fault. V.R.I. performed data preprocessing to train the CNN model and evaluated the trained model for fault diagnosis. Y.-M.H. and V.R.I. analyzed the experimental results with guidance from C.-C.K.; C.-C.K. and H.-C.C. revised the manuscript for submission.

Funding: This research received no external funding.

Conflicts of Interest: The authors declare no conflict of interest.

References

1. Albrecht, P.F.; Appiarius, J.C.; McCoy, R.M.; Owen, E.L.; Sharma, D.K. Assessment of the Reliability of Motors in Utility Applications—Updated. *IEEE Trans. Energy Convers.* **1986**, *EC-1*, 39–46. [CrossRef]
2. Bonnett, A.H.; Soukup, G.C. Cause and Analysis of Stator and Rotor Failures in Three-Phase Squirrel-Cage Induction Motors. *IEEE Trans. Ind. Appl.* **1992**, *28*, 921–937. [CrossRef]
3. Dai, X.; Gao, Z. From model signal to knowledge: A data-driven perspective of fault detection and diagnosis. *IEEE Trans. Ind. Inform.* **2013**, *9*, 2226–2238. [CrossRef]
4. Gao, Z.; Cecati, C.; Ding, S.X. A survey of fault diagnosis and fault-tolerant techniques-Part I: Fault diagnosis with model-based and signal-based approaches. *IEEE Trans. Ind. Electron.* **2015**, *62*, 3757–3767. [CrossRef]
5. Cecati, C. A survey of fault diagnosis and fault-tolerant techniques-Part II: Fault diagnosis with knowledge-based and hybrid/active approaches. *IEEE Trans. Ind. Electron.* **2015**, *62*, 3768–3774.
6. Lee, J.; Wu, F.; Zhao, W.; Ghaffari, M.; Liao, L.; Siegel, D. Prognostics and health management design for rotary machinery systems—Reviews, methodology and applications. *Mech. Syst. Signal Process.* **2014**, *42*, 314–334. [CrossRef]
7. Lamim Filho, P.C.M.; Pederiva, R.; Brito, J.N. Detection of stator winding faults in induction machines using flux and vibration analysis. *Mech. Syst. Signal Process.* **2014**, *42*, 377–387. [CrossRef]
8. Sun, W.; Chen, J.; Li, J. Decision tree and PCA-based fault diagnosis of rotating machinery. *Noise Vib. Worldw.* **2007**, *21*, 1300–1317. [CrossRef]
9. Ngaopitakkul, A.; Bunjongjit, S. An application of a discrete wavelet transform and a back-propagation neural network algorithm for fault diagnosis on single-circuit transmission line. *Int. J. Syst. Sci.* **2013**, *44*, 1745–1761. [CrossRef]
10. Yang, Y.; Yu, D.; Cheng, J. A fault diagnosis approach for roller bearing based on IMF envelope spectrum and SVM. *Measurement* **2007**, *40*, 943–950. [CrossRef]
11. Pandya, D.H.; Upadhyay, S.H.; Harsha, S.P. Fault diagnosis of rolling element bearing with intrinsic mode function of acoustic emission data using APF-KNN. *Expert Syst. Appl.* **2013**, *40*, 4137–4145. [CrossRef]
12. Lecun, Y.; Bengio, Y.; Hinton, G. Deep learning. *Nature* **2015**, *521*, 436–444. [CrossRef] [PubMed]
13. Qi, Y.; Shen, C.; Wang, D.; Shi, J.; Jiang, X.; Zhu, Z. Stacked Sparse Autoencoder-Based Deep Network for Fault Diagnosis of Rotating Machinery. *IEEE Access* **2017**, *5*, 15066–15079. [CrossRef]
14. Xia, M.; Li, T.; Xu, L.; Liu, L.; Silva, C.W. Fault Diagnosis for Rotating Machinery Using Multiple Sensors and Convolutional Neural Networks. *IEEE/ASME Trans. Mechatron.* **2018**, *23*, 101–110. [CrossRef]
15. Wen, L.; Gao, L.; Li, X. A New Deep Transfer Learning Based on Sparse Auto-Encoder for Fault Diagnosis. *IEEE Trans. Syst. Man Cybern. Syst.* **2017**, *99*, 1–9. [CrossRef]
16. Shao, H.; Jiang, H.; Zhang, X.; Niu, M. Rolling bearing fault diagnosis using an optimization deep belief network. *Meas. Sci. Technol.* **2015**, *26*, 115002. [CrossRef]
17. Shao, H.; Jiang, H.; Wang, F.; Zhao, H. An enhancement deep feature fusion method for rotating machinery fault diagnosis. *Knowl. Based Syst.* **2017**, *119*, 200–220. [CrossRef]

18. Lei, Y.; Jia, F.; Lin, J.; Xing, S.; Ding, S.X. An intelligent fault diagnosis method using unsupervised feature learning towards mechanical big data. *IEEE Trans. Ind. Electron.* **2016**, *63*, 3137–3147. [CrossRef]

19. Lee, K.B.; Cheon, S.; Chang, O.K. A Convolutional Neural Network for Fault Classification and Diagnosis in Semiconductor Manufacturing Processes. *IEEE Trans. Semicond. Manuf.* **2017**, *30*, 135–142. [CrossRef]

20. Wang, J.; Liu, P.; She, M.; Nahavandi, S.; Kouzani, A. Bag-of-words representation for biomedical time series classification. *Biomed. Signal Process. Control* **2016**, *8*, 634–644. [CrossRef]

21. Hatami, N.; Chira, C. Classifiers with a reject option for early time-series classification. In Proceedings of the 2013 IEEE Symposium on Computational Intelligence and Ensemble Learning (CIEL), Singapore, 16–19 April 2013; pp. 9–16.

22. Wang, Z.; Oates, T. Pooling sax-bop approaches with boosting to classify multivariate synchronous physio-logical time series data. In Proceedings of the 28th International FLAIRS Conference, Hollywood, FL, USA, 18–20 May 2015; pp. 335–341.

23. Ince, T.; Kiranyaz, S.; Eren, L.; Askar, M.; Gabbouj, M. Real-time motor fault detection by 1-D convolutional neural networks. *IEEE Trans. Ind. Electron.* **2016**, *63*, 7067–7075. [CrossRef]

24. Jerome, G. Empirical Wavelet Transform. *IEEE Trans. Signal Process.* **2013**, *61*, 3999–4010.

25. Wang, J.; Zhuang, J.; Duan, L.; Cheng, W. A multi-scale convolution neural network for featureless fault diagnosis. In Proceedings of the International Symposium on Flexible Automation (ISFA'16), Cleveland, OH, USA, 1–3 August 2016.

26. Lee, D.; Siu, V.; Cruz, R.; Yetman, C. Convolutional neural net and bearing fault analysis. In Proceedings of the International Conference on Data Mining Series (ICDM) Barcelona, San Diego, CA, USA, 12–15 December 2016; pp. 194–200.

27. Ge, H.; Chen, G.; Yu, H.; Chen, H.; An, F. Theoretical Analysis of Empirical Mode Decomposition. *Symmetry* **2018**, *10*, 623. [CrossRef]

28. Lei, Y.; He, Z.; Zi, Y. EEMD method and WNN for fault diagnosis of locomotive roller bearings. *Expert Syst. Appl.* **2011**, *38*, 7334–7341. [CrossRef]

29. Ma, F.; Zhan, L.; Li, C.; Li, Z.; Wang, T. Self-Adaptive Fault Feature Extraction of Rolling Bearings Based on Enhancing Mode Characteristic of Complete Ensemble Empirical Mode Decomposition with Adaptive Noise. *Symmetry* **2019**, *11*, 513. [CrossRef]

30. Ge, M.; Wang, J.; Xu, Y.; Zhang, F.; Bai, K.; Ren, X. Rolling Bearing Fault Diagnosis Based on EWT Sub-Modal Hypothesis Test and Ambiguity Correlation Classification. *Symmetry* **2018**, *10*, 730. [CrossRef]

31. Deng, W.; Zhao, H.; Yang, X.; Dong, C. A Fault Feature Extraction Method for Motor Bearing and Transmission Analysis. *Symmetry* **2017**, *9*, 60. [CrossRef]

32. Agrawal, P.; Jayaswal, P. Diagnosis and Classifications of Bearing Faults Using Artificial Neural Network and Support Vector Machine. *J. Inst. Eng. India Ser. C* **2019**, 1–12. [CrossRef]

33. Jayaswal, P.; Wadhwani, A.K. Application of artificial neural networks, fuzzy logic and wavelet transform in fault diagnosis via vibration signal analysis: A review. *Aust. J. Mech. Eng.* **2015**, *7*, 157–171. [CrossRef]

34. Bin, G.F.; Gao, J.J.; Li, X.J.; Dhillon, B.S. Early fault diagnosis of rotating machinery based on wavelet packets-Empirical mode decomposition feature extraction and neural network. *Mech. Syst. Signal Process.* **2012**, *27*, 696–711. [CrossRef]

35. Zhao, R.; Yan, R.; Chen, Z.; Mao, K.; Wang, P.; Gao, R.X. Deep learning and its applications to machine health monitoring: A survey. *arXiv* **2016**, arXiv:1612.07640. [CrossRef]

36. Jia, F.; Lei, Y.; Lin, J.; Zhou, X.; Lu, N. Deep neural networks: A promising tool for fault characteristic mining and intelligent diagnosis of rotating machinery with massive data. *Mech. Syst. Signal Process.* **2016**, *72*, 303–315. [CrossRef]

37. Cho, H.C.; Knowles, J.; Fadali, M.S.; Lee, K.S. Fault detection and isolation of induction motors using recurrent neural networks and dynamic Bayesian modeling. *IEEE Trans. Control Syst. Technol.* **2009**, *18*, 430–437. [CrossRef]

38. Xu, Y.; Liang, F.; Zhang, G.; Xu, H. Image Intelligent Detection Based on the Gabor Wavelet and the Neural Network. *Symmetry* **2016**, *8*, 130. [CrossRef]

39. OAbdeljaber; Avci, O.; Kiranyaz, S.; Gabbouj, M.; Inman, D.J. Real-time vibration-based structural damage detection using one-dimensional convolutional neural networks. *J. Sound Vib.* **2017**, *388*, 154–170. [CrossRef]

40. Chong, U.P. Signal model-based fault detection and diagnosis for induction motors using features of vibration signal in two-dimension domain. *Stroj. Vestn. J. Mech. Eng.* **2011**, *57*, 655–666.

41. Xu, G.; Liu, M.; Jiang, Z.; Söffker, D.; Shen, W. Bearing fault diagnosis method based on deep convolutional neural network and random forest ensemble learning. *Sensors* **2019**, *19*, 1088. [CrossRef] [PubMed]

42. Lu, C.; Wang, Y.; Ragulskis, M.; Cheng, Y. Fault diagnosis for rotating machinery: A method based on image processing. *PLoS ONE* **2016**, *11*, e0164111. [CrossRef]

43. Kang, M.; Kim, J.M. Reliable fault diagnosis of multiple induction motor defects using a 2-D representation of Shannon wavelets. *IEEE Trans. Magn.* **2014**, *50*, 1–13. [CrossRef]

44. Huang, N.E.; Shen, Z.; Long, S.R.; Wu, M.C.; Shih, H.H.; Zheng, Q.; Yen, N.-C.; Tung, C.C.; Liu, H.H. The empirical mode decomposition and the Hilbert spectrum for nonlinear and non-stationary time series analysis. *Proc. R. Soc. Lond. Ser. A Math. Phys. Eng. Sci.* **1971**, *454*, 903–995. [CrossRef]

45. Torres, M.E.; Colominas, M.A.; Schlotthauer, G.; Flandrin, P. A complete ensemble empirical mode decomposition with adaptive noise. In Proceedings of the 2011 IEEE International Conference on Acoustics, Speech and Signal Processing (ICASSP), Prague, Czech Republic, 22–27 May 2011; pp. 4144–4147.

46. Daubechies, I. *Ten Lectures on Wavelets*; ser. CBMS-NSF Regional Conference Series in Applied Mathematics; SIAM: Philadelphia, PA, USA, 1992.

47. Jaffard, S.; Meyer, Y.; Ryan, R.D. *Wavelets: Tools for Science and Technology*; SIAM: Philadelphia, PA, USA, 2001.

48. Mallat, S. *A Wavelet Tour of Signal Processing: The Sparse Way*, 3rd ed.; Elsevier/Academic: New York, NY, USA, 2009.

49. Huang, S.J.; Hsieh, C.T. High-impedance fault detection utilizing a Morlet wavelet transform approach. *IEEE Trans. Power Deliv.* **1999**, *14*, 1401–1410. [CrossRef]

50. Lin, J.; Qu, L. Feature extraction based on morlet wavelet and its application for mechanical fault diagnosis. *J. Sound Vib.* **2000**, *234*, 135–148. [CrossRef]

51. Peng, Z.K.; Chu, F.L. Application of the wavelet transform in machine condition monitoring and fault diagnostics: A review with bibliography. *Mech. Syst. Signal Process.* **2004**, *18*, 199–221. [CrossRef]

52. Verstraete, D.; Ferrada, A.; Droguett, E.L.; Meruane, V.; Modarres, M. Deep Learning Enabled Fault Diagnosis Using Time-Frequency Image Analysis of Rolling Element Bearings. *Shock Vib.* **2017**, *2017*, 5067651. [CrossRef]

53. Image Processing in Python. Available online: http://scikit-image.org/docs/dev/auto_examples/transform/plot_rescale.html (accessed on 12 May 2019).

54. Nwankpa, C.; Ijomah, W.; Gachagan, A.; Marshall, S. Activation Functions: Comparison of Trends in Practice and Research for Deep Learning. *arXiv* **2018**, arXiv:1811.03378.

55. Krizhevsky, A.; Sutskever, I.; Hinton, G.E. ImageNet classification with deep convolutional neural networks. In Proceedings of the Conference on Neural Information Processing Systems (NIPS12), Lake Tahoe, NV, USA, 3–6 December 2012; pp. 1097–1105.

56. Xu, B.; Wang, N.; Chen, T.; Li, M. Empirical Evaluation of Rectified Activations in Convolutional Network. *arXiv* **2015**, arXiv:1505.00853.

57. Park, S.; Kwak, N. Analysis on the Dropout Effect in Convolutional Neural Networks. In *Computer Vision—ACCV 2016*; Lecture Notes in Computer Science; Lai, S.H., Lepetit, V., Nishino, K., Sato, Y., Eds.; Springer: Cham, Switzerland, 2017; Volume 10112, pp. 189–204.

58. Kingma, D.P.; Ba, J. Adam: A method for stochastic optimization. *arXiv* **2014**, arXiv:1412.6980.

59. Chang, H.; Kuo, C.; Hsueh, Y.; Wang, Y.; Hsieh, C. Fuzzy-Based Fault Diagnosis System for Induction Motors on Smart Grid Structures. In Proceedings of the IEEE International Conference on Smart Energy Grid Engineering, Oshawa, ON, Canada, 14–17 August 2017.

60. The Dataset Used for This Study. Available online: https://drive.google.com/drive/u/0/folders/1Cvs-1LYagNmEQcj-s395dZflh3L4JWTO (accessed on 12 May 2019).

61. Zhang, X.; Liang, Y.; Zhou, J. A novel bearing fault diagnosis model integrated permutation entropy ensemble empirical mode decomposition and optimized SVM. *Measurement* **2015**, *69*, 164–179. [CrossRef]

62. Guo, X.; Chen, L.; Shen, C. Hierarchical adaptive deep convolution neural network and its application to bearing fault diagnosis. *Measurement* **2016**, *93*, 490–502. [CrossRef]

Article

Forecasting for Ultra-Short-Term Electric Power Load Based on Integrated Artificial Neural Networks

Horng-Lin Shieh * and **Fu-Hsien Chen**

St. John's University, 499, Sec. 4, Tam King Road, Tamsui District, New Taipei City 25135, Taiwan
* Correspondence: shieh@mail.sju.edu.tw

Received: 25 July 2019; Accepted: 14 August 2019; Published: 20 August 2019

Abstract: Energy efficiency and renewable energy are the two main research topics for sustainable energy. In the past ten years, countries around the world have invested a lot of manpower into new energy research. However, in addition to new energy development, energy efficiency technologies need to be emphasized to promote production efficiency and reduce environmental pollution. In order to improve power production efficiency, an integrated solution regarding the issue of electric power load forecasting was proposed in this study. The solution proposed was to, in combination with persistence and search algorithms, establish a new integrated ultra-short-term electric power load forecasting method based on the adaptive-network-based fuzzy inference system (ANFIS) and back-propagation neural network (BPN), which can be applied in forecasting electric power load in Taiwan. The research methodology used in this paper was mainly to acquire and process the all-day electric power load data of Taiwan Power and execute preliminary forecasting values of the electric power load by applying ANFIS, BPN and persistence. The preliminary forecasting values of the electric power load obtained therefrom were called suboptimal solutions and finally the optimal weighted value was determined by applying a search algorithm through integrating the above three methods by weighting. In this paper, the optimal electric power load value was forecasted based on the weighted value obtained therefrom. It was proven through experimental results that the solution proposed in this paper can be used to accurately forecast electric power load, with a minimal error.

Keywords: sustainable energy; power load forecasting; adaptive-network-based fuzzy inference system (ANFIS); back-propagation neural network (BPN); persistence; search algorithm

---

1. Introduction

Due to human over-exploitation, the global warming and energy crisis is a challenge that human beings must face. In order to achieve sustainable energy goals, developing renewable energy and improving energy efficiency are principal research methods. Power forecasting is important for power companies to improve energy efficiency. Power companies must closely monitor the supply and demand of electricity. Otherwise, whether the load demand is greater than the power supply capacity caused by a power jump, or the energy waste caused by an oversupply of electricity, the power cost for the power company will increase.

In recent years, the demand for electric power has been growing steadily along with booming economic growth in Taiwan. A stable electric power supply should be the basis for national economic development. A noticeable increase in demand for industrial and civil electricity has also been observed due to rapid economic development. A stable and sufficient electric power supply is most crucial to electric power companies. Electric power companies can reduce electric power operation costs and further improve the quality and stability of the electric power supply if the future load can be accurately forecasted.

Electric power load forecasting is of great importance. Therefore, electric power companies need to control the distribution of electric power. Excessive electric power supply will also cause grievous waste of energy, and on the other hand, a power trip will be caused if the load demand is greater than the electric power supply. In both cases, the electric power costs of electric power companies will be increased, such that civil electricity charges will be increased accordingly. The increase in civil electricity charges will lead to a rise in the costs of all things.

In view of the above reasons, it is necessary that effective electric power supply and maintenance of the appropriate reserve capacity and appropriate electric power distribution and dispatching are undertaken to determine the electric power demand increase due to various uncertainty factors. Moreover, reasonable electric power dispatching and distribution must be based on accurate load forecasting.

The quality of an electric power supply has a great influence on industrial development and living conditions. For the purpose of a stable electric power supply and eliminating heavy economic losses caused by power shortage, additional load increased by the demand for industrial development and civil electricity must be satisfied and an appropriate reserve capacity must be maintained. As a result, accurate load forecasting is extremely important.

Electric power load forecasting constitutes a part of an electric power system. Load forecasting can be classified into four categories according to time [1]: Long-term, mid-term, short-term and ultra-short-term, as set out below, respectively:

- Ultra-short-term load forecasting: The forecasting unit ranges from several minutes to several hours. Such a model is often used in flow control and used for detecting the stability of an electric power system and forecasting its reserve capacity, so as to prevent the occurrence of insufficient electric power dispatching.
- Short-term load forecasting: The forecasting unit ranges from 1 h to several weeks. Such a model is often used for adjusting the economic dispatching of electric power, analyzing electric power demand and supply and power flow, forecasting crisis in the case of accidents, and for corporate operations and equipment maintenance.
- Mid-term load forecasting: the forecasting unit ranges from several days to 1 month. Such a model is often used for estimating the peak electric power consumption and maintenance of power equipment. It can be used for detecting the time for maintenance and shutdown of power generators and is mainly used for decision making on energy management, such as electricity pricing and procurement of fuels used in power generation.
- Long-term load forecasting: The forecasting unit ranges from 1 year to several years. Such a model is often used for research on electric power policies. New generator sets can be developed or constructed thereby for the future electric power planning of power industry, planning of power transmission and distribution system, procurement of fuels, signing contracts on outsourced electricity, electricity pricing and price structure analysis, corporate operation management and earnings estimation, load management, etc. This load forecasting considers the economic growth, energy proportion planning, industrial structure, construction of electric power development, electric power demand management and other conditions, such as population, temperature, power saving effect of research organizations or government agencies.

2. Literature Review

In [2], Kuster et al. present a systematic review of electric power load forecasting. This paper reveals that regression methods are still popularity adopted and very efficient for long and very long-term electrical load forecasting. The machine-learning algorithms, such as artificial neural networks (ANN), support vector machines (SVM), and time series analysis are widely used for short and very short-term prediction.

Mi et al. [3] propose a short-term power load forecasting method based on the improved exponential smoothing grey model. Authors used grey correlation analysis to determine the importance factor affecting the power load, then conducted power load forecasting using the improved multivariable grey model. The results showed that proposed method has a satisfactory prediction effect and meets the requirements of short-term power load forecasting.

Minhas et al. [4] forecasted short-term electric power load based on the adaptive fuzzy neural system. In the ANN database, temperature and power load were used as a training model. The data of electric power load and temperature in 2015 were used for forecasting the electric power demand over the next few hours, and a fuzzy system was used for establishing membership functions. Subsequently, the probabilistic and stochastic hybrid adaptive fuzzy neural system was used for reducing the error in electric power load forecasting, in particular, the error on the weekend was much lower than that on weekdays.

Yin et al. [5] proposed the ultra-short-term load forecasting method based on the weighted average optimal local shape similarity model which was used for acquiring the preliminary forecasting values of ultra-short term load and rectifying the preliminary forecasting values of the ultra-short term load. Computation methods for various influence factors were proposed after analyzing other factors influencing the accuracy of ultra-short-term load forecasting. Human comfort and other influence factors were improved based on the impact of the improved air quality index on human behaviors. The second rectification was carried out on the forecasting value of the ultra-short-term load by applying the Super-Stable Adaptive Control Theory, based on the deviation value of actual data from the forecasting data. Such a method features a very good adaptability to the computation speed and the accuracy of large-scale ultra-short term load forecasting and satisfies the actual demands of site engineering.

Din and Marnerides [6] implemented applicability and comparison to the performance of the feed-forward deep neural network and recurrent deep neural network by leveraging the accuracy and computation capability of short-term forecasting. In that study, the data of 4 years were used for forecasting the load in several days or several weeks. The use of certain different input sources can accurately forecast the consumption of the short-term load. Such inputs included weather, time, official holidays and festivals. In addition, a higher accuracy can be obtained through feature analysis of the time frequency of collaborative use of the deep neural network.

Chen et al. [7] proposed a kind of nonlinear issue of partially connected neural network to be used for short-term load forecasting and have developed the group-based chaos genetic algorithm to produce various effective neural networks. In that study, a new pruning method was utilized to develop partially connected neural networks. In order to further improve forecasting accuracy, a non-linear partially connected neural network predictor based on the neural network has been developed in the paper. As a result of the application of this research, results in a PJM market dataset and an ISO New England dataset with errors of 1.76% and 1.29% have been obtained, respectively, which has proven that the network is an effective predictor.

Eljazzar et al. [8] introduced the main factors to short-term load forecasting, such as temperature, wind and humidity. They studied the relationship between inputs and peak load and measured the accuracy through comparison between actual value and forecasting value, residual error and the fitted model. Selection of correct parameters affecting forecasting was very important. Additional computation time would be required, and forecasting accuracy may not be improved if irrelevant parameters were selected. They introduced the impact of electric power load factors on short-term load forecasting, and applied certain factors (temperature, required temperature, wind and humidity) in ANN, so as to understand their impact on electric power load forecasting in the north of Cairo. The experimental results showed that MAPE, RMSE and MAE were reduced by more than a half as a result of the application of the model proposed.

López et al. [9] proposed the application of a linear hybrid model in short-term load forecasting. That study was in relation to load forecasting of a current Spanish Transport System Operator based on linear autoregressive techniques and neural networks. At present, the forecasting system forecasts load in each area in Spain respectively, to enable the load behaviors in each area to be subject to the effects of the same factors. Subsequently, certain areas have been integrated as a linear hybrid model, so as to utilize the information from other areas to understand the general behaviors of all areas, as well as to determine the individual deviation of each area. Such a technique is particularly useful for the modeling of impact of special days without sufficient information. Such a model has been applied in the three most relevant areas in the system, to collect the data of several years for the training of the model and forecast the demand load for the whole year. The proposed model provided a powerful database and average error was reduced by 4% through comparison between the experimental result and the original autoregressive model.

Tian et al. [10] proposed a hybrid deep learning model that integrates the hidden feature of the convolutional neural network (CNN) model and the long short-term memory (LSTM) model to improve the forecasting accuracy. The CNN extracts the local trend and captures the same pattern, and the LSTM learns the relationship in time steps. The performances of the hybrid model proposed by this paper were compared with the LSTM model and the CNN model; the experimental results showed that the proposed method can achieve a better and a more stable performance than either the CNN, or LSTM modes.

Semero et al. [11] proposed a hybrid technique for very short-term load forecasting in microgrids. The proposed method integrated genetic algorithm (GA), particle swarm optimization (PSO), and adaptive neuro fuzzy inference systems (ANFIS). The GA selects important predictors that significantly influence the load pattern among a number of candidate input variables. The PSO is used to optimize an ANFIS model for very short-term forecasting of load.

3. Proposed Method

In this study, preliminary forecasting values of electric power load were executed by integrating ANFIS and BPN and persistence. The preliminary forecasting values of the electric power load obtained therefrom were called suboptimal solutions and finally the optimal weighted value was determined by applying the search algorithm through integrating the above three methods into a forecast formula by weighting, in order to effectively improve forecasting accuracy.

3.1. Adaptive-Network-Based Fuzzy Inference System (ANFIS)

In this study, the IF-THEN rules of fuzzy system were developed systematically through input and output data by applying an artificial neural algorithm to adjust the parameters of ANFIS with the Sugeno model, as shown in Figure 1. In this paper, the architecture of ANFIS was utilized to forecast the electric power load value. A common IF-THEN rule of the Sugeno model is the flowing:

R_i: If x_1 is A_1^i and x_2 is A_2^i and ... x_k is A_k^i
Then $f^i = a_k^i x_k + a_{k-1}^i x_{k-1} + \ldots + a_0^i$

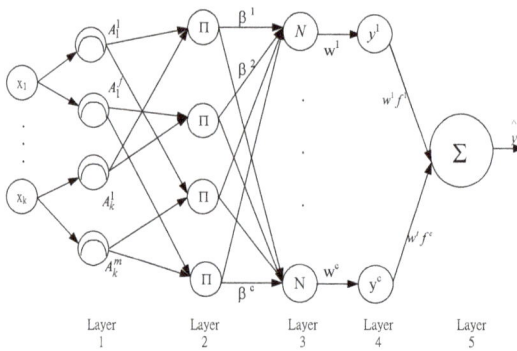

Figure 1. Architecture of the adaptive-network-based fuzzy inference system (ANFIS).

270

In this study, the ANFIS was applied with three input nodes and an output node. Let $\mathbf{X}= (x_1, x_2, x_3,)$ be input load data; from x_1 to x_3 in sequence, x_1 was the load value for the first period of time prior to the forecasting time point, x_2 was the load value for the second period of time; x_3 was the load value for the third period of time. The node at the output layer was the forecasting value of electric power load at the forecasting time point.

ANFIS was a five-layer artificial neural node. Set out below are the functions of each layer [12]:

Layer 1: In the architecture of ANFIS as shown in Figure 1, each node at the layer 1 was represented with i to indicate the membership function of antecedent input variables. Node function output was as shown in Equation (1), indicating the membership grade of output to its corresponding fuzzy set. The membership function could be any appropriate function, such as the Bell Function or Gaussian Function.

$$O_{1,i} = \mu_{A_j^i}(x_j), \; j = 1, 2, \ldots, k,$$ (1)

where x_j is the j-th input, $O_{1,i}$ is the output of Layer 1 and the $\mu_{A^i}(x_j)$ is the membership grade of a fuzzy set A_j^i.

Layer 2: Each node at the layer was marked with i to indicate that each node implemented arithmetic product computation to incoming signal. The node output was as shown in Equation (2), indicating the output of each rule.

$$O_{2,i} = \beta^i = \prod_{j=1}^{k} \mu_{A_j^i}(x_j),$$ (2)

where $\mu_{A^i}(x_j)$ is the membership grade of the input x_j that is fed to node i of Layer 2. The output of each node of layer 2 represents the firing strength of a rule.

Layer 3: Each node at the layer was marked with i to calculate the ratio of the output of the ith rule to the sum of outputs of all rules. It was called normalized fulfillment, as shown in Equation (3).

$$O_{3,i} = w^i = \frac{\beta^i}{\sum_{j=1}^{c} \beta^j}, \; i = 1, 2, \ldots, c,$$ (3)

where c is the number of rules.

Layer 4: Each adaptive node, marked with i, at the layer represented an output variable of consequent explicit function and was used for completing the computation as shown in Equation (4); where, c was the number of rules.

$$O_{4,i} = y^i = w^i f^i = w^i (a_2^i x_2 + a_1^i x_1 + a_0^i), \; i = 1, 2, \ldots, c,$$ (4)

where w^i is the output of layer 3 and a_j^i is the parameter of rule.

Layer 5: Only a single node at the layer was used for calculating the sum of outputs of all rules, as shown in Equation (5).

$$O_5 = \hat{y} = \sum_{i=1}^{c} y^i = \sum_{i=1}^{c} w^i f^i = \frac{\sum_{i=1}^{c} \beta^i f^i}{\sum_{i=1}^{c} \beta^i},$$ (5)

3.2. Ultra-Short Term Load Forecasting Based on Back-Propagation Neural Network

Operation of the BPN consisted of two parts: learning and recalling. Supervised learning was used in the learning part. Supervised learning acquired training data and target output value from the problem and imported training value into the system, and repeatedly adjusted the weighted value and bias through the Steepest slope method. In terms of the recalling part, through classification and forecasting, the network would inform us of the most possible forecasting result when a value has been imported. The BPN was applied to forecast electric power load value in this paper, as shown in Figure 2.

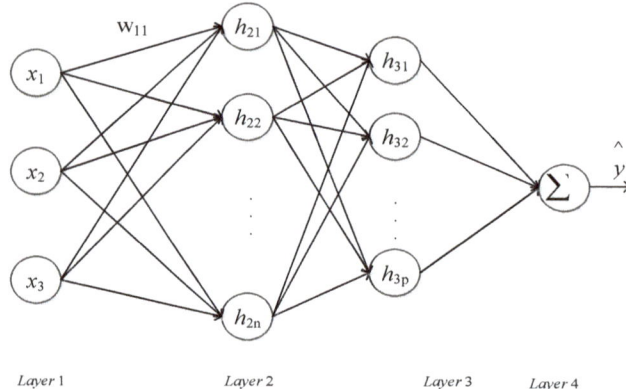

Figure 2. Architecture of the back-propagation neural network.

In this paper, the architecture of the BPN for electric power load forecasting used three input nodes and one output node. Nodes x_1 to x_3 at the input layer were the actual values of electric power load for the three periods of time prior to forecasting. The gradient descent approach is adopted for adjusting the parameters of the BPN. The node at the output layer was the forecasting value of the electric power load in the forecasting period of time.

3.3. Persistence

In terms of persistence, it was assumed that the future load was as the same as the load during forecasting [13]. In short-term load forecasting, persistence showed a higher accuracy than other load forecasting methods. The forecasting accuracy of persistence, however, reduced along with the increase of forecasting time [14].

Figure 3 is the curve graph of persistence. L_a is the current actual load value; L_f is forecasting value. As seen, the value of L_f was forecasted from the first period of time on the first day prior to the actual load value to the next period of time in sequence. However, in the case of bad weather or hot weather, such external conditions would lead to dramatic changes in the load line and hence forecasting based on persistence would not be so accurate. If the weather conditions and other external factors on that day are the same as that on the previous day, the forecasting based on the method will be quite accurate. Persistence is one of the most efficient methods in some cases.

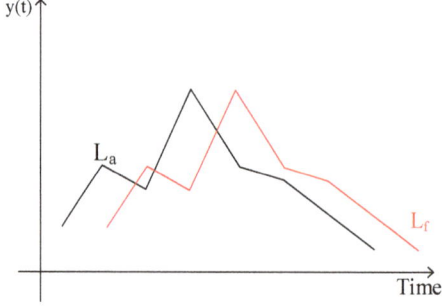

Figure 3. Curve graph of persistence.

Persistence was a method for forecasting the reference value and was applicable when there was a stable environment during electric power load forecasting, without any factor which would cause dramatic changes of load. In this case, persistence would be significantly useful. In the application of persistence in electric power load forecasting, a set of data of actual load values per 10 min in the previous period of time was regarded as the forecasting value of electric power load in the next period of time. Such a method is the traditional method for forecasting the reference value, as shown in Equation (6):

$$L_f(t+1) = L_a, \tag{6}$$

where, $L_f(t+1)$ is the forecasting value of time $t+1$, L_a is the actual value of time t.

3.4. Integrated Search Method

Figure 4 shows the diagram of the proposed predictor for an ultra-short-term power load forecasting. In Figure 4, this paper integrated three methods, ANFIS, BPN and persistence, into the predictor. Each method was a suboptimal solution, attached with a weight ($0 \le w_i \le 1$, $i = 1,2,3$); $w_1 + w_2 + w_3 = 1$; and the output y_{es} was evaluated as Equation (7).

$$y_{es} = w_1 y_1 + w_2 y_2 + w_3 y_3, \tag{7}$$

where y_1, y_2; y_3 are the outputs of ANFIS; BPN; and persistence, respectively. The optimal weight value was determined by applying the search algorithm, as shown in Algorithm 1.

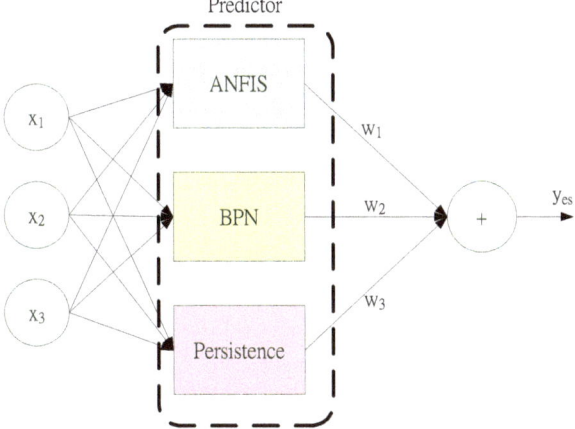

Figure 4. The diagram of proposed method.

In order to determine the best weighted value, Equation (8) was regarded as the criterion for evaluating weight in this paper.

$$\min_{w_i} \sum_{t=1}^{n} \left| (w_1 y_1(t) + w_2 y_2(t) + w_3 y_3(t) - y'(t)) \right|, \tag{8}$$

where, the restriction was $0 \le w_i \le 1$, $i = 1,2,3$, and $w_1+w_2+w_3 = 1$. y' was actual load value; n was data volume. In Equation (8), all training data were used for calculating the difference between the weight of load estimates of the three methods and the actual load value, and obtaining the w_i at the time of the minimal error value as the weight of training model.

Algorithm 1. Integrated search algorithm

Input: ANFIS, BPN, Persistence model, load dataset (X, y')

Output: $w_{1,opt}, w_{2,opt}, w_{3,opt}$

Step 1:   Input X, calculate ANFIS, BPN, Persistence output: y_1, y_2, y_3.

Step 2: $E_{min} = \infty$

Step 3:   For $w_1 = 0$: 1: step 0.01

Step 4:   For $w_2 = 0$: $(1 - w_1)$: step 0.01

Step 5: $w_3 = 1 - (w_1 + w_2)$

Step 6: $E_{new} = \sum_{t=1}^{n} |(w_1 y_1(t) + w_2 y_2(t) + w_3 y_3(t) - y'(t))|$

Step 7:   If $(E_{min} > E_{new})$ Then

Step 8 $E_{min} = E_{new}; w_{1,opt} = w_1; w_{2,opt} = w_2; w_{3,opt} = w_3;$

Step 9   End if

Step 10   Next w_2

Step 11   Next w_1

Step 12   End

---

4. Experimental Results

4.1. The Data Set

In this paper, electric power consumption in workdays and non-workdays in Taiwan was forecasted. Generally, work and rest hours in weekdays are roughly the same, therefore, the electric power load is stable. In holidays, the rest of certain industries causes a load different from that in weekdays. Therefore, forecasting was carried out in respect of two periods of time respectively, i.e., workdays and non-workdays, in this paper. Sample data derived from the daily electric power supply of the Taiwan Power Company. A set of data was acquired per 10 min, and a total of 144 sets of data were collected each day. In this paper, a total of seven modes were established on a weekly basis, and the simulation was implemented in one model on each day of each week. In the phase of model training, data on the same day in three consecutive weeks were used for training. For example, sample data on 31 May 2017, 7 June 2017 and 14 June 2017 were used as model training data. Upon the completion of model training, the model was used for forecasting the electric power consumption on the same day in the next week, i.e., 21 June 2017. There were three inputs for each model training, i.e., x_1, x_2, x_3. x_1 was the load value for the first period of time prior to the forecasting time point; x_2 was the load value for the second period of time, and x_3 was the load value for the third period of time. Output $y_i(t)$, $i = 1, 2, 3$ was the load at the forecasting time point, t; $y'(t)$ was the actual load. In model training, define $\varepsilon(t) = y(t) - y'(t)$ as the error value. For adjustment of parameters, please refer to [15].

Four methods were used in this study for ultra-short-term electric power load forecasting. The first method was ANFIS, the second method was BPN, the third method was persistence, and the fourth method was integrated search. In respect of all the four methods, actual values of electric power load were used as test data and error values were used to compare and judge the forecasting accuracy of the four methods.

4.2. Ultra-Short Term Electric Power Load Forecasting

In this paper, ANFIS used three input nodes and one output node. Inputs were the electric power load values for the first three periods of time prior to forecasting. Generally, load forecasting started from 00:00 each day to 23:55 next day. Data from Taiwan Power per 10 min were used as a set of training data. Therefore, data of the three periods of time prior to forecasting, i.e., the actual load values in the first period of time as at 11:50, the second period of time as at 11:40 and the third period of time as at 11:30, were used as training data.

Three input nodes and one output node were used in the architecture of BPN. The output layer was the forecasting value of electric power load during forecasting period of time.

In terms of persistence, a set of data of actual load values per 10 min in the previous period of time were used as the forecasting value of electric power load during forecasting period of time.

The integrated search has integrated the above three methods, i.e., BPN, ANFIS and persistence. Preliminary forecasting values of electric power load from the above three methods were called suboptimal solutions and the optimal weighted value was determined through search by integrating the three forecasting methods.

In order to express the error of each method, Mean Absolute Percentage Error (MAPE) and Maximum Absolute Percentage Error (MaxAPE) were used in this paper to calculate and compare error values, as shown below:

(1) Mean Absolute Percentage Error (MAPE, E_{ave}):

$$E_{ave} = \frac{1}{n}\left(\sum\nolimits_{t=1}^{n} \frac{|p_f(t) - p_L(t)|}{P_L(t)}\right) \times 100\%, \qquad (9)$$

where: P_L was the actual load value at the time t, P_f was the forecasting load value at the time t.

(2) Maximum Absolute Percentage Error (MaxAPE, E_{max}):

$$E_{max} = max\left(\frac{|p_f(t) - p_L(t)|}{P_L(t)}\right) \times 100\% , \; t = 1, 2, \ldots, n, \qquad (10)$$

4.2.1. Workday Ultra-Short-Term Electric Power Load Forecasting Experiment

In the experiment, forecasting object was the ultra-short-term electric power load in northern Taiwan on Wednesday, 21 June 2017. Comparison between the forecasting values of ultra-short-term electric power load by applying ANFIS, BPN, persistence, and integrated search and the actual load values are as shown in Figure 5. In Figure 5, the blue line represents the actual values; the red line represents the forecasting values of ANFIS; the yellow line represents the forecasting values of BPN, the purple line represents the forecasting values of persistence, and the green line represents the forecasting values of the integrated search. Table 1 sets out the comparison of errors in forecasting workday electric power load in northern Taiwan from 20 June to 26 June. Table 2 sets out the comparison of MAPE and MaxAPE in five workdays. As seen, the average error of the solution proposed in the five days, no matter absolute error or maximum error, is lower than that of ANFIS, BPN and persistence.

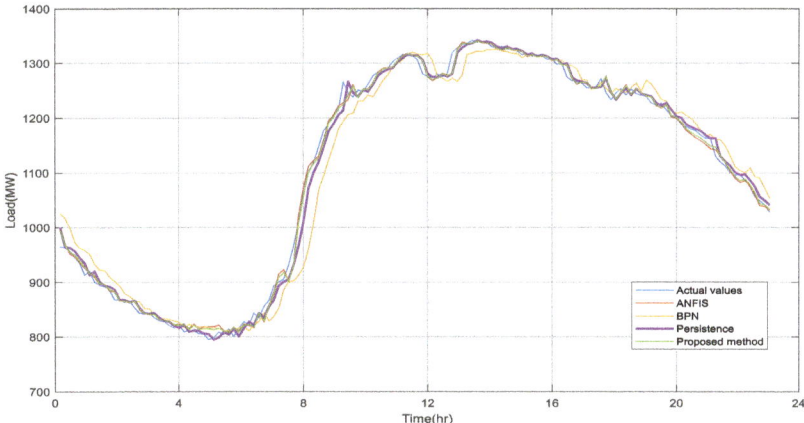

Figure 5. Ultra-short-term electric power load forecasting chart in northern Taiwan on 21 June.

Table 1. Table of forecasting errors in workdays in northern Taiwan in June.

Date	Method	E_{ave} (%)	E_{max} (%)
20	ANFIS	0.78	4.73
	BPN	2.33	12.79
	Persistence	0.95	6.03
	Proposed Method	0.75	4.7
21	ANFIS	0.74	3.19
	BPN	2.3	12.56
	Persistence	0.93	5.308
	Proposed Method	0.73	3.09
22	ANFIS	0.79	3.67
	BPN	2.17	13.4
22	Persistence	0.94	5.69
	Proposed Method	0.74	3.63
23	ANFIS	0.66	3.26
	BPN	2.22	12.08
	Persistence	0.9	5.09
	Proposed Method	0.65	3.23
26	ANFIS	0.68	3.65
	BPN	2.31	15.8
	Persistence	0.92	5.31
	Proposed Method	0.67	3.64

Table 2. Comparison of average forecasting errors in workdays in northern Taiwan in June.

Methods	E_{ave} (%)	E_{max} (%)
ANFIS	0.73	3.70
BPN	2.27	13.32
Persistence	0.93	5.48
Proposed Method	0.71	3.65

4.2.2. Non-workday Ultra-Short-Term Electric Power Load Forecasting Experiment

Figure 6 shows the comparison of load forecasting in northern Taiwan on a non-workday, i.e., Saturday, 24 June 2017.

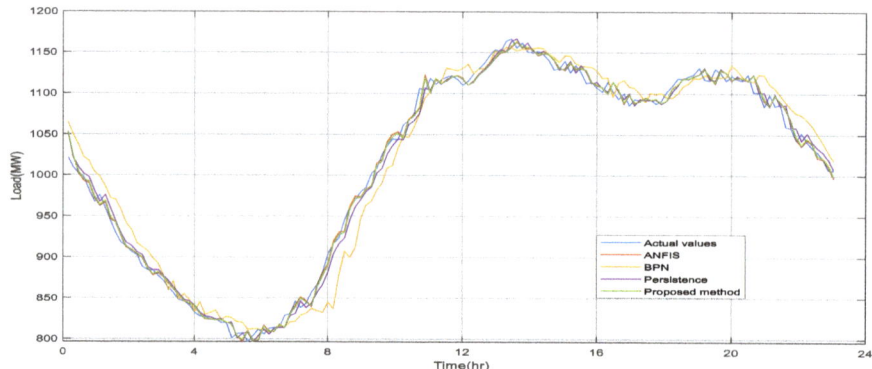

Figure 6. Ultra-Short-Term electric power load forecasting chart in northern Taiwan on 24 June.

In Figure 6, the blue line represents the actual values; the red line represents the forecasting values of ANFIS; the yellow line represents the forecasting values of BPN, the purple line represents the forecasting values of persistence, and the green line represents the forecasting values of integrated search. Table 3 sets out the comparison of load forecasting errors in non-workdays in northern Taiwan on 24 June and 24 June. Table 4 shows the comparison of MAPE and MaxAPE in non-workdays. As seen, the average error of the solution proposed in non-workdays, no matter absolute error or maximum error, is lower than that of ANFIS, BPN and persistence.

Table 3. Table of forecasting errors in non-workdays in northern Taiwan on 24 June and 25 June.

Date	Methods	E_{ave} (%)	E_{max} (%)
24 June	ANFIS	0.72	3.12
	BPN	1.73	8.69
	Persistence	0.83	3.10
	Proposed Method	0.71	3.05
25 June	ANFIS	0.81	3.54
	BPN	1.94	6.99
	Persistence	0.86	3.53
	Proposed Method	0.79	3.34

Table 4. Comparison of average forecasting errors in non-workdays in northern Taiwan in June.

Methods	E_{ave} (%)	E_{max} (%)
ANFIS	0.77	3.33
BPN	2.18	7.84
Persistence	0.86	3.32
Proposed Method	0.75	3.19

4.2.3. National Workday Ultra-Short-Term Electric Power Load Forecasting Experiment

In order to forecast workday ultra-short-term electric power load across Taiwan, ANFIS, BPN and persistence, and integrated search were used in this paper to forecast electric power consumption load across Taiwan on 20 June 2017. Comparison results are as shown in Figure 7. The representation of lines in Figure 7 is the same as that in above experiments.

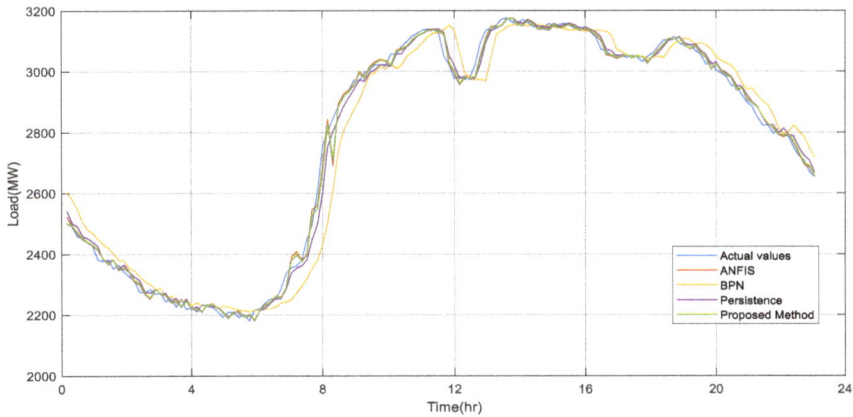

Figure 7. National workday ultra-short-term electric power load forecasting chart on 20 June.

Table 5 sets out the comparison between the solution proposed and the other three methods in workdays across Taiwan. Table 6 sets out average errors. As seen, the average error of the solution proposed is lower than that of the other three methods.

Table 5. Table of forecasting errors in workdays across Taiwan.

Date	Method	E_{ave} (%)	E_{max} (%)
20	ANFIS	0.61	5.4
	BPN	1.74	11.63
	Persistence	0.74	5.42
	Proposed Method	0.59	4.6
21	ANFIS	0.58	2.86
	BPN	1.76	12.44
	Persistence	0.73	5.51
	Proposed Method	0.56	2.86
22	ANFIS	0.66	3.53
	BPN	1.87	1.87
	Persistence	0.707	4.74
	Proposed Method	0.587	2.46
23	ANFIS	0.51	2.2
	BPN	1.78	10.82
	Persistence	0.7	4.62
	Proposed Method	0.5	2.2
26	ANFIS	0.51	2.15
	BPN	1.86	12.05
	Persistence	0.73	5.06
	Proposed Method	0.5	2.08

Table 6. Comparison of average forecasting errors in workdays across Taiwan.

Methods	E_{ave} (%)	E_{max} (%)
ANFIS	0.57	3.17
BPN	1.80	9.76
Persistence	0.72	5.07
Proposed Method	0.54	2.84

4.2.4. National Non-Workday Ultra-Short-Term Electric Power Load Forecasting Experiment

Non-workday electric power load across Taiwan was also forecasted in this paper. Figure 8 shows the comparison of ultra-short-term electric power load forecasting on 24 June 2017 between the solution proposed in this paper and the other three methods. The representation of lines in Figure 7 is the same as that in above experiments.

Table 7 sets out forecasting errors in non-workdays across Taiwan in June.

Table 7. Table of forecasting errors in non-workdays across Taiwan on 24 June and 25 June.

Date	Methods	E_{ave} (%)	E_{max} (%)
24 June	ANFIS	0.59	2.05
	BPN	1.44	4.24
	Persistence	0.62	2.17
	Proposed Method	0.52	2.05
25 June	ANFIS	0.63	2.48
	BPN	1.27	4.65
	Persistence	0.62	2.80
	Proposed Method	0.59	2.50

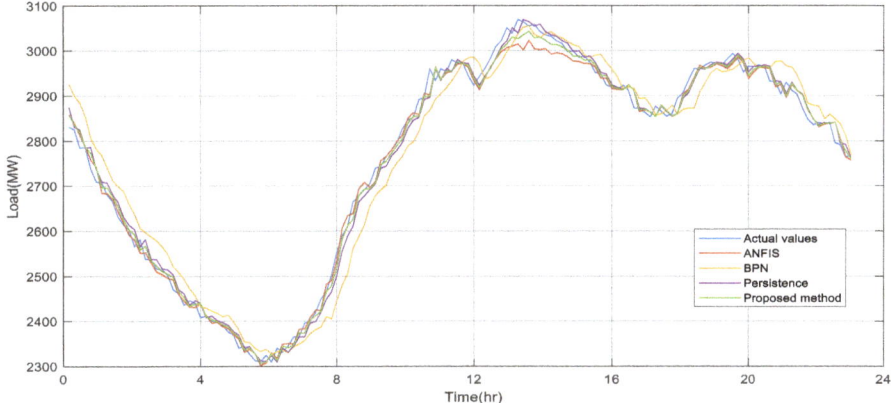

Figure 8. National non-workday ultra-short-term electric power load forecasting chart on 24 June.

Table 8 sets out average errors in non-workdays across Taiwan. As seen, the average error of the solution proposed is lower than that of the other three methods.

Table 8. Comparison of average forecasting errors in non-workdays across Taiwan on 24 June and 25 June.

Methods	E_{ave} (%)	E_{max}(%)
ANFIS	0.61	2.27
BPN	1.36	4.45
Persistence	0.62	2.49
Proposed Method	0.56	2.27

5. Discussion

An integrated electric power load forecasting method was proposed in this paper. ANFIS, BPN and persistence were integrated by weighting, and the optimal weighted value was determined by applying the search algorithm. The MAPE and MaxAPE analysis of experimental results indicated that among the three methods of ANFIS, BPN and persistence, ANFIS features a better forecasting accuracy. But the result through the weighted integrated search proposed in this paper showed that the solution proposed in this paper showed a better forecasting accuracy than ANFIS. The optimal weighted value can be determined by applying integrated search. In addition, significant effect will be observed if integrating more efficient forecasting methods in the future. It can be effectively applied by practitioners in certain relevant fields.

In this paper, actual load data in the three periods of time prior to forecasting were used to forecast electric power load. Such a practice is relatively not subject to the great influence of weather or other factors. In the future, data in relation to load, such as temperature, weather, apparent temperature, relative humidity, wind force and daylight hours can be utilized to increase the accuracy of the system in forecasting. In this paper, the search algorithm was employed to determine the optimal weighted value. In future study, the genetic algorithm, bee colony algorithm and other optimization theories can be used to determine the optimal weighted portfolio, so as to achieve better forecasting performance.

The search algorithm determines the w_1, w_2, and w_3 parameter values. Since there are only three parameters to decide, using the search algorithm can find the best value in a short time. Since the current load value is estimated using the power load values of the previous three 10-min periods, the experimental results show that the predicted value of this paper is more accurate than the other

Symmetry **2019**, *11*, 1063

three methods. When future research considers long-term or mid-term electric power load forecasting, the search algorithm requires more calculation time because of the longer estimation time and the need to consider the weather and temperature. In future research, gene algorithms, bee colony algorithms, and PSO algorithms can be used to estimate parameters for long-term or mid-term electric power load forecasting.

Author Contributions: This research article contains two authors, including H.L.S. and F.H.C. F.H.C. and H.L.S. jointly design the overall architecture and related algorithms, and also conceived and designed the experiments, however, H.L.S. coordinated the overall plan and direction of the experiments and related skills; F.H.C. and H.L.S. not only contributed analysis tools, but also analyzed the data; F.H.C. performed the experiments; and H.L.S. wrote this paper and related reply.

Funding: This research was funded by Ministry of Science and Technology, Taiwan, grant number "MOST 107-2622-E-129 -003 -CC3" and "MOST 107-2632-E-129 -001".

Conflicts of Interest: The authors declare no conflict of interest.

References

1. Luis, H.; Carlos, B.; Javier, M.A.; Belen, C.; Antonio, J.S.-E.; Jaime, L.; Joaquim, M. A Survey on Electric Power Demand Forecasting: Future Trends in Smart Grids, Microgrids and Smart Buildings. *IEEE Commun. Surv. Tutor.* **2014**, *16*, 1460–1495.
2. Kuster, C.; Rezgui, Y.; Mourshed, M. Electrical load forecasting models: A critical systematic review. *Sustain. Cities Soc.* **2017**, *35*, 257–270. [CrossRef]
3. Jianwei, M.; Libin, F.; Xuechao, D.; Yuanying, Q. Short-Term Power Load Forecasting Method Based on Improved Exponential Smoothing Grey Model. *Math. Probl. Eng.* **2018**, *2018*, 1–11.
4. Daud, M.M.; Raja, R.K.; Georg, F. Short term load forecasting using hybrid adaptive fuzzy neural system: The performance evaluation. In Proceedings of the 2017 IEEE PES PowerAfrica, Accra, Ghana, 27–30 June 2017; pp. 468–473.
5. Yin, Z.; Chen, Y.; Zhang, W.; Li, J. An ultra-short term load forecasting method based on improved human comfort index. In Proceedings of the 2017 4th International Conference on Electrical and Electronic Engineering (ICEEE), Ankara, Turkey, 8–10 April 2017; pp. 468–473.
6. Ghulam, M.U.D.; Angelos, K.M. Short term power load forecasting using Deep Neural Networks. In Proceedings of the 2017 International Conference on Computing, Networking and Communications (ICNC), Santa Clara, CA, USA, 26–29 January 2017; pp. 594–598.
7. Chen, L.-G.; Chiang, H.-D.; Dong, N.; Liu, R.-P. Group-based chaos genetic algorithm and non-linear ensemble of neural networks for short-term load forecasting. *IET Gener. Transm. Distrib.* **2016**, *10*, 1440–1447. [CrossRef]
8. Maged, M.E.; Elsayed, E.H. Feature selection and optimization of artificial neural network for short term load forecasting. In Proceedings of the 2016 Eighteenth International Middle East Power Systems Conference (MEPCON), Cairo, Egypt, 27–29 December 2016; pp. 827–831.
9. López, M.; Valero, S.; Senabre, C. Short-Term Load Forecasting of Multiregion Systems Using Mixed Effects Models. In Proceedings of the 2017 14th International Conference on the European Energy Market (EEM), Dresden, Germany, 6–9 June 2017; pp. 1–5.
10. Tian, C.; Ma, J.; Zhang, C.; Zhan, P. A Deep Neural Network Model for Short-Term Load Forecast Based on Long Short-Term Memory Network and Convolutional Neural Network. *Energies* **2018**, *11*, 3493. [CrossRef]
11. Yordanos, K.S.; Zhang, J.; Zheng, D.; Wei, D. An Accurate Very Short-Term Electric Load Forecasting Model with Binary Genetic Algorithm Based Feature Selection for Microgrid Applications. *Electr. Power Compon. Syst.* **2018**, *46*, 1570–1579. [CrossRef]
12. Wayan, S.; Kemal Maulana, A. A comparison of ANFIS and MLP models for the prediction of precipitable water vapor. In Proceedings of the 2013 IEEE International Conference on Space Science and Communication (IconSpace), Melaka, Malaysia, 1–3 July 2013; pp. 243–248.
13. Zhao, X.; Wang, S.; Li, T. Review of Evaluation Criteria and Main Methods of Wind Power Forecasting. *Energy Procedia* **2011**, *12*, 761–769. [CrossRef]

14. Wu, Y.-K.; Hong, J.-S. A Literature Review of Wind Forecasting Technology in the World. In Proceedings of the IEEE Conference on Power Tech, Lausanne, Switzerland, 1–5 July 2007; pp. 504–509.
15. Jang, J.-S.R.; Sun, C.-T.; Mizutani, E. *Neruo-Fuzzy and Soft Computing*; Prentice Hall: Englewood Cliffs, NJ, USA, 1997.

Article

Behavior Modality of Internet Technology on Reliability Analysis and Trust Perception for International Purchase Behavior

Shiow-Luan Wang *, Yung-Tsung Hou and Sarawut Kankham

Department of Information Management, National Formosa University, Yunlin 632, Taiwan
* Correspondence: slwang@nfu.edu.tw; Tel.: +886-933-173-886

Received: 17 July 2019; Accepted: 29 July 2019; Published: 2 August 2019

Abstract: The proliferation of Internet technology and balance of composition in major feature of many visual products have been advantageous for businesses and changed the distribution channels through which industries reach their consumers. The intensive development of Internet technology and the increasing popularity of online shopping have further changed customers' purchasing behaviors and the methods by which companies disseminate their video advertisements. The main research question that this study intends to answer is, "What do users do when a YouTube advertisement appears? Do they avoid or confront them?" The aim of this study is to explore the perceptions and related behaviors of international purchasing and consumers' trust of YouTube advertisements. Statistical analyses focus on the demographics of a sample population in Thailand. The findings are based on data obtained by a questionnaire, the results of which were analyzed by *t*-test and multiple regression. The results indicate that YouTube advertising has a significant effect on behavioral trends. Moreover, the subjects in the sample reported that they are more likely to avoid YouTube ads than confront them. The study subjects have low satisfaction with YouTube advertising, and males have significantly lower satisfaction than females. This study also analyzes the reliability of trust perception toward purchasing. The results indicate that the reliability is greater than 90% at an α level of 5% and a 95% confidence interval.

Keywords: YouTube advertising; uses and gratifications approach; demographic characteristics; behavioral modality; reliability

1. Introduction

Since its appearance, the Internet has become increasingly influential by facilitating the connection of people all over the world. In particular, the introduction of the World Wide Web (www) has allowed people to easily discover information about any topic by simply searching and clicking. Symmetry's expertise in connecting engineers to the technology allows consumers to save time and money by getting their design products of complex interactions among various stimuli and perceptions. The Internet revolution has led to the advancement of online business all over the world [1]. Users rely on websites to serve their practical demands, such as finding entertainment, enjoying social media, making or chatting with friends, watching movies, doing business, receiving news updates, reading articles, checking on product stock in stores, and so on.

Nevertheless, nothing in this world is free, including the Internet. Users generally pay monthly fees for Internet access through their smartphones or computers. Many highly popular websites are free to users but include many advertisements. By profiting from advertisers who purchase ad space on these websites, website creators are able to maintain the quality of their websites while keeping them up to date. The development of computer technology has accelerated economic development while creating new opportunities and sectors of activity amidst an increasingly competitive environment and easily accessible information could greatly affect the online consumption decision [2,3].

Im2market [4] explained that advertisers focus on providing information that generates or encourages customers' enjoyment. Advertising media can be a tool that promotes a company's product to customers and target groups, and its success is dependent on the chosen media type.

The Digital Advertising Association in Thailand (DAAT) [5] has shown that Facebook (www. facebook.com) has continued to be a predominant platform preferred by companies for advertising and communicating their brand to customers. Advertising on Facebook was valued at 4084 million baht (Thai monetary unit) or one third of digital media advertising revenue in 2017. YouTube (www.youtube.com) was in second place in advertising revenue, which amounted to 2105 million baht, and the third highest was attributed to the display of advertising banners or website banners, with a value of 1340 million baht.

YouTube is a website for watching videos that users can easily upload and share with people around the world. The website is a collection of countless video types of different topics and various content. Globally, more than 1000 million people per day watch videos on YouTube, with hundreds of millions of hours watched per day [6].

Marketers insert advertising into regular media in an attempt to deploy attention-grabbing facts about their goods or services. Consequently, customers are increasingly shunning such advertising. Customers can avoid advertisements on television using a range of actions, such as changing channels or muting the TV when advertising appears. Internet users tend to avoid advertisements in online media as well; for example, users can close banners and pop-up ads that appear on websites by clicking the window cross-box.

In this study, we concentrate on the behavioral trends of Thai individuals to investigate the effects of YouTube advertising, with a particular focus on the "uses and gratifications" approach and the demographic characteristics of Thai people. We used these methods to determine whether the behavioral trend of Thai Internet users is toward avoiding or confronting YouTube advertisements.

2. Literature Review

2.1. YouTube Website

Sanook [7] suggested that YouTube could serve as an effective video-sharing website on which users upload videos, view existing videos, and share videos. These videos are available free of charge to anybody on the YouTube website, although much of the content includes advertising. The types of video content include short film clips, TV shows, music videos, video blogs, and so on. Videos on YouTube are mostly short clips with a duration that ranges from about one minute to more than one hour, and they are recorded by the people in the general public or content creators. The videos are sorted by the website under categories such as recent videos, most viewed videos, most liked videos, and so on. YouTube is available on various platforms: such as a website on PCs or as an application on smartphones, smart TVs, and tablets. In this study, we focus on the PC platform because YouTube started as a website that is accessible through browsers, such as Google Chrome, Internet Explorer, Opera, Safari, and so on.

2.2. Types of Advertising on the Desktop Version of YouTube

Nuttaputch [8] explained that the desktop version of YouTube has six types of advertisements.

(1) Mastheads (Scheme 1) are oversized banners that appear at the top of YouTube's homepage. The standard banner size is 970 × 250 pixels or 970 × 500 pixels as an expandable size. It has multiple formats (schemes or video advertising) and is a fixed banner that is viewed multiple times every day.

Scheme 1. Mastheads.

(2) Display ads (Scheme 2) are banner ads that are displayed as slides or animated gifs. They are typically 300 × 250 pixels and appear on the right side of the playing video.

Scheme 2. Display ads.

(3) Overlay-in-video ads (Scheme 3) are banner ads beneath the video content to discourage the audience from hiding it.

Scheme 3. Overlay-in-video ads

The next two types of ads are forms of TrueView advertising. TrueView is advertising for which YouTube can only charge the advertiser who placed the ad once users view the ad. TrueView advertisements can automatically find their target audiences through the system.

(4) TrueView in-stream ads (Scheme 4) are video ads that play before the main video that the user intends to watch. The cost of advertising to the ad owner depends on whether the video is watched until the end without the user skipping it. However, users who want to skip it can only do so five seconds after the advertisement appears.

Scheme 4. TrueView in-stream ads.

(5) TrueView in-search ads (Scheme 5) are recommended video clips at the top of YouTube's search results. The ad owner is charged once a link is clicked to view the ad.

Scheme 5. TrueView in-search ads.

(6) Non-skippable in-stream ads (Scheme 6) are also called reserve videos. They are mostly video clips similar to TrueView in-stream ads, but this type of ad forces the user to view the video until the end before they can access the main video. The video ad is no longer than 20 s. The ad owner is only charged per 1000 views.

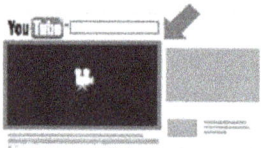

Scheme 6. Non-skippable in-stream ads.

2.3. Demographic Characteristics

Bryant, and Oliver, [9] indicated that because of the influential role the mass media play in society, understanding the psychosocial mechanisms through which symbolic communication influences human thought, affect, and action is of considerable import. To disseminate information (such as news, advertising, and so on), advertisers apply market segmentation to define their target market using demographic characteristics such as gender, age, occupation, social and economic status, education, and so on.

2.4. Uses and Gratifications Approach

When mass society theory was first proposed, it claimed that humanity is the victim of influential media. Later, the credibility of this theory decreased as a result of social studies and observations that illustrated that advertising media do not have a direct effect on everybody or affect everyone in the same way.

Pira Jirasopon [10] recommended the "uses and gratifications theory" as a potential social and psychological theory that views human communication activities as being driven by wants and motivations. By this theory, the receiver looks for specific media and content to satisfy their wants and motivations to obtain self-satisfaction.

Dainton and Zelley [11] asserted that communication constitutes giving and receiving meaning and includes the concept of interaction followed by sharing with others. The media are considered to play a role in the production and dissemination of content that corresponds to the wants of the receiver.

Hence, the uses and gratification approach is based on the view that receivers are enthusiastically active rather than passive in their search for media that is in line with their wants and needs.

2.5. Exit-Voice

As described by Albert O. Hirschman [12], Exit-Voice Theory pertains to the response of customers to the product or service of a company when they feel dissatisfied with that product or service. For instance, dissatisfaction can arise if a product's quality deteriorates but still sells at its original price or if the quality of the item does not match what is advertised. Consumers react to this situation in two major ways:

(1) The "exit option" is an expression of consumer dissatisfaction with a product or service in the form of behaviors such as no longer purchasing that product or service, changing brands, and so on.

(2) Consumers who exercise the "voice option" continue to purchase such products or services and employ other expressions of dissatisfaction that are either direct or indirect. An example is issuing a complaint to the company by writing an e-mail, criticizing the company on social media, and so on.

In cases of the voice option, consumers are likely to spend time or money to express their dissatisfaction to others, such as the time spent writing e-mails to complain to the company. Therefore, the exit option is often a simpler and more convenient expression of dissatisfaction than the voice option.

The theoretical exit-voice concepts are applied to this research to study the behavioral trends of Thai people in response to YouTube advertising. Their possible responses are divided into two types of expression:

(1) Avoidance behavior is the use of measures to avoid viewing the ad on the YouTube website. There are many ways in which advertising can be avoided: closing the YouTube website, refreshing the page (reloading the YouTube website), returning to a previous page, waiting for five seconds to skip the advertisement, watching other videos, installing Web browser extensions (such as AdBlock), and so on.

(2) Confrontation behavior is viewing the ad on YouTube rather than avoiding it, and the video is allowed to play to completion. Such confrontation actions include watching the advertisement, leaving the YouTube ad open, and so on.

3. The Conceptual Framework of Trust Perception-Based Purchasing Behavior

This study analyzes the international purchasing behavior by Thai individuals. The structure of perception-based behavior is shown in Figure 1.

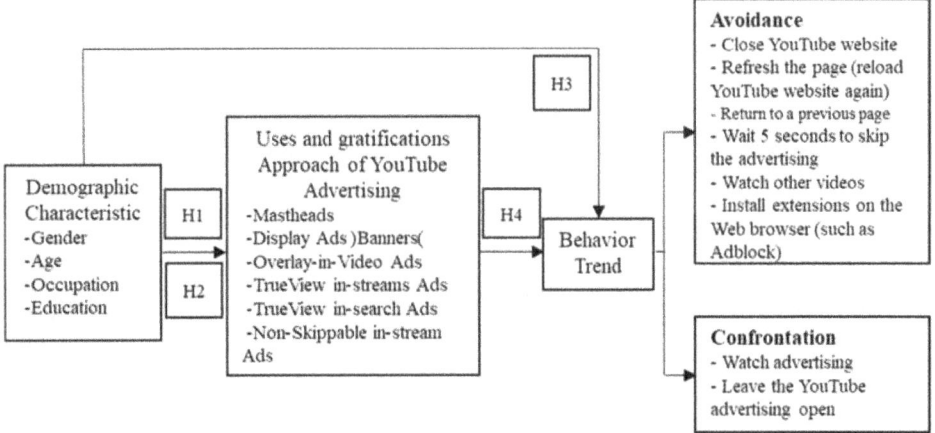

Figure 1. Conceptual framework and hypothesis.

The four hypotheses in this study are as follows:

Hypothesis 1. *Thai subjects have low satisfaction with YouTube advertising.*

Hypothesis 2. *Male subjects have lower satisfaction with YouTube advertising than Female subjects.*

Hypothesis 3. *The behavioral trend of Thai subjects in response to YouTube advertising is that of avoidance rather than confrontation.*

Hypothesis 4. *YouTube advertising has a significant effect on behavioral trends.*

4. Research Methodology and Behavioral Modality Establishment

This research used a quantitative approach through the collection of data from Thai subjects using a questionnaire and sample population are presented on Table 1.

Table 1. Sample Thai population.

Population	Details
Thai Individuals	- Equal number of males and females. - Age range of 11–60 years old. - Living in Chiang Mai, Thailand. - Used the YouTube website for at least one month via laptop or personal computer.

4.1. Sample Selection

There are roughly 960,906 individuals living in Chiang Mai, Thailand [13]. Therefore, we determined the sample size for a precision level defined by a 95% confidence level and a degree of accuracy of 0.05. Thus [14],

$$\text{Size} = \frac{X^2 NP\ (1-P)}{d^2\ (N-1)+X^2 P\ (1-P)},$$

$$\text{Size} = \frac{3.84 \times 1{,}728{,}242 \times 0.50 \times (1-0.50)}{0.05^2\ (1{,}728{,}242-1)+3.84 * 0.50 * (1-0.50)}, \tag{1}$$

$$\text{Size} = 384,$$

where X^2 is the Chi-Square value for df (degrees of freedom) = 1 for the desired confidence level of 95%, i.e., X^2 = 3.84; N is the population size; P is the population proportion (defined as 0.50); and d is the degree of accuracy (expressed as a proportion), these are shown on Table 1.

According to the above formula and the desired confidence level and accuracy, the sample Thai population size for this study was 384 Thai subjects. For easier analysis, we added 16 more subjects. Therefore, there were 400 subjects in the sample.

4.2. Random Sampling

We used probability sampling by multi-stage sampling. The steps were:

Step 1: From Thai sample (400 subjects, these data are presented as Table 2), we conducted non-probability sampling by using quota sampling with respect to the age of Thai subjects.

Table 2. Sample population of Thai subjects.

Group	Gender	Number (Samples)
1	Male	200
2	Female	200

Step 2: We used non-probability sampling, accident sampling, snowball sampling, and convenience sampling to distribute the questionnaire until the number of completed questionnaires was that of the calculated sample size.

4.3. The Study Instrument

The questionnaire was written in two languages, English and Thai, and consisted of three parts.

- Part 1: Information about the demographic characteristics of the study subjects. The requested information included gender, age, education, and occupation. There were four questions in this part.
- Part 2: Information about Thai subjects' application of the uses and gratification approach to YouTube advertising. This part of the questionnaire consisted of closed-ended questions so that subjects could choose a certain answer from six scaled-response questions, which asked about satisfaction with six types of YouTube advertising: mastheads, display ads (banners), overlay-in-video ads, TrueView in-stream ads, TrueView in-search ads, and non-skippable in-stream ads. Table 3 shows the five possible responses on the importance scale.
- Part 3: Behavioral trend of Thai subjects in response to YouTube advertising. This part of the questionnaire had one closed-ended question, with more than one possible answer (i.e., checklist question), as well as one open-ended question, the answer to which respondents could write whatever they chose. Importance scale are presented here as Table 3.

Table 3. Importance scale.

Number	Meaning
(Number) 5	Most
(Number) 4	More
(Number) 3	Moderate
(Number) 2	Low
(Number) 1	Very Low

4.4. Data Analysis and Statistics for Analyzing Data

Data analysis for analyzing data and classification of satisfaction are presented here as Tables 4 and 5. We used the standardized satisfaction of subjects by following the formula below [15]:

$$\text{Class Interval} = \frac{\text{Upper Class Limit} - \text{Lower Class Limit}}{\text{Amount of Class}} = \frac{5-1}{5} = 0.80. \tag{2}$$

Table 4. Data Analysis Statistics for Analyzing Data

Data	Scale	Variable	Statistic
Demographic Characteristics	Nominal	Gender, Occupation	number, percentage, t-test
	Ordinal	Age, Education	
uses and gratifications Approach to Media of Thai Subjects in Response to YouTube Advertising	Internal	- Mastheads - Display ads (banner) - Overlay-in-video ads - TrueView in-stream ads - TrueView in-search ads - Non-Skippable in-stream ads	percentage, mean, standard deviation
The Behavioral Trend of Thai Subjects in Response to YouTube Advertising	Internal	**Avoidance** - Return to a previous page - Wait for 5 s to skip the advertising - Watch other videos - Install Web browser extensions (such as AdBlock) **Confrontation** - Watch advertising - Leave the YouTube advertising open	number, percentage, multiple regression

289

Table 5. Average range of points for classification of satisfaction.

Average Points	Meaning
4.21–5.00	Most Satisfaction
3.41–4.20	More Satisfaction
2.61–3.40	Moderate Satisfaction
1.81–2.60	Low Satisfaction
1.00–1.80	Very Low Satisfaction

4.5. Perception-Based Behavior toward Purchasing: Reliability Analysis

This study analyzed the purchaser's perception-based behavior and the reliability of purchasing perception for relevant purchases in the future. Reliability is defined as the probability that an element (that is, a component, subsystem, or full system) will accomplish its assigned task within a specified time, which is designated by the interval t = [0, tM] [16]. Reliability is closely related to the following four factors: (1) probability value; (2) predetermined function; (3) predetermined life; and (4) prescribed environment. The probability function of reliability allocation is defined in the next subsection.

4.6. Exponential Distribution

Hazard rate:

$$h(x) = (f(x))/(R(x)),\qquad(3)$$

where $f(x)$ is the probability density function of exponential distribution,

$$f(x) = \lambda\, e^{-\lambda x}, x \geq 0\qquad(4)$$

where λ is the failure rate.

The mean time between failures (MTTF) is calculated by the following calculations.

Let X be a random variable that indicates the expiration time. Then, the probability of the product failing at a specific time x is

$$P(X \leq x) = F(x), x \geq 0,\qquad(5)$$

where $F(x)$ is the failure distribution function.

If the product still functions as intended at time x, then

$$R(x) = P(X > x) = 1 - F(x).\qquad(6)$$

4.7. Weibull Distribution Probability Density Function

The probability density function of a Weibull is

$$f(x) = \frac{\beta}{\lambda}\left(\frac{x}{\lambda}\right)^{\beta-1} \exp\left[-\left(\frac{x}{\lambda}\right)^{\beta}\right], x \geq 0\qquad(7)$$

The cumulative distribution function is:

$$F(x) = 1 - \exp\left[-\left(\frac{x}{\lambda}\right)^{\beta}\right], x \geq 0\qquad(8)$$

Reliability is:

$$R(x) = 1 - F(x) = \exp\left[-\left(\frac{x}{\lambda}\right)^{\beta}\right], x \geq 0\qquad(9)$$

The average time to failure is:

$$\mathrm{MTTF} = \lambda\Gamma\left(1 + \frac{1}{\beta}\right)\qquad(10)$$

The failure rate function is:

$$h(x) = \frac{\beta}{\lambda}\left(\frac{x}{\lambda}\right)^{\beta-1}, \ x \geq 0 \tag{11}$$

where the conditions are as follows: when $\beta < 1$, the failure rate decreases with time (early stage); when $\beta = 1$, the failure rate is constant (opportunity period); when $\beta > 1$, the failure rate increases with time (loss period).

Reliability defines the reliability of a product or system. This study had two key computations: the internal series calculation and the Internet of Things (IOT) system. The R_A is the A-system reliability, R_B is the B-system reliability, R_C is the C-system reliability, and R_D is the D-system reliability.

$$R_S = (R_A)(R_B)(R_C)(R_D) \tag{12}$$

In the internal parallel calculation, the internal components of the system are connected in series, and the Internet of Things (IoT) system is also connected in series. The parallel equation is as follows:

$$R_P = 1 - (1 - R_A)(1 - R_B)(1 - R_C)(1 - R_D) \tag{13}$$

The above scenario and calculations were applied in this experiment by using 400 subjects from Thailand, as shown in Table 6.

Table 6. Thai sample population.

Group	Gender	Number (Samples)
1	Male	200
2	Female	200

From the Thai sample population (400 subjects), we conducted non-probability sampling by using quota sampling with respect to the age of Thai subjects.

The purpose of this study was to determine the reliability of using perception-based behavior to predict an individual's final decision on their willingness to purchase the product. The results are shown in Table 7. The reliability of the predicted purchasing behavior (i.e., whether the individual will purchase) was low when it was based on a single use of YouTube. With multiple uses of YouTube, the predicted purchasing behavior was highly reliable. Overall, the reliability of determining purchasing behavior on the basis of YouTube use was greater than 90%.

Table 7. Statistical analysis of reliability on purchaser and abandonment purchase behavior.

Group	Purchaser		Abandonment Purchaser		Reliability of Purchasing Behavior $(R = R_A * R_B * R_C * R_D)$
	Single Use of YouTube	Multiple Uses of YouTube	Single Use of YouTube	Multiple Uses of YouTube	
Male (200 subjects)	0.86	0.96	0.88	0.98	0.712
Female (200 subjects)	0.84	0.97	0.96	0.94	0.735
Reliability of purchasing behavior $R = 1-(1 - R_1) * (1 - R_2)$			0.923		

5. Summary of Findings

The data from the sample population are summarized in Table 8.

Table 8. Demographic characteristics of sample population.

Demographic Characteristics	Number (Subjects)	Percentage
Gender		
Male	200	50%
Female	200	50%
Age		
11–20 years old	56	14%
21–30 years old	328	82%
31–40 years old	12	3%
41–50 years old	4	1%
51–60 years old	0	0%
Occupation		
Self-Employed	8	2%
Bureaucrat	8	2%
Student	344	86%
Unemployed/Retirement	8	2%
Other (includes Professor and Librarian)	32	8%
Education		
Junior high school	12	3%
Senior high school	24	6%
Bachelor's degree	348	87%
Higher than bachelor's degree	16	4%

The results indicate that most of the Thai subjects were students with a bachelor's degree, and the age range of the majority was 21–30 years old.

6. Hypothesis Testing

6.1. Hypothesis 1: Thai Subjects Have Low Satisfaction with YouTube Advertising

Table 9 reports the satisfaction of the subjects with the six types of YouTube advertising. The results reveal that Thai subjects had a moderate satisfaction with mastheads, display ads (banners), and TrueView in-search ads. Overlay-in-video ads and TrueView in-stream ads were scored as low satisfaction, and Thai subjects had very low satisfaction with non-skippable in-stream ads.

Table 9. Satisfaction with types of YouTube advertising.

Variable	Mean	St. Dev.	Description
Mastheads	2.96	1.132	Moderate Satisfaction
Display ads/banners	3.08	1.145	Moderate Satisfaction
Overlay-in-video ads	2.20	1.218	Low Satisfaction
TrueView in-stream ads	2.37	1.455	Low Satisfaction
TrueView in-search ads	3.03	1.139	Moderate Satisfaction
Non-skippable in-stream ads	1.72	1.195	Very Low Satisfaction
Total	2.54	0.835	Low Satisfaction

We next used the ranges of average points to classify satisfaction, as specified in Table 5. The results reveal that Thai subjects have low satisfaction with YouTube advertising, with a mean score of 2.54.

6.2. Hypothesis 2: Male Subjects Have Lower Satisfaction with YouTube Advertising than Female Subjects

Table 10 compares the mean scores given by male and female subjects. These results demonstrate that the difference in satisfaction with YouTube advertising between genders was not significant at the 0.05 level, which means that males and females had the same opinion toward YouTube advertising.

Table 10. Satisfaction with YouTube advertising according to the gender of Thai subjects (independent samples *t*-test).

Gender	N	$\bar{x}$	S.D.	t	df	Sig. (Two-Tailed)	p
Male	200	2.48	0.825	1.519	398	0.129 *	0.098
Female	200	2.60	0.842				

* Significant at the 0.05 level.

The mean satisfaction score given by male subjects was 2.48, which was lower than that given by female subjects, whose mean was 2.60. In other words, male samples have lower satisfaction with YouTube advertising than female samples.

6.3. Hypothesis 3: The Behavioral Trend of Thai Subjects in Response to YouTube Advertising Is that of Avoidance Rather Than Confrontation

The results show that most Thai subjects engaged in Avoidance behavior toward YouTube advertising. This option was selected 564 times, which was 76.2% of the total. The most common avoidance behavior was waiting for five seconds and then skipping the ad. This behavior was selected 328 times, which is 44.3% of the total.

Confrontation behavior was reported 176 times (23.8% of the total). The most common confrontation behavior was leaving the YouTube advertisement open, which was selected 124 times, which was 16.8% of the total.

This implies that the behavioral trend of Thai subjects was more avoidance than confrontation in response to YouTube advertising. These data are presented here as Table 11.

Table 11. The number and percentage of behaviors toward YouTube advertising reported by Thai subjects.

Behavioral Trend of Subjects toward YouTube Advertising	Number	Percentage
Avoidance	564	76.2%
Close YouTube website	56	7.6%
Refresh the page (reload the YouTube website)	72	9.7%
Return to a previous page	28	3.8%
Wait for 5 s to skip the advertisement	328	44.3%
Watch other videos	64	8.6%
Install Web browser extensions (such as AdBlock)	16	2.2%
Confrontation	176	23.8%
Watch advertising	52	7%
Leave the YouTube advertisement open	124	16.8%
Total	740	100%

6.4. Hypothesis 4: YouTube Advertising Has a Significant Effect on Behavioral Trends

Table 12 provides the regression results with avoidance as the dependent variable. The result indicates that variation in masthead ads and display ads (banner) explained 4.2% of the variation in the avoidance behavior of Thai subjects at a significance level of 0.01 for these two independent variables. Thus, the independent variables that led to the most variation in avoidance behavior were display ads (banner) and mastheads.

Table 12. Regression results (dependent variable = avoidance).

Independent Variable	B	S.E.	Beta
Mastheads	−0.024	0.007	−0.201 **
Display ads (Banner)	0.026	0.007	0.227 **
Overlay-in-video ads	00.001	0.007	0.005
TrueView in-stream ads	0.006	0.006	0.066
TrueView in-search ads	0.003	0.007	0.024
Non-skippable in-stream ads	0.002	0.008	0.021
$R^2 = 0.042$ SEE $= 0.130$ F $= 3.851$ **			

** Significant at the 0.01 level.

This result shows that Thai subjects tended to avoid mastheads and display ads (banners) the most among all types of YouTube advertising. The results for the remaining independent variables imply that Thai subjects were consistent in their accepted level of advertising.

Table 13 reports the regression results with confrontation as the dependent variable. The result indicates that variation in mastheads, TrueView in-search ads, and non-skippable ads explained 5.9% of the variation in the confrontation behavior of Thai subjects at the 0.05 significance level for mastheads and at the 0.01 level for the two other independent variables. Thus, the independent variables that led to the most variation were TrueView in-search ads, non-skippable in-stream ads, and mastheads, from highest to lowest significance.

Table 13. Regression results (dependent variable = confrontation).

Independent Variable	B	S.E.	Beta
Mastheads	−0.035	0.015	−0.142 *
Display ads (banners)	0.005	0.014	0.021
Overlay-in-video ads	0.020	0.015	0.089
TrueView in-stream ads	−0.017	0.012	−0.086
TrueView in-search ads	0.060	0.014	0.244 **
Non-skippable in-stream ads	−0.041	0.016	−0.174 **
$R^2 = 0.059$ SEE $= 0.270$ F $= 5.048$ **			

** Significant at the 0.01 level; * Significant at 0.05 level.

The result implies that Thai subjects tended to confront mastheads, TrueView in-search ads, and non-skippable in-stream ads. This means that they accepted watching these types of YouTube ads.

Note that mastheads were a special case: Thai subjects tended to both confront and avoid them. However, comparing significance levels revealed that Thai subjects tended to avoid more than confront mastheads since the former behavior was associated with a significance level of 0.01. These data are presented here as Table 14.

Table 14. Summary of the hypotheses.

Hypothesis	Result
H1: Thai subjects have low satisfaction with YouTube advertising.	Supported
H2: Male subjects have lower satisfaction with YouTube advertising than female subjects.	Supported
H3: The behavioral trend of Thai subjects in response to YouTube advertising is that of avoidance rather than confrontation.	Supported
H4: YouTube advertising has a significant effect on behavioral trends.	Supported

7. Conclusions

The results indicate that Thai subjects using the desktop version of YouTube had low satisfaction with YouTube advertising, and the YouTube advertising type that subjects felt the most displeasure encountering was non-skippable in-stream ads. This was likely because it is the only kind of YouTube advertising on the desktop version of the website that users cannot avoid unless they install specific Web browser adds-on (i.e., extensions, such as AdBlock). Further, male subjects had lower satisfaction with YouTube advertising than female subjects. Demographic characteristic theories [6] argue that individuals with different demographic characteristics (such as gender, age, occupation, education, and so on) behave differently.

Avoidance or confrontation? In fact, the presented results indicated that subjects were more likely to avoid YouTube advertising than confront it. The most common way to avoid ads was waiting for five seconds and then skipping the ad, which is applicable to TrueView in-stream ads (video ads that play before the main video that the user intends to watch). Research on why people avoid advertising on the Internet has been conducted, and it was suggested in Reference [17] that the more that advertising interrupts the Internet user's activity, the more they avoid the advertising website. Some comments in subjects' responses to the questionnaire included, "I do not like YouTube advertising that waste[s] my time", "we can install an extension to block advertising", "advertising should be suitable with video", "YouTube advertising should not have non-skip[p]able in-stream ads because I cannot skip", and so on.

Riedl and Kenning's study found that perceived trustworthiness of Internet offers is affected by neurobiology [18]. Organizational trust can be divided into intra- and inter-organizational trust. Inter-organizational trust refers to the extent to which organizational members have a collectively held trust orientation towards the partner firm [19]. We have to emphasis on this point in the future study.

What should YouTube do? Understanding how IT impacts consumer behavior can serve as a critical foundation for businesses to identify and develop effective and sustainable marketing communication strategies [20]. To enhance user satisfaction, YouTube ought to improve its advertising format by providing a choice to users, such as monthly or yearly subscriptions to the website for an ad-free experience, or reducing non-skippable advertising on the website. Future work may need to consider other countries and apply a similar method for comparison with Thailand, as well as study avoidance behavior for other social media platforms, such as Instagram, Line, and Facebook.

Author Contributions: Conceptualization, all authors; methodology, S.K.; software, Y.-T.H.; formal analysis and data extraction, S.K., Y.-T.H.; writing—original draft preparation, S.K.; writing—review and editing, S.-L.W.; visualization, Y.-T.H.; supervision, S.-L.W.

Funding: This research received no external funding.

Conflicts of Interest: The authors declare no conflicts of interest.

References

1. Oláh, J.; Kitukutha, N.; Haddad, H.; Pakurár, M.; Máté, D.; Popp, J. Achieving sustainable e-Commerce in environmental, social and economic dimensions by taking possible trade-offs. *Sustainability* **2018**, *11*, 89. [CrossRef]
2. Chin, A.J.; Wafa, S.A.; Ooi, A.Y. The effect of internet trust and social influence towards willingness to purchase online in Labuan, Malaysia. *Int. Bus. Res.* **2009**, *2*, 72–81. [CrossRef]
3. Cheung, C.M.; Lee, M.K.; Rabjohn, N. The impact of electronic word-of-mouth: The adoption of online opinions in online customer communities. *Internet Res.* **2008**, *18*, 229–247. [CrossRef]
4. Im2market. Advertisement. Available online: https://www.im2market.com/2017/02/11/4290 (accessed on 7 August 2018).
5. Digital Advertising Association (Thailand). Available online: www.daat.in.th/index.php/thailand-digital-advertising-spend-by-daat2018/ (accessed on 7 August 2018).
6. Pudpong Woradech. Media Exposure, Attitudes and Advertising Avoidance Behaviors in YouTube of Thai Teenagers. Master's Thesis, Burapha University, Chon Buri, Thailand, 2015.

7. Sanook. YouTube. Available online: https://guru.sanook.com/2292/ (accessed on 7 August 2018).
8. Nuttaputch. Six Ad Formats on YouTube that Marketers and Viewers Should Know. Available online: https://www.nuttaputch.com/5-major-youtube-ad-types/ (accessed on 10 August 2018).
9. Bryant, J.; Oliver, M.B. *Media Effects: Advances in Theory and Research*, 3rd ed.; Routledge: New York, NY, USA, 2009; p. 31.
10. Pira Jirasopon. *Paradigm about the Theory of Results from Mass Communication (Instructor's Handbook: Strategic Communication Theory)*; Bangkok: Bangkok, Thailand, 2013.
11. Dainton, M.; Zelley, E.D. *Applying Communication Theory for Professional Life: A Practical Introduction*, 2nd ed.; Sage: Singapore, 2011; p. 167.
12. Hirschman, A.O. *Exit, Voice, and Loyalty: Responses to Decline in Firms, Organizations, and States*; Harvard University Press: Cambridge, MA, USA, 1970.
13. Wikipedia. Chiang Mai Province. Available online: https://en.wikipedia.org/wiki/Chiang_Mai_Province (accessed on 10 September 2018).
14. Krejcie, R.V.; Morgan, D.W. Determining sample size for research activities. *Educ. Psychol. Meas.* **1970**, *30*, 607–610. [CrossRef]
15. Suta Phachareeya. Perception and Attitude of Consumers towards Goods Purchasing through QR Code Payment in Bangkok Metropolitan Area. Master's Thesis, Rajamangala University of Technology Thanyaburi, Pathum Thani, Thailand, 2012.
16. Myers, A. *Complex System Reliability: Multichannel Systems with Imperfect Fault Coverage*, 2nd ed.; Springer: London, UK, 2010.
17. Cho, C.H.; Cheon, H.J. Why Do People Avoid Advertising on the Internet? *J. Advert.* **2004**, *33*, 89–97. [CrossRef]
18. Riedl, R.; Hubert, M.; Kenning, P. Are there neural gender differences in online trust? An fMRI study on the perceived trustworthiness of eBay offers. *MIS Q.* **2010**, *34*, 397428. [CrossRef]
19. Oláh, J.; Karmazin, G.; Fekete, M.F.; Popp, J. An examination of trust as a strategical factor of success in logistical firms. *Bus. Theory Pract.* **2017**, *18*, 171–177. [CrossRef]
20. Xiang, Z.; Magnini, V.P.; Fesenmaier, D.R. Information technology and consumer behavior in travel and tourism: Insights from travel planning using the internet. *J. Retail. Consum. Serv.* **2015**, *22*, 244–249. [CrossRef]

Article

Application of the Symmetric Model to the Design Optimization of Fan Outlet Grills

Hsin-Hung Lin [1,2,*] and Jui-Hung Cheng [3]

1   Department of Creative Product Design, Asia University, Taichung City 41354, Taiwan
2   Department of Medical Research, China Medical University Hospital, China Medical University, Taichung 404, Taiwan
3   Department of Mold and Die Engineering, National Kaohsiung University of Science and Technology, Kaohsiung 80778, Taiwan
*   Correspondence: hhlin@asia.edu.tw or a123lin0@gmail.com; Tel.: +886-04-2332-3456-1051

Received: 1 July 2019; Accepted: 21 July 2019; Published: 30 July 2019

Abstract: In this study, different designs of the opening pattern of computer fan grills were investigated. The objective of this study was to propose a simulation analysis and compare it to the experimental results for a set of optimized fan designs. The FLUENT computational fluid dynamics (CFD) simulation software was used to analyze the fan blade flow. The experimental results obtained by the simulation analysis of the optimized fan designs were analyzed and compared. The effect of different opening pattern designs on the resulting airflow rate was investigated. Six types of fans with different grills were analyzed. The airflow velocity distribution in the simulated flow channel indicated that the wind speed efficiency of the fan and its influence were comparable with the experimental model. The air was forced by the fan into the air duct. The flow path was separately measured by analog instruments. The three-dimensional flow field was determined by performing a wind speed comparison on nine planes containing the mainstream velocity vector. Moreover, the three-dimensional curved surface flow field at the outlet position and the highest fan rotation speed were investigated. The air velocity distribution at the inlet and the outlet of the fan indicated that among the air outlet opening designs, the honeycomb shaped air outlet displayed the optimal performance by investigating the fan characteristics and the estimated wind speed efficiency. These optimized designs were the most ideal configurations to compare these results. The air flow rate was evenly distributed at the fan inlet.

Keywords: fan design; numerical simulation; fan experiments; axial flow fan; electronics cooling

1. Introduction

The opening pattern of an axial fan grill is one of the most important factors of the resulting airflow rate. The fan grill is provided with a number of significant characteristics. From the standpoint of aerodynamic performance, the pressure rise will decrease depending on the different types of opening patterns on the axial fan grill. The increased density of the boundary layer of the fan grill opening pattern affects the exhaust flow of an axial. The resulting phenomenon is the influence of the flowing vortex when the airflow passes the fan grill. On the other hand, the style of the fan grill also affects the overall efficiency of the intake airflow when the blade design remains unchanged. A smaller aspect ratio and a higher opening ratio allow for an axial fan to generate better outflow performance. The overall airflow rate can be increased by optimizing the opening pattern of the outflow when considering the complex aerodynamic parameters [1].

The influence of different fan grill patterns on the resulting airflow velocity has been investigated. In the simulation, the airflow velocity distribution indicated the efficiency of the air velocity. The three-dimensional flow field of the experimental model was also verified from the verification of the airflow velocity. One of the most important factors that affect the resulting airflow velocity of an axial fan grill is its characteristics. Sergio Marinetti et al. 2001 carried out the investigation of the rotational speed of a fan by forcing airflow through an evaporator and two fans. The three-dimensional flow field was investigated by measuring the airflow distribution at different elevations at the evaporator inlet and outlet. The results indicated a uniform air velocity distribution at the evaporator inlet [1]. Gebrehiwot et al. 2010 applied the CFD approach to the investigation of the performance of a cross-flow fan by using three fans with similar geometry at the cross-flow opening. The fan load can be determined from the impedance curve of the perforated plates with different openings. The results of the wind tunnel testing indicated that the non-uniform airflow distribution at the air inlet along the width of a cross-flow fan was an important factor for the CFD simulation [2]. The findings from Chen's et al. 2009 investigation indicated that the design of a distorted stator of a transonic fan presented better aerodynamic characteristics [3]. Betta et al. 2010 studied the fluid dynamic performance by comparing it to the conventional axial ventilation system and by applying the CFD analysis to the system by the *k*–*e* model [4]. Li et al. 2008 carried out the analysis and experimental investigation of the aerodynamics of forward inclined impellers and radial low pressure axial impellers. The design of the forward-inclined blades was optimized by CFD techniques and by measuring the aerodynamics and aeroacoustics of the two blades. In comparison to typical radial blades, a forward-inclined blade was proven to be able to improve the efficiency by carrying out a detailed flow rate measurement and the calculated exhaust flow field. The results indicated that a forward-inclined blade could trigger the redistribution along the radial direction and reduce the tip overload [5]. Delele et al. 2005 A CFD model was developed for the investigation of the three-dimensional airflow pattern of the ground velocity of the cross-flow air sprayer. The researchers conducted a simulation of the rotational speed of two different fans, and the results of the instant cross-sectional velocity distribution and simulation of the maximum vertical exhaust velocity and the change in directions indicated good consistency. This is due to the fact that the magnitude of the exhaust velocity of the jet flow is larger and the change in direction is greater [6]. The flow allocation and performance investigation were carried out by CFD approaches and anemometers in order to determine the effect of the balanced split vortex walls between two grid points. The results indicated that a two-dimensional CFD model could be used to predict fan performance to an acceptable level, especially at the portion between two outlets. The modification indicated a more uniform air vortex distribution at the walls and the mass flow rate between the two outlets affected the rotor exhaust air to a greater extent [7]. Li et al. 2011 carried out the comparison between two impellers with diameters 5% and 10% larger than the original impeller by both numerical analysis and experimental study. The numerical simulation of the internal characteristics indicated that the flow speed and total pressure increased, so the axial power and sound pressure also increased. When the efficiency decreased, the impeller needed to have a larger diameter for a better operating point [8]. The fan assembly could keep the average level of the components of the heat transfer coefficient. The results indicated that if a fan design could resolve the problem of cross-flow environment, the heat transfer efficiency could be improved by 30% [9]. Other studies have implemented CFD approaches in the flow field of axial fans and their performance and characteristics. The deviation of the fan performance curve obtained by experimental results reached the horizontal DFR method, and the use of static meshes and moving meshes was less than 3% and 1.5%, respectively. This presents significant improvement to the conventional approach, which has a significant deviation of 26% [10,11]. Hu et al. 2013 the calculation includes three steps: firstly, the unsteady viscous flow around the blades is calculated using the CFD method to acquire the noise source information; secondly, the radiated sound pressure is calculated using the acoustic analogy Curle equation in the frequency domain; lastly, the scattering effect of the duct wall on the propagation of the sound wave is expressed using the thin-body BEM method [12]. Owen 2013 a comparison of test data collected at an existing ACC (cooling performance of an air-cooled condenser) and numerical data generated in a CFD analysis of the flow around the same ACC shows a discrepancy in the predicted effects of wind on fan inlet temperature. Careful analysis of the test data indicates the potential involvement of atmospheric temperature distributions in fan inlet temperature deviations. A numerical case study is conducted considering four differing atmospheric temperature distributions [13]. Detailed flow measurement and computation were performed for outlet flow field for investigating the responsible flow mechanisms. Yang et al. 2008 The results show the forward-skewed blade can cause a spanwise redistribution of flow toward the blade mid-span and reduce tip loading [14]. Carolus et al. 2007 In the limits of the necessary assumptions the SEM (a simple semi-empirical noise prediction model) predicts the noise spectra and the overall sound power surprisingly well without any further tuning of parameters; the influence of the fan operating point and the nature of the inflow is obtained. Naturally, the predicted spectra appear unrealistically "smooth", since the empirical input data are averaged and modeled in the frequency domain. By way of contrast the LES (The numerical large eddy simulation) yields the fluctuating forces on the blades in the time domain. Details of the source characteristics and their origin are obtained rather clearly. The predicted effects of the ingested turbulence on the fluctuating blade forces and the fan noise compare favorably with experiments [15].

The objective of this study is to investigate the performance of different gap designs of fan grills. The flow field of an axial fan was investigated by CFD simulation and the result was compared to experimental one. The airflow velocity of the resulting airflow was determined in the three-dimensional flow field. The maximum rotating speed at the fan outlet was also measured. The distribution of the airflow velocity at the fan inlet and outlet was analyzed in order to determine the effect of different gap designs. The objective is to generate uniform airflow velocity at the fan inlet so that an optimal configuration can be created.

2. Research Model

Configuration of Model Parameters

The dimension of the target fan model was 35 mm (L) × 95 mm (W) × 146 mm (H), as shown in Figure 1. The main components of the external structure included a casing or housing, impeller, and exhaust housing. A fan operates by creating a pressure difference by the rotating blades so that the surrounding fluid is forced to move. Therefore, its energy is transferred to the surrounding fluid in a dynamic way. The main effect is to overcome the system impedance by the air pressure that is created by the fan.

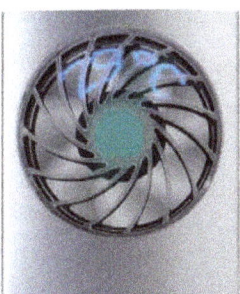

Figure 1. New fan model.

Typically, when designing an impeller, various parameters and equations that affect the fan performance need to be considered since the geometry of the impeller deals with three-dimensional surfaces. A fan designer is required to design based on the criteria of geometrical form design and needs to carry out the design process again if a resulting impeller does not meet the performance criteria. Axial fan parameters were collected for the investigation in this study with the relevant theories and equations summarized for further review to determine the optimal impeller design and parameters such as the inner and outer diameters. With further configuration of other detailed parameters, a designer can quickly generate the required impeller without much effort in the calculation and modification. The final impeller design can be determined by the curve-fitting results and its geometric parameters, as shown in Table 1.

Table 1. Configuration parameters of the new fan model.

Airfoil Type	Airfoil Name	Blade No.	Radius of Hub	Tip of Blade	Radius of Shroud
NACA65	NACA65-Parabolic	7	8	36	40
Thickness of hub	Section No.	Tip clearance	Incidence angle of the blade at the hub	Incidence angle of the blade at the tip	Blade width at the hub
23.5	31	0.75	50	35	11

One of the most important factors that affect the flow rate of a fan is the opening pattern. In order to make the overall evaluation framework more complete, the parameters and curve equations that affect the fan performance should be considered when designing the blade profile. The conditions of the geometric shapes were listed, and the simulation results of the new fan models were compared with the real test results as shown in Figure 1. Therefore, after analyzing the fans that are available on the market, a variety of new fan grill designs were created in this study. These new grill designs were screened out according to the ergonomic design principle that human fingers do not penetrate the grill gaps. Moreover, the qualified designs need to present symmetric and regular curve patterns. A total of six grill patterns were determined to be the qualified ones and were analyzed by CFD simulation in order to determine the most optimal fan grill design. The geometric parameters of the new fan blade designs are shown in Table 1. Of all the new opening pattern designs, six were selected for further simulation. The six different opening pattern designs are shown in Figure 2 as follows.

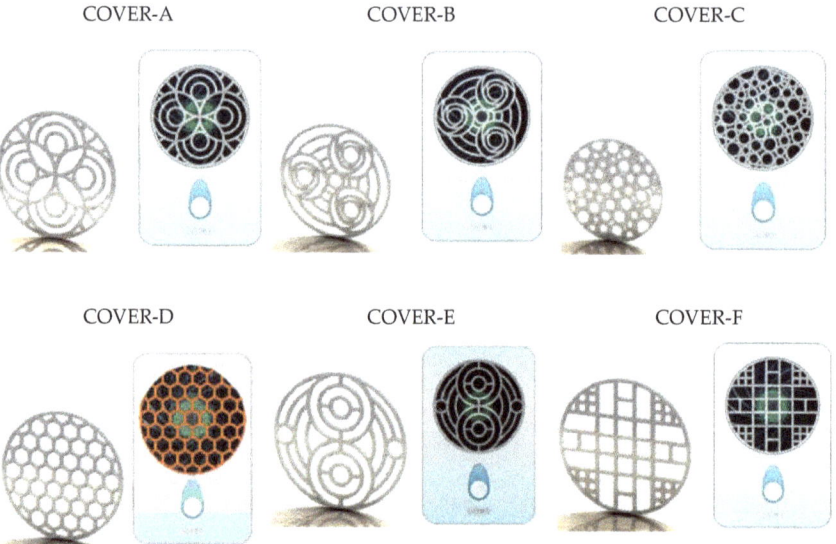

Figure 2. The six different opening patterns.

The purpose of this study was to investigate the design of the opening patterns. A fan rotates to move air to the opening, and the domain of the upstream and downstream of the fan should be included into the numerical simulation to obtain more accurate numerical results. The symmetric air velocity measurement points on the fan model is shown in Figure 3. Along the centerline of the fan model, there were a total of nine measurement points marked as Points A to I, among which Points C to F were the points located within the fan itself. Points A and I were located at the inlet and outlet of this flow field to determine the flow pattern of the region close to the opening. The velocity component V can be determined as these points are located at the boundary of this solution domain. The velocity along the vertical axis can be obtained from the numerical analysis and can be later compared to the real measurement results. The numerical model was based on the standard k–ε model. The vertical component V indicates a larger difference than the component along the flow direction. This is the main purpose of optimizing the opening pattern design in order to obtain satisfactory results.

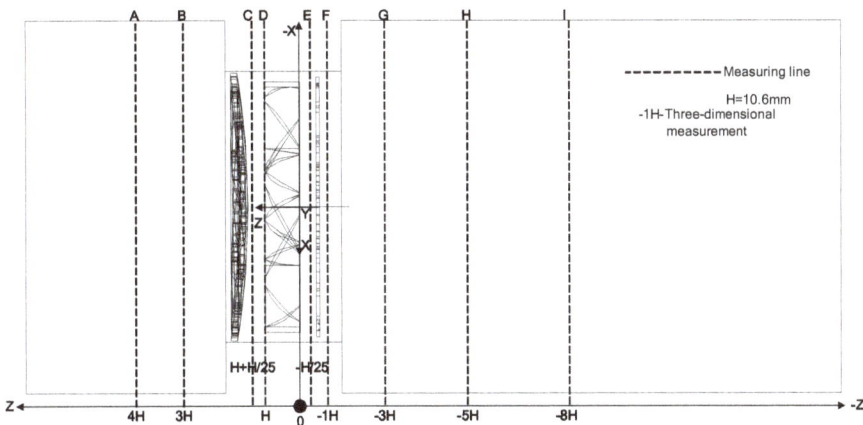

Figure 3. Symmetric air velocity measurement points on the model.

3. Research Methodology

3.1. Numerical Analysis

When a numerical method is used for the simulation and analysis, some fundamental and reasonable assumptions need to be made in order to simplify the complexity of the numerical simulation. These assumptions are described as follows.

3.1.1. Governing Equations

In the 3D Cartesian coordinate system, the governing equations are as follows [16–18].
Continuity equation:

$$\frac{\partial u}{\partial x} + \frac{\partial v}{\partial y} + \frac{\partial w}{\partial z} = 0 \tag{1}$$

Momentum equation:
X direction:

$$\frac{\partial u}{\partial t} + \frac{\partial (u^2)}{\partial x} + \frac{\partial (uv)}{\partial y} + \frac{\partial (uw)}{\partial z} = -\frac{1}{\rho}\frac{\partial P}{\partial x} + v\left[\frac{\partial^2 u}{\partial x^2} + \frac{\partial^2 u}{\partial y^2} + \frac{\partial^2 u}{\partial z^2}\right] \tag{2}$$

Y direction:

$$\frac{\partial v}{\partial t} + \frac{\partial (uv)}{\partial x} + \frac{\partial (v^2)}{\partial y} + \frac{\partial (vw)}{\partial z} = -\frac{1}{\rho}\frac{\partial P}{\partial y} + v\left[\frac{\partial^2 v}{\partial x^2} + \frac{\partial^2 v}{\partial y^2} + \frac{\partial^2 v}{\partial z^2}\right] \tag{3}$$

Z direction:

$$\frac{\partial w}{\partial t} + \frac{\partial(uw)}{\partial x} + \frac{\partial(vw)}{\partial y} + \frac{\partial(w^2)}{\partial z} = -\frac{1}{\rho}\frac{\partial P}{\partial z} + v\left[\frac{\partial^2 w}{\partial x^2} + \frac{\partial^2 w}{\partial y^2} + \frac{\partial^2 w}{\partial z^2}\right] \tag{4}$$

Energy equation:

$$\frac{\partial T}{\partial t} + \frac{\partial(uT)}{\partial x} + \frac{\partial(vT)}{\partial y} + \frac{\partial(wT)}{\partial z} = \alpha\left(\frac{\partial^2 T}{\partial x^2} + \frac{\partial^2 T}{\partial y^2} + \frac{\partial^2 T}{\partial z^2}\right) + \frac{q}{\rho C_P}. \tag{5}$$

The governing equations can be represented by the general equations as follows:

$$\frac{\partial(\rho\phi)}{\partial t} + \frac{\partial(\rho\phi u)}{\partial x} + \frac{\partial(\rho\phi v)}{\partial y} + \frac{\partial(\rho\phi w)}{\partial z} = \frac{\partial}{\partial x}\left(\Gamma\frac{\partial\phi}{\partial x}\right) + \frac{\partial}{\partial y}\left(\Gamma\frac{\partial\phi}{\partial y}\right) + \frac{\partial}{\partial z}\left(\Gamma\frac{\partial\phi}{\partial z}\right) + s. \tag{6}$$

$$\frac{\partial(\rho\phi u)}{\partial x} + \frac{\partial(\rho\phi v)}{\partial y} + \frac{\partial(\rho\phi w)}{\partial z}$$

is the convective term;

$$\frac{\partial}{\partial x}\left(\Gamma\frac{\partial\phi}{\partial x}\right) + \frac{\partial}{\partial y}\left(\Gamma\frac{\partial\phi}{\partial y}\right) + \frac{\partial}{\partial z}\left(\Gamma\frac{\partial\phi}{\partial z}\right)$$

is the diffusive term; S is the source term; and $\frac{\partial(\rho\phi)}{\partial t}$ is the unsteady term [19]. This term is not considered under the steady-state assumption. The symbol φ represents dependent variables such as u, v, w, k, ε, and T in Table 2 [20]. Γ is the corresponding diffusivity of each physical variable [12]. u, v, and w are the velocity components in the x, y, and z directions, respectively.

Table 2. List of independent variables.

Equation	ψ
Continuity	1
X-momentum	u
Y-momentum	v
Z-momentum	w
Energy	I or T

Based on the fundamentals of the finite-volume method, the computational domain must be partitioned into many small control volumes. After a volume integral, the equations of the mass, energy, and momentum of fluids can then be transformed into algebraic equations as follows:

$$\frac{\partial}{\partial t}\int_v(\rho\varphi)dV + \int_A \vec{n}\cdot\left(\rho\varphi\vec{V}\right)dA = \oint_A \vec{n}\cdot\left(\Gamma_\varphi\nabla\varphi\right)dA + \int_V S_\varphi\cdot dV. \tag{7}$$

where $\oint_A \vec{n}\cdot(\rho\varphi\vec{V})dA$ is the convective term; $\oint_A \vec{n}\cdot(\Gamma_\varphi\nabla\varphi)dA$ is the diffusive term; $\int_V S_\varphi\cdot dV$ is the generation term; and $\frac{\partial}{\partial t}\int_v(\rho\varphi)dV$ is the unsteady term. This term is not considered under the steady-state assumption.

3.1.2. Theory of Turbulence Model

Since turbulence causes the exchange of momentum, energy, and concentration variations between the fluid medium, it causes quite a few fluctuations. Such fluctuations are of a small scale and with high frequency. Therefore, for real engineering calculations, a direct simulation requires very high-end computer hardware [13]. Therefore, when simulating turbulent flows, manipulations of the control equations are required to filter out turbulence components that are at an extremely high frequency or of extremely small scale. However, the modified equations may comprise variables that are unknown to us, while the turbulence model requires the use of known variables to confirm these variables [21].

3.1.3. Standard k–ε Turbulence Model

The standard k–ε is a type of semi-empirical turbulence model. It is mostly based on basic physical equations to derive the transport equations that describe the turbulent flow transmission of turbulence kinetic energy (k) and dissipation (ε) [22]. These equations are as follows [23]:

Turbulence kinetic energy equation (k):

$$\frac{\partial}{\partial t}(\rho k) + \frac{\partial}{\partial x_i}(\rho k u_i) = \frac{\partial}{\partial x_j}\left[\left(\mu + \frac{\mu_t}{\sigma_k}\right)\frac{\partial k}{\partial x_j}\right] + G_k + G_b - \rho\varepsilon - Y_M; \tag{8}$$

Dissipation equation (ε):

$$\frac{\partial}{\partial t}(\rho\varepsilon) + \frac{\partial}{\partial x_i}(\rho\varepsilon u_i) = \frac{\partial}{\partial x_j}\left[\left(\mu + \frac{\mu_t}{\sigma_\varepsilon}\right)\frac{\partial\varepsilon}{\partial x_j}\right] + C_{1s}\frac{\varepsilon}{k}(G_k + C_{3\varepsilon}G_b) - C_{2g}\rho\frac{\varepsilon^2}{k}; \tag{9}$$

(3) Coefficient of turbulent viscosity (μ_t):

$$\mu_t = \rho C_\mu \frac{k^2}{\varepsilon}. \tag{10}$$

In the equation, G_k indicates the turbulence kinetic energy generated by the velocity gradient of laminar flow. G_b is the turbulence kinetic energy generated by the buoyancy. In compressible turbulent flows, Y_M is the fluctuation generated by the excessive diffusion. σ_k and σ_ε are the turbulent Prandtl numbers of turbulence kinetic energy and turbulent dissipation; and $C_{1\varepsilon}$, $C_{2\varepsilon}$, and $C_{3\varepsilon}$ are the empirical constants. The recommended values of these coefficients are shown in Table 3 [24].

Table 3. Coefficients of the standard k–ε turbulence model.

$C_{1\varepsilon}$	$C_{2\varepsilon}$	C_μ	C_k	$C_{3\varepsilon}$
1.44	1.92	0.09	1.0	1.3

The k–ε model is based on the resulting equations by assuming that the flow field is completely at the turbulence state and the condition in which the molecular viscosity is negligible. Therefore, the standard k–ε model provided a better result for calculating the fully turbulent flow fields [25].

3.1.4. RNG k–ε Turbulence Model

The RNG k–ε model is derived from the mathematical method of renormalization group in combination with the transient Navier-Stokes equations (N-S equations). This model is similar to the standard model and its analytical characteristics directly evolved from the standard model. The main difference between the RNG model and the standard one is due to the consideration of the turbulent vortex by the RNG model. This consideration enhances the calculation precision on the vortex. Moreover, the turbulent Prandtl number also provides a comprehensive analytical equation. A condition is also included into the turbulent diffusion equation in order to improve the precision of the standard model. The flow field can be presented in a more precise way. The equation of the RNG model is described as follows [18].

Equation of turbulence kinetic energy (k)

$$\frac{\partial}{\partial t}(\rho k) + \frac{\partial}{\partial x_i}(\rho k u_i) = \frac{\partial}{\partial x_j}\left(\alpha_k \mu_{eff}\frac{\partial k}{\partial x_j}\right) + C_k + G_b - \rho\varepsilon - Y_M \tag{11}$$

Equation of diffusivity (ε)

$$\frac{\partial}{\partial t}(\rho\varepsilon) + \frac{\partial}{\partial x_i}(\rho\varepsilon u_i) = \frac{\partial}{\partial x_j}\left(\alpha_k \mu_{eff}\frac{\partial\varepsilon}{\partial x_j}\right) + C_{1\varepsilon}\frac{\varepsilon}{k} + (G_k + C_{3\varepsilon}G_b) - C_{2\varepsilon}\rho\frac{\varepsilon^2}{k} - R_\varepsilon \tag{12}$$

where α_K and α_ε are the turbulence Prandtl number of the turbulence kinetic energy and the turbulence diffusivity, μ_{eff} is the coefficient of equivalent turbulence viscosity, R_ε is the parameter of the modified turbulence viscosity, the constants are $C_{1\varepsilon} = 1.42$, $C_{2\varepsilon} = 1.68$ respectively.

The main difference between the RNG model and the standard model is described as follows. The RNG model is used to build up a new equation based on the condition of low Reynolds number. The equation is as follows.

$$d\left(\frac{\rho^{2k}}{\sqrt{\varepsilon\mu}}\right) = 1.72\frac{\hat{v}}{\sqrt{\hat{v}^3 - 1} + C_V}\tag{13}$$

where $C_V \approx 100$, $\hat{v} = \frac{\mu_{eff}}{\mu}$.

The equation depicts how the Reynolds number affects the coefficient of equivalent turbulence viscosity so that a model could perform better at a low Reynolds number. At conditions with a higher Reynolds number, the turbulence velocity equation of the standard model is still used, except that the C_μ parameter is set as 0.0845 according to the RNG theoretical calculation.

Moreover, since the turbulence in a uniform flow is also affected by the vortex, the turbulence viscosity is also modified to compensate for this influence. The equation is as follows.

$$\mu_t = \mu_{t0}f\left(\alpha_s, \Omega, \frac{k}{\varepsilon}\right)\tag{14}$$

Here μ_{t0} is the quantity that is not modified from the original equation of the turbulence viscosity coefficient. Ω is a characteristic parameter that is used by the FLUENT software. α_s is a vortex constant and its value is determined form the vortex intensity of the flow condition. For a moderate vortex flow state, α_s is set as 0.05. For a stronger vortex flow, a larger value can be used.

For the turbulence Prandtl number, the RNG theory supplies a comprehensive analytical equation that can be used to calculate α_k and α_ε as follows.

$$\left|\frac{\alpha - 1.3929}{\alpha_0 - 1.3929}\right|^{0.6321}\left|\frac{\alpha + 2.3929}{\alpha + 2.3929}\right|^{0.3679} = \frac{\mu_{mol}}{\mu_{eff}}\tag{15}$$

where $\alpha_0 = 1.0$ and $\alpha_k = \alpha_\varepsilon \approx 1.393$ for a larger Reynolds number.

Finally, the conditional parameter of R_ε is also included into the diffusivity equation. This parameter leads to the main difference from the standard model and the equation is as follows.

$$R_\varepsilon = \frac{C_\mu \rho \eta^3 (1 - \eta/\eta_0)}{1 + \beta\eta^3}\frac{\varepsilon^2}{k}\tag{16}$$

where $\eta = \frac{Sk}{\varepsilon}$, $\eta_0 = 4.38$, $\beta = 0.0 = 0.012$.

Since the RNG model provides a comprehensive definition and correction to several parameters, the RNG model can react to the flow field with immediate changes and curved streamlines. This is also the reason why the RNG can present better performance in this type of flow field.

3.2. Performance Testing Equipment for Fans

From the aspect of performance measurements in this experiment, the detailed configuration of the measuring equipment and apparatus and the corresponding operations were described separately as shown in Figure 4. The main device of the performance testing equipment for the fans used adopted the outlet-chamber wind tunnel according to the AMCA 210-99 standard. This type of wind tunnel is composed of a main body, flow setting means, multiple nuzzles, and a flow-rate regulating device [14]. Its major function is to simulate various types of the air-flow condition downstream of the fan and supply a good and stable flow field for measurement in order to obtain the complete performance curves [26].

1	Test fan	12	Thermocouple
2	Auxiliary fan	13	Fiber-optic tachometer
3	Flow-rate adjusting device	14	Multi-functional interface card for extraction
4	Flow setting means	15	Multi-functional signal converter card
5	Multiple nuzzles	16	Desktop computer
6	Pressure tap upstream of the nozzle plate	17	Laser printer
7	Pressure tap downstream of the nozzle plate	18	Thermometer and hygrometer
8	Pressure tap at the inlet	19	Barometer
9	Pressure transducer #1	20	Digital inverter
10	Pressure transducer #2	21	Power supply
11	Adjusting device for fiber-optic tachometer	22	Access hole

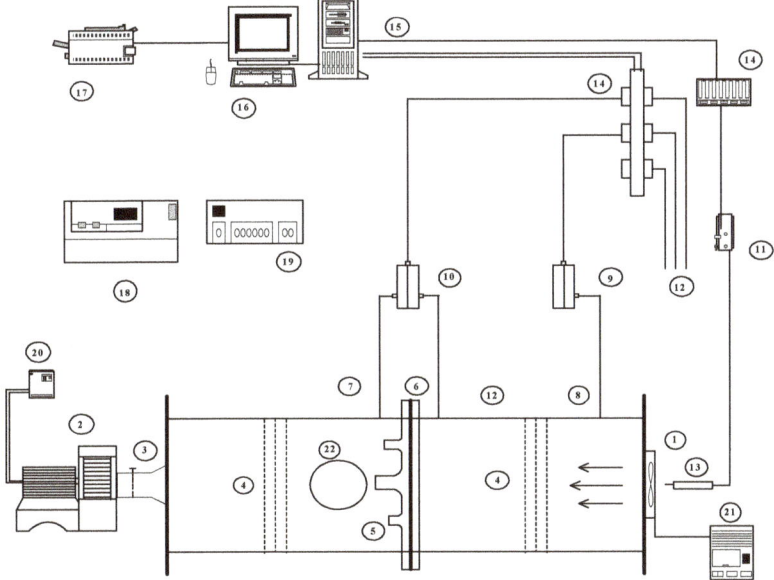

Figure 4. Configuration of the fan performance testing system and specifications of the major instruments.

3.2.1. Calculation of Flow Rates

According to the calculation by the standard equations and flow-rate measurements by the National Laboratory, the errors can be obtained through a comparison with the measurement readings. The standard equations are as follows [27].

The pressure difference between the nozzle outlet and inlet PL_5 and PL_6 can be obtained. The flow rates on the cross-sections of the nozzles can be determined with varying nozzle coefficients as shown in Figure 5. If there is a need to calculate the outlet flow rate of the fan under test, then the effect of density variations must be considered, the measurement of which is as follows [28].

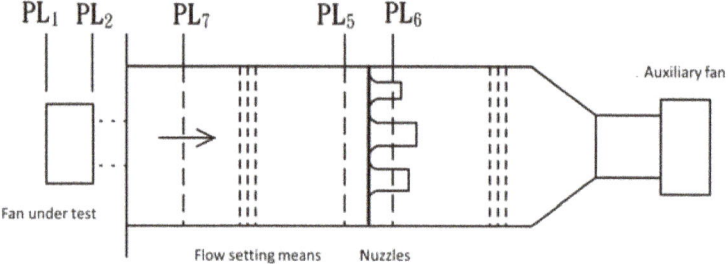

Figure 5. Definition of various measurement planes.

The equation for the calculation of flow rates in a test chamber with multiple nuzzles is

$$Q_5 = 265.7Y \sqrt{\Delta P / \rho_5} \sum_n (C_n A_{6n}) \tag{17}$$

where Q_5 is the total flow rate measured by a bank of nozzles, CMM; ΔP is the pressure difference across the nozzles, mm-Aq; ρ_5 is the air density upstream of the nozzles, kg/m³; is the expansion factor; C_n is the discharge coefficient of the nth nozzle; and A_{6n} is the cross-sectional area of the nth nozzle's throat, m².

3.2.2. Calculation of Air Pressures

Typical pressure readings can be directly measured by instruments, but requires understanding the definition of a fan's static pressure (ΔP_s) and total pressure (ΔP_t). The static pressure, defined as the difference between the fan's static pressure at typical pressure readings, can be directly measured by instruments, but understanding (P_{s_2}) and the static pressure at inlet (P_{s_1}) is required. The total pressure is the difference between the fan's total pressure at outlet (P_{t_2}) and the total pressure at inlet (P_{t_1}). The equations for measurement and calculation are explained respectively as follows.

Since the outlet and inlet planes of the fan under test are PL_2 and PL_1, respectively, they can be defined as follows [29]:

$$P_s = P_t - P_v \tag{18}$$

$$P_t = P_{t_2} - P_{t_1} \tag{19}$$

where P_s is the static pressure of the fan under test; P_t is the total pressure of the fan under test; P_v is the dynamic pressure of the fan under test; P_{t_2} is the total pressure at the fan's outlet (or plane PL_2); and P_{t_1} is the total pressure at the fan's inlet (or plane PL_1).

Since in this experiment there was no duct at the inlet of the fan under test, therefore $P_{t_1} = 0$. On the other hand, the measured static pressure at the outlet is the same as the static pressures measured at the measuring plane PL_7. Therefore, $P_{s_2} = P_{s_7}$.

$$P_{t_2} = P_{s_7} + P_v \tag{20}$$

$$P_s = P_{s_7} \tag{21}$$

It can be concluded from the above equation that the static pressure of the fan under test happened to be equal to the static pressure obtained at the outlet test chamber P_{t_7}. The Type A method (a test method with no duct at either the outlet or the inlet) that was carried out at the outlet test box is a special case of testing. When carrying out different types of tests or different equipment, the equation of the static pressure of the fan under test is thus more complicated. The calculation of dynamic pressure is as follows [30]:

$$P_{v_2} = \frac{\rho_2 V_2^2}{19.6} \tag{22}$$

where P_{v_2} is the outlet dynamic pressure of the fan under test, mm-Aq; V_2 is the outlet air velocity of the fan under test, m/s; ρ_2 is the outlet air density of the fan under test, kg/m^3; and $V_2 = \frac{Q_2}{60 A_2} = \frac{Q}{60 A_2} \cdot \frac{\rho}{\rho_2} = \frac{Q}{50 \rho_2 A_2}$ where Q_2 is the outlet flow rate of the fan under test, CMM; Q is the standard flow rate of the fan under test, CMM; A_2 is the outlet cross-sectional area of the fan under test, m^2; ρ is the density of air at STP (1.2 kg/m^3); and $P_t = P_s + P_v = P_s + P_{v_2}$.

$$P_t = P_s + \frac{\rho_2 V_2^2}{19.6} \tag{23}$$

3.2.3. Fan Performance Power and Efficiency

The calculation of power can be obtained from torque and rotation speed. By measuring the torque of a fan by a torque gauge and measuring the rotation speed by a fiber-optic tachometer, the fan input power (W) can be obtained. The fan efficiency can also be obtained from the air pressure and the flow rate. It can be estimated by the equations as follows:

$$W = \frac{2\pi \times T \times n}{33,000 \times 12} \tag{24}$$

$$\eta_s = \frac{P_s \cdot Q}{4500 W} \tag{25}$$

$$\eta_t = \frac{P_t \cdot Q}{4500 W}. \tag{26}$$

4. Numerical Simulation

The flow passage of the numerical model of the target axial fan is shown in Figure 6. Within the entire range of the flow rate, the model was composed of not only one flow passage. It is known that the flow field behavior follows the continuity equation and the momentum equation. The axial fan is composed of the inlet cone, impeller, and the fan housing. The resulting mesh structure contains structured meshes as the majority and unstructured meshes as the minority. After further mesh refinement and coupling, the final number of the mesh contained 750,000 cells as shown in Figure 7. The mesh structure contained a rotating mesh system between the inlet and the outlet. The outlet boundary had a uniform pressure of 1 atm. The rotation speed of the rotating mesh was 2000 RPM. The pressure and velocity coupling wind-facing difference method was selected as the simple algorithm. The maximum residual was defined as $<10^{-3}$ for convergence.

Since this is a problem for rotating machinery, the commercial CFD software FLUENT6.3 was used for the simulation. The selected region was configured to rotate against an axis so the momentum equation could be rectified automatically. The source terms were automatically added into the relevant equations for calculation. The configuration of the boundary conditions needs to consider the operating condition of a real object, i.e., to comply with the physical phenomenon. Otherwise, the accuracy of the calculated result might be affected. The boundary conditions of this simulation include the inlet boundary condition, outlet boundary condition, and wall boundary condition. Their descriptions are listed in Table 4 as follows as shown in Figure 8 [15,28,31].

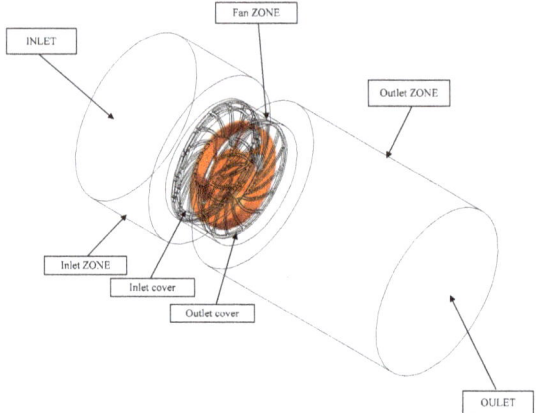

Figure 6. Numerical model of the target axial fan.

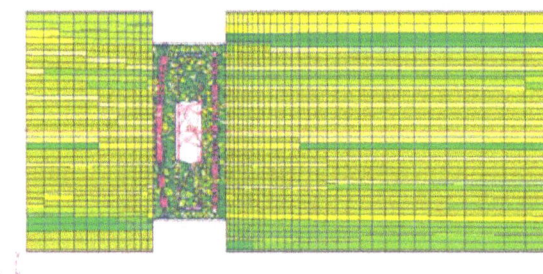

Figure 7. Mesh structure along the cross-section of the numerical model of the axial fan.

Table 4. Boundary conditions of the fan model for simulation.

Inlet boundary condition [9,32]	The inlet condition is for the initial calculation, this research simulates the fan in an infinite domain condition, therefore, at the inlet, it selects and adopts normal atmospheric pressure P0.
Outlet boundary condition [33]	The flow generated by the rotation of the fan is the simulated flow toward the ambient atmosphere. Therefore, the outlet boundary condition of the normal atmospheric pressure P0 is also adopted.
Wall boundary condition	Except for the non-permeable condition to be satisfied when a fluid flows through a wall, it also needs to satisfy the no-slip condition), i.e., $u = v = w = 0$. k and ε are determined by the near-wall model.
Assumption that is made to reduce the complexity of flow field calculation [34]	The flow field is at the steady state and the fluid is incompressible air. The turbulence model is the standard $k-\varepsilon$ model with eddy rectification. The influence of gravity is neglected. Related fluid properties such as viscosity, density, and specific heat are all constants. A rotation speed of 2000 RPM is set for the MRF fluid rotating region. The relative velocity between the solid surface and the fluid is zero, which is the no-slip condition. The influence of radiation and buoyancy is neglected. Moreover, physical properties do not vary with temperature.
Rotating speed of the fan	Configured to be 2000 RPM.

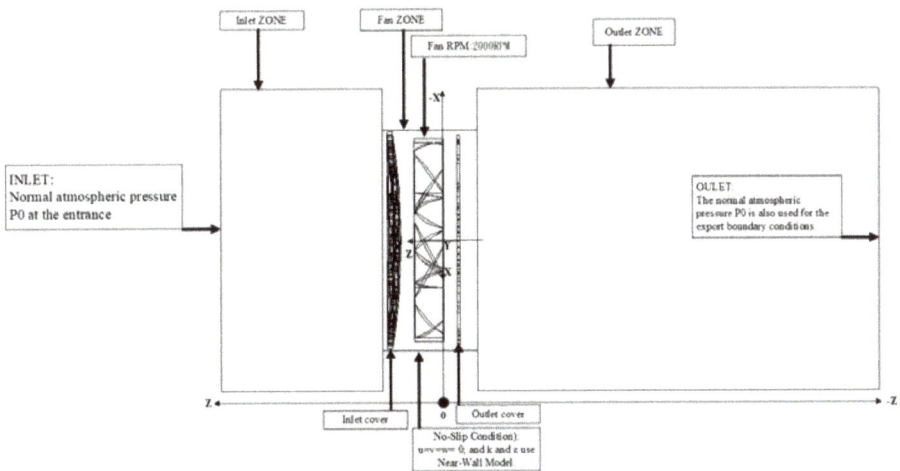

Figure 8. Boundary conditions.

5. Verification of the Case Study and Numerical Analysis

5.1. Verification between Numerical Simulation and Experiment Testing

After the completion of the numerical simulation, we compared the results to the experiment in order to determine whether the trend corresponded to each other. Furthermore, the accuracy of the numerical simulation in this study was also verified as shown in Figure 9. First, to verify the accuracy of the numerical simulation, it is known from the comparison of the numerical results to the experimental ones in Table 5 that the numerical results were on average 3% smaller than the experimental ones. After exploring the reason, it was found that the deviations occurred most from the difference between the configuration of the real test environment and that of the simulated environment. This is due to the fact that when testing a real fan, the air impedance varies according to different operating points. It is also known from the experimental results that the average fan speed during the real tests was 3% larger than the design fan speed, so that it could be closer to the real test result [34].

Table 5. Comparison of the numerical simulation result and the experimental result.

	Numerical Simulation	Experimental Result	Deviation
Air flow rate	16.8CFM	16.3CFM	3%
Static pressure	1.75 mm-H_2O	1.71 mm-H_2O	2%

By comparing the numerical simulation results to the experimental results, the non-dimensional air velocity at the inlet, as shown in Figure 10, served as a good reference for analyzing the flow field distribution of these six opening patterns. From the average value at the inlet and the pulse velocity distribution, the correct air velocity can be determined from Figure 10. The results indicated that the deviation of the non-dimensional velocity at the inlet 4H was the same for the six opening patterns on the boundary. Therefore, the mesh structure was valid for the simulation in this study.

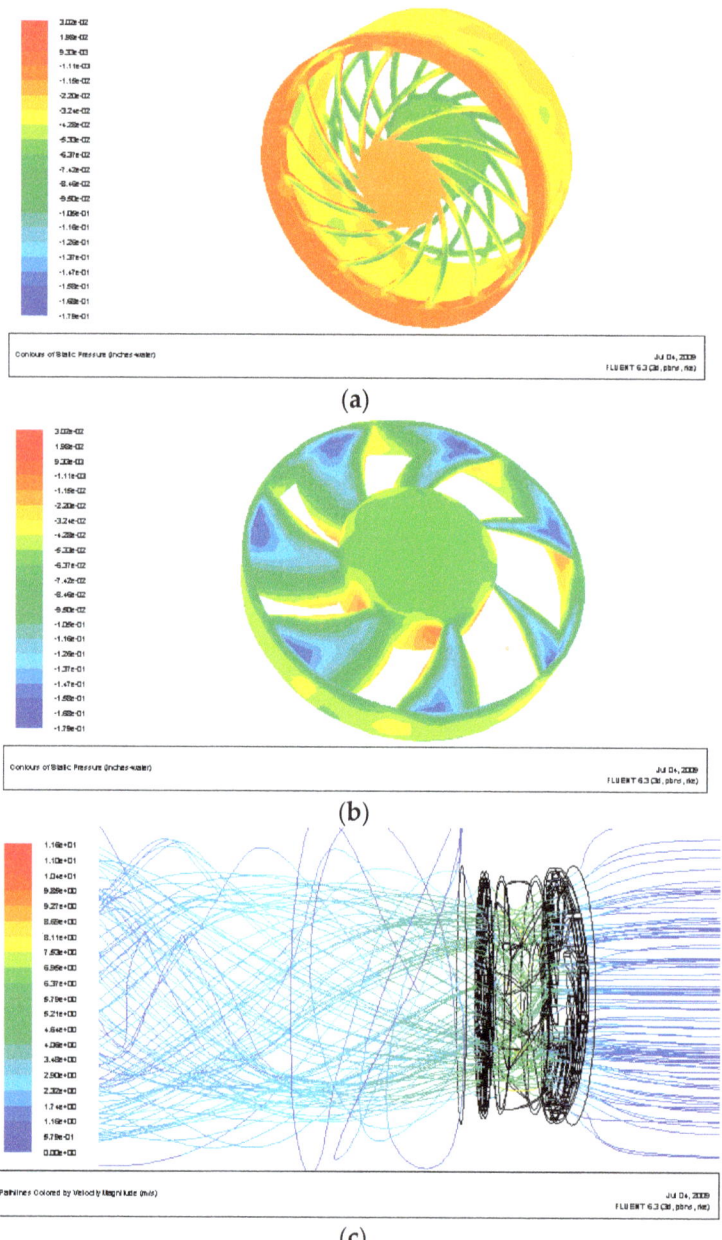

Figure 9. *Cont.*

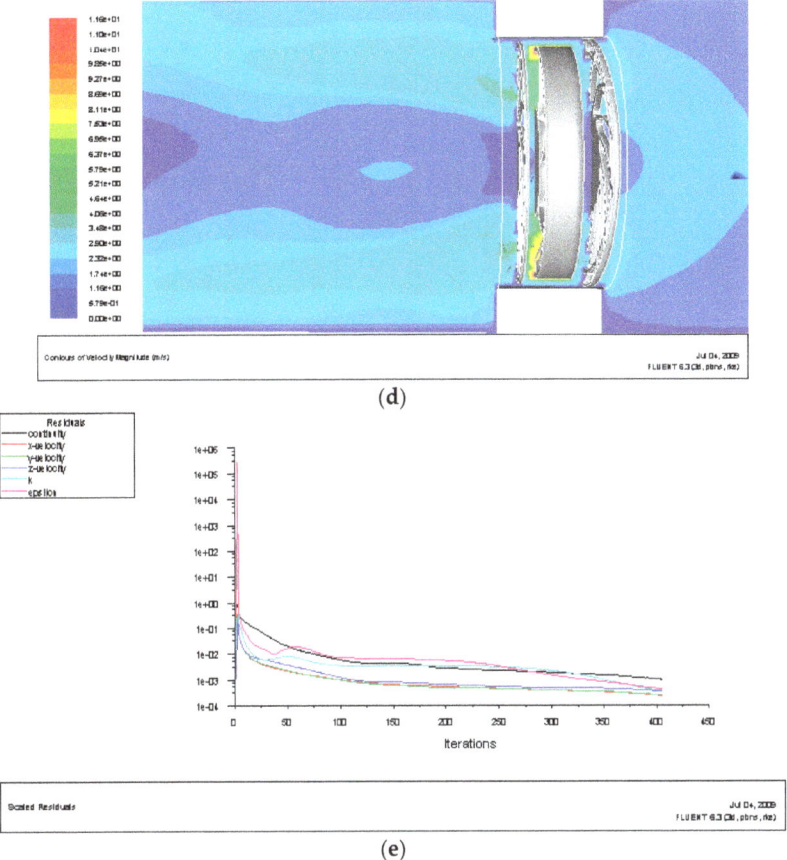

(d)

(e)

Figure 9. Resulting plots of CFD simulation for the numerical fan model. (**a**) Pressure distribution on top and bottom covers; (**b**) pressure distribution on the fan; (**c**) distribution of stream lines; (**d**) distribution for flow field; and (**e**) program convergence plot.

Figure 11 indicates the non-dimensional velocity V along the vertical direction from A to I. Since the air flow rate of a fan is affected by the gaps at the outlet, the relative location of the intake channel is different according to its air velocity. The velocity component of the six curves on 4H in Figure 11 was consistent with those in Figure 10. The velocity component V at 10 mm in front of the fan was almost uniform. The variation in the air velocity was due to the interference of the direction of the rear gaps. At a distance of −1H, the velocity started to vary and the change in the air velocity was in the same direction. A similar trend could be observed at the eight vertical lines at the rear end of −3H. The V component was larger at both the left and right sides and was negative at the centerline of the fan. It is known that the magnitude of the velocity component at −5H is relatively smaller than the left and right sides. The results also indicated that the velocity component of the Idea-D magnitude remained at the highest air velocity from the measurement line between E and I. Under this condition, the boundary layer effect of the velocity component could be clearly observed. The magnitude of the velocity component was gradually increased and the Idea-D can be viewed as the optimal factor of the opening pattern design.

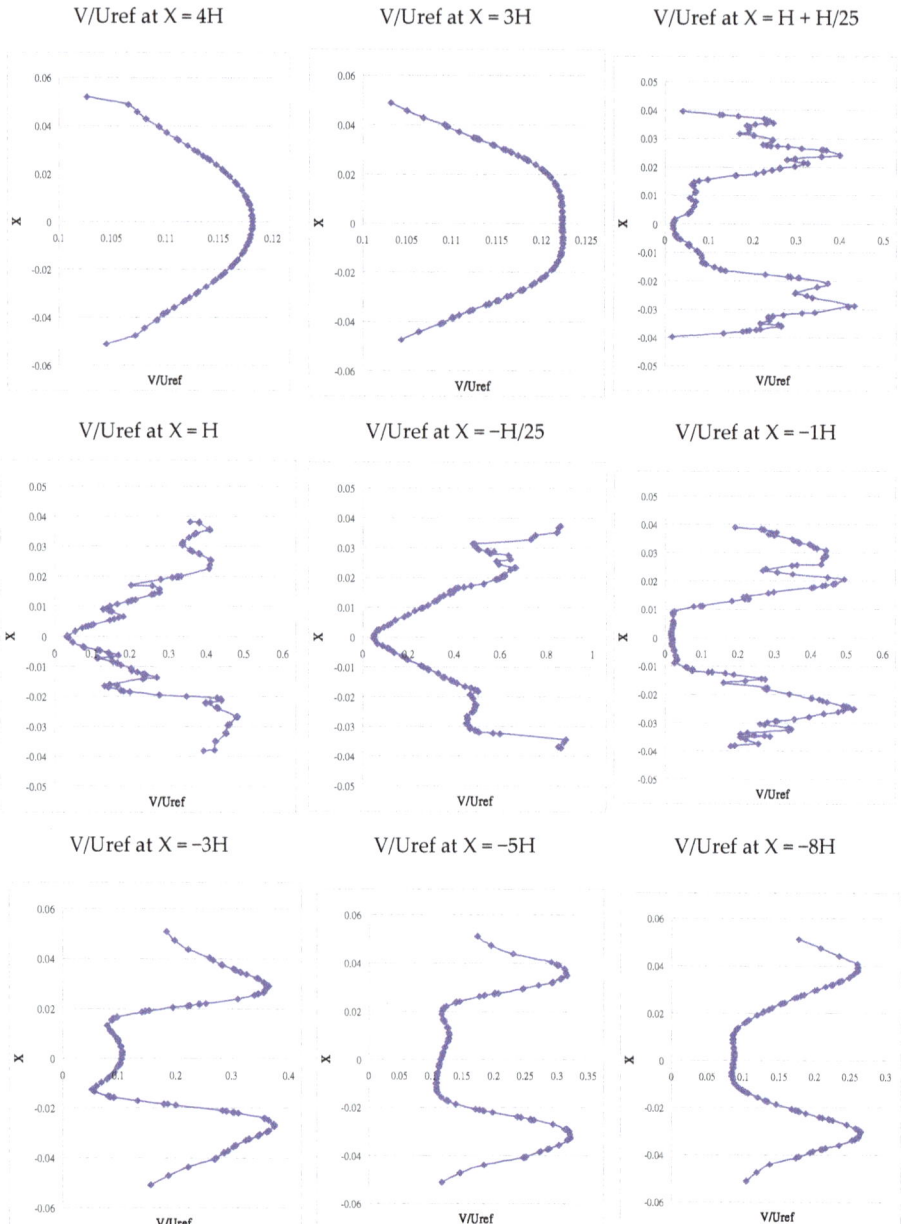

Figure 10. Non-dimensional velocity V/Uref at nine planes along the vertical centerline of the model for the numerical simulation.

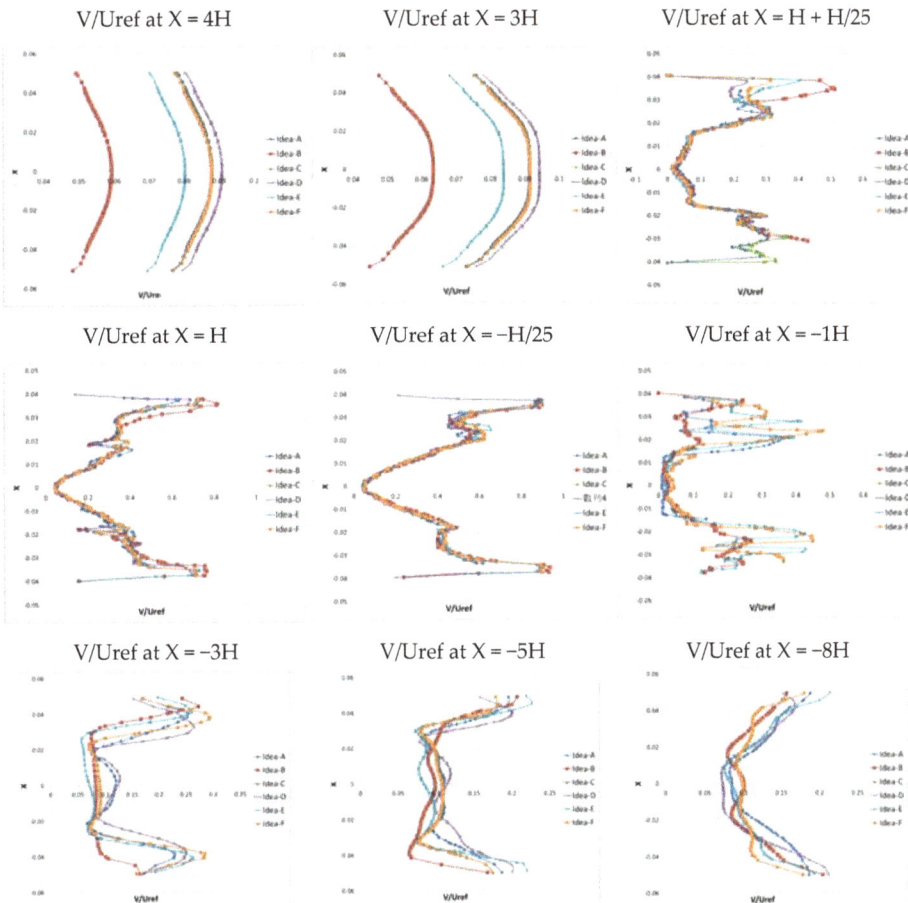

Figure 11. Non-dimensional velocity V/Uref at nine planes along the vertical centerline of the model for the numerical simulation for these six opening patterns.

5.2. Design Cases and Comparison between Simulation Results

Figure 12 indicates the qualitative velocity distribution on the three-dimensional plane at location F of the outlet opening. The purpose of this figure is to understand the velocity pattern on these six planar regions. As the results clearly indicated a cone-shaped distribution, the velocity distribution at the outlet region was the most apparent. Moreover, Figure 12 also indicates that the out ring of the distribution of Idea-C revealed a low air velocity. In contrast, the air velocity at the peak was the highest. The Idea-D had the largest number of peaks and this indicates that this region was the region with the highest air velocity and the distribution had a more focused region in the central region. For the velocity distribution of Idea-B, E, and F, since they were affected by the non-uniform gaps, the air flow velocity was reduced. When the fan was rotating, the air layers also rotated. The air at the central region was affected by the viscosity and molecular attraction, so the rotating effect was transferred to the air on the out ring of the flow field.

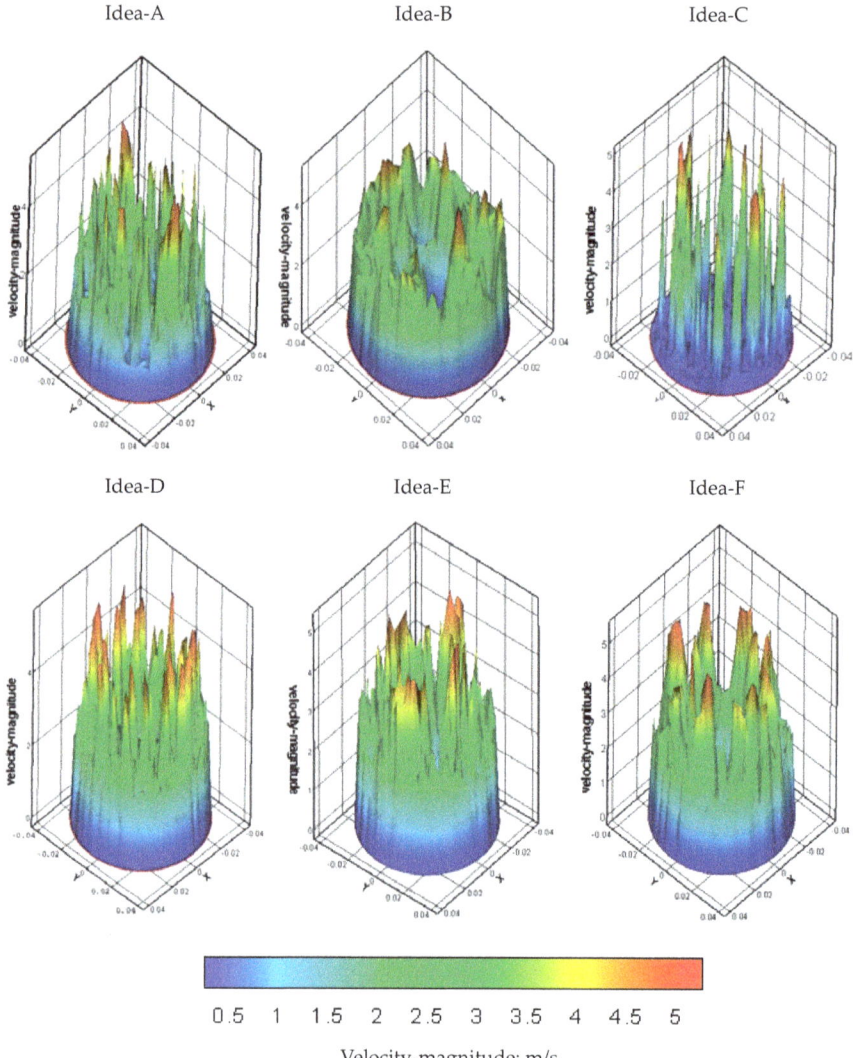

Figure 12. Three-dimensional distribution of the air velocity at the outlet.

The results of the simulation for the six opening patterns are shown in Table 6. The effect of different opening patterns can be evaluated by a comparison of the numerical results. The variation in the flow field of six different opening patterns and the change in the performance were reviewed from the inlet and outlet of each streamline. Table 6 reveals that the high velocity regions were on the left and right sides of the flow field. This phenomenon also directly affected the outflow of the air between the blades. This is the same phenomenon that reduced the range of higher air velocity on the downstream of the fan. For Idea-A, the expansion of the opening pattern provided some improvement and became more and more apparent. The region of low air velocity extended inward so that the velocity along the inner diameter increased gradually. For Idea-B and E, the region of the highest air velocity at the outlet reduced gradually. The region of low air velocity also expanded along the downstream direction. The downstream flow field of the Idea-D honeycomb type opening still remained smooth, but not affected by the opening gap. The flow field revealed high velocity at the middle portion and downstream portion. On the overall distribution, the outlet of the impeller still remained at the high air velocity.

Table 6. Comparison of the numerical simulation results for six different cases.

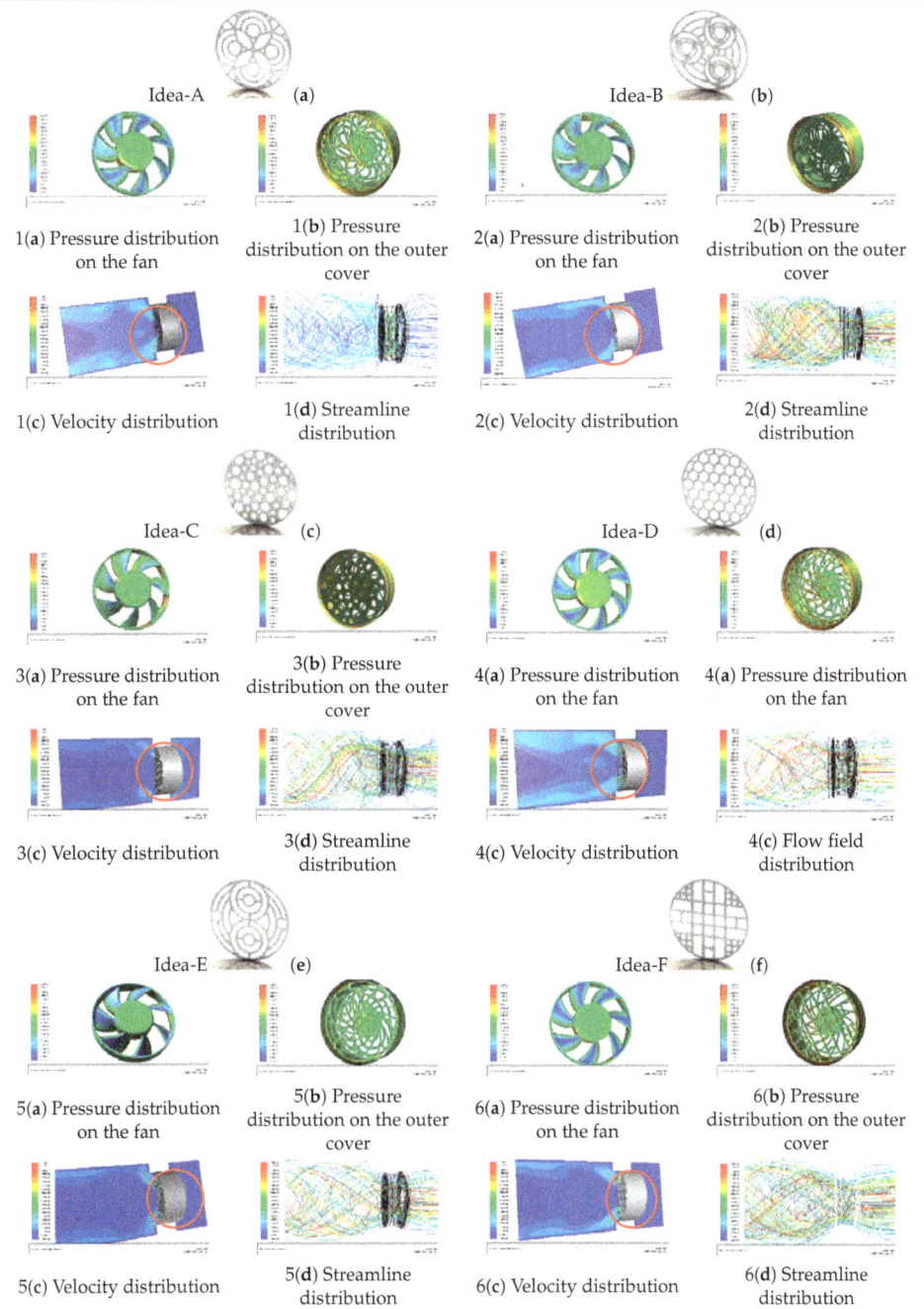

Moreover, the flow field of those six opening patterns indicated a trend of moving downstream toward the outlet. A non-ideal velocity distribution at the outlet could severely affect the total air flow rate. It is known from Table 6 that Idea-B presented a very non-uniform velocity distribution. Most of the fluid aggregated to the left and right sides of the circular region. Reverse airflow and low air velocity can be found at the central region. The total air flow rate of the fan was not greatly affected by this phenomenon, and therefore this portion could be excluded.

The comparison of the resulting flow rate of the six opening patterns is shown in Figure 13. At the outlet of Idea-D with the honeycomb shape, the flow rate obtained from the numerical simulation was 18CFM. Similarly, the resulting flow rate of Idea-A was 16CFM. Figure 13 also reveals that the air flow rate of Idea-D was 2CFM larger than Idea-A. For the static pressure, the A4 model had the highest static pressure of 1.87 mm-H_2O. The result static pressure of Idea-A was 1.74 mm-H_2O. From the aspect of overall air flow rate, the simulation results indicated that Idea-D had the largest air flow rate.

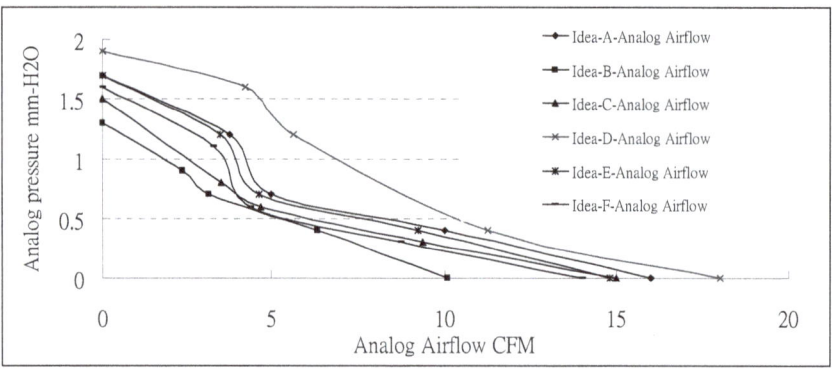

Figure 13. Comparison of air flow rate and static pressure of new fan designs by simulation.

6. Conclusions

In this study, the aerodynamic performance of axial fans was investigated to determine the effect of different opening patterns. As the air flow rate decreases with increasing pressure, the fan's aerodynamic performance is affected if the density of the opening pattern is increased, so the boundary layer is affected. The cross-sectional flow field of an axial fan was reviewed and investigated. The flow field could be improved by making the streamlines smoother. Some reverse flows were also observed at the outlet of the fan. Conclusions of the investigation of the flow field along the radial direction and at the outlet are as follows.

1. The distribution at the inlet can be smoother.
2. After passing through the inlet, the air pressure will increase at some portions about 1/2 of the impeller height.
3. The maximum velocity at each cross-sectional plane occurs closer to the outlet.
4. The change in the outlet location makes the air velocity increase and move toward the outlet direction.
5. At the outlet plane of Idea-B, C, and E, many regions were found to have a low air velocity and recirculation. This phenomenon indicates inferior outlet conditions.
6. Idea-D, with the honeycomb shape, had the most uniform air velocity among the six opening patterns. From the aspect of the leaving flow rate, it was also the most optimal opening pattern.
7. If a fan is not improved by the airfoil design, the assessment can be carried out on the design factors of the outlet pattern.

There are many design parameters when desigrning the opening patterns of a fan outlet. The analysis in this study investigated six factors in order to provide future fan designs with a good reference for the optimization of the opening pattern. The results indicate that the honeycomb-shaped opening pattern can be beneficial to the performance enhancement of an axial fan. The parameters of the inlet design can also be included for further investigation on design optimization.

Author Contributions: The author contributed to the paper. H.-H.L. Collects and organizes data and acts as the corresponding author, J.-H.C. and the authors propose methods.

Funding: This research received no external funding.

Conflicts of Interest: The authors declare no conflict of interest.

References

1. Marinetti, S.; Cavazzini, G.; Fedele, L.; De Zan, F.; Schiesaro, P. Air velocity distribution analysis in the air duct of a display cabinet by PIV technique. *Int. J. Refrig.* **2012**, *35*, 2321–2331. [CrossRef]
2. Gebrehiwot, M.G.; De Baerdemaeker, J.; Baelmans, M. Effect of a cross-flow opening on the performance of a centrifugal fan in a combine harvester: Computational and experimental study. *Biosyst. Eng.* **2010**, *105*, 247–256. [CrossRef]
3. Chen, F.; Li, S.; Su, J.; Wang, Z. Experimental Study of Bowed-twisted Stators in an Axial Transonic Fan Stage. *Chin. J. Aeronaut.* **2009**, *22*, 364–370.
4. Betta, V.; Cascetta, F.; Musto, M.; Rotondo, G. Fluid dynamic performances of traditional and alternative jet fans in tunnel longitudinal ventilation systems. *Tunn. Undergr. Space Technol.* **2010**, *25*, 415–422. [CrossRef]
5. Li, Y.; Liu, J.; Ouyang, H.; Du, Z.-H. Internal flow Mechanism and Experimental Research of low Pressure Axial fan with Forward-Skewed Blades. *J. Hydrodyn.* **2008**, *20*, 299–305. [CrossRef]
6. Delele, M.; De Moor, A.; Sonck, B.; Ramon, H.; Nicolaï, B.; Verboven, P. Modelling and Validation of the Air Flow generated by a Cross Flow Air Sprayer as affected by Travel Speed and Fan Speed. *Biosyst. Eng.* **2005**, *92*, 165–174. [CrossRef]
7. Gebrehiwot, M.G.; De Baerdemaeker, J.; Baelmans, M. Numerical and experimental study of a cross-flow fan for combine cleaning shoes. *Biosyst. Eng.* **2010**, *106*, 448–457. [CrossRef]
8. Chunxi, L.; Ling, W.S.; Yakui, J. The performance of a centrifugal fan with enlarged impeller. *Energy Convers. Manag.* **2011**, *52*, 2902–2910. [CrossRef]
9. Stafford, J.; Walsh, E.; Egan, V. The effect of global cross flows on the flow field and local heat transfer performance of miniature centrifugal fans. *Int. J. Heat Mass Transf.* **2012**, *55*, 1970–1985. [CrossRef]
10. Liu, S.H.; Huang, R.F.; Lin, C.A. Computational and experimental investigations of performance curve of an axial flow fan using downstream flow resistance method. *Exp. Therm. Fluid Sci.* **2010**, *34*, 827–837. [CrossRef]
11. Zhao, X.; Sun, J.; Zhang, Z. Prediction and measurement of axial flow fan aerodynamic and aeroacoustic performance in a split-type air-conditioner outdoor unit. *Int. J. Refrig.* **2013**, *36*, 1098–1108. [CrossRef]
12. Hu, B.-B.; Ouyang, H.; Wu, Y.-D.; Jin, G.-Y.; Qiang, X.-Q.; Du, Z.-H. Numerical prediction of the interaction noise radiated from an axial fan. *Appl. Acoust.* **2013**, *74*, 544–552. [CrossRef]
13. Owen, M.; Kröger, D.G. Contributors to increased fan inlet temperature at an air-cooled steam condenser. *Appl. Ther. Eng.* **2013**, *50*, 1149–1156. [CrossRef]
14. Moonen, P.; Blocken, B.; Roels, S.; Carmeliet, J. Numerical modeling of the flow conditions in a closed-circuit low-speed wind tunnel. *J. Wind Eng. Ind. Aerodyn.* **2006**, *94*, 699–723. [CrossRef]
15. Carolus, T.; Schneider, M.; Reese, H. Axial flow fan broad-band noise and prediction. *J. Sound Vib.* **2007**, *300*, 50–70. [CrossRef]
16. Bredell, J.; Kröger, D.; Thiart, G. Numerical investigation of fan performance in a forced draft air-cooled steam condenser. *Appl. Eng.* **2006**, *26*, 846–852. [CrossRef]
17. Hsiao, S.-W.; Lin, H.-H.; Lo, C.-H. A study of thermal comfort enhancement by the optimization of airflow induced by a ceiling fan. *J. Interdiscip. Math.* **2016**, *19*, 859–891. [CrossRef]
18. Hsiao, S.W.; Lin, H.H.; Lo, C.H.; Ko, Y.C. Automobile shape formation and simulation by a computer-aided systematic method. *Concurr. Eng. Res. Appl.* **2016**, *24*, 290–301. [CrossRef]

19. Lin, H.H.; Huang, Y.Y. Application of ergonomics to the design of suction fans. In Proceedings of the 1st IEEE International Conference on Knowledge Innovation and Invention, Jeju Island, South Korea, 23–27 July 2018; pp. 203–206.
20. Andrea, T. On the theoretical link between design parameters and performance in cross-flow fans. A numerical and experimental study. *Comput. Fluids* **2005**, *34*, 49–66.
21. Hurault, J.; Kouidri, S.; Bakir, F. Experimental investigations on the wall pressure measurement on the blade of axial flow fans. *Exp. Fluid Sci.* **2012**, *40*, 29–37. [CrossRef]
22. Lin, H.H.; Hsiao, S.W. A Study of the Evaluation of Products by Industrial Design Students. *Eurasia J. Math. Sci. Technol. Educ.* **2018**, *14*, 239–254. [CrossRef]
23. Shih, Y.-C.; Hou, H.-C.; Chiang, H. On similitude of the cross flow fan in a split-type air-conditioner. *Appl. Eng.* **2008**, *28*, 1853–1864. [CrossRef]
24. Kim, J.-H.; Hur, N.; Kim, W. Development of algorithm based on the coupling method with CFD and motor test results to predict performance and efficiency of a fuel cell air fan. *Renew. Energy* **2012**, *42*, 157–162. [CrossRef]
25. Jason, S.; Walsh, E.; Egan, V.; Grimes, R. Flat plate heat transfer with impinging axial fan flows. *Int. J. Heat Mass Transfer.* **2010**, *53*, 5629–5638.
26. Greenblatt, D.; Avraham, T.; Golan, M. Computer fan performance enhancement via acoustic perturbations. *Int. J. Heat Fluid Flow* **2012**, *34*, 28–35. [CrossRef]
27. Lin, H.-H. Improvement of Human Thermal Comfort by Optimizing the Airflow Induced by a Ceiling Fan. *Sustainability* **2019**, *11*, 3370. [CrossRef]
28. Lin, S.-C.; Tsai, M.-L. An integrated performance analysis for a backward-inclined centrifugal fan. *Comput. Fluids* **2012**, *56*, 24–38. [CrossRef]
29. Stinnes, W.H. Effect of cross-flow on the performance of air-cooled heat exchanger fans. *Appl. Ther. Eng.* **2002**, *22*, 1403–1415. [CrossRef]
30. Casarsa, L.; Giannattasio, P. Experimental study of the three-dimensional flow field in cross-flow fans. *Exp. Fluid Sci.* **2011**, *35*, 948–959. [CrossRef]
31. Liu, G.; Liu, M. Development of simplified in-situ fan curve measurement method using the manufacturers fan curve. *Build. Environ.* **2013**, *48*, 77–83. [CrossRef]
32. Hurault, J.; Kouidri, S.; Bakir, F.; Rey, R. Experimental and numerical study of the sweep effect on three-dimensional flow downstream of axial flow fans. *Flow Meas. Instrum.* **2010**, *21*, 155–165. [CrossRef]
33. Chen, T.Y.; Shu, H.T. Flow structures and heat transfer characteristics in fan flows with and without delta-wing vortex generators. *Exp. Ther. Fluid Sci.* **2004**, *28*, 273–282. [CrossRef]
34. Stafford, J.; Walsh, E.; Egan, V. Local heat transfer performance and exit flow characteristics of a miniature axial fan. *Int. J. Heat Fluid Flow* **2010**, *31*, 952–960. [CrossRef]

Article

Energy Consumption Load Forecasting Using a Level-Based Random Forest Classifier

Yu-Tung Chen [1], Eduardo Piedad Jr. [2] and Cheng-Chien Kuo [1,*]

[1]   Department of Electrical Engineering, National Taiwan University of Science and Technology, Taipei City 10607, Taiwan

[2]   Department of Electrical Engineering, University of San Jose-Recoletos, Cebu City 6000, Philippines

*   Correspondence: cckuo@mail.ntust.edu.tw; Tel.: +886-02-27333141 (ext. 7710)

Received: 1 June 2019; Accepted: 13 July 2019; Published: 29 July 2019

Abstract: Energy consumers may not know whether their next-hour forecasted load is either high or low based on the actual value predicted from their historical data. A conventional method of level prediction with a pattern recognition approach was performed by first predicting the actual numerical values using typical pattern-based regression models, hen classifying them into pattern levels (e.g., low, average, and high). A proposed prediction with pattern recognition scheme was developed to directly predict the desired levels using simpler classifier models without undergoing regression. The proposed pattern recognition classifier was compared to its regression method using a similar algorithm applied to a real-world energy dataset. A random forest (RF) algorithm which outperformed other widely used machine learning (ML) techniques in previous research was used in both methods. Both schemes used similar parameters for training and testing simulations. After 10-time cross training validation and five averaged repeated runs with random permutation per data splitting, the proposed classifier shows better computation speed and higher classification accuracy than the conventional method. However, when the number of its desired levels increases, its prediction accuracy seems to decrease and approaches the accuracy of the conventional method. The developed energy level prediction, which is computationally inexpensive and has a good classification performance, can serve as an alternative forecasting scheme.

Keywords: energy level consumption; pattern recognition; random forest; machine learning; load forecasting; level classification

---

1. Introduction

Energy load forecasting is becoming one of the latest trends due to advancements in energy and power systems and management. As a result, emerging techniques in the field of artificial intelligence (AI) have recently come into play. One particular study reviews various prediction techniques for energy consumption prediction in buildings [1]. Energy regression models are studied in [2]. Machine learning (ML) techniques such as artificial neural networks (ANNs) and support vector machines (SVMs) are employed to predict energy consumption and draw energy-saving mechanisms [3]. Another study reviews the use of a probabilistic approach in load forecasting [4]. Other studies analyze the effectiveness of AI tools applied in smart grid and commercial buildings [5–7]. Most of the studied AI tools focus primarily on actual value forecasting. For example, consumers may not know whether the next-hour forecasted load value based on these models is either high or low. The conventional way is to categorize the forecasted value into reasonable levels, such as low, average, or high, which consumers can understand. This study proposes an alternative method which can be applied when estimated levels instead of actual values are already sufficient for a load forecasting application.

Short-term forecasting of energy consumption load uses the most important historical data ranging from a few hours even up to a number of weeks before the forecasted day. Recently, short-term load forecasting research studies employed advance machine learning such as artificial neural networks [8], fuzzy logic algorithms and wavelet transform techniques integrated in a neural network system [9], and an extreme learning machine [10]. Studies on short-term forecasting also cover various settings such as residential [11], non-residential [12], and micro-grid [13] buildings.

In residential houses, a typical energy consumption forecasting is driven by data generated from humidity and temperature sensors [14]. Occupant behavior assessment can also predict building consumption [15]. A number of research papers study short- and long-term energy consumption both in residential and small commercial establishments. The emergence of algorithms and an increasing computational capability have encouraged the development of different prediction models. Stochastic models can reliably predict the energy consumption of buildings and identify areas of possible energy waste [15–17]. Standard engineering regression and a statistical approach still have good prediction results [1,7,14]. A combination with genetic programming is also effective [18]. Various machine learning tools such as support vector machines and neural network algorithms provide an acceptable energy prediction performance [19,20]. Random forest outperforms other widely used classifiers such as artificial neural networks and support vector machines in energy consumption forecasting [21].

2. Machine Learning Methodology

This section introduces the machine learning models and presents their implementation. This covers two parts—the pipeline and implementation of ML models, and the random forest classifier as the ML model used in this study.

2.1. Machine Learning Pipeline and Implementation

Figure 1 shows the typical implementation flow of machine learning (ML) algorithms. Two main stages of an ML algorithm are the training and testing phases. First, the training phase creates the ML model using a training dataset based on the chosen ML classifier models. The three most commonly known ML models, namely, artificial neural network (ANN), support vector machine (SVM), and random forest (RF), are employed. The performance validation of the training stage guarantees the general performance of the classifier model and is used to avoid the overfitting issue. Then, verification is performed on the trained model in the testing stage using the testing dataset as input to the trained classifier. This testing dataset is the other partitioned data from the original dataset; therefore, it has identical characteristics to the training dataset. The original dataset is partitioned into 70% and 30% for training and testing, respectively. Performance metrics are used to evaluate both stages. By comparing the performance of both training and testing, any overfitting issue can be determined. It occurs when the training performance is relatively higher than the testing performance.

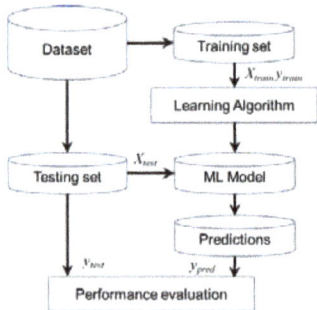

Figure 1. Pseudocode of a usual machine learning implementation with training and testing phases, and final evaluation stage.

Algorithm 1 presents the implemented program ML classifier similar to the pseudocode of [21]. A *k* number of times of cross-validation was performed. In this study, 10-time cross-validation was used. This cycle was repeated for another 10 times. The overall average performance and random data shuffling were taken. This verification helped avoid any overfitting issue. This was performed by taking the loss function of random forest present the difference between the training and testing results. For the evaluation of the conventional method, which is a regression-type problem, the root-mean-square error (RMSE) function in Equation (1) is used. F-score, classification accuracy, and confusion matrix are the metrics used for the proposed classifier. F-score accuracy metrics in Equation (2) weigh the significance of both precision and recall performance of the ML model. Precision measures its positive predictive value, whereas recall measures its sensitivity.

Algorithm 1. Machine Learning Implementation

\# Initialization

In the initialization stage, data pre-processing is performed such as the loading and shuffle-splitting of the dataset into feature X and predictor y, and the importation of the necessary python-based libraries.

\# Repeat n times the training and testing of the model

for i=1:n

Shuffle-splitting of dataset into training, validation, and testing datasets

\# k-time Training Cross-Validation

for j=1:k-time

Training of the model using an ML algorithm using the training dataset

Performance evaluation of the trained model using the validation dataset

\# Testing the model

Testing of the trained model using the testing dataset

Performance evaluation of the tested model

\# Display Performance Results

Compute classification accuracy and F-score

Compute classification confusion matrix

Another measure, classification accuracy in (3), was also taken. This metric is the percent of correctly classified levels over the total number of taken levels.

$$\text{RMSE} = \sqrt{\sum_{i=1}^{n} \frac{(w^T x(i) - y(i))^2}{n}} \tag{1}$$

$$\text{F score} = \frac{2*\text{Precision} * \text{Recall}}{\text{Precision} * \text{Recall}} \tag{2}$$

$$\text{Accuracy} = \frac{\text{No. of correctly classified energy levels}}{\text{Total number of classified energy levels}} \tag{3}$$

Finally, a confusion matrix normalized between zero and one helps visualize the classification performance of the model. scikit-learn in [22] is an open source platform that provides Python libraries and support. This was used to implement the three ML model classifiers—ANN, SVM, and RF.

2.2. Random Forest Classifier

A decision tree was used as the predictive model. The model predicts from the subject observations up to the model decision on which the subject's target value is based. The subject observations are also called branches while subject's target values are also known as leaves. Bagging is a technique of estimated prediction by reducing its variance which is suitable for decision trees [23]. For its regression application, a recursive fit of a similar regression tree was performed to produce bootstrap-sampled versions of training data taking its mean value. For classification, a predicted class was chosen by the majority vote from each committee of trees. Random forest (RF) is a modified bagging that produces a large collection of independent trees and averages their results [24]. Each of the trees generated from bagging is identically distributed, making it hard to improve other than achieving variance reduction. RF performs the tree-growing process by random input variable selection, thereby improving bagging by the correlation reduction between trees without an excessive variance increase.

3. Energy Data Processing

This section presents the implementation of the proposed machine learning classifier using a real energy dataset. The dataset was processed according to the state-of-the-art data class interval method. It was then compared with the conventional forecasting technique.

A 12-month energy dataset of [25] from a large hypermarket was used in this study. An hourly energy consumption collected via a smart metering device and hourly temperature records retrieved from meteorological sensors are shown in Figures 2 and 3, respectively. During the sunny days of the year between June to September, the energy load consumption is relatively high due to the prevalent use of air conditioning units in response to the high temperature.

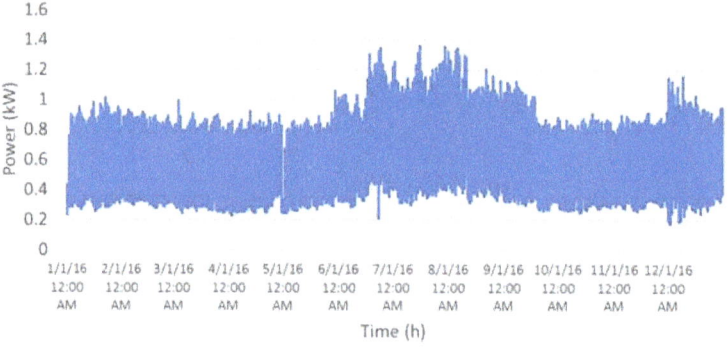

Figure 2. Whole-year time series energy consumption data of a commercial entity.

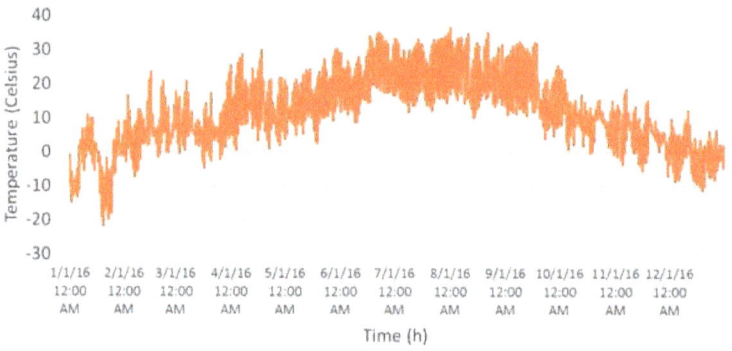

Figure 3. Whole-year time series temperature data of a commercial entity.

The conventional method and the new proposed scheme of predicting energy level are shown in Figure 4. The conventional method of energy level prediction is performed by first predicting the actual numerical values using typical regression models and then classifying them into consumer-preferred levels (e.g., low, average, and high). Since the regression model becomes computationally expensive as its model becomes more complicated, a proposed prediction scheme was developed to directly predict the desired levels using simpler classifier models without undergoing regression.

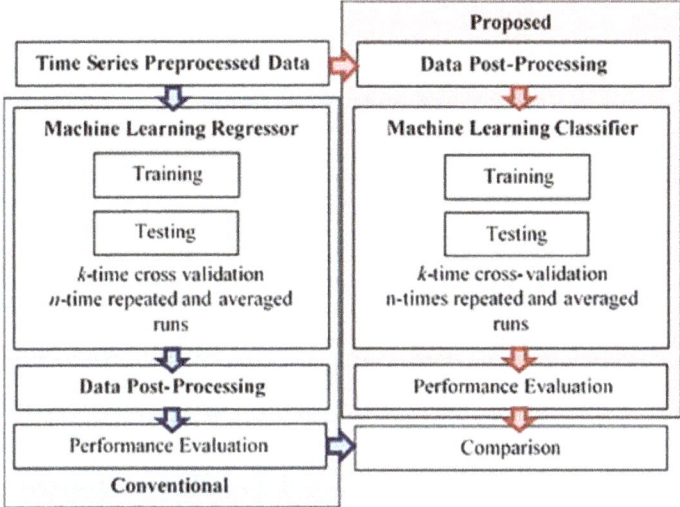

Figure 4. Methodology and comparison of the conventional and the proposed time series machine learning classifiers (source: authors' own conception).

In the proposed scheme, the energy consumption values are classified into ordinal bins using a general percentile statistic. Ordinal bin partitioning has an approximately equal number of data points as shown. For example, five bins representing five energy levels (very low, low, mid, high, very high) can be created using five even percentile ranges of the energy consumption data resulting in the [0.174, 0.366), [0.366, 0.634), [0.634, 0.782), [0.782, 0.874), [0.874, 1.36] energy value ranges, respectively. For prediction implementation, energy levels were converted into their respective ordinal values (1, 2, 3, 4, 5) for the machine learning implementation. The dataset contains three energy level cases—three, five and seven classes, as shown in Table 1. The modified dataset can be found in [26]. With these, three prediction cases were conducted using a machine learning random forest classifier.

Table 1. Three *n*-level cases for a real dataset.

n-Level Cases	Interval	Data Points
3-level	[0, 0.525)	2927
	[0.525, 0.807)	2917
	[0.807, 1.36]	2940
5-level	[0, 0.366)	1755
	[0.366, 0.634)	1756
	[0.634, 0.7816)	1759
	[0.7816, 0.874)	1752
	[0.874, 1.36]	1762
7-level	[0, 0.3366)	1255
	[0.3366, 0.427)	1249
	[0.427, 0.675)	1260
	[0.675, 0.771)	1236
	[0.771, 0.827)	1271
	[0.827, 0.935)	1257
	[0.935, 1.36]	1256

4. Results and Discussion

This section presents the implementation results of the previous proposed random forest classifier with the previous preprocessed energy data. The results were compared with those of the conventional forecasting classifier.

A brute-force simulation was performed to tune the hyperparameters of both the conventional classifier and the proposed random forest classifier. The training and testing loss function differences of both the conventional and the proposed classifiers are shown in Figure 5. Based on three-level cases, it can be observed that, most of the time, the proposed RF classifier has a lower train–test difference, indicating a better model performance to avoid overfitting compared to the conventional classifier. In addition, the proposed method tends to converge in less than 2% train–test loss function difference as the parameters increase, whereas the conventional method deviates. Furthermore, the average standard deviation on the classification accuracy of the proposed method is lower than the conventional one in all three cases, as shown in Table 2. Accordingly, the lower minimum and maximum standard deviations of the proposed method suggest a more precise prediction than the performance of the conventional method.

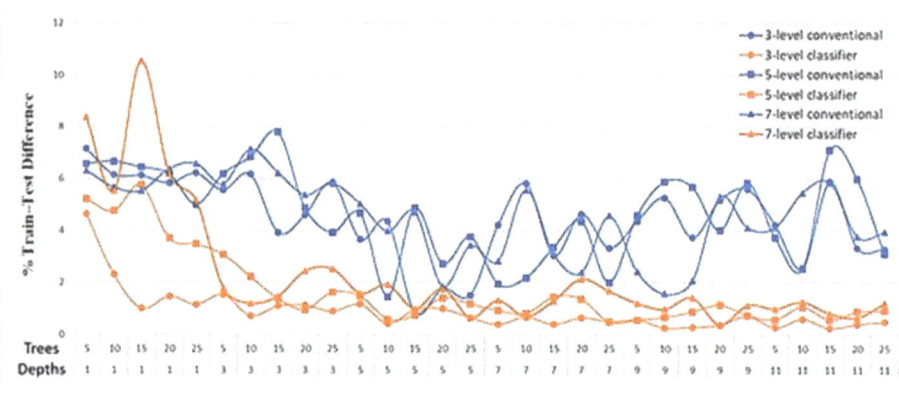

Figure 5. Training and testing difference from the loss function graph of the random forest (RF) classifiers.

Table 2. Classification performance of the conventional and proposed methods in three cases.

Classifier Models		Classification Accuracy		
		std_min	std_ave	std_max
3-level	conventional	0.0032	0.0106	0.9131
	proposed	0.0012	0.0048	0.0100
5-level	conventional	0.0024	0.0093	0.0206
	proposed	0.0023	0.0068	0.0148
7-level	conventional	0.0049	0.0100	0.0175
	proposed	0.0033	0.0070	0.0123

The proposed RF classifier tends to perform better with a lower number of energy levels and compared with the conventional method. Based on the F-score in Table 3, the proposed classifier deviates further as the number of levels increases. For example, seven-energy level prediction suggests two times deviation as compared with the three-energy level prediction.

Table 3. F-score performance of the proposed method in three cases.

Proposed	F Score			
	min	std_ave	max	std
3-level	0.0054	0.6491	3.3674	0.6388
5-level	0.0085	1.0993	7.1509	0.9413
7-level	0.0058	1.3333	5.2442	1.1107

Parameter simulations of both the conventional and the proposed classifiers in three cases are compared in Figure 6a–c. This was conducted to determine the classification performance and the execution time of both methods as the respective parameters become more complicated. Both classifiers seem to perform better with lower energy levels. Both classifiers converge to a classification accuracy around 90% in three-level prediction in Figure 6a, while reaching around 83% and 75% for five- and seven-level predictions in Figure 6b,c, respectively. However, the proposed classifier is observed to outperform the conventional classifier based on a higher classification accuracy performance and a lower execution time in all three cases. Initially, the execution time of the proposed model takes almost the same time as the conventional one using simpler parameters. With the increasing complexity of the parameters, the former does not change significantly, whereas the latter changes abruptly. It seems that this is due to the fact that the conventional method has a regression model structure which is more complicated than the classification model of the proposed method. The performance of the conventional method approaches that of the proposed method in terms of classification accuracy at the expense of computation time.

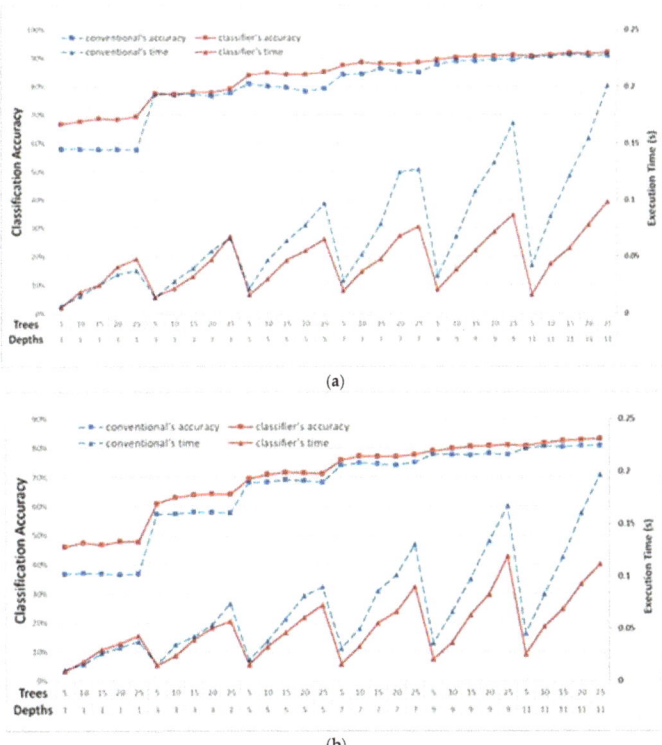

(a)

(b)

Figure 6. *Cont.*

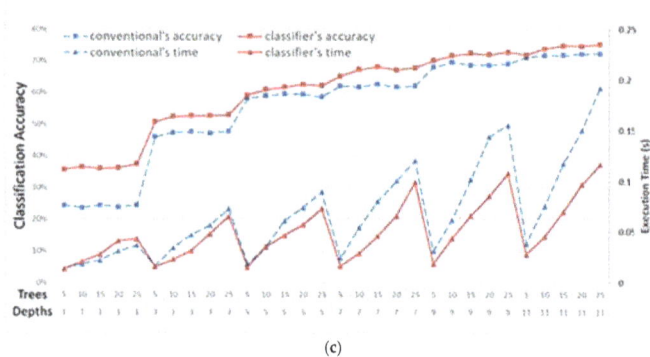

(c)

Figure 6. Parameter simulation of three cases, using (**a**) three-energy level, (**b**) five-energy level, and (**c**) seven-energy level prediction of both methods.

5. Conclusions

Energy level prediction was performed using a developed random forest classifier. Instead of undergoing regression-based load forecasting from the conventional method, the developed classifier preprocessed the numerical-valued data into levels and then later predicted them using a simpler classification process. Both classifiers perform better with a lower number of energy levels. The developed classifier outperforms the conventional classifier based on its classification accuracy and execution time when simulating 3, 5 and 7 level cases –. However, the performance of the conventional classifier approaches that of the proposed method in terms of classification accuracy but at the expense of computation time. The proposed random forest classifier serves as an alternative to regression-based problems not only for energy consumption forecasting but also for other similar applications. This study was limited to only a single real dataset. Further studies can use other real datasets.

Author Contributions: Conceptualization, C.-C.K.; Data curation, Y.-T.C.; Formal analysis, Y.-T.C. and E.P.J.; Investigation, E.P.J.; Methodology, E.P.J. and C.-C.K.; Project administration, C.-C.K.; Software, Y.-T.C.; Supervision, C.-C.K.

Conflicts of Interest: The authors declare no conflict of interest.

References

1. Zhao, H.X.; Magoulès, F. A review on the prediction of building energy consumption. *Renew. Sustain. Energy Rev.* **2012**, *16*, 3586–3592. [CrossRef]
2. Fumo, N.; Rafe Biswas, M.A. Regression analysis for prediction of residential energy consumption. *Renew. Sustain. Energy Rev.* **2015**, *47*, 332–343. [CrossRef]
3. Ahmad, A.S.; Hassan, M.Y.; Abdullah, M.P.; Rahman, H.A.; Hussin, F.; Abdullah, H.; Saidur, R. A review on applications of ANN and SVM for building electrical energy consumption forecasting. *Renew. Sustain. Energy Rev.* **2014**, *33*, 102–109. [CrossRef]
4. Hong, T.; Fan, S. Probabilistic electric load forecasting: A tutorial review. *Int. J. Forecast.* **2016**, *32*, 914–938. [CrossRef]
5. Raza, M.Q.; Khosravi, A. A review on artificial intelligence based load demand forecasting techniques for smart grid and buildings. *Renew. Sustain. Energy Rev.* **2015**, *50*, 1352–1372. [CrossRef]
6. Yildiz, B.; Bilbao, J.I.; Sproul, A.B. A review and analysis of regression and machine learning models on commercial building electricity load forecasting. *Renew. Sustain. Energy Rev.* **2017**, *73*, 1104–1122. [CrossRef]
7. Menezes, A.C.; Cripps, A.; Buswell, R.A.; Wright, J.; Bouchlaghem, D. Estimating the energy consumption and power demand of small power equipment in office buildings. *Energy Build.* **2014**, *75*, 199–209. [CrossRef]

8. Tsekouras, G.J.; Kanellos, F.D.; Mastorakis, N. Short term load forecasting in electric power systems with artificial neural networks. In *Computational Problems in Science and Engineering*; Springer: Berlin, Germany, 2015; pp. 19–58.

9. Chaturvedi, D.K.; Sinha, A.P.; Malik, O.P. Short term load forecast using fuzzy logic and wavelet transform integrated generalized neural network. *Int. J. Electr. Power Energy Syst.* **2015**, *67*, 230–237. [CrossRef]

10. Li, S.; Wang, P.; Goel, L. Short-term load forecasting by wavelet transform and evolutionary extreme learning machine. *Electr. Power Syst. Res.* **2015**, *122*, 96–103. [CrossRef]

11. Jain, R.K.; Smith, K.M.; Culligan, P.J.; Taylor, J.E. Forecasting energy consumption of multi-family residential buildings using support vector regression: Investigating the impact of temporal and spatial monitoring granularity on performance accuracy. *Appl. Energy* **2014**, *123*, 168–178. [CrossRef]

12. Massana, J.; Pous, C.; Burgas, L.; Melendez, J.; Colomer, J. Short-term load forecasting in a non-residential building contrasting models and attributes. *Energy Build.* **2015**, *92*, 322–330. [CrossRef]

13. Chitsaz, H.; Shaker, H.; Zareipour, H.; Wood, D.; Amjady, N. Short-term electricity load forecasting of buildings in microgrids. *Energy Build.* **2015**, *99*, 50–60. [CrossRef]

14. Candanedo, L.M.; Feldheim, V.; Deramaix, D. Data driven prediction models of energy use of appliances in a low-energy house. *Energy Build.* **2017**, *140*, 81–97. [CrossRef]

15. Virote, J.; Neves-Silva, R. Stochastic models for building energy prediction based on occupant behavior assessment. *Energy Build.* **2012**, *53*, 183–193. [CrossRef]

16. Oldewurtel, F.; Parisio, A.; Jones, C.N.; Morari, M.; Gyalistras, D.; Gwerder, M.; Stauch, V.; Lehmann, B.; Wirth, K. Energy efficient building climate control using Stochastic Model Predictive Control and weather predictions. In Proceedings of the 2010 American Control Conference, Baltimore, MD, USA, 30 June–2 July 2010; pp. 5100–5105.

17. Arghira, N.; Hawarah, L.; Ploix, S.; Jacomino, M. Prediction of appliances energy use in smart homes. *Energy* **2012**, *48*, 128–134. [CrossRef]

18. Castelli, M.; Trujillo, L.; Vanneschi, L. Prediction of energy performance of residential buildings: A genetic programming approach. *Energy Build.* **2015**, *102*, 67–74. [CrossRef]

19. Tsanas, A.; Xifara, A. Accurate quantitative estimation of energy performance of residential buildings using statistical machine learning tools. *Energy Build.* **2012**, *49*, 560–567. [CrossRef]

20. Li, K.; Su, H.; Chu, J. Forecasting building energy consumption using neural networks and hybrid neuro-fuzzy system: A comparative study. *Energy Build.* **2011**, *43*, 2893–2899. [CrossRef]

21. Chang, H.-C.; Kuo, C.-C.; Chen, Y.-T.; Wu, W.-B.; Piedad, E.J. Energy Consumption Level Prediction Based on Classification Approach with Machine Learning Technique. In Proceedings of the 4th World Congress on New Technologies (NewTech'18), Madrid, Spain, 19–21 August 2018; pp. 1–8.

22. Pedregosa, F.; Varoquaux, G.; Gramfort, A.; Michel, V.; Thirion, B.; Grisel, O.; Blondel, M.; Müller, A.; Nothman, J.; Louppe, G.; et al. Scikit-learn: Machine Learning in Python. *J. Mach. Learn. Res.* **2011**, *12*, 2825–2830.

23. Bickel, P.; Diggle, P.; Fienberg, S.; Gather, U.; Olkin, I.; Zeger, S. *Springer Series in Statistics*; Springer: New York, NY, USA, 2009.

24. Breiman, L. Random forests. *Mach. Learn.* **2001**, *45*, 5–32. [CrossRef]

25. Pîrjan, A.; Oprea, S.V.; Carutasu, G.; Petrosanu, D.M.; Bâra, A.; Coculescu, C. Devising hourly forecasting solutions regarding electricity consumption in the case of commercial center type consumers. *Energies* **2017**, *10*, 1727.

26. Piedad, E.J.; Kuo, C.-C. A 12-Month Data of Hourly Energy Consumption Levels from a Commercial-Type Consumer. Available online: https://data.mendeley.com/datasets/n85kwcgt7t/1/files/6cfc7434-315c-4a2d-8d8c-ce6a2bb80a01/energy_consumption_levels.csv?dl=1 (accessed on 25 June 2018).

 *symmetry*

Article

The Computer Course Correlation between Learning Satisfaction and Learning Effectiveness of Vocational College in Taiwan

Ru-Yan Chen [1] and Jih-Fu Tu [2,*]

1   Department of Computer Science and Information Engineering, Vanung University, ZhongliDist.,
    Taoyuan City 32061, Taiwan; cjchen088@gmail.com
2   Department of Industrial Management, St. John's University, Tamsui District, New Taipei 25135, Taiwan
*   Correspondence: tu@mail.sju.edu.tw; Tel.: +886-930038679

Received: 20 May 2019; Accepted: 10 June 2019; Published: 21 June 2019

Abstract: In this paper, we surveyed the influence of learn effectiveness in a computer course under the factors of learning attitude and learning problems for students in senior-high school. We followed the formula for a regression line as $R = A + BX + \varepsilon$ and simulated on SPSS platform with symmetry to obtained the results as follows: (1) In learning attitude, both the cognitive-level and behavior-level, are positively correlated with satisfaction. This means the students have cognitive-level and behavior-level more positively correlated with satisfaction in computer subjects and have a high degree of self-learning effectiveness. (2) In learning problems, the female students had higher learning effectiveness than male students, and the students who practiced on the computer on their own initiative long-term each week had higher learning effectiveness.

Keywords: learning attitude; computer subject; learning effectiveness

1. Introduction

Education should focus on cultivating professional skills and combining practice with practice to enable students to have the ability to work in response to the rapid changes in the current industrial structure. Therefore, in addition to paying attention to students' professional and practical abilities, students should also have professional abilities in order to improve their graduation employment rate and employment competitiveness [1].

The ultimate goal of school is to assist students in acquiring the skills of employment. Therefore, the content of the curriculum should complement the workplace and assist students in obtaining professional skills related to their work [2].

How to make students have the motivation to learn, and then generate interest in active learning is an important question. A teacher should understand the behavioral motivation of students in a timely manner while teaching [3], as well as the learning attitude (including cognition, emotion, and behavior) that students hold while studying computer courses, and the troubles encountered in learning.

We also analyze the "learning attitude" and "learning problems" of the learning effectiveness frame for high vocational students to understand the student's personal background variables for learning attitude, learning problems, and learning effectiveness, and the relationship between learning attitude, learning problems, and learning effectiveness [4,5].

This paper is based on the symmetry subject of computer courses with students of the Information Process Department and International Trade Department to survey student learning attitudes and learning problems in the subject of computers. This paper aims to analyze the learning effectiveness and related issues of computer courses for higher vocational students through the questionnaire approach.

The rest of this paper is structured as follows. Section 2 describes the design structure and progress of the research. Section 3 presents the implemented research. Section 4 documents symmetry regressing on analysis between learning attitude and learning effectiveness. The regression analysis of learning problems to learning effectiveness is described in Section 5. Finally, we conclude the paper in Section 6.

2. Research Design and Progress

In this paper, we use a frame work to measure learning attitudes and learning problems for analyzing the learning effectiveness influences on students studying a computer course. In order to implement the above framework, firstly, we used a questionnaire survey to search the symmetry relationship between variables, and statistical methods for analyzing empirical data and verification the hypothesis.

2.1. Research Object and QuestionnaireResponse

The research objects of this paper were aimed at the three degree students of the Information Process Department and the International Trade Department. We issued in total 219 questionnaires through on-site distribution to students. The number of responses of valid questionnaires was 196 (89.49% response rate).

2.2. Research Tools

The measurement tool was a questionnaire of "learning attitudes and learning problems for learning effectiveness influences of computer courses", which included the following four categories: learning attitude, learning problems, learning effectiveness, and personal basic information. They are described as follows.

(1) Personal basic information includes sex, age, department, and practice computer time every week.
(2) Learning effectiveness factors include the learning attitude frame and the learning problems frame. The learning attitude frame has three parts as cognitive, emotional, and behavioral. The learning problems frame has four parts: personal, family, school, and course content. The learning effectiveness frame has four parts: class schedule, teacher teaching, learning environment, and learning results.

2.3. Assessment Method of Questionnaire

The questionnaire used a scale from 1 to 4, with "4" indicating very much agree, "3" indicating agree, "2" indicating disagree, and "1" indicating very much disagree. For the questionnaire, we set some reverse test questions, which meant that the sampled students had the attention item content when answering "not applicable".

2.4. Questionnaire Pretest

The pretest questionnaire was processed by random to 25 third grade students of high vocational education school. The main purpose of this pretest was to test the internal consistency and consistency when answering the questionnaire questions. We found α coefficient with high reliability for values larger than 0.898. The results of α coefficient are listed as Table 1.

Table 1. Questionnaire Cronbach α reliability analysis.

Frame Classification	Numbers	Cronbach α Coefficient
Learning attitude-cognitive	8	0.877
Learning attitude-emotional	8	0.805
Learning attitude-behavioral	11	0.871
Learning problems frame	14	0.887
Learning effectiveness-class schedule	3	0.755
Learning effectiveness-teacher teaching	8	0.929
Learning effectiveness-learning environment	5	0.850
Learning effectiveness-learning results	4	0.791
Total Reliability	61	0.898

3. Research Implementation

The samples of this questionnaire were third grade students of the International Trade Department and Information Process Department. As for the initial issued samples, the number was 219 and the number of valid questionnaires was 196. The details of these are shown in Table 2.

Table 2. Questionnaire response information.

Items	Issues	Sort	Response Questionnaire	Ratio of Useful Questionnaire
International Trade Dept. A	45	valid invalid	43 2	95.56%
International Trade Dept. B	36	valid invalid	30 6	83.33%
Information Process Dept. A	34	valid invalid	30 4	88.24%
Information Process Dept. B	34	valid invalid	33 1	97.06%
Information Process Dept. C	35	valid invalid	33 2	94.29%
Information Process Dept. D	35	valid invalid	27 8	77.14%
Total	219	valid invalid	196 23	89.49%

3.1. Data Analysis and Results

After administering the response questionnaires, we deleted the invalid questionnaire, and then carried out statistics and analyzed the valid questionnaire using SPSS 15.0. The methods of statistical analysis are described as follows.

(1) Narrative statistics was used to organize and present the characteristics of the existing data using statistical description groups. The methods used in this paper were: average, standard deviation, number of times, and percentage to understand the situation of the subject [6].

(2) Variance analysis was used to test whether there was a difference in the average of the maternal group. If there was a significant difference in variance analysis, then carried out a T-test to confirm whether there was a difference in the average of each selected group. In this paper, single factor variance analysis was used to test whether the student's personal background variables had a significant level of learning attitude, learning problems, and learning effectiveness [7,8].

(3) Regression analysis (RA) was used as a symmetry statistical method for analyzing data, mainly to determine specific relationships between the survey data.

Regression analysis is a relation model that was used to create the symmetry dependent variable Y and independent variable X. In this paper, we exploited multi regression analysis to create the frame of learning attitude and learning problems to use as a pre-prediction model for learning effectiveness, and then survey the influence level between those two frames and learning effectiveness [9].

3.2. Learning Attitude and Learning Problems for Learning Effectiveness

For the learning attitude, if a higher score was obtained, the more positive the learning attitude was than if a lower score was obtained, and vices versa. The average score of the fourth-order scale was 2.5 points. However, for learning problems, if the average scale was lower than 2.5 points which indicated those had less learning and those who had higher learning.

For learning effectiveness, an average scores for each question higher than 2.5 indicated that the training effectiveness was slightly positive. We found that if the teacher patiently guided students through computer related issues, the average score was 3.01 points. This meant students had satiable faction of learning because the teacher was able to patiently teach the students and the students were more satisfied.

4. Regression Analysis between Learning Attitude and Learning Effectiveness

In this paper, we used the symmetry regression analysis (R), coefficient of determination (R^2), adjusted R^2, and F test. Regression analysis is a powerful statistical method that allows the examination of the relationship between two or more variables of interest. The formula for a regression line can be given as $R = A + BX + \varepsilon$, where A and B are coefficients (also are constants). X and R are the independent and dependent variables, and ε is the error term [10].

We also showed the p value, which referred to the probability that the statistical symmetry summary (such as the difference between the two groups of samples) was the same as the actual observation data, or even larger, in a probability model.

4.1. Learning Attitude and Learning Effectiveness under Class Schedule

Table 3 shows the regression analysis of computer subject for learning attitude to learning effectiveness under class schedule. For the class schedule relation, the cognitive coefficient of learning attitude to learning effectiveness was 0.578 and was significantly positively correlated. This meant that with students in computer courses, the learning attitude was higher and more positive, and the active students had more satisfaction with the class schedule.

Table 3. Attitude to class schedule regression analysis.

Item		Value		
Regression factor (R)		0.656		
R^2		0.430		
AdjustedR^2		0.421		
F-test		48.36		
Class schedule Learning attitude	Cognitive	Emotion	Behavior	
Coefficient	0.578	0.205	0.327	
p value	<0.000***	0.144	<0.01**	

For the class schedule correlation, the coefficient of learning attitude behavior to learning effectiveness was 0.327 and was also significantly positively correlated. This indicated that the computer course students had a higher number and more positive learning attitude, and the positive students were more satisfied with the class schedule.

The learning effectiveness correlation coefficient of the emotion of learning attitude was 0.205, but the p value was 0.144, which indicted that there was no significant difference. Inferred from the above, the possible reasons may be that the students were just beginning to experience the computer subject or if just contact with the course was not enough to influence class schedule satisfaction by the aspect of learning emotion.

4.2. Regression Analysis between Learning Attitude and Teacher Teaching

Table 4 shows the symmetry regression analysis of teacher teaching under the computer subject of learning attitude cognitive to learning effectiveness with a *p* value <0.000***, so the equation was valid. As seen in Table 4, we found the correlative coefficient between cognitive of learning attitude cognitive and teacher teaching of learning effectiveness was 0.287 and had a positive correlation ($P < 0.005**$).

Table 4. Subject of learning attitude to teacher teaching regression analysis.

Item	Value		
Regression factor (R)	0.586		
R^2	0.344		
AdjustedR^2	0.333		
F-test	33.58		
Teacher teaching　　Learning attitude	Cognitive	Emotion	Behavior
Coefficient	0.287	0.041	0.406
p value	<0.005**	0.717	<0.000***

Based on the above results, if the learning attitude of students in the computer course was higher and more positive, the positive students had higher satisfaction with teacher teaching.

Regression analysis of teacher teaching of the computer subject of learning attitude behavior to learning effectiveness had a *p* value <0.000***, so the equation was also valid. As seen in Table 4, the correlation coefficient between cognitive of learning attitude cognitive and teacher teaching of learning effectiveness was 0.406 and had appositive correlation. This indicated that the students' learning behavior in the computer course was higher and more positive, and the positive students were more satisfied with teacher teaching.

Finally, we conducted symmetry regression analysis of teacher teaching under the computer subject of learning attitude emotion to learning effectiveness. We found the correlation coefficient between emotion of learning attitude cognitive and teacher teaching of learning effectiveness was 0.041, but the *p* value was 0.717, which meant there was no significant difference. The possible reason was that students were beginning to understand the computer subject, and they had a preliminary feeling about learning emotion in the course. There were not enough factors to influence their satisfaction with teacher teaching.

4.3. Regression Analysis between Learning Attitude and Learning Environment

Table 5 shows the regression analysis of learning attitude to learning environment in the computer course. As seen in Table 5, we found the correlation coefficient between cognitive of learning attitude cognitive and learning environment of learning effectiveness to be 0.390, which was a positive correlation. This indicates that the computer course students' subject of learning attitude cognitive was higher and more positive, and the positive students had higher satisfaction with the learning environment.

The symmetry correlation coefficient between emotion of learning attitude cognitive and learning environment of learning effectiveness was 0.326, which was a positive correlation. This indicates that the computer course students' learning attitude emotion was higher and more positive, and the positive students were more satisfied with the learning environment.

Furthermore, the symmetry correlative coefficient between behavior of learning attitude cognitive and learning environment of learning effectiveness was 0.446 and had a positive correlation. This indicated that the computer course students' learning attitude behavior was higher and more positive, and the positive students had higher satisfaction with the learning environment.

Table 5. Subject of learning attitude to learning environment regression analysis.

Item	Value
Regression factor (R)	0.692
R^2	0.479
AdjustedR^2	0.470
F-test	58.86

Learning environment / Learning attitude	Cognitive	Emotional	Behavioral
Coefficient	0.390	0.326	0.446
p value	<0.001***	<0.01**	<0.000***

4.4. Regression Analysis between Learning Attitude and Learning Results

Table 6 shows the regression question for the computer subject of learning attitude to learning results. As seen in Table 6, we found the correlation coefficient between cognitive of learning attitude cognitive and learning environment of learning effectiveness to be 0.573 with a positive correlative. This indicates that the computer course students' learning attitude cognitive was higher and more positive, and the positive students had higher satisfaction with the learning results.

The correlative coefficient between emotions of learning attitude cognitive and learning environment of learning effectiveness was 0.126, but the p value was 0.717, which meant that there was no significant difference. The results may indicate that students were beginning to understand the computer subject, and they had a preliminary feeling about learning emotion in the course. There were not enough factors to influence their satisfaction with learning results.

Furthermore, the correlative coefficient between behavior of learning attitude cognitive and learning environment of learning effectiveness was 0.382 and was appositive correlation. This indicates the computer course students' learning attitude behaviors were higher and more positive, and the positive students had the higher satisfaction with the learning results.

Table 6. Subject of learning attitude to learning results regression analysis.

Item	Value
Regression factor (R)	0.655
R^2	0.429
AdjustedR^2	0.420

Learning results / Learning attitude	Cognitive	Emotional	Behavioral
Coefficient	0.573	0.126	0.382
p value	<0.000***	0.360	<0.000***

5. Regression Analysis of Learning Problems to Learning Effectiveness

5.1. Regression Analysis of Class Schedule between Learning Problems and Learning Effectiveness

As seen in Table 7, which shows the regression question between the computer subject of learning problems and class schedule, we also found the p value <0.000***, which indicated the regression question was valid. We also found the correlation coefficient between learning problems and learning effectiveness of class schedule was –0.479, which was a negative correlation. This represents the lower degree of computer subject learning problems of student with higher satisfaction in the class schedule.

Table 7. Computer subject of learning problems to class schedule regression analysis.

Item	Value
Regression factor (R)	0.378
R^2	0.143
AdjustedR^2	0.138
F-test	32.442
Coefficient	−0.479

5.2. Regression Analysis between Learning Problems and Teacher Teaching

As seen Table 8, which shows the regression question between the computer subject of learning problems and teacher teaching, we also found the p value <0.000***, which indicated the regression question was valid. We also found the correlation coefficient between learning problems and learning effectiveness was −0.309, which was a negative correlation. This represents the lower the degree of computer subject learning problems of student with higher satisfaction in teacher teaching.

Table 8. Computer subject of learning problems to teacher teaching.

Item	Value
Regression factor (R)	0.319
R^2	0.102
AdjustedR^2	0.097
F-test	22.129
Coefficient	−0.309

5.3. Regression Analysis between Learning Problems and Learning Environment

As seen in Table 9, which shows the regression question between the computer subject of learning problems and learning environment, we also found the p value <0.000***, which indicated the regression question was valid. We also found the correlation coefficient between learning problems and learning effectiveness was −0.510, which was a negative correlation. This represents the lower the degree of computer subject learning problems of students with higher satisfaction in the learning environment.

Table 9. Computer subject of learning problems to learning environment.

Item	Value
Regression factor (R)	0.408
R^2	0.166
AdjustedR^2	0.162
F-test	38.777
Coefficient	-0.510

5.4. Regression Analysis of Learning Problems to Learning Results

In Table 10, which shows the regression question (=0.465) between the computer subject of learning problems and learning results, we also found the p value <0.000***, which indicated the regression question was valid. We also found a negative correlation coefficient between learning problems and learning effectiveness of −0.580. It represents the lower degree of computer subject learning problems of student with higher satisfaction in learning results.

Table 10. Computer subject of learning problems to learning results regression analysis.

Item	Value
Regression factor (R)	0.465
R^2	0.216
AdjustedR^2	0.212
F-test	53.64
Coefficient	−0.580

6. Conclusions

In this paper, symmetry we surveyed a computer course of learning attitudes and learning problems to learning effectiveness as influences. We summarized the results as follows.

1. For students in computer course, learning attitude to learning effectiveness had a higher difference. However, for the level of cognitive and behavioral with positive correlative satisfaction results showed students with more positive learning attitude cognitive and learning behavior than with higher satisfaction of class schedule, teacher teaching, teaching environment, and learning results.
2. For the emotion of learning attitude to learning effectiveness, there are significant differences in learning environment. The most students had more approval in the learning environment in emotion.
3. For the computer course analysis of learning problems to learning effectiveness, the learning problems to learning effectiveness symmetry as class schedule, teacher teaching, teaching environment, and learning had results with negative correlations. If they have higher satisfaction of class schedule, teacher teaching learning environment, and learning results.

Author Contributions: The authors contributed equally to the conception of the idea, implementing and analyzing the experimental results, and writing the manuscript.

Funding: This research received no external funding.

Conflicts of Interest: The authors declare no conflict of interest.

References

1. Campos, A.M. Analyzing the Effectiveness of Learning Objects and Designs. In Proceedings of the 2011 IEEE 11th International Conference on Advanced Learning Technologies, Athens, GA, USA, 6–8 July 2011; pp. 650–651.
2. Chatzara, K.; Karagiannidis, C.; Stamatis, D. Students Attitude and Learning Effectiveness of Emotional Agents. In Proceedings of the 2010 10th IEEE International Conference on Advanced Learning Technologies, Sousse, Tunisia, 5–7 July 2010; pp. 558–559.
3. Yang, Y.; Wang, Y.; Yuan, X. Bidirectional Extreme Learning Machine for Regression Problem and Its Learning Effectiveness. *IEEE Trans. Neural Netw. Learn. Syst.* **2012**, *23*, 1498–1505. [CrossRef] [PubMed]
4. Mishra, N.R.; Chavhan, R.K. Effectiveness of mobile learning on awareness about learning disability among student teachers. In Proceedings of the 2012 IEEE International Conference on Technology Enhanced Education (ICTEE), Kerala, India, 3–5 January 2012; pp. 1–6.
5. Krikun, I. Applying learning analytics methods to enhance learning quality and effectiveness in virtual learning environments. In Proceedings of the 2017 5th IEEE Workshop on Advances in Information, Electronic and Electrical Engineering (AIEEE), Riga, Latvia, 24–25 November 2017; pp. 1–6.
6. Shadiev, R.; Hwang, W.; Huang, Y.; Liu, A. Cognitive Diffusion Model: Facilitating EFL Learning in an Authentic Environment. 2017. Available online: https://ieeexplore.ieee.org/abstract/document/7497494 (accessed on 17 June 2019).
7. Joseph, N.; Pradeesh, N.; Chatterjee, S.; Bijlani, K. A novel approach for group formation in collaborative learning using learner preferences. In Proceedings of the 2017 International Conference on Advances in Computing, Communications and Informatics (ICACCI), Karnataka, India, 13–16 September 2017; pp. 1564–1568.

8.   Boicu, C.; Tecuci, G.; Boicu, M. Learning complex problem solving expertise from failures. In Proceedings of the Sixth International Conference on Machine Learning and Applications (ICMLA 2007), Cincinnati, OH, USA, 13–15 December 2007.
9.   Dehghani-Pilehvarani, A.; Karimaghaee, P.; Khayatian, A. Combined gradient and Iterative Learning Control method for magnetostatic inverse problem. In Proceedings of the 3rd International Conference on Control, Instrumentation, and Automation, Tehran, Iran, 28–30 December 2013; pp. 334–339.
10. Krishnan, M.; Muhammad, R.; Ruhizan, Y. Problem based learning in Engineering Education at Malaysian polytechnics: A proposal. In Proceedings of the 2009 International Conference on Engineering Education (ICEED), Kuala Lumpur, Malaysia, 7–8 December 2009; pp. 122–124.

Review

Locality Sensitive Discriminative Unsupervised Dimensionality Reduction

Yun-Long Gao [1], Si-Zhe Luo [1], Zhi-Hao Wang [1], Chih-Cheng Chen [2],* and Jin-Yan Pan [2],*

1   Department of Automation, Xiamen University, Xiamen 361005, China
2   School of Information Engineering, Jimei University, Xiamen 361021, China
*   Correspondence: 201761000018l@jmu.edu.cn (C.-C.C.); jypan@jmu.edu.tw (J.-Y.P.)

Received: 15 July 2019; Accepted: 7 August 2019; Published: 12 August 2019

Abstract: Graph-based embedding methods receive much attention due to the use of graph and manifold information. However, conventional graph-based embedding methods may not always be effective if the data have high dimensions and have complex distributions. First, the similarity matrix only considers local distance measurement in the original space, which cannot reflect a wide variety of data structures. Second, separation of graph construction and dimensionality reduction leads to the similarity matrix not being fully relied on because the original data usually contain lots of noise samples and features. In this paper, we address these problems by constructing two adjacency graphs to stand for the original structure featuring similarity and diversity of the data, and then impose a rank constraint on the corresponding Laplacian matrix to build a novel adaptive graph learning method, namely locality sensitive discriminative unsupervised dimensionality reduction (LSDUDR). As a result, the learned graph shows a clear block diagonal structure so that the clustering structure of data can be preserved. Experimental results on synthetic datasets and real-world benchmark data sets demonstrate the effectiveness of our approach.

Keywords: machine learning; graph embedding method; dimensionality reduction; diversity learning; adaptive neighbors

1. Introduction

Due to the large number of data generated by the advancements of science and technology, dimensionality reduction has become an important task in data mining and machine learning research with many applications [1–4]. These data have such characteristics as high dimensionality, nonlinearity, and extreme complexity, which bring a lot of problems to the subsequent data processing. However, the intrinsic structure of data are often suspected to be much lower due to the redundant information hidden in the original space [5]. Therefore, revealing the potential low-dimensional representation involved in the corresponding high-dimensional structure is an essential preprocessing step for various applications. Under the background, a lot of supervised and unsupervised dimensionality reduction methods are proposed, such as principal component analysis (PCA) [6], linear discriminant analysis (LDA) [7], Laplacian embedding(LE) [8–10], local linear embedding (LLE) [11], locality preserving projections (LPP) [12], neighborhood minmax projections (NMMP) [13], isometric feature mapping (IsoMAP) [14], discriminant sparsity neighborhood preserving embedding (DSNPE) [15], and multiple empirical kernel learning with locality preserving constraint (MEKL-LPC) [16], etc. Obviously, the unsupervised dimensionality reduction method is more challenging than other methods due to the lack of label information. Among them, the graph embedding method exhibits significant performance because it captures the structural information of high-dimensional space. The graph embedding method is built on the basis of manifold assumption, which means the data are formed according to a certain manifold structure and the nearby data points tend to have the same labels.

The commonly used graph-based algorithms, such as LPP [12], IsoMAP [14], local graph based correlation clustering (LGBACC) [17], and locality weighted sparse representation (LWSR) [18] generally have the same steps—for instance, (1) build adjacency graph for each neighborhood; (2) construct pairwise feature (similarity) for each neighborhood to describe the intrinsic manifold structure; and (3) convert the problem into an eigenvalue problem. Thus, we can find the traditional graph-based algorithms mentioned above are all established independently of the subsequent processes, i.e., cluster indicators need to be extracted through post-processing, such dimensionality reduction results are highly dependent on the input pairwise feature matrix [19]. For graph-based algorithms taken, only local distances' account in the original space cannot adequately eliminate noise and capture the underlying manifold structure [20], in that it is an insufficient description for data similarity. Moreover, it is usually difficult to explicitly capture the intrinsic structure of data only by using pairwise data during the graph construction process [21]. In fact, for pairwise data, the similarity is dependent on the adjacency graph constructed by a pair of data individually, without consideration for the local environment of pairwise data. It can be seen from Figure 1, though the distance between A and B is shorter than that between A and C in the original space, and, clearly, $S(A, B)$ is called a similarity, one pairwise feature is bigger than $S(A, C)$, hence point A and point C should be sorted out to one class, and B to the other class. However, point A and point C could get more similar in regular classification or clustering tasks because there exists a dense distribution of many points which link A and C, resulting from a big gap between A and B, which are regarded as less similar in some traditional methods with two more manifold and consequently divided into different class. Therefore, the traditional definition of similarity does not sufficiently describe the structure.

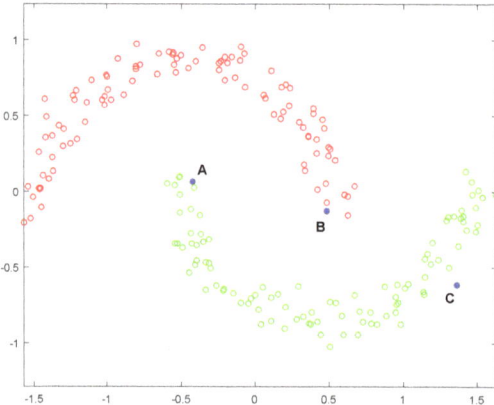

Figure 1. A data point map (point A and point B are closer, but point A and point C get bigger similarity in two more manifold structures.)

In recent years, there has been a lot of research devoted to solving these problems. For example, the constrained Laplacian rank (CLR) method [22] learns a block diagonal similarity matrix so that the clustering indicators can be immediately extracted. For Cauchy graph embedding [23], a new objective is proposed to preserve the similarity between the original data for the embedded space, and emphasize the closer two nodes in the embedding space, those that are more similar. Projected clustering with adaptive neighbors (PCAN) [24] designs a similarity matrix and is assigned adaptive and optimal neighbors to every piece of data on the basis of the local distances to learn instead of learning a probabilistic affinity matrix before dimensionality reduction. Stable semi-supervised discriminant learning (SSDL) [25] is worked out to learn the intrinsic structure of a constructed adjacency graphs which could extract the local topology characteristics, and get the geometrical properties as well. Nonetheless, these methods only focus on parts of the problems mentioned above, and the challenge in reasonably representing underlying data structure or adaptively adjusting the similarity graph still exists. As a consequence, it is quite necessary and challenging to develop an algorithm to address these problems.

In this paper, we propose a novel adaptive graph learning method, namely locality sensitive discriminative unsupervised dimensionality reduction (LSDUDR), which aims to uncover the intrinsic topology structures of data by proposing two objective functions. In the first step, one of the objective functions is aimed at guaranteeing the mapping of all points close to each other in the subspace, while the other one is with the purpose of excluding points with a large distance from the subspace. Furthermore, a data similarity matrix is learned to adaptively adjust the initial input data graph according to the basis of the projected local distances, that is to say, we adjust the projection jointly with graph learning. Moreover, we constrain the similarity matrix by imposing a rank constraint to make it contain more explicit data structure information. It is worthwhile to emphasize the main contributions of our method: (1) LSDUDR can construct a discriminative linear embedded representation that can deal with high-dimensional data and characterize the intrinsic geometrical structure among data; (2) compared with traditional two-stage graph embedding methods, which need an independent affinity graph to be constructed in advance in LSDUDR, and a clustering-oriented graph can be learned and the clustering indicators are extracted with no post-processing needed for the graph; (3) comprehensive experiments were performed on both synthetic data sets and real world benchmark data sets and better effectiveness of the proposed LSDUDR was demonstrated.

2. Related Work

2.1. Principal Component Analysis (PCA)

PCA is one of the most representative unsupervised dimensionality reduction methods. The main idea of PCA is to seek a projection transformation to maximise the variance of data. Assume that we have a data matrix $X \in R^{d \times n}$, where $x_i \in R^{d \times 1}$ denotes the i-th sample. For better generality, the samples in the data set are centralized, i.e., $\sum_{i=1}^{n} x_i = 0$. PCA aims to solve the following problem:

$$\max_{W^T W = I} \sum_{i,j=1}^{n} \left\| W^T x_i - W^T x_j \right\|_2^2, \tag{1}$$

where $W \in R^{d \times m}$ is the projection matrix, and m is the dimensionality of the linear subspace. When data points lie in a low-dimensional manifold and the manifold is linear or nearly-linear, the low-dimensional structure of data can be effectively captured by a linear subspace spanned by the principal PCA directions; the property provides a basis for utilizing the global scatter of samples as regularization in many applications.

2.2. Locality Preserving Projections (LPP)

LPP is very popular to substitute algorithms in linear manifold learning in which the data are projected responding to the direction of maximal variance, and the adjacent graph was employed to extract the structure properties of high-dimensional data, and structure properties were transplanted into low-dimensional subspace. The objective function of LPP is

$$\min_{W^T W = I} \sum_{i,j=1}^{n} \left\| W^T x_i - W^T x_j \right\|_2^2 s_{ij}, \tag{2}$$

where s_{ij} is defined as the similarity between samples x_i and x_j. As we can see, LPP is a linear version of Laplacian Eigenmaps that uses linear model approximation to nonlinear dimensionality reduction, Thus, it shares many of the data representation properties of nonlinear techniques such as Laplacian eigenmaps or locally linear embedding.

2.3. Clustering and Projected Clustering with Adaptive Neighbors (PCAN)

The PCAN algorithm performs subspace learning and clustering simultaneously instead of learning an initial pairwise feature matrix that is constructed before dimensionality reduction. The goal of PCAN is to assign the optimal and adaptive neighbors for each data point according to the local distances so that it can learn a new data similarity matrix. Therefore, it can be used as a clustering method, and can also be used as a unsupervised dimensionality reduction method. Denote the total scatter matrix by $S_t = XHX^T$, where H is the centering matrix defined as $H = I - \frac{1}{n}\mathbf{1}\mathbf{1}^T$, and $\mathbf{1}$ is a column vector whose elements are all 1. PCAN constrains the subspace with $W^T S_t W = I$ so that the data in the subspace has no statistical correlation. In [24], the definition of PCAN is:

$$\min_{S,W} \sum_{i,j=1}^{n} \left(\|W^T x_i - W^T x_j\|_2^2 s_{ij} + \theta s_{ij}^2 \right)$$
$$s.t. \forall i, \mathbf{s}_i^T \mathbf{1} = 1, 0 \le s_{ij} \le 1, W^T S_t W = I,$$
$$rank(L) = n - c,$$

(3)

where $L = D - (S + S^T)/2$ is called Laplacian matrix in graph theory, and the i-th diagonal element of the degree matrix $D \in R^{n \times n}$ is $\sum_j (s_{ij} + s_{ji})/2$. Then, by assigning the adaptive neighbors according to the local distances, the neighbors assignment divides the data points into c clusters based on the learned similarity matrix S, which can be directly used for clustering without having to perform other post-procedures.

3. Locality Sensitive Discriminative Unsupervised Dimensionality Reduction

3.1. Intrinsic Structure Representation

The proposed method needs a pre-defined affinity matrix S as the initial graph. While learning the affinity values of S, we get a smaller distance by adopting the the square of Euclidean distance $\|x_i - x_j\|_2^2$, which is related to a larger affinity value s_{ij}. Thus, determining the value of s_{ij} can be seen as solving the following problem:

$$\min_{\mathbf{s}_i^T \mathbf{1}=1, s_i \ge 0, s_{ii}=0} \sum_{j=1}^{n} \left(\|x_i - x_j\|_2^2 s_{ij} + \theta s_{ij}^2 \right),$$

(4)

where θ is the regularization parameter. The affinities are learned using a suitable θ in formula (4) so that we can get the optimal solution s_i with k nonzero values, i.e., the number of neighbors k. Let us define $e_{ij} = \|x_i - x_j\|_2^2$ and denote e_i as a vector and e_{ij} as j-th element; formula (4) can be simplified as

$$\min_{\mathbf{s}_i^T \mathbf{1}=1, s_i \ge 0, s_{ii}=0} \frac{1}{2} \left\| \mathbf{s}_i + \frac{1}{2\theta} e_i \right\|_2^2 .$$

(5)

According to [22], we can get the optimal affinities $\hat{s}_{ij}$ as follows:

$$\hat{s}_{ij} = \begin{cases} \frac{e_{i,k+1} - e_{ij}}{k e_{i,k+1} - \sum_{h=1}^{k} e_{ih}}, j \le k, \\ 0, j > k. \end{cases}$$

(6)

Next, we define two adjacency graphs $M_s = \{X, S\}$ and $M_d = \{X, V\}$ in order to characterize the intrinsic structure of data. Among them, the elements in matrix S represent the similarity between nearby points, and the elements in matrix V represent the diversity between nearby points. We define the elements v_{ij} in V as follows:

$$v_{ij} = \begin{cases} 1 - s_{ij}, j \le k, \\ 0, j > k. \end{cases}$$

(7)

Following the above work, we still did not get a clear and simple intrinsic structure if only using similarity or diversity. Therefore, two objective functions simultaneously proposed to emphasize the local intrinsic structure. One objective function is proposed to guarantee that nearby data points should be embedded to be close to each other in the subspace and mainly focuses on preserving the similarity relationships among nearby data; the other objective function mainly focuses on the shape of a manifold and guarantees that nearby data with large distance are not embedding to be very close to each other in the subspace and effectively preserves the diversity relationships of data. By integrating this two objective functions, the local topology are guaranteed, that is to say, similarity property and diversity property of the data can be perfectly preserved. Based on the above conclusions, we employ the following objective functions to capture the local intrinsic structure:

$$\min_{W^T W=I} \sum_{i,j=1}^{n} \left\| W^T x_i - W^T x_j \right\|_2^2 s_{ij}, \tag{8}$$

$$\max_{W^T W=I} \sum_{i,j=1}^{n} \left\| W^T x_i - W^T x_j \right\|_2^2 v_{ij}. \tag{9}$$

By simple algebra, we have:

$$\sum_{i,j=1}^{n} \left\| W^T x_i - W^T x_j \right\|_2^2 s_{ij} = tr\left(W^T X L_S X^T W \right), \tag{10}$$

$$\sum_{i,j=1}^{n} \left\| W^T x_i - W^T x_j \right\|_2^2 v_{ij} = tr\left(W^T X L_V X^T W \right), \tag{11}$$

where $L_S = D - (S + S^T)/2$ and $L_V = P - (V + V^T)/2$, $P \in R^{n \times n}$ is a diagonal matrix and its entries are column sum of V. Furthermore, in order to consider the global geometric structure information of data, we introduce the third objective function, i.e., preserving as much information as possible by maximizing overall variance of the input data. Then, inspired by LDA, we can construct a concise discriminant criterion by combining the three objective functions, which contain both local and global geometrical structures information for dimensionality reduction:

$$\min_{W} \frac{tr\left(W^T X \left(L_S - \beta L_V \right) X^T W \right)}{tr\left(W^T X H X^T W \right)}, \tag{12}$$
$$s.t. W^T W = I.$$

Bringing the definitions of L_S and L_V into Equation (12), we have:

$$tr\left(W^T X \left(L_S - \beta L_V \right) X^T W \right) = \sum_{i,j=1}^{n} \left\| W^T x_i - W^T x_j \right\|_2^2 s_{ij} - \beta \sum_{i,j=1}^{n} \left\| W^T x_i - W^T x_j \right\|_2^2 v_{ij}$$
$$= (1 + \beta) \sum_{i,j=1}^{n} \left\| W^T x_i - W^T x_j \right\|_2^2 s_{ij} - \beta \sum_{i=1}^{n} \sum_{j=1}^{k} \left\| W^T x_i - W^T x_j \right\|_2^2. \tag{13}$$

According to the definition of s_{ij}, when $j > k$, we have $s_{ij} = 0$. Therefore, $\sum_{i,j=1}^{n} \left\| W^T x_i - W^T x_j \right\|_2^2 s_{ij}$ models the local geometric structure, while $\sum_{i=1}^{n} \sum_{j=1}^{k} \left\| W^T x_i - W^T x_j \right\|_2^2$ represents the total scatter in the local region. Thus, we call this model locality sensitive discriminative unsupervised dimensionality reduction.

3.2. Analysis of Optimal Graph Learning

When the data contain a large number of noise samples, the similarity matrix S obtained by Equation (6) is virtually impossible to be the ideal state. The desired situation is that we map the data to a low-dimensional subspace in which the elements of similarity matrix within a cluster is nonzero and evenly distributed while the values of elements between clusters are zero. Based the above considerations, we adopt a novel and feasible way to achieve the desired state:

$$\min_{W,S} \frac{tr\left(W^T X\left(L_S - \beta L_V\right) X^T W\right)}{tr\left(W^T X H X^T W\right)} + \theta \left\|S\right\|_F^2 \tag{14}$$
$$s.t. \forall i, \mathbf{s}_i^T \mathbf{1} = 1, 0 \leq s_{ij} \leq 1, W^T W = I, rank(L_S) = n - c.$$

In order to exclude the situation of trivial solution, we add the regularization term $\theta \left\|S\right\|_F^2$. The first and second constraints are added according to the definition of graph weights, which is defined for a vertex as the sum of the distance between one vertex and the members and is non-negative. In addition, we also add the rank constraint to the problem. If S is non-negative, the Laplacian matrix has a significant property:

Theorem 1. *A graph S with $s_{ij} \geq 0 (\forall i, j)$ has c connected components if and only if the algebraic multiplicity of eigenvalue 0 for the corresponding Laplacian matrix L_S is c [26].*

Theorem 1 reveals that, when $rank(L_S) = n - c$, the obtained graph could distinctly divide the data set into exactly c clusters based on the block diagonal structure of similarity matrix S. It is worth mentioning that Equation (14) can simultaneously learn the projection matrix W and the similarity matrix S, which is significantly different from previous works. However, it is hard to tackle it directly, especially when there are several strict constraints. In order to solve the question, an iterative optimization algorithm is proposed.

4. Optimization

4.1. Determine the Value of S, W, F

Without loss of generality, suppose $\sigma_i(L_S)$ is the i-th smallest eigenvalue of L_S. It is clearly seen that $\sigma_i(L_S) \geq 0$ since L_S is positive semi-definite. Then, if λ is big enough, Equation (14) can be rewritten as:

$$\min_{W,S} \frac{tr\left(W^T X\left(L_S - \beta L_V\right) X^T W\right)}{tr\left(W^T X H X^T W\right)} + \theta \left\|S\right\|_F^2 + 2\lambda \sum_{i=1}^{c} \sigma_i\left(L_S\right) \tag{15}$$
$$s.t. \forall i, \mathbf{s}_i^T \mathbf{1} = 1, 0 \leq s_{ij} \leq 1, W^T W = I.$$

Hyperparameter λ here can be used to trade balance between the rank of the graph Laplacian and consistency of the data structure. The rank constraint of the graph Laplacian is usually satisfied with a large enough λ. Meanwhile, given a rank-enforcing matrix $F \in R^{n \times c}$, suppose that node i is assigned a function value as $f_i \in R^{1 \times c}$. According to the Ky Fan's Theorem [27], the rank constraint term in Equation (15) can be seen as the optimization of the smallest c eigenvalues of the Laplacian matrix. Thus, we can transform Equation (15) into the following form:

$$\min_{W,S,F} \frac{tr\left(W^T X\left(L_S - \beta L_V\right) X^T W\right)}{tr\left(W^T X H X^T W\right)} + \theta \left\|S\right\|_F^2 + 2\lambda tr\left(F^T L_S F\right), \tag{16}$$
$$s.t. \forall i, \mathbf{s}_i^T \mathbf{1} = 1, 0 \leq s_i \leq 1, W^T W = I, F^T F = I.$$

When S and F are fixed, problem (16) can be rewritten as:

$$\min_{W} \frac{tr\left(W^T X\left(L_S - \beta L_V\right) X^T W\right)}{tr\left(W^T X H X^T W\right)} \tag{17}$$
$$s.t. W^T W = I.$$

We can use the iterative method introduced in [28] to solve W from Equation (17), and the Lagrangian function is constructed according to Equation (17):

$$L(W, \eta) = \frac{tr\left(W^T X (L_S - \beta L_V) X^T W\right)}{tr\left(W^T X H X^T W\right)} - \eta tr\left(W^T W - I\right),\tag{18}$$

where η is a scalar. Then, taking the derivative of W and letting the result be zero, we have

$$\left(X(L_S - \beta L_V)X^T - \frac{tr\left(W^T X (L_S - \beta L_V) X^T W\right)}{tr\left(W^T X H X^T W\right)} X H X^T\right) W = \tilde{\eta} W,\tag{19}$$

where $\tilde{\eta} = \eta tr\left(W^T X H X^T W\right)$. The optimal solution of W in Equation (19) is formed by the m eigenvectors corresponding to the m smallest eigenvalues of the matrix:

$$\left(X(L_S - \beta L_V)X^T - \frac{tr\left(W^T X (L_S - \beta L_V) X^T W\right)}{tr\left(W^T X H X^T W\right)} X H X^T\right).\tag{20}$$

When W and S are fixed, problem (16) becomes

$$\min_{F} 2\lambda tr(F^T L_S F)\tag{21}$$
$$s.t. F^T F = I.$$

Since λ is a constant, the optimal solution of rank-enforcing matrix F in Equation (21) is composed of c eigenvectors, which are derived from c smallest eigenvalues of Laplacian matrix L_S.

When we fix W and F, problem (16) was written as:

$$\min_{S} \sum_{i,j=1}^{n} \left(\frac{(1+\beta)\left\|W^T x_i - W^T x_j\right\|_2^2 s_{ij}}{tr\left(W^T X H X^T W\right)} + \theta s_{ij}^2 + \lambda \left\|f_i - f_j\right\|_2^2 s_{ij}\right)\tag{22}$$
$$s.t. \forall i, s_i^T \mathbf{1} = 1, 0 \le s_{ij} \le 1.$$

Note that problem (22) can be solved independently for different s_i, so that the following problem can be solved separately for each i:

$$\min_{s_i} \sum_{i=1}^{n} (\Gamma_{ij} s_{ij} + \theta s_{ij}^2 + \lambda \Psi_{ij} s_{ij})\tag{23}$$
$$s.t. \forall i, s_i^T \mathbf{1} = 1, 0 \le s_{ij} \le 1,$$

where $\Gamma_{ij} = \frac{(1+\beta)\left\|W^T x_i - W^T x_j\right\|_2^2}{tr\left(W^T X H X^T W\right)}$ and $\Psi_{ij} = \left\|f_i - f_j\right\|_2^2$. Then, Equation (23) can be rewritten as:

$$\min_{s_i} \left\|s_i + \frac{1}{2\theta}(\Gamma_i + \lambda \Psi_i)\right\|_2^2\tag{24}$$
$$s.t. \forall i, s_i^T \mathbf{1} = 1, 0 \le s_{ij} \le 1.$$

Thus, Equation (24) can be solved easily with a close form solution. Denote vector $d_i \in R^{n \times 1}$ with $d_{ij} = \Gamma_{ij} + \lambda \Psi_{ij}$. For each i, Lagrange functions can be obtained:

$$L(W, \varsigma, \gamma_i) = \frac{1}{2}\left\|s_i + \frac{1}{2\theta} d_i\right\|_2^2 - \varsigma(s_i^T \mathbf{1} - 1) - \gamma_i^T s_i,\tag{25}$$

where ς and $\gamma_i^T \geq 0$ are the Lagrangian multipliers. Take a partial derivative for each $\mathbf{s}_i$ and set it to zero; then, according to **K.K.T.** conditions:

$$
\begin{aligned}
(\mathbf{s}_i)_j - (\mathbf{d}_i)_j + \varsigma - \gamma_i &= 0, \\
(\mathbf{s}_i)_j &\geq 0, \\
\gamma_i &\geq 0, \\
(\mathbf{s}_i)_j \gamma_i &\geq 0, \\
\mathbf{s}_i^T \mathbf{1} - 1 &= 0.
\end{aligned}
\tag{26}
$$

Then, we can obtain $\mathbf{s}_i$ that should be:

$$
s_{ij} = \left[-\tfrac{1}{2\theta_i} \mathbf{d}_i + \varsigma \right]_+ .
\tag{27}
$$

4.2. Approach to Determine the Initial Value of θ, λ

In actual experiments, regularization parameters are difficult to tune because their values may range from zero to infinity. In this section, we propose an efficient way to determine the regularization parameter θ and λ as follows:

$$
\lambda = \theta = \frac{1}{n} \sum_{j=1}^{n} \left[\tfrac{k}{2} d_{i,k+1} - \tfrac{1}{2} \sum_{j=1}^{k} d_{ij} \right].
\tag{28}
$$

k is a pre-defined parameter. In this way, we only need to set the number of neighbors we prefer rather than setting two hyper-parameters of θ and λ. The number of neighbors is usually easy to set according to the number of samples and locality of the data set. The rationality of deciding θ and λ using the distance gaps between k-th neighbor and $(k+1)$-th neighbor lies in the fact that, to achieve a desired similarity where the top k-neighbor similarities are kept and the rest are set to zeros, we should approximately achieve

$$
\tfrac{k}{2} d_{ik} - \tfrac{1}{2} \sum_{j=1}^{k} d_{ij} < \theta_i \leq \tfrac{k}{2} d_{i,k+1} - \tfrac{1}{2} \sum_{j=1}^{k} d_{ij},
\tag{29}
$$

where $d_{i1}, d_{i2}, ..., d_{in}$ are sorted in ascending order. If we set the inequality to equality, we can get an estimation of θ:

$$
\theta \sim \frac{1}{n} \sum_{j=1}^{n} \left[\tfrac{k}{2} d_{i,k+1} - \tfrac{1}{2} \sum_{j=1}^{k} d_{ij} \right].
\tag{30}
$$

Similarly, λ is set to be equal to θ as follows:

$$
\lambda = \theta \sim \frac{1}{n} \sum_{j=1}^{n} \left[\tfrac{k}{2} d_{i,k+1} - \tfrac{1}{2} \sum_{j=1}^{k} d_{ij} \right].
\tag{31}
$$

Since these two parameters control the regularization strength, we adaptively update the parameters during each iteration:

1. When the connected components are insufficient, i.e., the number of zero eigenvalues is smaller than c, we multiply λ by 2.
2. The number of connected components could be overrun, i.e., the number of zero eigenvalues is larger than c. We divide λ by 2.
3. If the graph has exact c connected components, then we stop the algorithm in this case and return the result.

The detailed steps are summarized in Algorithm 1.

Algorithm 1 Framework of the LSDUDR method.

Require: Data $X \in R^{d \times n}$, cluster number c, projection dimension m.

Initialize S and V according to Equations (6) and (7). Initialize parameter θ and λ by the Equation (28).
If algorithm 1 not converge:

repeat

1. Construct the Laplacian matrix $L_S = D - (S + S^T)/2$ and $L_V = P - (V + V^T)/2$.

2. Calculate F, columns of F are c eigenvectors of L_S and are derived from the c samllest eigenvalues.

3. Calculate the projection matrix W by the m eigenvectors corresponding to the m smallest eigenvalues of the matrix:

$$\left(X \left(L_S - \beta L_V \right) X^T - \frac{tr \left(W^T X \left(L_S - \beta L_V \right) X^T W \right)}{tr \left(W^T X H X^T W \right)} X H X^T \right).$$

4. Compute S by updating s_i according to Equation (27).

5. Calculate the number of connected components of the graph, if it is smaller than c, then multiply λ by 2; if larger than c, then divide λ by 2.

until Convergence

End if

return

Projection matrix $W \in R^{d \times m}$ and similarity matrix $S \in R^{n \times n}$.

---

5. Discussion

5.1. Analysis

As previously discussed, LSDUDR represents the local intrinsic structure of data set based on Equations (8) and (9). Then, we integrate the two objective functions as follows:

$$\sum_{i,j=1}^{n} \left\| W^T x_i - W^T x_j \right\|_2^2 s_{ij} - \beta \sum_{i,j=1}^{n} \left\| W^T x_i - W^T x_j \right\|_2^2 v_{ij}$$

$$= \sum_{i=1}^{n} \sum_{j=1}^{k} \left\| W^T x_i - W^T x_j \right\|_2^2 \left(s_{ij} + \beta s_{ij} - 1 \right) \tag{32}$$

$$= \sum_{i,j=1}^{n} \left\| W^T x_i - W^T x_j \right\|_2^2 z_{ij},$$

where the elements z_{ij} are defined as follows:

$$z_{ij} = \begin{cases} s_{ij} + \beta s_{ij} - 1, j \le k, \\ 0, j > k. \end{cases} \tag{33}$$

It is easy to see that Equation (32) is very similar to Equation (8). However, they are completely different when they express the intrinsic geometrical structure of the data. Without loss of generality, we set the weight elements s_{ij} in Equation (8) as a heat kernel function. Figure 2 shows their weight change process with a distance between two points x_i and x_j.

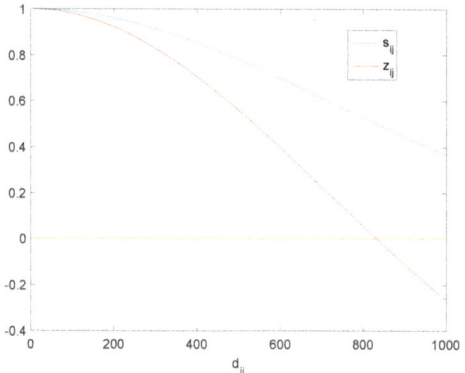

Figure 2. Difference between s_{ij} and z_{ij}.

As we know, the real-world data are usually unbalanced and complex, thus some points may be distributed in sparse areas while other data points are distributed in compact areas. As shown in Figure 2, z_{ij} is positive for data points in compact regions, thus Equation (32) maps these data points to be very close in the subspace, and mainly preserves the similarity of data. If data points lie in sparse regions, z_{ij} is negative, and Equation (32) mainly characterizes diversity of data in this case, i.e., the shape of a manifold structure. However, the difference among points in a neighborhood is not considered in Equation (8), and always projects the neighborhood points to be close into subspace, which ignores the intrinsic geometrical structure of data.

It is noteworthy that three updating rules are included in the proposed algorithm, which are computationally efficient. In fact, [29] has already proven the convergence of the alternative optimization method. In our algorithm, the main cost lies in each iteration being the eigen-decomposition step for Equations (7) and (21). The time computational complex of the proposed method is $O((d^2m + n^2c)t)$, where t is the number of iterations.

5.2. Convergence Study

The method proposed by Algorithm 1 can be used to find a locally optimal solution of problem (14). The convergence of Algorithm 1 is given through Theorem (2).

Theorem 2. *The alternate updating rules in Algorithm 1 monotonically decrease the objective function value of optimization problem (14) in each iteration until convergence.*

Proof. In the procedure of iteration, we get the global optimal selective matrix W_{t+1} by solving optimization problem $W_{t+1} = \arg\min\limits_{W^T W = I} \dfrac{tr(W^T X(L_S - \beta L_V)X^T W)}{tr(W^T X H X^T W)}$. As a result, we have the following inequality:

$$\frac{tr(W_{t+1}^T X(L_S - \beta L_V)X^T W_{t+1}^T)}{tr(W_{t+1}^T X H X^T W_{t+1}^T)} \leq \frac{tr(W_t^T X(L_S - \beta L_V)X^T W_t^T)}{tr(W_t^T X H X^T W_t^T)}. \tag{34}$$

Since variable F_{t+1} is updated by solving problem $F_{t+1}^T = \arg\min\limits_{F^T F = I} 2\lambda tr(F_t^T L F_t^T)$, we obtain the following inequality:

$$tr(F_{t+1}^T L F_{t+1}^T) \leq tr(F_t^T L F_t^T). \tag{35}$$

Consequently, we have the following inequality:

$$\frac{tr(W_{t+1}^T X(L_S - \beta L_V)X^T W_{t+1}^T)}{tr(W_{t+1}^T X H X^T W_{t+1}^T)} + tr(F_{t+1}^T L F_{t+1}^T) \leq \frac{tr(W_t^T X(L_S - \beta L_V)X^T W_t^T)}{tr(W_t^T X H X^T W_t^T)} + tr(F_t^T L F_t^T). \tag{36}$$

In addition, **K.K.T.** conditions (26) illustrate that the converged solution of Algorithm 1 is at least a stationary point of Equation (25). Because the updating of weights matrix $S_{t+1} \in R^{n \times n}$ can be divided into n independently sub-optimization problem with respect to n-dimensional vector. Consequently, the objective function value of optimization problem (14) decreases monotonically in each iteration until the algorithm convergence. □

6. Experiment

In the experiment, the following two metrics are used to evaluate the performance of the proposed LSDUDR algorithm: Accuracy (ACC) and Normalized Mutual Information (NMI) [30]. Accuracy is defined as

$$ACC = \frac{\sum\limits_{i=1}^{n} \delta\left(t_i, map\left(t_i^g\right)\right)}{n}, \tag{37}$$

where t_i is the label of the clustering result and t_i^g is the known label of x_i. $map\left(t_i^g\right)$ is the optimal mapping function that permutes the label set of the clustering results and the known label set of samples. $\delta\left(t_i, map\left(t_i^g\right)\right)$ is an indicator function. Normalized Mutual Information is defined as

$$NMI = \frac{\sum\limits_{i,j=1}^{c} t_{ij} \log \frac{n \times t_{ij}}{t_i \hat{t}_j}}{\sqrt{\left(\sum\limits_{i=1}^{c} t_i \log \frac{t_i}{n}\right)\left(\sum\limits_{j=1}^{c} \hat{t}_j \log \frac{\hat{t}_j}{n}\right)}}, \tag{38}$$

where t_i is the number of samples in the i-th cluster C_i according to clustering results and $\hat{t}_j$ is the number of samples in the j-th ground truth class G_j. t_{ij} is the number of overlap between C_i and G_j.

We compare the performance of LSDUDR with K-Means [31], Ratio Cut [32], Normalized Cut [33] and PCAN methods, since they are closely related to LSDUDR, i.e., the information contained in the eigenvectors of an affinity matrix is used to detect the similarity. We made comparisons with Ratio Cut, Normalized Cut to show that LSDUDR can effectively mitigate the influence of outliers by inducing robustness and adaptive neighbors. To emphasize the importance of describing the intrinsic manifold structure, we compared the results of PCAN with LSDUDR which concatenates to uncover the intrinsic topology structures of data by proposing two objective functions and performs discriminatively embedded K-Means clustering.

6.1. Experiment on the Synthetic Data Sets

To verify the robust performances and strong discriminating power of the proposed LSDUDR, two simple synthetic examples (two-Gaussian and multi-cluster data) are given in this experiment.

In this first synthetic data set, we deliberately set a point away from the two-Gaussian distribution as an outlier so that a one-dimensional linear manifold representation was obtained to clearly divide two clusters. LSDUDR and PCAN were demonstrated on the synthetic examples respectively and the results are shown in Figure 3. It is clear that one cluster shown in pink almost submerges in another one as blue in the one-dimensional representation using PCAN, while it is separated distinctly out using LSDUDR, so we can conclude that LSDUDR has more discriminating power than PCAN. Furthermore, LSDUDR is less sensitive to outliers than PCAN because the objective function of LSDUDR will bring a heavy penalty to two points when they are embedded to be close in the subspace but with large distance in the origin space.

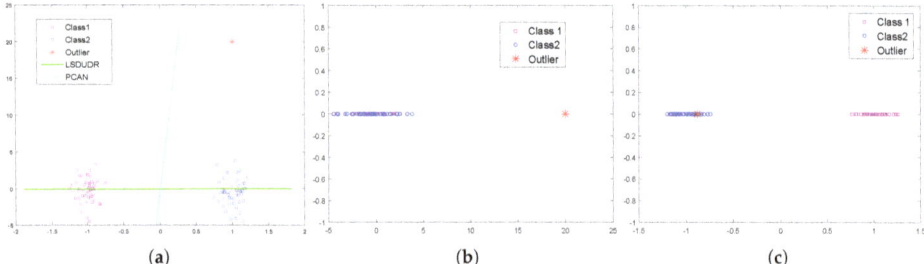

Figure 3. (**a**) two-Gaussian synthetic data projection results; (**b**) one-dimensional representation obtained by Projected clustering with adaptive neighbors. (PCAN); (**c**) one-dimensional representation obtained by Locality Sensitive Discriminative Unsupervised Dimensionality Reduction. (LSDUDR).

The second synthetic data set is a multi-cluster data, which contains 196 randomly generated clusters that are distributed in a spherical manner. We compared LSDUDR with K-means and PCAN. Due to the fact that K-means is sensitive to initialization [34], we repeatedly run K-means 100 times and use the minimal K-means objective value as the result. To be fair, the parameters of PCAN are adjusted to report the best performance of PCAN. As for LSDUDR, we run LSDUDR once to generate a clustering result and use it as initialization for K-means and report the best performance. Table 1 and Figure 4 show the experiment results of LSDUDR and other two algorithms on multi-cluster data. As can be seen from Table 1, LSDUDR obtained better performance than those of other methods according to the the the minimal K-means objective value and clustering accuracy. Thus, LSDUDR has stronger discriminating power than PCAN and K-means especially when the data distribution is complex.

Table 1. Compare results on multi-cluster synthetic data sets.

Methods	ACC%	Minimal K-Means Objective
K-Means	66.94	336.46
PCAN	98.62	107.33
LSDUDR	99.49	106.21

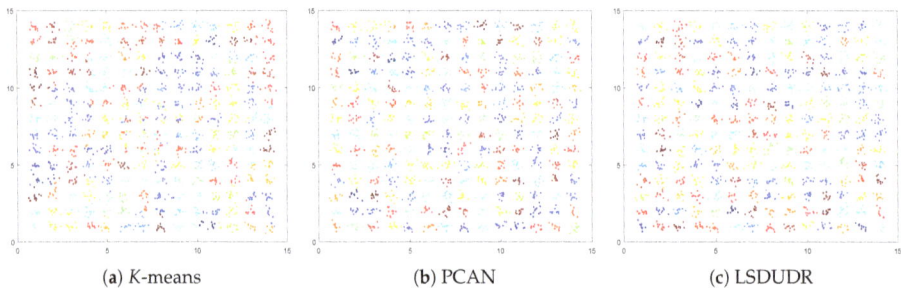

Figure 4. Clustering results of three algorithms.

6.2. Experiment on Low-Dimensional Benchmark Data Sets

In this subsection, we evaluate the performance of the proposed LSDUDR on ten low-dimensional benchmark data sets with comparison to four related methods, including K-Means, Ratio Cut, Normalized Cut and PCAN methods. Description of these data sets is summarized in Table 2, including four synthetic data sets and six University of CaliforniaIrvine (UCI) datasets [35]. In low-dimensional data, we set the projection dimension in PCAN and LSDUDR to be $c - 1$. For the methods that require a fixed input data graph, we use the self-turn Gaussian method [34,36] to build the graph. For the methods involving K-Means to extract the clustering labels, we repeatedly ran K-Means 100 times with the same settings and chose the best performance. As for PCAN and LSDUDR, we only ran it once and reported the result directly from the learned graph. The experimental results are shown in Tables 3 and 4.

Table 2. Specifications of the data sets.

Data Set	#Classes (c)	#Data Points (n)	#Dimensions (d)
Spiral	3	312	2
Pathbased	3	300	2
Compound	6	399	2
Movements	15	360	90
Iris	3	150	4
Cars	3	392	8
Glass	6	214	9
Vote	2	435	16
Diabetes	2	768	8
Dermatology	6	366	34

Table 3. ACC(%) on low-dimensional benchmark data sets.

ACC%	K-Means	RatioCut	NormalizedCut	PCAN	LSDUDR
Spiral	33.97	99.68	99.68	**100**	100
Pathbased	74.33	77.33	76.67	**87.00**	87.00
Compound	80.20	76.69	65.91	78.95	**88.22**
Movements	10.56	5.83	10.56	**56.11**	55.56
Iris	66.67	68.00	66.39	77.33	**92.00**
Cars	44.90	53.27	47.70	48.98	**58.42**
Glass	52.21	36.45	51.87	49.07	**52.80**
Vote	83.45	61.61	83.68	67.36	**85.75**
Diabetes	56.02	64.71	61.98	58.46	**65.10**
Dermatology	85.25	54.92	93.72	94.81	**95.90**

Table 4. NMI(%) on low-dimensional benchmark data sets.

NMI%	K-Means	RatioCut	NormalizedCut	PCAN	LSDUDR
Spiral	12.52	98.35	98.35	**100**	100
Pathbased	51.33	54.96	53.10	75.63	**79.27**
Compound	79.74	71.60	66.32	77.48	**85.16**
Movements	44.11	15.85	44.91	**84.95**	84.17
Iris	61.68	61.3	59.05	61.85	**77.52**
Cars	39.10	21.61	39.06	39.03	**39.39**
Glass	35.83	35.33	34.88	35.76	**35.90**
Vote	36.58	30.17	35.66	35.23	**39.37**
Diabetes	52.67	50.02	61.98	64.01	**68.10**
Dermatology	85.20	41.24	88.43	91.83	**93.53**

In this experiment, we can observe that PCAN and LSDUDR are much better than those of fixed graph-based methods. This observation confirms that separation of graph construction and dimensionality reduction leads the similarity matrix to not being able to be fully relied on and the experimental results will seriously deteriorate. In addition, LSDUDR outperforms other methods in nine data sets on account of preserving locality structure among data.

6.3. Embedding of Noise 3D Manifold Benchmarks

To confirm the ability of robustly characterizing the manifold structure of LSDUDR, we use three typical 3D manifold benchmark data sets [37], i.e., Guassian, Toroidal Helix and Swiss Roll. In this experiment, we tried to map these 3D manifold benchmarks to 2D in order to find out a low-dimensional embedding but with the most manifold structure information. The experimental results are shown in Figure 5.

(**a**) Guassian original data   (**b**) PCAN   (**c**) LSDUDR

(**d**) Toroidal Helix original data   (**e**) PCAN   (**f**) LSDUDR

(**g**) Swiss Roll original data   (**h**) PCAN   (**i**) LSDUDR

Figure 5. Projection results on 3D manifold benchmarks by the PCAN and LSDUDR methods.

Under the same conditions, PCAN method is also tested for comparison. Figure 5 shows the 2D embedding results of PCAN and LSDUDR which each row is related to on the manifold benchmark. It is obvious that PCAN did not find a suitable projection direction. This is because PCAN only considers the similarity between data points, which is not enough to characterize the intrinsic structure of data and even causes the destruction of a manifold structure. However, LSDUDR considers both similarity and diversity of the data set, and thus has strong sensitivity to local topology of data.

6.4. Experiment on the Image Data Sets

6.4.1. Visualization for Handwritten Digits

To further test the low-dimensional embedding applicability of the proposed LSDUDR algorithm, another experiment is carried out on a Binary Alphadigits data set [38], as shown in Figure 6. We select four letters ("C", "P", "X", "Z") and four digits ("0", "3", "6", "9") from the Binary Alphadigits data set, which comprises binary digits from "0" to "9" and capital "A" to "Z". The embedding results are drawn in Figure 7.

Figure 6. Some image samples of the handwritten digits.

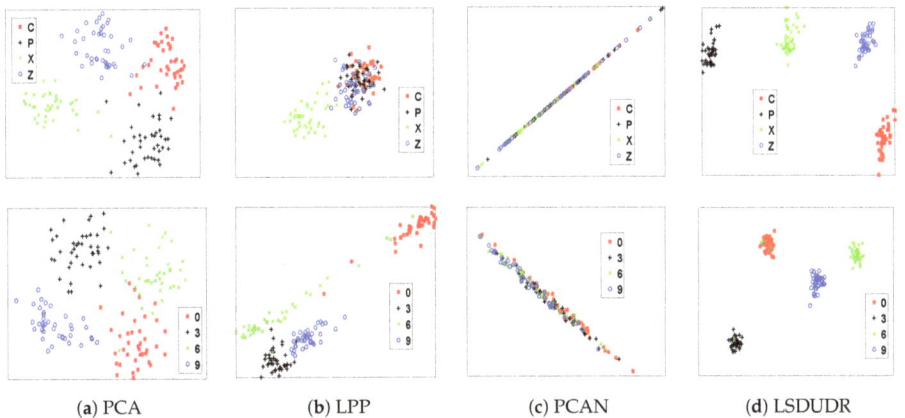

| (a) PCA | (b) LPP | (c) PCAN | (d) LSDUDR |

Figure 7. Experiment on the Alphadigits data set.

It can be seen from Figure 7a,b that there are overlaps in clusters of "C", "P" and "Z", digits "0" and "6" when we use PCA and LPP. In addition, worse results are obtained from PCAN and it is shown in Figure 7c that almost all points are tangled for all clusters. However, for LSDUDR, results in Figure 7d show that classes are separated clearly, which reflects that diversity plays an important role in representing the intrinsic structure of data.

6.4.2. Face Benchmark Data Sets

We use four image benchmark data sets in this section for experiments on projection, since these data typically have high dimensionality. We summarize the four face image benchmark data sets in Table 5. To study the data-adaptiveness and noise-robustness of the proposed LSDUDR algorithm, we use a range of data sets contaminated by different kinds of noise based on the face data sets, as shown in Figure 8. Similar to the above-mentioned experiment, three algorithms, including PCAN, PCA and LPP, are used for comparison.

Table 5. The description of the face image benchmark data sets.

Data Sets	#Classes (c)	#Data Points (n)	#Dimensions (d)
YaleA	15	165	3456
Jaffe	10	213	1024
CBCI	120	840	7668
UMIST	120	840	768

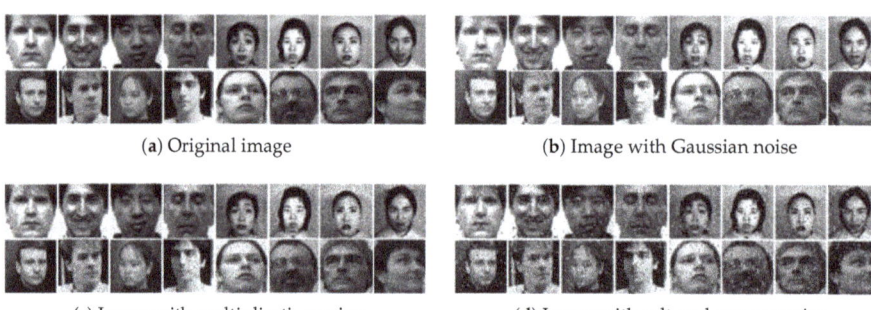

(**a**) Original image   (**b**) Image with Gaussian noise

(**c**) Image with multiplicative noise   (**d**) Image with salt-and-pepper noise

Figure 8. Some image samples of the data sets with different kinds of noise.

The experimental results about face benchmark data sets are shown in Figure 9, from which we get a convincing observation that the experimental results obtained by adaptive graph learning algorithms are usually more outstanding, especially when the dimensionality of projection space increases. This is because adaptive graph learning algorithms can use the embedded information that are obtained in the previous step to update the similarity matrix, hence the dimensionality reduction results are more accurate. In addition, we observe that PCA and LPP are more sensitive to the dimensionality of embedded space while the curve of LSDUDR is basically stable with the change of dimensionality. Furthermore, LSDUDR is capable of projecting the data into a subspace with a relatively small dimension $c - 1$; such subspace with low dimensionality obtained by our method would be even better than the subspaces obtained by PCA and LPP with higher dimensionality. It indicates that local topology and geometrical properties were taken into account for the similarity and diversity of data when using LSDUDR, and thus have better performance and achieved higher accuracy than PCAN when the images reserve sufficient spatial information.

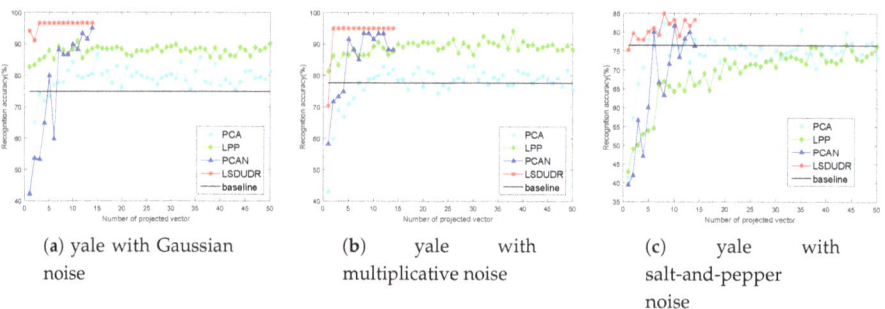

(**a**) yale with Gaussian noise

(**b**) yale with multiplicative noise

(**c**) yale with salt-and-pepper noise

Figure 9. *Cont.*

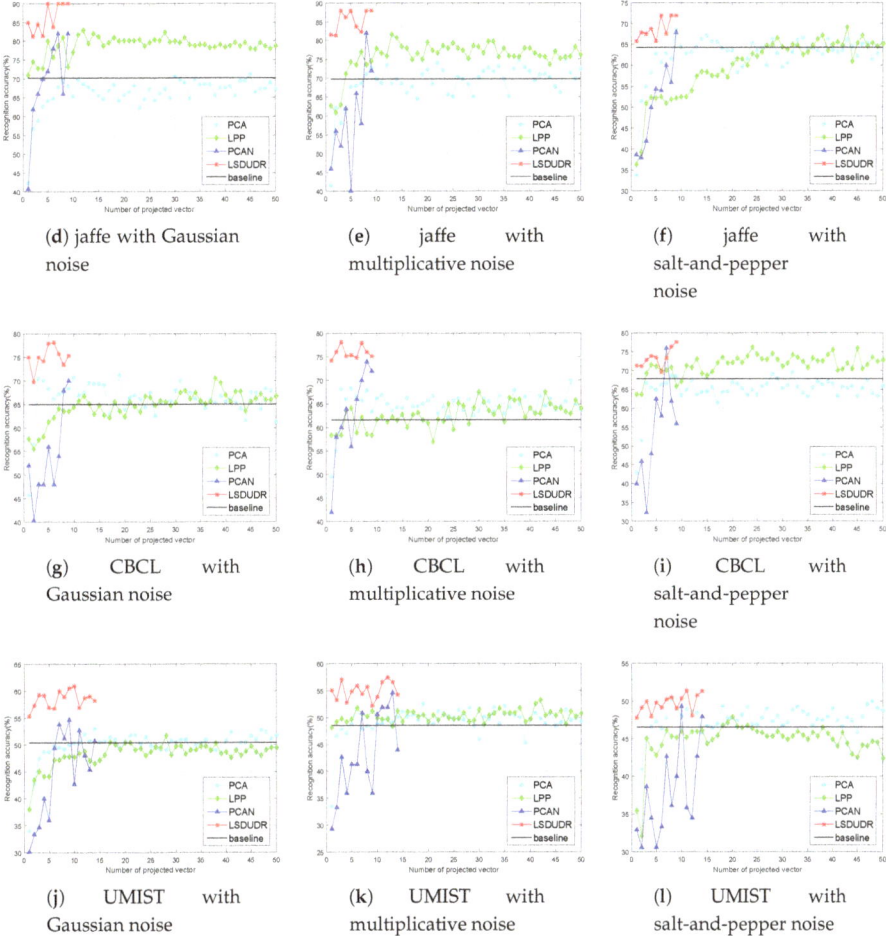

(**d**) jaffe with Gaussian noise

(**e**) jaffe with multiplicative noise

(**f**) jaffe with salt-and-pepper noise

(**g**) CBCL with Gaussian noise

(**h**) CBCL with multiplicative noise

(**i**) CBCL with salt-and-pepper noise

(**j**) UMIST with Gaussian noise

(**k**) UMIST with multiplicative noise

(**l**) UMIST with salt-and-pepper noise

Figure 9. Projection results on face image benchmark data sets with different kinds of noise.

7. Conclusions

In this paper, a novel adaptive graph learning method (LSDUDR) is proposed from a new perspective by integrating a similarity graph and diversity graph to learn a discriminative subspace where data can be easily separated. Meanwhile, LSDUDR performs dimensionality reduction and local structure learning simultaneously based on the high quality Laplacian matrix. Different from previous graph-based models, LSDUDR constructs two adjacency graphs that could represent the intrinsic structure of data well in learning the local sensitivity of the data. Furthermore, LSDUDR doesn't require other clustering methods to obtain cluster indicators but extracts label information from a similarity graph or diversity graph, which adaptively updates in a reconstruction manner. We also discuss the convergence of the proposed algorithm as well as the value of trade-off parameters. Experimental results on the synthetic data, face image databases and several benchmark data illustrate the effectiveness and superiority of the proposed method.

In this paper, we focus on the scenario of construction of two adjacency graphs to represent the original structure with data similarity and diversity. Our method can be used to remove irrelevant and correlated features involved in high-dimensional feature space and convert data represented in subspaces [39]. In our future work, it is potentially interesting to extend the proposed methods to unsupervised feature selection of data points with multiview and multitask.

Author Contributions: Y.-L.G. and S.-Z.L. conceived and designed the experiments; S.-Z.L. performed the experiments; C.-C.C. and Z.-H.W. analyzed the data; J.-Y.P. contributed analysis tools.

Funding: This research was funded by [National Natural Science Foundation of China] grant number [61203176] and [Fujian Provincial Natural Science Foundation] grant number [2013J05098, 2016J01756].

Conflicts of Interest: The authors declare no conflicts of interest.

References

1. Everitt, B.S.; Dunn, G.; Everitt, B.S.; Dunn, G. *Cluster Analysis*; Wiley: Hoboken, NJ, USA, 2011; pp. 115–136.
2. Xu, R.; Wunsch, D. Survey of clustering algorithms. *IEEE Trans. Neural Netw.* **2005**, *16*, 645–678. [CrossRef] [PubMed]
3. Cheng, Q.; Zhou, H.; Cheng, J. The Fisher-Markov Selector: Fast Selecting Maximally Separable Feature Subset for Multiclass Classification with Applications to High-Dimensional Data. *IEEE Trans. Pattern Anal. Mach. Intell.* **2011**, *33*, 1217–1233. [CrossRef] [PubMed]
4. Dash, M.; Liu, H. Feature selection for clustering. In Proceedings of the Pacific-Asia Conference on Knowledge Discovery and Data Mining, Keihanna Plaza, Japan, 17–20 April 2000; pp. 110–121.
5. Ben-Bassat, M. Pattern recognition and reduction of dimensionality. *Handb. Stat.* **1982**, *2*, 773–910.
6. Wold, S.; Esbensen, K.; Geladi, P. Principal component analysis. *Chemom. Intell. Lab. Syst.* **1987**, *2*, 37–52. [CrossRef]
7. FISHER, R.A. The use of multiple measurements in taxonomic problems. *Ann. Eugen.* **1936**, *7*, 179–188. [CrossRef]
8. Hall, K.M. Anr-Dimensional Quadratic Placement Algorithm. *Manag. Sci.* **1970**, *17*, 219–229. [CrossRef]
9. Belkin, M.; Niyogi, P. Laplacian Eigenmaps for Dimensionality Reduction and Data Representation. *Neural Comput.* **2003**, *15*, 1373–1396. [CrossRef]
10. Luo, D.; Ding, C.; Huang, H.; Li, T. Non-negative Laplacian Embedding. In Proceedings of the 2009 Ninth IEEE International Conference on Data Mining, Miami, FL, USA, 6–9 December 2009. [CrossRef]
11. Roweis, S.T. Nonlinear Dimensionality Reduction by Locally Linear Embedding. *Science* **2000**, *290*, 2323–2326. [CrossRef]
12. He, X.; Niyogi, P. Locality preserving projections. In *Advances in Neural Information Processing Systems*; MIT Press: Vancouver, BC, Canada, 2004; pp. 153–160.
13. Nie, F.; Xiang, S.; Zhang, C. Neighborhood MinMax Projections. In Proceedings of the International Joint Conference on Artificial Intelligence (IJCAI), Hyderabad, India, 8 January 2007; pp. 993–998.
14. Tenenbaum, J.B. A Global Geometric Framework for Nonlinear Dimensionality Reduction. *Science* **2000**, *290*, 2319–2323. [CrossRef]
15. Lu, G.F.; Jin, Z.; Zou, J. Face recognition using discriminant sparsity neighborhood preserving embedding. *Knowl.-Based Syst.* **2012**, *31*, 119–127. [CrossRef]
16. Fan, Q.; Gao, D.; Wang, Z. Multiple empirical kernel learning with locality preserving constraint. *Knowl.-Based Syst.* **2016**, *105*, 107–118. [CrossRef]
17. Pandove, D.; Rani, R.; Goel, S. Local graph based correlation clustering. *Knowl.-Based Syst.* **2017**, *138*, 155–175. [CrossRef]
18. Feng, X.; Wu, S.; Zhou, W.; Min, Q. Efficient Locality Weighted Sparse Representation for Graph-Based Learning. *Knowl.-Based Syst.* **2017**, *121*, 129–141. [CrossRef]
19. Du, L.; Shen, Y.D. Unsupervised feature selection with adaptive structure learning. In Proceedings of the 21th ACM SIGKDD International Conference on Knowledge Discovery and Data Mining, Sydney, Australia, 10–13 August 2015; pp. 209–218. [CrossRef]
20. Nie, F.; Wang, H.; Huang, H.; Ding, C. Unsupervised and semi-supervised learning via L1-norm graph. In Proceedings of the 2011 International Conference on Computer Vision, Barcelona, Spain, 6–13 November 2011. [CrossRef]
21. Gao, Q.; Liu, J.; Zhang, H.; Gao, X.; Li, K. Joint Global and Local Structure Discriminant Analysis. *IEEE Trans. Inf. Forensics Secur.* **2013**, *8*, 626–635. [CrossRef]

22. Nie, F.; Wang, X.; Jordan, M.I.; Huang, H. The Constrained Laplacian Rank Algorithm for Graph-Based Clustering. In Proceedings of the Thirtieth AAAI Conference on Artificial Intelligence, Phoenix, AZ, USA, 12–17 February 2016; pp. 1969–1976.

23. Luo, D.; Nie, F.; Huang, H.; Ding, C.H. Cauchy graph embedding. In Proceedings of the 28th International Conference on Machine Learning (ICML-11), Bellevue, WA, USA, 28 June–2 July 2011; pp. 553–560.

24. Nie, F.; Wang, X.; Huang, H. Clustering and projected clustering with adaptive neighbors. In Proceedings of the 20th ACM SIGKDD international conference on Knowledge discovery and data mining, New York, NY, USA, 24–27 August 2014; pp. 977–986. [CrossRef]

25. Gao, Q.; Huang, Y.; Gao, X.; Shen, W.; Zhang, H. A novel semi-supervised learning for face recognition. *Neurocomputing* **2015**, *152*, 69–76. [CrossRef]

26. Mohar, B.; Alavi, Y.; Chartrand, G.; Oellermann, O. The Laplacian spectrum of graphs. *Graph Theory Comb. Appl.* **1991**, *2*, 12.

27. Fan, K. On a Theorem of Weyl Concerning Eigenvalues of Linear Transformations I. *Proc. Natl. Acad. Sci. USA* **1949**, *35*, 652–655. [CrossRef]

28. Nie, F.; Xiang, S.; Jia, Y.; Zhang, C. Semi-supervised orthogonal discriminant analysis via label propagation. *Pattern Recognit.* **2009**, *42*, 2615–2627. [CrossRef]

29. Bezdek, J.C.; Hathaway, R.J. Convergence of alternating optimization. *Neural Parallel Sci. Comput.* **2003**, *11*, 351–368.

30. Chen, W.; Feng, G. Spectral clustering: A semi-supervised approach. *Neurocomputing* **2012**, *77*, 229–242. [CrossRef]

31. Macqueen, J. Some Methods for Classification and Analysis of MultiVariate Observations. In Proceedings of the Fifth Berkeley Symposium on Mathematical Statistics and Probability, Oakland, CA, USA, 21 June–18 July 1965; pp. 281–297.

32. Hagen, L.; Kahng, A. New spectral methods for ratio cut partitioning and clustering. *IEEE Trans. Comput.-Aided Des. Integr. Circuits Syst.* **1992**, *11*, 1074–1085. [CrossRef]

33. Shi, J.; Malik, J. Normalized cuts and image segmentation. *IEEE Trans. Pattern Anal. Mach. Intell.* **2000**, *22*, 888–905.

34. Nie, F.; Xu, D.; Li, X. Initialization Independent Clustering With Actively Self-Training Method. *IEEE Trans. Syst. Man Cybern. Part B Cybern.* **2012**, *42*, 17–27. [CrossRef]

35. Dua, D.; Graff, C. UCI Machine Learning Repository. 2017. Available online: http://archive.ics.uci.edu/ml (accessed on 1 August 2018).

36. Zelnik-Manor, L.; Perona, P. Self-tuning spectral clustering. In *Advances in Neural Information Processing Systems*; MIT Press: Vancouver, BC, Canada, 2005; pp. 1601–1608.

37. Chen, S.B.; Ding, C.H.; Luo, B. Similarity learning of manifold data. *IEEE Trans. Cybern.* **2015**, *45*, 1744–1756. [CrossRef]

38. Roweis, S. Binary Alphadigits. Available online: https://cs.nyu.edu/~roweis/data.html (accessed on 1 August 2018).

39. Har, M.T.; Conrad, S.; Shirazi, S.; Lovell, B.C. Graph embedding discriminant analysis on Grassmannian manifolds for improved image set matching. In Proceedings of the IEEE Conference on Computer Vision and Pattern Recognition, Colorado Springs, CO, USA, 20–25 June 2011.

MDPI

St. Alban-Anlage 66

4052 Basel

Switzerland

Tel. +41 61 683 77 34

Fax +41 61 302 89 18

www.mdpi.com

Symmetry Editorial Office

E-mail: symmetry@mdpi.com

www.mdpi.com/journal/symmetry